Recent Advances in Microbial Biotechnology

Recent Advances in Microbial Biotechnology

Editor: Elsa Cooper

R CALLISTO REFERENCE

www.callistoreference.com

Callisto Reference,
118-35 Queens Blvd., Suite 400,
Forest Hills, NY 11375, USA

Visit us on the World Wide Web at:
www.callistoreference.com

ISBN: 978-1-64116-116-9 (Hardback)

Cataloging-in-Publication Data

Recent advances in microbial biotechnology / edited by Elsa Cooper.
p. cm.
Includes bibliographical references and index.
ISBN 978-1-64116-116-9
1. Microbial biotechnology. 2. Biotechnological microorganisms.
3. Biotechnology. I. Cooper, Elsa.
TP248.27.M53 R43 2019
660.62--dc23

Table of Contents

Preface.. VII

Chapter 1   Metabolite labelling reveals hierarchies in *Clostridium acetobutylicum* that
             selectively channel carbons from sugar mixtures towards biofuel precursors................1
             Ludmilla Aristilde

Chapter 2   Wine microbiology is driven by vineyard and winery anthropogenic factors............14
             Cédric Grangeteau, Chloé Roullier-Gall, Sandrine Rousseaux, Régis D. Gougeon,
             Philippe Schmitt-Kopplin, Hervé Alexandre and Michèle Guilloux-Benatier

Chapter 3   Mechanism and regulation of sorbicillin biosynthesis by *Penicillium
             chrysogenum*..30
             Fernando Guzmán-Chávez, Oleksandr Salo, Yvonne Nygård, Peter P. Lankhorst,
             Roel A. L. Bovenberg and Arnold J. M. Driessen

Chapter 4   Engineering *Ashbya gossypii* strains for *de novo* lipid production using
             industrial by-products...41
             Patricia Lozano-Martínez, Rubén M. Buey, Rodrigo Ledesma-Amaro,
             Alberto Jiménez and José Luis Revuelta

Chapter 5   Metallic bionanocatalysts: potential applications as green catalysts and energy
             materials...50
             Lynne E. Macaskie, Iryna P. Mikheenko, Jacob B. Omajai, Alan J. Stephen and
             Joseph Wood

Chapter 6   Dynamics of mono-and dual-species biofilm formation and interactions
             between *Staphylococcus aureus* and Gram-negative bacteria.................................60
             Jitka Makovcova, Vladimir Babak, Pavel Kulich, Josef Masek, Michal Slany and
             Lenka Cincarova

Chapter 7   Community structure of partial nitritation-anammox biofilms at decreasing
             substrate concentrations and low temperature..74
             Frank Persson, Carolina Suarez, Malte Hermansson, Elzbieta Plaza, Razia Sultana
             and Britt-Marie Wilén

Chapter 8   Antimicrobial activity of biogenically produced spherical Se-nanomaterials
             embedded in organic material against *Pseudomonas aeruginosa* and
             Staphylococcus aureus strains on hydroxyapatite-coated surfaces............................85
             Elena Piacenza, Alessandro Presentato, Emanuele Zonaro, Joseph A. Lemire,
             Marc Demeter, Giovanni Vallini, Raymond J. Turner and Silvia Lampis

Chapter 9   Synthetic extreme environments: overlooked sources of potential
             biotechnologically relevant microorganisms...100
             Timothy Sibanda, Ramganesh Selvarajan and Memory Tekere

Chapter 10   **Advances and bottlenecks in microbial hydrogen production**..**116**
Alan J. Stephen, Sophie A. Archer, Rafael L. Orozco and Lynne E. Macaskie

Chapter 11   **Diversity and functions of the sheep faecal microbiota: a multi-omic
characterization**..**124**
Alessandro Tanca, Cristina Fraumene, Valeria Manghina, Antonio Palomba,
Marcello Abbondio, Massimo Deligios, Daniela Pagnozzi, Maria Filippa Addis
and Sergio Uzzau

Chapter 12   **Engineering *Mycobacterium smegmatis* for testosterone production** **138**
Lorena Fernández-Cabezón, Beatriz Galán and José L. García

Chapter 13   **Biosynthesis of micro- and nanocrystals of Pb (II), Hg (II) and Cd (II) sulfides
in four *Candida* species: a comparative study of *in vivo* and *in vitro* approaches**.........................**149**
Mayra Cuéllar-Cruz, Daniela Lucio-Hernández, Isabel Martínez-Ángeles,
Nicola Demitri, Maurizio Polentarutti, María J. Rosales-Hoz and Abel Moreno

Chapter 14   **Effect of temperature and colonization of *Legionella pneumophila* and
Vermamoeba vermiformis on bacterial community composition of copper
drinking water biofilms**...**169**
Helen Y. Buse, Pan Ji, Vicente Gomez-Alvarez, Amy Pruden, Marc A. Edwards
and Nicholas J. Ashbolt

Chapter 15   **Production of a novel medium chain length poly(3- hydroxyalkanoate) using
unprocessed biodiesel waste and its evaluation as a tissue engineering scaffold**.........................**184**
Pooja Basnett, Barbara Lukasiewicz, Elena Marcello, Harpreet K. Gura,
Jonathan C. Knowles and Ipsita Roy

Permissions

List of Contributors

Index

Preface

Microbes are microscopic organisms like bacteria, virus, fungi, protozoa, etc. which are either unicellular, multicellular or acellular. Microbiology studies different classes of these microbes and branches out into bacteriology, mycology, protozoology and phycology. Microbiology also studies microbes for use in medicine, enzyme production, DNA transfer tools, industrial fermentation, bioremediation, microbial biodegradation, etc. Biopolymers like polysaccharides, polyesters, polyamides, etc. are also produced by microorganisms. Advanced medical applications of tissue engineering and drug delivery are also aided by microorganisms. The ever growing need of advanced technology is the reason that has fueled the research in the field of microbial biotechnology in recent times. This book strives to explain the concepts and theories central to the development of microbial biotechnology. It also presents researches and studies performed by experts across the globe. The book is appropriate for students seeking detailed information in this area as well as for experts.

This book is a result of research of several months to collate the most relevant data in the field.

When I was approached with the idea of this book and the proposal to edit it, I was overwhelmed. It gave me an opportunity to reach out to all those who share a common interest with me in this field. I had 3 main parameters for editing this text:

1. Accuracy – The data and information provided in this book should be up-to-date and valuable to the readers.

2. Structure – The data must be presented in a structured format for easy understanding and better grasping of the readers.

3. Universal Approach – This book not only targets students but also experts and innovators in the field, thus my aim was to present topics which are of use to all.

Thus, it took me a couple of months to finish the editing of this book.

I would like to make a special mention of my publisher who considered me worthy of this opportunity and also supported me throughout the editing process. I would also like to thank the editing team at the back-end who extended their help whenever required.

Editor

Metabolite labelling reveals hierarchies in *Clostridium acetobutylicum* that selectively channel carbons from sugar mixtures towards biofuel precursors

Ludmilla Aristilde*

Department of Biological and Environmental Engineering, College of Agriculture and Life Sciences, Cornell University, Ithaca, NY 14853, USA.

Summary

Clostridial fermentation of cellulose and hemicellulose relies on the cellular physiology controlling the metabolism of the cellulosic hexose sugar (glucose) with respect to the hemicellulosic pentose sugars (xylose and arabinose) and the hemicellulosic hexose sugars (galactose and mannose). Here, liquid chromatography–mass spectrometry and stable isotope tracers in *Clostridium acetobutylicum* were applied to investigate the metabolic hierarchy of glucose relative to the different hemicellulosic sugars towards two important biofuel precursors, acetyl-coenzyme A and butyryl-coenzyme A. The findings revealed constitutive metabolic hierarchies in *C. acetobutylicum* that facilitate (i) selective investment of hemicellulosic pentoses towards ribonucleotide biosynthesis without substantial investment into biofuel production and (ii) selective contribution of hemicellulosic hexoses through the glycolytic pathway towards biofuel precursors. Long-term isotopic enrichment demonstrated incorporation of both pentose sugars into pentose-phosphates and ribonucleotides in the presence of glucose. Kinetic labelling data, however, showed that xylose was not routed towards the biofuel precursors but there was minor contribution from arabinose. Glucose hierarchy over the hemicellulosic hexoses was substrate-dependent. Kinetic labelling of hexose-phosphates and triose-phosphates indicated that mannose was assimilated but not galactose. Labelling of both biofuel precursors confirmed this metabolic preference. These results highlight important metabolic considerations in the accounting of clostridial mixed-sugar utilization.

*For correspondence. E-mail ludmilla@cornell.edu

Introduction

Decomposition of lignocellulosic wastes by anaerobic bacteria, including the *Clostridium* species, is an important component in the turnover of organic carbons in soils. Several of the soil-dwelling *Clostridium* species, including notably *Clostridium acetobutylicum*, have been exploited for biofuel production due to their ability to ferment sugars from polysaccharides and produce hydrogen gas, short-chain carboxylic acids (butyrate, acetate), alcohols (ethanol and butanol) and ketones (acetone) (Grupe and Gottschalk, 1992; Dürre, 1998; Desai *et al.*, 1999; Gheshlagi *et al.*, 2009; Ren *et al.*, 2010; Amador-Noguez *et al.*, 2011; Hu *et al.*, 2011; Servinsky *et al.*, 2012; Aristilde *et al.*, 2015; Dash *et al.*, 2016). Polysaccharides from lignocellulosic wastes are composed of a mixture of different types of sugars, which include primarily the following: the hexose glucose from cellulose, the pentoses xylose and arabinose from hemicellulose, and other hexoses (mannose, galactose, in addition to glucose) from hemicellulose. The different sugars in the hemicellulosic component are dependent on the source material (Scheller and Ulvskov, 2010). Of particular interest for optimizing the conversion of plant waste materials to biofuel products in clostridial species is a comprehensive understanding of the cellular metabolism of glucose with respect to hemicellulosic hexoses and pentoses.

Transcriptional analysis of *C. acetobutylicum* grown on each hexose or pentose sugar as a sole carbon source revealed the expression of all the relevant sugar uptake transporters (Servinsky *et al.*, 2010; Mitchell, 2016; Fig. 1A). In accordance with these findings, growth of *C. acetobutylicum* on each sugar as the single carbon source was reported for glucose (Amador-Noguez *et al.*, 2011; Aristilde *et al.*, 2015), galactose (Guiterrez and Maddox, 1996; Raganati *et al.*, 2012), mannose (Raganati *et al.*, 2012; Voigt *et al.*, 2014), xylose (Raganati *et al.*, 2012; Aristilde *et al.*, 2015; Kudahettige-Nilsson *et al.*, 2015) and arabinose (Raganati *et al.*, 2012; Servinsky *et al.*, 2012; Zhang *et al.*, 2012; Aristilde *et al.*, 2015). The genes encoding for the two mannose transporters were shown to be expressed within the same

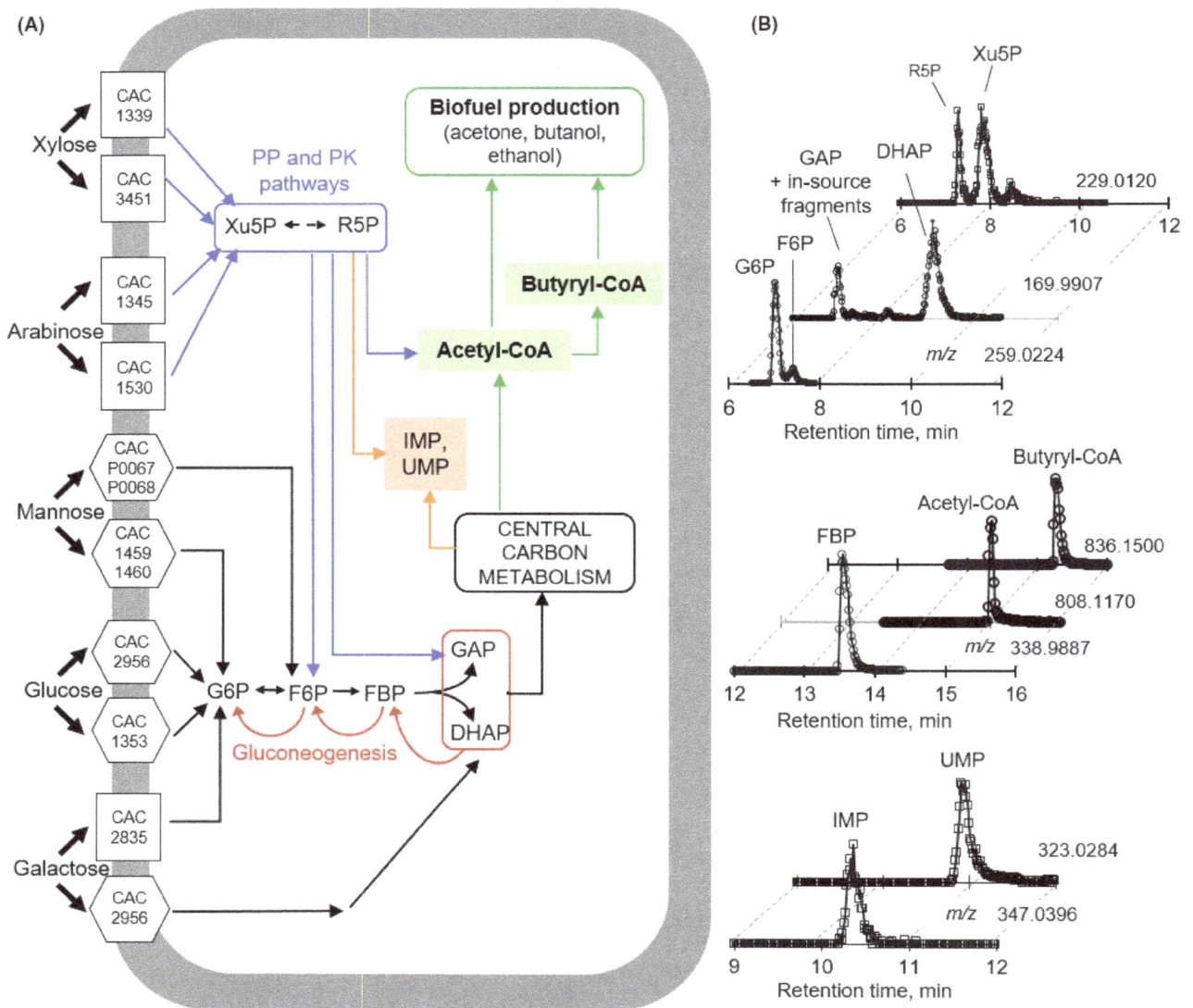

Fig. 1. Schematic central carbon metabolism and metabolite identification.
(A) Schematic routing of sugar catabolism towards biofuel production following sugar uptake in *Clostridium acetobutylicum*. The black lines represent catabolic pathways for the hexose sugars and in the central carbon metabolism, the blue lines represent catabolic pathways of pentose sugars through the pentose-phosphate (PP) and phosphoketolase (PK) pathways, the dark red lines show reactions involved in gluconeogenesis; the genes reported to encode the sugar uptake transporters are shown (Servinsky *et al.*, 2010).
(B) Liquid chromatography–mass spectrometry chromatogram of important metabolites isolated from glucose-grown cells.
Abbreviations: G6P, glucose 6-phosphate; F6P, fructose-6-phosphate; DHAP, dihydroxyacetone-phosphate; GAP, glyceraldehyde-3-phosphate; Xu5P, xylulose-5-phosphate; R5P, ribose-5-phosphate; FBP, fructose-1,6-bisphosphate; Acetyl-CoA, acetyl-coenzyme A; butyryl-CoA, butyryl-coenzyme A; IMP, inosine monophosphate; UMP, uridine monophosphate.

order of magnitude during growth on mannose alone versus glucose alone, with slightly higher expression during growth on mannose alone (Servinsky *et al.*, 2010; Fig. 1A). By contrast, the corresponding genes for galactose transport were completely suppressed when cells were grown on glucose alone (Servinsky *et al.*, 2010; Fig. 1A). The genes encoding the four uptake transporters, two each for xylose and arabinose, were highly expressed in cells grown on xylose alone or arabinose

alone but not in glucose-grown cells (Servinsky *et al.*, 2010). Thus, these transcriptional results implied that glucose-grown *C. acetobutylicum* may accommodate uptake of mannose but not of galactose, xylose or arabinose. However, much still remains unknown regarding the simultaneous utilization of both glucose and another sugar in *C. acetobutylicum*.

Following uptake, pentose and hexose sugars follow distinct paths through the central carbon metabolism and towards two important biofuel precursors: acetyl-coenzyme A (acetyl-CoA) and butyryl-coenzyme A (butyryl-CoA) (Fig. 1A). Acetyl-CoA is a precursor to acetate and ethanol (Lee et al., 2008; Aristilde et al., 2015; Dash et al., 2016). Acetyl-CoA combines with another acetyl-CoA to produce acetoacetyl-CoA, which is a precursor to acetone and butyryl-CoA; the latter is used to synthesize butyrate and butanol (Lee et al., 2008; Aristilde et al., 2015; Dash et al., 2016). Glucose, the primary hexose, is metabolized through the glycolytic pathway, which involves glucose phosphorylation to hexose-phosphates [glucose 6-phosphate (G6P) and fructose-6-phosphate (F6P)] followed by splitting of a bis-phosphorylated hexose, fructose-1,6-bisphosphate (FBP), to triose-phosphates [glyceraldehyde-3-phosphate (GAP) and dihydroxyacetone-3-phosphate (DHAP)] (Fig. 1A). These triose-phosphates subsequently feed into the central carbon metabolism, which connects to the production of biofuels (Fig. 1A).

Mannose-derived carbons can enter metabolism either via G6P, similar to glucose, or via F6P (Servinsky et al., 2010); galactose-derived carbons are catabolized either via the Leloir pathway that generates G6P or the tagatose-6P pathway that directly produces GAP and DHAP (Sund et al., 2013; Fig. 1A). The gene for the enzyme that connects phosphorylated mannose to glycolysis via F6P was similarly expressed in both glucose-grown and mannose-grown C. acetobutylicum (Servinsky et al., 2010). On the other hand, the genes responsible for galactose assimilation were minimally expressed when cells were grown on glucose alone (Servinsky et al., 2010) (Fig. 1A). Preferential consumption of glucose over galactose in a C. acetobutylicum during growth on both substrates implied that inhibition of galactose assimilation by the presence of glucose may persist even when galactose is also present (Guiterrez and Maddox, 1996). It is not yet clear how mannose transport and assimilation are influenced by the presence of glucose.

With respect to pentose sugar metabolism, there are two paths for their catabolic route to biofuel production. Following phosphorylation to xylulose-5-phosphate (Xu5P) and ribose-5-phosphate (R5P), pentose sugars get fed into the pentose-phosphate (PP) pathway, which connects to the glycolytic pathway by generating the hexose-phosphate F6P and the triose-phosphate GAP. Alternatively, metabolites in the PP pathway can bypass most of glycolysis to produce acetyl-CoA directly through the phosphoketolase (PK) pathway (Fig. 1A; Servinsky et al., 2012; Liu et al., 2012; Aristilde et al., 2015). The PK reactions involve the cleavage of the PP pathway metabolite Xu5P or the glycolytic metabolite F6P to produce, respectively, GAP and acetyl-phosphate (acetyl-P) or erythrose-4-phosphate and acetyl-P. The metabolite acetyl-P can be converted directly to

either acetate or acetyl-CoA. Genes involved in pentose catabolism in both the PP and PK pathways were found to be still expressed in glucose-grown cells, though less abundantly (Servinsky et al., 2010, 2012). The inhibition of pentose metabolism in the presence of glucose has been well documented in C. acetobutylicum (Ounine et al., 1985; Fond et al., 1986; Gu et al., 2009; Grimmler et al., 2010; Aristilde et al., 2015). Despite the seemingly minimal pentose utilization in C. acetobutylicum in the presence of glucose based on extracellular substrate depletion, ^{13}C-labelling experiments recently revealed that assimilated pentose carbons from glucose:pentose mixtures can be accumulated in PP pathway intermediates, leaving glucose as the dominant sugar incorporated into glycolytic metabolites (Fig. 1A; Aristilde et al., 2015). Regarding the involvement of the PK pathway, the labelling patterns of acetyl-P in tracer experiments revealed an increased participation of the PK pathway in the presence of arabinose whereby arabinose-derived carbons were routed through the PK pathway during feeding on glucose:arabinose mixture (Aristilde et al., 2015). Whether the PK pathway may also provide a connection between arabinose catabolism and biofuel precursors in the presence of glucose was not determined.

In addition to connecting assimilated pentoses to the glycolytic and PK pathways, the PP pathway provides the ribose sugar backbone for inosine monophosphate (IMP) and uridine monophosphate (UMP). These two metabolites are required for de novo ribonucleotide biosynthesis: IMP is a precursor to purines and UMP is a precursor to pyrimidines (Fig. 1A). Pentose accumulation in the PP pathway in the presence of glucose was thus proposed to serve as a strategy to route pentoses towards ribonucleotide synthesis during growth on glucose:pentose mixtures (Aristilde et al., 2015). Confirmation of this metabolic strategy has not yet been reported in C. acetobutylicum or other clostridium species.

Building on the aforementioned studies, this study employs a metabolomics approach to investigate the following four hypotheses regarding the co-metabolism of glucose and each hemicellulosic sugar in C. acetobutylicum: (1) glucose inhibits galactose metabolism and its subsequent contribution to biofuel precursors; (2) glucose does not compromise mannose metabolism towards biofuel precursors, (3) pentoses are routed from the PP pathway towards ribonucleotide biosynthesis but not glycolysis, and (4) pentoses are routed from the PP pathway to biofuel precursors through the PK pathway in the presence of glucose. High-resolution liquid chromatography–mass spectrometry (LC-MS) was applied to track the simultaneous incorporation of stable isotope-labelled and unlabelled substrates from mixed-sugar mixtures into intracellular metabolites in glycolysis, the

PP pathway, ribonucleotide biosynthesis and biofuel precursors. The results unravelled the metabolic hierarchies of glucose with respect to each of the hemicellulosic sugars. Contrary to galactose metabolism, which was subjected to near-complete inhibition by glucose, mannose was well incorporated into glycolysis as well as biofuel precursors. Investment of both pentoses into ribonucleotide precursors was evident in the presence of glucose, despite the minimal contribution of the pentoses to biofuel precursors via glycolysis. The data also indicated that the PK pathway may connect specifically arabinose to biofuel production, albeit to a relatively smaller contribution compared to glucose. These findings shed light into the constitutive metabolic hierarchy that underpins the channelling of sugar mixtures towards biofuel-generating pathways in *C. acetobutylicum*.

Results and discussion

Proof-of-concept labelling experiments

Essential to tracking cellular metabolism is the identification of metabolites in cellular extracts (Fig. 1A). All the relevant metabolites were detected using established methods applying high-performance LC followed by electrospray ionization and detection using high-resolution MS in negative mode (Kimball and Rabinowitz, 2006; Rabinowitz and Kimball, 2007; Lu *et al.*, 2010; Xu *et al.*, 2015; Fig. 1B). The four pentose- and hexose-monophosphates (Xu5P, R5P, G6P, F6P) were detected between retention time (RT) of 7 and 8.5 min and FBP at RT of 13.6 min (Fig. 1B). As a result of the chromatographic separation, the PP metabolites, Xu5P and R5P, were detected despite their common *m/z* value at 229.0120 (Fig. 1B). In a similar fashion, the chromatographic separation allowed for simultaneous detection of G6P and F6P at the same *m/z* channel of 259.0024 (Fig. 1B). The metabolite FBP was detected at *m/z* 338.9887 and chromatographic separation facilitated the detections of both GAP and DHAP at *m/z* 169.9907 (Fig. 1B). It was recently pointed out that, even with soft ionization such as electrospray ionization, in-source fragmentation can interfere with distinct metabolite detection (Xu *et al.*, 2015). One such example is the fragmentation of F6P into GAP (Xu *et al.*, 2015) (Fig. 1B). As a result, only the isotopic enrichment of DHAP was monitored here when investigating the metabolism of the different sugar mixtures, which will be discussed in the next sections.

In addition to the phosphorylated metabolites in both the PP and glycolytic pathways, the detection of important precursors to both biofuel and nucleic acid biosynthesis was achieved (Fig. 1A and B). Acetyl-CoA is a direct precursor to ethanol and, combined with another acetyl-CoA, to yield acetone as well as butyryl-CoA, a precursor to butanol (Aristilde *et al.*, 2015). At very close

RT of 15–15.2 min, acetyl-coA and butyryl-CoA were detected, respectively, at *m/z* of 808.1170 and 836.1500 (Fig. 1B). Precursors to *de novo* biosynthesis of purines and pyrimidines, respectively IMP and UMP, were captured at close RT (10.9 and 10.3 min, respectively) but at their distinct *m/z* of 347.0396 and 323.0284 respectively (Fig. 1B). The findings above confirmed that the pertinent metabolites in the metabolic pathways of interest can be detected well by the LC-MS approach applied here (Fig. 1A and B).

Next, proof-of-concept labelling experiments were conducted with *C. acetobutylicum* fed on [U-$^{13}C_6$]-glucose alone or with equimolar unlabelled glucose (Fig. 2). Preliminary kinetic experiments indicated no significant changes in the isotopic enrichment in the glycolytic metabolites after 30 min (Fig. S1). Therefore, only the 30 min labelling data are presented in Fig. 2. When the cells were fed only the labelled glucose, about 90% of both G6P and F6P were fully labelled whereas only up to 40–50% of these metabolites were fully labelled when the cells were fed simultaneously labelled glucose and unlabelled glucose (Fig. 2), in accordance with near equal incorporation of the labelled and unlabelled glucose by 30 min. The slightly less than 50% of fully labelled F6P is due to up to 10% of triply labelled F6P, suggesting gluconeogenic flux from FBP to F6P (Fig. 2). Indeed, the FBP labelling shows clear evidence of gluconeogenic flux (Fig. 2). Forward glycolytic flux would only result in either non-labelled or fully ^{13}C-labelled FBP but triply ^{13}C-labelled FBP was measured in both glucose labelling schemes, indicating reverse flux of nonlabelled and fully labelled triose-phosphates (GAP and DHAP) combined to form FBP (Fig. 2). Specifically, 15% and 44% of FBP on average were triply ^{13}C-labelled, respectively, in the cells grown on labelled glucose alone or with unlabelled glucose (Fig. 2). Accordingly, DHAP was found to be over 92% fully labelled in cells fed labelled glucose and over 42% fully labelled in cells fed the mixture of labelled with unlabelled glucose (Fig. 2).

The metabolites (Xu5P and R5P) in the PP pathway also exhibited differences in the labelling, in agreement with the labelling of the hexose- and triose-phosphates as discussed above (Fig. 2 and Fig. S2). When cells were fed only labelled glucose, both Xu5P and R5P were about 82% fully labelled, 6–7% triply ^{13}C-labelled and 5–10% doubly ^{13}C-labelled (Fig. 2); the doubly and triply ^{13}C-labelled are due to reactions involving the minor fractions of nonlabelled hexose- and triose-phosphates remaining at the 30 min labelling time (Figs 2 and S2). Due to higher fraction of nonlabelled hexose- and triose-phosphates in cells grown on both labelled and unlabelled labelled glucose compared to the growth condition with labelled glucose alone, the nonlabelled

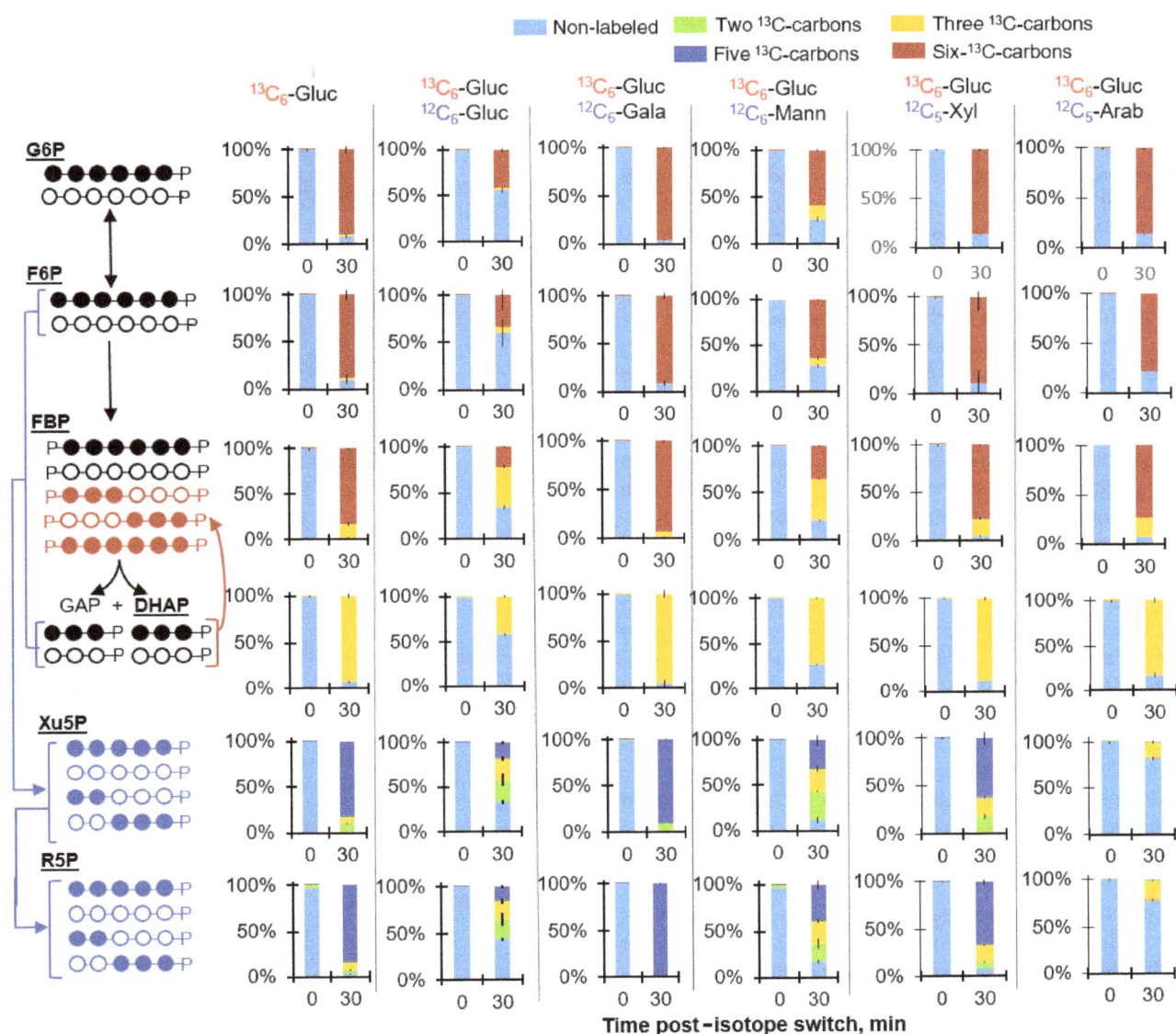

Fig. 2. Mixed-sugar catabolism of stable isotope-labelled glucose with unlabelled glucose, galactose, mannose, xylose or arabinose. The carbon mapping on the left illustrates the different labeling forms of the metabolites based on isotopic enrichment from substrate feeding. The filled circles and open circles represent, respectively, ^{13}C-labeled carbons and unlabeled carbons; black, red, and blue circles are assigned to labeling schemes in glycolytic, gluconeogenic, and pentose-phosphate pathways respectively. Labeling of glycolytic and pentose-phosphate pathway metabolites following 30 min incorporation of fully labelled glucose ([U-^{13}C$_6$]-Gluc) alone or with unlabelled Gluc, galactose (Gala), mannose (Mann), xylose (Xyl) or arabinose (Arab) respectively. The carbon mapping on the left illustrates the different labelling forms of the metabolites based on the fed substrate(s). Colour codes for the labelling isotopologues: non-labelled carbon (light blue), two ^{13}C-carbons (light green), three ^{13}C-carbons (yellow); five ^{13}C-carbons (dark blue), and six ^{13}C-carbons (dark red). The measured data (average ± standard deviation) were from biological replicates (n = 2–3). Non-noticeable error bars were in cases where standard deviation values were small.

doubly and triply ^{13}C-labelled fractions for R5P and Xu5P were higher at 31–42%, 20–25% and 19–23% respectively (Figs 2 and S2).

The aforementioned proof-of-concept results were used as a guide to determine the hierarchy in the co-metabolism of glucose with hemicellulosic sugars such that labelling results that match the ^{13}C-glucose-alone data would indicate complete repression of the accompanying substrate in the presence of glucose whereas labelling results that match the data obtained with the mixture of ^{13}C-glucose and unlabelled glucose would be consistent with equal metabolism of glucose and the hemicellulosic sugar. The following sections detailed the co-metabolism of labelled glucose with the unlabelled form of each hemicellulosic sugar of interest: galactose, mannose, xylose or arabinose.

Glucose metabolism with respect to a hemicellulosic hexose: mannose or galactose

Figure 2 also illustrates 30 min labelling patterns of the intracellular metabolites obtained following feeding on [U-$^{13}C_6$]-glucose with an unlabelled hemicellulosic hexose: mannose or galactose (Fig. 2). Kinetic experiments showed that within 15 min, the isotopic enrichment of G6P, F6P and the triose-phosphate DHAP had reached equilibrium (Fig. S3). Therefore, the 30 min labelling data shown in Fig. 2 represented near steady-state labelling of these metabolites by the assimilated substrates.

During simultaneous feeding on glucose and galactose, the labelling patterns of metabolites in the glycolytic and PP pathways were about identical to the metabolite labelling during feeding on glucose alone (Fig. 2). These labelling data thus indicated the exclusion of galactose catabolism in the presence of glucose, in agreement with a previous report that glucose was preferred over galactose in *C. acetobutylicum* P262 (Guiterrez and Maddox, 1996). By contrast, the metabolite labelling patterns during simultaneous feeding on glucose and mannose revealed incorporation of both nonlabelled carbons from mannose and the labelled carbons from glucose (Fig. 2). By comparing the specific isotopologues in cells fed ^{13}C-glucose with unlabelled mannose versus those fed ^{13}C-glucose with unlabelled glucose, it was clear that mannose catabolism was not identical to glucose catabolism (Fig. 2).

In the presence of unlabelled mannose and labelled glucose, fully ^{13}C-labelled fractions of G6P, F6P and DHAP were, on average, 58%, 60%, and 73% respectively (Fig. 2). These results implied higher incorporation of glucose than mannose (Fig. 2). The persistence of triply ^{13}C-labelled FBP (on average, 43%) was consistent with the occurrence of gluconeogenic flux in the presence of mannose, similar to the glucose-only condition (Fig. 2). However, slightly higher fractions of triply ^{13}C-labelled G6P and F6P (on average, up to 15%) implied greater flux of the gluconeogenic flux in the presence of mannose than in the presence of glucose alone (Fig. 2). The presence of partially labelled PP metabolites (Xu5P and R5P) with doubly (up to 24%) and triply (up to 32%) ^{13}C-labelled fractions were also in agreement with the assimilation of mannose in the presence of glucose (Fig. 2). It was previously reported that glucose-grown *C. cellulotycium*, a less efficient biofuel producer than glucose-grown *C. acetobutylicum*, exhibited a reversal glycolytic pathway whereby feeding on 50% labelled glucose led to about 12% of triply ^{13}C-labeled F6P (Rabinowitz *et al.*, 2015). Therefore, *C. acetobutylicum* grown on a glucose:mannose mixture had a gluconeogenic flux that was comparable to glucose-grown *C. cellulolitycum*. This gluconeogenic flux was proposed to impair glycolytic flux towards biofuel production in *C.*

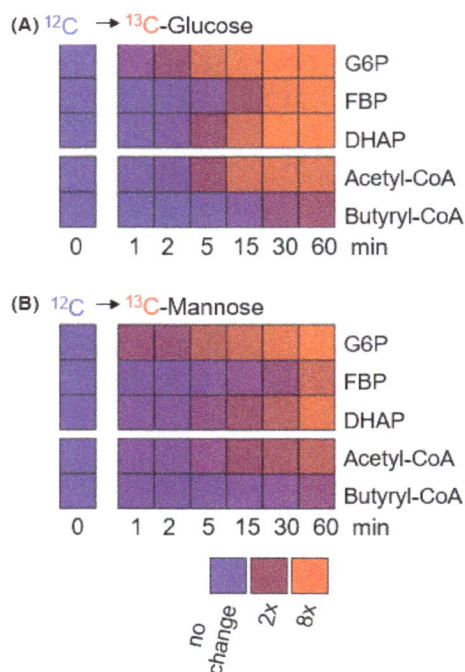

Fig. 3. Mannose substitution for glucose in *Clostridium acetobutylicum*.
(A) Kinetic metabolite labelling when cells were switched from unlabelled (^{12}C) glucose to fully labelled (^{13}C) glucose. (B) Kinetic metabolite labelling when cells were switched from unlabelled (^{12}C) glucose to fully labelled (^{13}C) mannose.
Levels of nonlabelled and labelled fractions are shown in blue and red respectively. The measured data used for the heat map are averages from two biological replicates.

cellulolitycum (Rabinowitz *et al.*, 2015). Whether the same phenomenon can contribute to decreased biosynthesis of biofuel precursors in *C. acetobutylicum* grown on mixed substrates remains to be determined.

A previous study reported similar expression levels of genes encoding for mannose transport and metabolism in glucose-grown cells versus mannose-grown cells (Servinsky *et al.*, 2010). Therefore, for the cells grown on the glucose:mannose mixture, mannose assimilation may be facilitated by constitutive transporters present in glucose-grown cells. In fact, it was determined that mannose transport was primarily via glucose transporters in two human cells lines (Rodríguez *et al.*, 2005). Here, switch-substrate labelling experiments demonstrated that *C. acetobutylicum* can fully substitute glucose by rapidly catabolizing mannose (Fig. 3). Within only 1 min following glucose removal and introduction of ^{13}C-labelled mannose, there was already incorporation of mannose-derived labelled carbons into both glycolytic intermediates and the biofuel precursor acetyl-CoA (Fig. 3). Within 15 min after the isotope switch, assimilated mannose fully populated the metabolites and, by 60 min, the isotopic enrichment was nearly the same as in cells incorporating labelled glucose (Fig. 3). Therefore, in agreement with results from the transcriptional analysis of glucose-grown *C. acetobutylicum* (Servinsky *et al.*, 2010), the results presented here are consistent with constitutive flexibility for mannose uptake and metabolism in glucose-grown cells.

Glucose metabolism with respect to a hemicellulosic pentose: xylose or arabinose

In regard to the co-metabolism of glucose with hemicellulosic pentose sugars in *C. acetobutylicum*, Fig. 2 illustrates 30 min labelling of the intracellular metabolites obtained following feeding of *C. acetobutylicum* on [U-^{13}C$_6$]-glucose with unlabelled xylose or unlabelled arabinose. In a previous study (Aristilde *et al.*, 2015), data were provided to show that near steady-state isotopic enrichment in glycolytic and PP intermediates was achieved by 30 min during growth on labelled glucose alone and labelled xylose alone. Here, additional experiments were conducted to confirm that a time period of 30 min was sufficient to achieve near steady-state isotopic enrichment in the glycolytic and PP pathway metabolites during growth on the glucose:xylose and glucose:arabinose mixtures wherein glucose was fully labelled (Fig. S4). The data indicated no appreciable changes in the labelling patterns of the metabolites extracted at 30 min versus 60 min, thus confirming a 30 min isotopic enrichment was also sufficient (Fig. S4).

The 30 min labelling patterns of glycolytic metabolites indicated that these metabolites were largely exclusively populated by glucose-derived carbons similar to glucose-alone conditions (Fig. 2). And, measurements of both non-labelled and partially labelled Xu5P and R5P compared to the fully labelled PP pathway metabolites measured during growth on labelled glucose alone were consistent with assimilation of the pentose sugars into PP pathway (Fig. 2). These results are thus in agreement with previous results, which showed that the pentose sugars were incorporated into PP pathway metabolites after long-term isotopic enrichment during cell growth for multiple doubling time on labelled glucose and unlabelled pentose sugar (Aristilde *et al.*, 2015). However, the kinetic data obtained here revealed more rapid kinetic incorporation of arabinose than xylose (Fig. 2). Whereas the pentose-phosphates were about 80% nonlabelled and the remaining fraction triply ^{13}C-labelled in the presence of unlabelled arabinose and labelled glucose, the corresponding labelling pattern in the presence of unlabelled xylose and labelled glucose was, on average, 60–65% quintuply ^{13}C-labelled, 18–20% triply ^{13}C-labelled, and 8–6% doubly ^{13}C-labelled (Fig. 2). This difference indicated that the assimilation of the pentose sugar was more prioritized over the contribution of glucose-derived carbons for the biosynthesis of the pentose-

phosphates in the presence of arabinose than in the presence of xylose (Fig. 2). These results stressed the preference of arabinose over xylose for uptake and metabolism in the PP pathway, as previously reported (Ezeji and Blaschek, 2007; Aristilde *et al.*, 2015).

It was proposed that the assimilation of pentoses into PP pathway in *C. acetobutylicum* with little subsequent contribution towards glycolytic intermediates may be a metabolic strategy to invest pentose-derived carbons specifically into ribonucleotide biosynthesis (Aristilde *et al.*, 2015). The biosynthesis of these ribonucleotides combines metabolites from different metabolic pathways: PPP, glycolysis and TCA cycle (Fig. 4A). The biosynthesis of the purine IMP combines the PP pathway metabolite R5P with glycine [an amino acid derived from the TCA cycle intermediate oxaloacetate (Amador-Noguez *et al.*, 2010)], formate [one-carbon unit derived from the glycolytic metabolite pyruvate (Amador-Noguez *et al.*, 2010)] and dissolved bicarbonate species (Fig. 4A). And, the biosynthesis of the pyrimidine UMP stems from the combination of R5P with aspartate (a TCA cycle-derived amino acid) and dissolved carbonate species (Fig. 4A). A preliminary kinetic isotopic flux experiment with fully labelled glucose revealed that the labelling patterns of both IMP and UMP at 60 min were significantly different from those obtained at 30 min (SI, Fig. S5). Therefore, long-term isotopic enrichment experiments of the ribonucleotides IMP and UMP were conducted wherein the cells were subjected to growth for several hours in minimal medium solution containing unlabelled glucose with either [1,2-^{13}C$_2$]-xylose or [1-^{13}C$_1$]-arabinose in order to evaluate the proposal that assimilated pentoses were routed to ribonucleotide biosynthesis (Fig. 4B and Fig. 4C).

In accordance with the above discussion that glucose dominated glycolysis and downstream metabolic pathways, only R5P from the PP pathway populated by the assimilated pentoses was expected to contribute labelled carbons to the ribonucleotides. Indeed, the labelled forms of both IMP and UMP reflected the labelled forms of R5P under each growth condition: doubly ^{13}C-labelled IMP (50%) and UMP (42%) in cells grown on the glucose:xylose mixture with unlabelled glucose and ^{13}C$_2$-xylose; singly ^{13}C-labelled IMP (84%) and UMP (61%) in cells grown on the glucose:arabinose mixture with unlabelled glucose and ^{13}C$_1$-arabinose (Fig. 4B and C; SI, Fig. S6). These data also highlight two differences between the two glucose:pentose growth conditions. First, in both conditions, there was a higher isotopic enrichment (by 15–27%, on average) of IMP than of UMP (Fig. 4B and C), implying less *de novo* biosynthesis of the latter than the former. In agreement with more recycling (thus less *de novo* biosynthesis) of UMP than IMP, the kinetic data with fully labelled glucose indicated primarily labelled forms of IMP in accordance with incorporation of glucose-derived

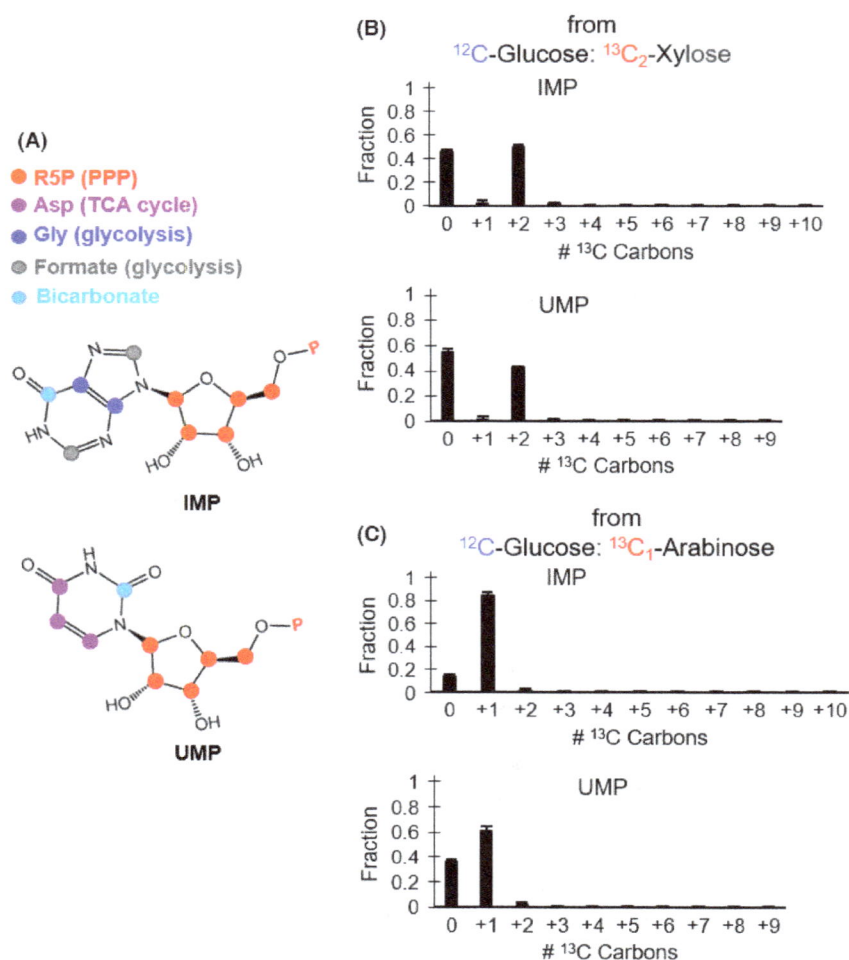

Fig. 4. Incorporation of pentoses into the ribonucleotides inosine monophosphate (IMP) and uridine monophosphate (UMP) in the presence of glucose.
(A) Metabolic pathways linked to ribonucleotide biosynthesis.
(B) Isotopic enrichment of IMP (top) and UMP (bottom) following growth on unlabelled glucose with $[1,2-^{13}C_2]$-xylose.
(C) Isotopic enrichment of IMP (top) and UMP (bottom) following growth on unlabelled glucose with $[1-^{13}C_1]$-arabinose.
The measured data (average $\pm$ standard deviation) were from biological replicates (n = 2–3). Non-noticeable error bars were in cases where standard deviation values were small.

carbons whereas the labelling pattern of UMP still contained, on average, about 34% nonlabelled forms (Fig. S5). Second, there was a higher fraction (by up to 50%, on average) of the labelled forms of both IMP and UMP in the presence of arabinose than xylose (Fig. 4), implying higher rate of *de novo* ribonucleotide biosynthesis in the presence of arabinose. This would be consistent with the aforementioned metabolic preference of arabinose over xylose in *C. acetobutylicum* (Aristilde *et al.*, 2015).

Contributions of hemicellulosic sugars to biofuel precursors in the presence of glucose

To determine the consequence of the different metabolic hierarchies on the routing of carbons towards biofuel

precursors, the labelling patterns of both acetyl-CoA and butyryl-CoA were obtained following feeding on fully ^{13}C-labelled glucose alone or with unlabelled glucose, galactose, mannose, xylose or arabinose (Fig. 5). The triose-phosphates generated in the glycolytic pathway ultimately produce the two-carbon acetyl moiety in acetyl-CoA following a decarboxylation step; the butyryl moiety in butyryl-CoA is the joining of two moles of acetyl moiety (Fig. 5) – the CoA moiety is generated from secondary metabolism that combines ATP with metabolites derived from glycolysis. The short-term 30 min labelling data obtained here focused on profiling the kinetic labelling of the acetyl and butyryl moieties when labelling of the CoA component would be relatively minor (Fig. 5). Accordingly, during growth on fully labelled glucose alone, acetyl-CoA was primarily doubly ^{13}C-labelled

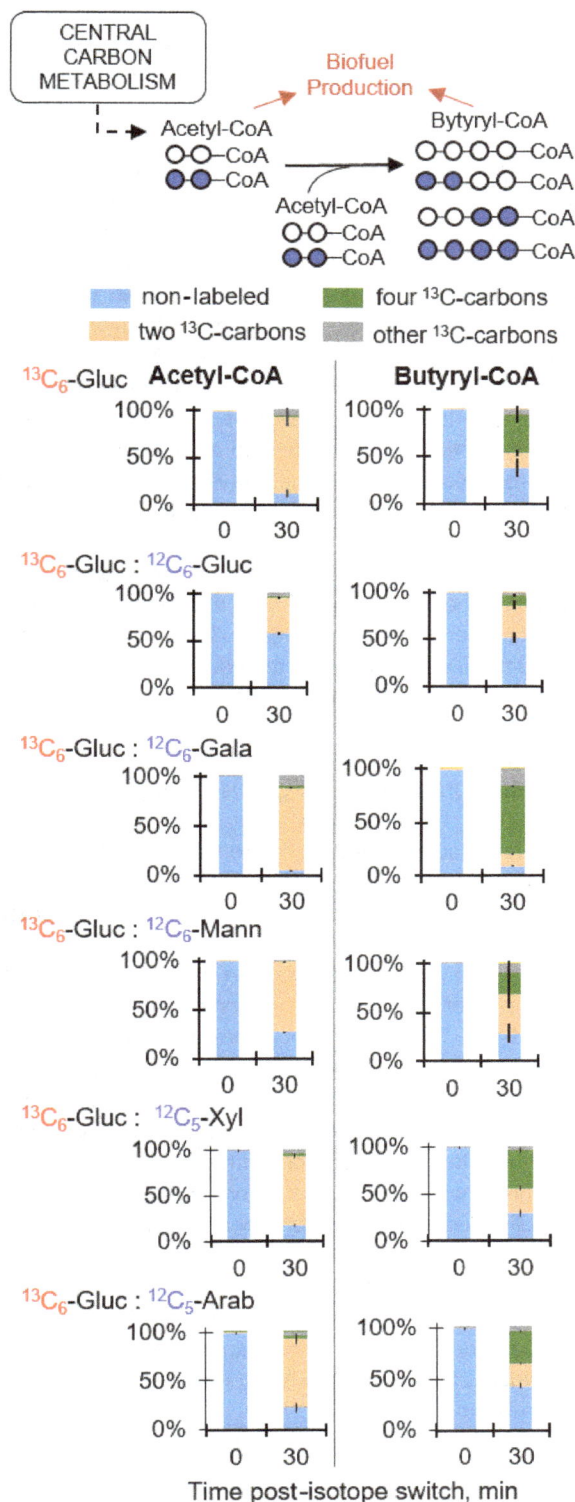

Fig. 5. Sugar investment in the biofuel precursors acetyl-coenzyme A (acetyl-CoA) and butyryl-coenzyme A (butyryl-CoA). Labelling patterns were obtained following 30 min incorporation of substrates as detailed in Fig. 2 legend. The measured data (average ± standard deviation) were from biological replicates (n = 2–3). Non-noticeable error bars were in cases where standard deviation values were small.

(on average, greater than 82%) in accordance with the decarboxylation of the triply ^{13}C-labelled fraction (about 90%, on average) of triose-phosphates as previously described (Fig. 2 and Fig. 5). Interestingly, only about 60% of butyryl-CoA was labelled with, on average, 17% doubly and 41% quadruply ^{13}C-labelled fractions (Fig. 5). The difference (~20% less) between the ^{13}C-labelled fraction in butyryl-CoA compared to acetyl-CoA implied a delay in the metabolic flux to synthesize butyryl-CoA downstream of acetyl-CoA (Fig. 5). In the 50:50 mixture with labelled glucose and unlabelled glucose, the labelling pattern of acetyl-CoA was approaching a near equal fraction of nonlabelled and doubly ^{13}C-labelled fractions, consistent with the incorporation of both fully labelled and nonlabelled glucose-derived carbons (Fig. 2 and Fig. 5). The subsequent labelling of butyryl-CoA confirmed the delay in its biosynthetic flux as there was very little quadruply ^{13}C-labelled fraction (Fig. 5).

Next, the contribution of the hemicellulosic hexose sugars to the biofuel precursors in the presence of glucose was determined (Fig. 5). During feeding on labelled glucose and unlabelled galactose, the labelling of acetyl-CoA, on average at 83% doubly ^{13}C-labelled, was nearly identical to the labelling during feeding on labelled glucose alone, but there was an unexpected higher isotopic enrichment in butyryl-CoA, with about a 58% increase in the quadruply ^{13}C-labelled fraction when compared to the glucose-alone condition (Fig. 5). Thus, the labelling patterns of both acetyl-CoA and butyryl-CoA during growth on the glucose:galactose mixture were consistent with the complete exclusion of galactose assimilation from the glucose:galactose mixture (Fig. 5). In addition, the higher isotopic enrichment of butyryl-CoA implied a higher biosynthetic flux of butyryl-CoA from feeding on the glucose:galactose mixture than feeding on glucose alone (Fig. 5). The significance of this phenomenon warrants further investigation. In terms of mannose contribution to the biofuel precursors during feeding on the glucose:mannose mixture, there was a 12% decrease in the doubly ^{13}C-labelled in acetyl-CoA and a 50% decrease in the quadruply ^{13}C-labelled in butyryl-CoA when compared to feeding on glucose alone (Fig. 5). These results indicated that, in contrast to feeding on the glucose:galactose mixture, there was simultaneous routing of carbons from both mannose and glucose towards biofuel precursors during feeding on the glucose:mannose mixture (Fig. 5), although glucose was still preferred over mannose.

With respect to the pentose contribution to the biofuel precursors in the presence of glucose, the contribution was not the same from xylose and arabinose (Fig. 5). During feeding on the glucose:xylose mixture, the labelling patterns of both acetyl-CoA and butyryl-CoA were comparable to those obtained in the presence of glucose alone, indicating little contribution of the xylose assimilated in the PP pathway towards the biofuel precursors (Figs 2 and 5). Accordingly, the reported yields of acids and solvents obtained during *C. acetobutylicum* growth on the glucose: xylose mixture were comparable to those obtained in the presence of glucose alone (Aristilde *et al.*, 2015).

Compared to cells grown on the glucose:xylose mixture, the labelling data obtained during feeding on the glucose:arabinose mixture demonstrated an increase in the incorporation of nonlabelled carbons into both acetyl-coA and butyryl-CoA by, on average, 28% and 43% greater, respectively (Fig. 5). This investment of arabinose into acetyl-CoA (Fig. 5), combined with the lack of arabinose-derived carbons in upper glycolytic metabolites (G6P, F6P, FBP) (Fig. 2), during growth on the glucose:arabinose mixture was in agreement with the previously reported generation of acetyl-P, a precursor to acetyl-CoA, equally from glycolysis and the PK pathway (Aristilde *et al.*, 2015). It was also shown previously that there was a higher yield (by up to 20%) of acetate following growth on glucose:arabinose compared to growth on glucose:xylose; there was no change, however, in the yield of alcohols and acetone (Aristilde *et al.*, 2015). Therefore, the acetyl-P generated to produce acetyl-CoA in the presence of arabinose seemed to be discarded as acetate instead of being invested into solvent production (Aristilde *et al.*, 2015).

Concluding remarks

Clostridial species are important in the fermentation of cellulosic and hemicellulosic sugars in environmental matrices and engineered bioreactors. The present study sought to gain metabolic insights into the sugar hierarchies in the notable biofuel producer *C. acetobutylicum* (Rabinowitz *el al.*, 2015). The following four hypothesized hierarchies regarding the metabolism of glucose in relation to four different hemicellulosic sugars were evaluated here: (1) glucose inhibition of galactose metabolism, (2) uncompromised co-metabolism of mannose in the presence of glucose, (3) contribution of pentoses to ribonucleotides and not biofuel production and (4) connection of pentoses to biofuel precursors via the PK pathway. Using ^{13}C tracer experiments, intracellular metabolite labelling was monitored to unravel these metabolic hierarchies.

Galactose incorporation into intracellular metabolism was not observed. Thus, the repression of the genes that encode galactose catabolism in glucose-grown cells (Servinsky *et al.*, 2010) persisted during growth on glucose:galactose mixtures. In contrast to galactose metabolism, mannose metabolism was not inhibited by the presence of glucose. Furthermore, it was found that mannose metabolism is analogous to glucose metabolism such that mannose can fully substitute glucose upon glucose absence. Subsequent biochemical studies are needed to determine whether mannose uptake exploits constitutive transporters of glucose in addition to mannose transporters in clostridial species. With respect to the metabolism of the glucose:pentose mixtures, the results revealed that both pentose sugars contributed to *de novo* ribonucleotide biosynthesis. The data were also in agreement with previously reported preference of arabinose over xylose for both consumption (Ezeji and Blaschek, 2007) and assimilation into the PP and PK pathways (Aristilde *et al.*, 2015). Moreover, the results here demonstrated appreciable contribution of arabinose to the biofuel precursors via the PK pathway, with potential contribution to acetate but not acetone production according to previous reports of acid and solvent yields from glucose:pentose mixtures (Aristilde *et al.*, 2015).

Two important factors should be considered when evaluating the relevance of the metabolic hierarchies presented here for mixed-sugar utilization in *C. acetobutylicum*. First, plant waste materials are composed of glucose with multiple hemicellulosic sugars simultaneously present. Based on substrate consumption rates during growth of the same strain of *C. acetobutylicum* (strain 824) on a mixture of glucose with multiple hemicellulosic sugars, a previous study (Ezeji and Blaschek, 2007) reported substrate preference of glucose over mannose, glucose over both pentoses, and arabinose over xylose. The last two substrate hierarchies agreed with the metabolomics results presented here, but the first was not consistent with the non-preferential co-metabolism of glucose and mannose (Ezeji and Blaschek, 2007). This discrepancy may be due to the 5:1 glucose:mannose ratio in the mixture composition of the previous study (Ezeji and Blaschek, 2007) compared to the 1:1 glucose:mannose mixture used here. It is important to note that the scope of the metabolomics analysis performed here was focused on monitoring glucose metabolism with respect to one hemicellulosic hexose or pentose sugar. Therefore, a metabolomics investigation of *C. acetobutylicum* fed simultaneously on glucose with a complete suite of hemicellulosic sugars is needed to shed light on how the metabolic hierarchies revealed here would manifest in the presence of more complex sugar mixtures.

Second, in order to design optimal engineering strategies for enhancing mixed-sugar metabolism towards biofuel production, it is important to distinguish between metabolic regulation versus transcriptional regulation (Liao *et al.*, 2015; Dash *et al.*, 2016; Richter *et al.*, 2016). Therefore, as was conducted for *C. acetobutylicum* fed on a single hexose or pentose substrate (Servinsky *et al.*, 2010), a detailed transcriptional analysis of *C. acetobutylicum* fed on sugar mixtures is warranted. As a necessary complement to this analysis, the present findings provide metabolic evidence for the hierarchical investment of different sugars through central carbon metabolism and towards the biosynthesis of nucleic acids and biofuel precursors.

Experimental procedures

Culturing conditions

Batch growth experiments of *C. acetobutylicum* (strain 824, American Type Culture Collection) were conducted in 250 ml Erlenmeyer flasks inside a Bactron IV SHEL LAB (Cornelius, OR, USA) anaerobic chamber (atmosphere: 90% N_2, 5% H_2 and 5% CO_2) at 37°C. An attached sensor continuously monitored the chemical composition of the air inside the chamber. Cells (two to three biological replicates) were grown in a minimal medium solution consisting of 14.7 mM KH_2PO_4, 11.5 mM K_2HPO_4, 0.81 mM $MgSO_4 \cdot 7H_2O$, 28.0 mM NH_4Cl, 1.6 mM $CaCl_2 \cdot 2H_2O$, 52.7 μM $FeSO_4 \cdot 7H_2O$, 16.0 nM $CuSO_4 \cdot 2H_2O$, 0.80 nM $MnCl_2$, 1.46 μM $CoCl_2$, 0.15 nM $Na_2MoO_4 \cdot 2H_2O$, 89.8 nM $NiSO_4 \cdot 2H_2O$, 0.26 nM $ZnCl_2$, 48.5 nM H_3BO_3, 532 nM biotin and 1.17 μM 4-aminobenzoic acid. For the carbon source, the minimal medium was supplemented with a total of 333 mmol C l^{-1} for glucose (i.e., 55.5 mM glucose or 10 g/L glucose) alone or with (at equimolar amount) galactose, mannose, xylose or arabinose. All chemicals were obtained from Fisher or Sigma-Aldrich (analytical grade). Cell growth was monitored by measuring the optical density at 650 nm (OD_{650}).

Stable isotope tracer experiments

Stable isotope-labelled sugars were purchased from Cambridge Isotopes (Tewskbury, MA, USA) or Omicron Biochemicals (South Bend, IN, USA). Intracellular kinetic labelling of metabolites in glycolysis, PP pathway, acetyl-CoA and butyryl-CoA at each growth condition was conducted following established protocols (Yuan *et al.*, 2008; Sasnow *et al.*, 2016). Briefly, 3 ml aliquots liquid cultures (three biological replicates) at early exponentially growth phase under each growth condition as described above were filtered (0.45 μm pore size) and the cell-containing filters were placed on top of plates containing agar-solidified medium of the same substrate composition. To determine when the cells reached logarithmic growth on the plates, the cells from parallel plates subjected to the same preparation at the same growth condition were rinsed off into a 3 ml suspension for OD_{650} reading. At the early onset of logarithmic growth, the filters containing the cells were switched from the unlabelled media plates to media plates with fully labelled glucose ([U-$^{13}C_6$]-glucose) combined with either unlabelled glucose or the hemicellulosic sugar. Therefore, the experiments with isotopic switch were the following: from unlabelled glucose to labelled glucose, from unlabelled glucose to 1:1 labelled glucose:unlabelled glucose, from 1:1 unlabelled glucose:unlabelled galactose to 1:1 labelled glucose:unlabelled galactose, from 1:1 unlabelled glucose:unlabelled mannose to 1:1 labelled glucose:unlabelled mannose, from 1:1 unlabelled glucose:unlabelled xylose to 1:1 labelled glucose:unlabelled xylose, and from 1:1 unlabelled glucose:unlabelled arabinose to 1:1 labelled glucose:unlabelled arabinose. Metabolism was quenched (see details in the next section) after specific time points: 1, 2, 5, 15, 30 or 60 min. Cells that were only grown on unlabelled media were used as a control for time 0 min.

Faster rate of isotopic enrichment was found for central carbon metabolites than for ribonucleotides during kinetic isotopic enrichment with fully labelled glucose (Fig. S5). Therefore, to monitor incorporation of pentoses into ribonucleotides during growth on the glucose:pentose mixtures, long-term isotopic enrichment experiments were performed using liquid cultures (three biological replicates) grown for at least two doubling times on unlabelled glucose with either doubly ^{13}C-labelled xylose ([1,2-$^{13}C_2$]-xylose) or singly ^{13}C-labelled arabinose ([1-$^{13}C_1$]-arabinose).

Monitoring intracellular metabolite labelling

Cellular metabolism for each of the tracer experiments described above (two to three biological replicates) was quenched by quickly submerging cell-containing filters from media plates or filtered cells from the liquid cultures into a cold (−20°C) solvent mixture composed of 40:40:20 methanol:acetonitrile:water as previously described (Kimball and Rabinowitz, 2006; Sasnow *et al.*, 2016). Metabolites were isolated by reverse-phase high-performance LC with high-accurate orbitrap MS operated in negative mode on a Thermo Exactive mass spectrometer following established methods (Lu *et al.*, 2010; Xu *et al.*, 2015). Using standards, the detection of the following metabolites was verified: R5P, Xu5P, G6P, F6P, FBP, IMP, UMP, acetyl-CoA, butyryl-CoA. Using the MAVEN software package (Clasquin *et al.*, 2012), the multiple isotopologues (different labelled forms of the same compound with the same number of ^{13}C-labelled carbons) resulting from the stable isotope tracer experiments were determined. The ^{13}C-labelled fractions were corrected for the natural abundance of ^{13}C.

Acknowledgements

The author is grateful to Joshua D. Rabinowitz (Princeton University) for providing laboratory facilities and

Daniel Amador-Noguez (University of Wisconsin-Madison) for sharing technical insights during the initial stages of this project. This work was supported in part by the U.S. National Science Foundation (Division of Molecular and Cellular Biosciences, MCB 1337292) and a start-up package from Cornell University.

Conflict of Interest

None declared.

References

Amador-Noguez, D., Feng, X.J., Fan, J., Roquet, N., Rabitz, H., and Rabinowitz, J.D. (2010) Systems-level metabolic flux profiling elucidates a complete, bifurcated tricarboxylic acid cycle in *Clostridium acetobutylicum*. *J Bacteriol* **192:** 4452–4461.

Amador-Noguez, D., Brasg, I.A., Feng, X.-J., Roquet, N., and Rabinowitz, J.D. (2011) Metabolome remodeling during the acidogenic-solventogenic transition in *Clostridium acetobutylicum*. *Appl Environ Microbiol* **77:** 7984–7997.

Aristilde, L., Lewis, I.A., Park, J.O., and Rabinowitz, J.D. (2015) Hierarchy in pentose sugar metabolism in *Clostridium acetobutylicum*. *Appl Environ Microbiol* **81:** 1452–1462.

Clasquin, M., Melamud, E., and Rabinowitz, J.D. (2012) LC-MS data processing with MAVEN: a metabolomic analysis and visualization engine. *Curr Protoc Bioinformatics* **14:** 14.11.

Dash, S., Ng, C.Y. and Maranas, C.D. (2016) Metabolic modeling of clostridia: current developments and applications. *FEMS Microbiol Lett* **363:** pii: fnw004. doi: 10.1093/femsle/fnw004

Desai, R.P., Harris, L.M., Welker, N.E., and Papoutsakis, E.T. (1999) Metabolic flux analysis elucidates the importance of the acid formation pathways in regulating solvent production by *Clostridium acetobutylicum*. *Metab Eng* **1:** 206–213.

Dürre, P. (1998) New insights and novel developments in clostridial acetone/butanol/isopropanol fermentation. *Appl Microbiol Biotechnol* **49:** 639–648.

Ezeji, T., and Blaschek, H.P. (2007) Fermentation of dried distillers' grains and solubles (DDGS) hydrosylates to solvents and value-added products by solventogenic clostridia. *Bioresour Technol* **99:** 5232–5242.

Fond, O., Engasser, J.-M., Matta-El-Amouri, G., and Petitdemange, H. (1986) The acetone butanol fermentation on glucose and xylose. I. Regulation and kinetics in batch cultures. *Biotechnol Bioeng* **28:** 160–166.

Gheshlagi, R., Scharer, J.M., Moo-Young, M., and Chou, C.P. (2009) Metabolic pathways of clostridia for producing butanol. *Biotechnol Adv* **27:** 764–781.

Grimmler, C., Held, C., Liebl, W., and Ehrenreich, A. (2010) Transcriptional analysis of catabolite repression in *Clostridium acetobutylicum* growing on mixtures of D-glucose and D-xylose. *J Biotechnol* **150:** 315–323.

Grupe, H., and Gottschalk, G. (1992) Physiological events in *Clostridium acetobutylicum* during the shift from acidogenesis to solventogenesis in continuous culture and presentation of a model for shift induction. *Appl Environ Microbiol* **58:** 3896–3902.

Gu, Y., Li, J., Zhang, L., Chen, J., Niu, L., Yang, Y., *et al.* (2009) Improvement of xylose utilization in *Clostridium acetobutylicum* via expression of the *talA* gene encoding transaldolase from *Escherichia coli*. *J Biotechnol* **43:** 284–287.

Guiterrez, N.A., and Maddox, I.S. (1996) Galactose transport in *Clostridium acetobutylicum* P262. *Lett Appl Microbiol* **23:** 97–100.

Hu, S., Zheng, H., Gu, Y., Zhao, J., Zhang, W., Yang, Y., *et al.* (2011) Comparative genomic and transcriptomic analysis revealed genetic characteristics related to solvent formation and xylose utilization in *Clostridium acetobutylicum* EA 2018. *BMC Genom* **12:** 93.

Kimball, E., and Rabinowitz, J.D. (2006) Identifying decomposition products in extracts of cellular metabolites. *Anal Biochem* **358:** 273–280.

Kudahettige-Nilsson, R., Helmerius, J., Nilsson, R.T., Sjöblom, M., Hodge, D.B., and Rova, U. (2015) Biobutanol production by *C. acetobutylicum* using xylose recovered from birch Kraft black liquor. *Bioresour Technol* **176:** 71–79.

Lee, S.Y., Park, J.H., Jang, S.H., Nielsen, L.K., Kim, J., and Jung, K.S. (2008) Fermentative butanol production by clostridia. *Biotechnol Bioeng* **101:** 209–228.

Liao, C., Seo, S.-O., Celik, V., Liu, H., Kong, W., Wang, Y., *et al.* (2015) Integrated, systems metabolic picture of acetone-butanol-ethanol fermentation by *Clostridium acetobutylicum*. *Proc Natl Acad Sci USA* **112:** 8505–8510.

Liu, L., Zhang, L., Tang, W., Gu, Y., Hua, W., Yang, S., *et al.* (2012) Phosphoketolase pathway for xylose catabolism in *Clostridium acetobutylicum* revealed by ^{13}C metabolic flux analysis. *J Bacteriol* **194:** 5413–5422.

Lu, W., Clasquin, M.F., Melamud, E., Amador-Noguez, D., Caudy, A.A., and Rabinowitz, J.D. (2010) Metabolomic analysis via reversed-phase ion-pairing liquid chromatography coupled to a stand alone OrbiTrap mass spectrometer. *Anal Chem* **82:** 3212–3221.

Mitchell, W.J. (2016) Sugar uptake by the solventogenic clostridia. *World J Microbiol Biotechnol* **32:** 32.

Ounine, K., Petitdemange, H., Raval, G., and Gay, R. (1985) Regulation and butanol inhibition of D-xylose and D-glucose uptake in *Clostridium acetobutylicum*. *Appl Environ Microbiol* **49:** 874–878.

Rabinowitz, JD, Aristilde, L and Amador-Noguez, D. (2015) Metabolomics of clostridial biofuel production. Technical Report, SC0006839. United States Department of Energy, doi: 10.2172/1213974

Rabinowitz, J.D., and Kimball, E. (2007) Acidic acetonitrile for cellular metabolome extraction from *Escherichia coli*. *Anal Chem* **79:** 6167–6173.

Raganati, F., Curth, S., Goetz, P., Olivieri, G., and Marzocchelia, A. (2012) Butanol production from lignocellulosic-based hexoses and pentoses by fermentation of *Clostridium acetobutylicum*. *Chem Eng Trans* **27:** 91–96.

Ren, C., Gu, Y., Hu, S., Wu, Y., Wang, P., Yang, Y., *et al.* (2010) Identification and inactivation of pleiotropic regulator CcpA to eliminate glucose repression of xylose utilization in *Clostridium acetobutylicum*. *Metab Eng* **12:** 446–454.

Richter, H., Molitor, B., Wei, H., Chen, W., Aristilde, L., and Angenent, L.T. (2016) Ethanol production in syngas-fermenting *Clostridium ljungdahlii* is controlled by thermodynamics rather than by enzyme expression. *Energy Environ Sci* **9:** 2392–2399.

Rodríguez, P., Rivas, C.I., Godoy, A., Villanueva, M., Fischbarg, J., Vera, J.C., and Reyes, A.M. (2005) Redefining the facilitated transport of mannose in human cells: absence of glucose-insensitive, high-affinity facilitated mannose transport. *Biochemistry* **44:** 313–320.

Sasnow, S.S., Wei, H., and Aristilde, L. (2016) Bypasses of glucose metabolism in iron-limited *Pseudomonas putida*. *Microbiologyopen* **5:** 3–20.

Scheller, H.V., and Ulvskov, P. (2010) Hemicelluloses. *Annu Rev Plant Biol* **61:** 263–289.

Servinsky, M.D., Kiel, J.T., Dupuy, N.F., and Sund, C.J. (2010) Transcriptional analysis of differential carbohydrate utilization by *Clostridium acetobutylicum*. *Microbiology* **156:** 3478–3491.

Servinsky, M.D., Germane, K.L., Liu, S., Kiel, J.T., Clark, A.M., Shankar, J., and Sund, C.J. (2012) Arabinose is metabolized via a phosphoketolase pathway in *Clostridium acetobutylicum* ATCC 824. *J Ind Microbiol Biotechnol* **39:** 1859–1867.

Sund, C.J., Servinsky, M.D., and Gerlach, E.S. (2013) Differing roles for *Clostridium acetobutylicum*'s galactose utilization pathways. *Adv Microbiol* **3:** 490–497.

Voigt, C., Bahl, H., and Fischer, R.-J. (2014) Identification of PTS[Fru] as the major fructose uptake system of *Clostridium acetobutylicum*. *Appl Microbiol Biotechnol* **98:** 7161–7172.

Xu, Y.-F., Lu, W., and Rabinowitz, J.D. (2015) Avoiding misannotation of in-source fragmentation products as cellular metabolites in liquid chromatography–mass spectrometry-based metabolomics. *Anal Chem* **87:** 2273–2281.

Yuan, J., Bennett, B.D. and Rabinowitz, J.D. (2008) Kinetic flux profiling for quantitation of cellular metabolic fluxes. *Nat Protoc* **3:** 31328–31340.

Zhang, L., Leyn, S.A., Gu, Y., Jiang, W., Rodionov, D.A., and Yang, C. (2012) Ribulokinase and transcriptional regulation of arabinose metabolism in *Clostridium acetobutylicum*. *J Bacteriol* **194:** 1055–1064.

Wine microbiology is driven by vineyard and winery anthropogenic factors

Cédric Grangeteau,[1,2] Chloé Roullier-Gall,[3,4]
Sandrine Rousseaux,[1,2,*] Régis D. Gougeon,[1,5]
Philippe Schmitt-Kopplin,[3,4] Hervé Alexandre[1,2] and
Michèle Guilloux-Benatier[1,2]

[1]*Univ. Bourgogne Franche-Comté, AgroSup Dijon, PAM UMR A 02.102, F-21000 Dijon, France.*
[2]*IUVV Equipe VAlMiS, rue Claude Ladrey, BP 27877, 21078 Dijon Cedex, France.*
[3]*Chair of Analytical Food Chemistry, Technische Universität München, Alte Akademie 10, 85354 Freising-Weihenstephan, Germany.*
[4]*Research Unit Analytical BioGeoChemistry, Department of Environmental Sciences, Helmholtz Zentrum München, Ingolstaedter Landstrasse 1, 85764 Neuherberg, Germany.*
[5]*IUVV Equipe PAPC, rue Claude Ladrey, BP 27877, 21078 Dijon Cedex, France.*

Summary

The effects of different anthropic activities (vineyard: phytosanitary protection; winery: pressing and sulfiting) on the fungal populations of grape berries were studied. The global diversity of fungal populations (moulds and yeasts) was performed by pyrosequencing. The anthropic activities studied modified fungal diversity. Thus, a decrease in biodiversity was measured for three successive vintages for the grapes of the plot cultivated with Organic protection compared to plots treated with Conventional and Ecophyto protections. The fungal populations were then considerably modified by the pressing-clarification step. The addition of sulfur dioxide also modified population dynamics and favoured the domination of the species *Saccharomyces cerevisiae* during fermentation. The non-targeted chemical analysis of musts and wines by FT-ICR-MS showed that the wines could be discriminated at the end of alcoholic fermentation as a function of adding SO_2 or not, but also and above all as a function of phytosanitary protection, regardless of whether these fermentations took place in the presence of SO_2 or not. Thus, the existence of signatures in wines of chemical diversity and microbiology linked to vineyard protection has been highlighted.

*For correspondence. E-mail sandrine.rousseaux@u-bourgogne.fr

Funding Information
This work was funded by the Regional Council of Burgundy and the Interprofessional Office of Burgundy wines with technical support of Vinipôle South Burgundy.

Introduction

For over 7500 years, humans have sought to control vine development, grape berry maturation and alcoholic fermentation to produce wine (McGovern *et al.*, 1996). Over the last 20 years, the emergence of various vineyard management methods has been observed, particularly with the increasing number of vineyards practising organic viticulture (Zafrilla *et al.*, 2003). This diversity and especially the use of chemical or organic phytosanitary products could affect the biodiversity of grape microorganisms. Various studies have been conducted to compare the effects of these different systems and more particularly the non-target effects of phytosanitary treatments on fungal populations present on berries. Although all these studies show an effect of plant protection on the diversity of yeasts present in grape berries, the results cannot be generalized and are very often contradictory. Cordero-Bueso *et al.* (2011) and Martins *et al.* (2014) observed a wider diversity of yeasts for organic plots compared with conventional plots. Regarding the study by Milanović *et al.* (2013), the lowest diversity of isolated yeasts was observed for the organic modality. At genus or species level, Guerra *et al.* (1999) observed that the species *Saccharomyces cerevisiae* was not isolated in the conventional modality compared with the organic modality. The fermentative yeast genera such as *Hanseniaspora* and *Metschnikowia* have been isolated mainly in control (untreated) and organic plots, whereas *Aureobasidium pullulans* was the majority species isolated in grape berries from conventional plots (Comitini and Ciani, 2008). But in another study, the species *A. pullulans* was described as the majority species isolated in grape berries from organic plots (Martins *et al.*, 2014) and the species *Metschnikowia pulcherrima* was isolated more in samples obtained from the conventional modality (Milanović *et al.*, 2013). Significant variability between studies is likely due to differences in grape varieties, the geographical location of the vineyard, the sampling method, identification techniques and finally intra-vine variation (plot level) (Hierro *et al.*, 2006; Xufre *et al.*, 2006; Nisiotou *et al.*, 2007; Barata *et al.*, 2008, 2012; Setati *et al.*, 2012, 2015). Furthermore, *in vitro* studies are still required to determine the sensitivity of different fungal genera to these products as well as studies in the vineyard to determine the direct effect of the products used at the time of application. For Cadez *et al.* (2010), the presence of fungicides has a minor impact on yeast communities associated with grape berries because, after the safety interval, colonization with yeast is possible.

In addition, such works do not consider whether the differences of yeast biodiversity observed as a function of plant protection are maintained in the musts and during alcoholic fermentation. Indeed the pre-fermentation operations carried out to ensure the quality of the final product could reduce or on the contrary amplify the differences observed for yeast biodiversity in the vineyard. For musts obtained from white grapes (Chenin Blanc and Prensal White), cold settling reduces the overall yeast population and particularly affects the growth of certain species such as *Hansenula anomala*, *Issatchenkia terricola* and *S. cerevisiae*, instead of other species such as *Candida zemplinina* and *Hanseniaspora uvarum*, not very sensitive to this process and which become or remain the major species after racking (Mora and Mulet, 1991). Sturm *et al.* (2006) observed for Riesling grape must that non-*Saccharomyces* yeasts persist longer during fermentation if pressing is preceded by crushing or maceration. The temperature during pre-fermentation maceration of red grape varieties (Cabernet sauvignon and Malbec) also seems to play a role in the evolution of yeast populations (Maturano *et al.*, 2015). Thus, maceration carried out at 14°C resulted in the development of yeast populations with a high proportion of *H. uvarum*. For maceration performed at 2.5 or 8°C, yeast populations did not increase but the proportions of *S. cerevisiae* and *C. zemplinina* increased at 8 and 2.5°C respectively. These results show the strong preference of the species *H. uvarum* for temperatures around 15°C and confirm the psychrotolerant characteristic of the species *C. zemplinina* (Sipiczki, 2003). The addition of SO_2 in red and white wine promotes the establishment of *S. cerevisiae*, often to the detriment of non-*Saccharomyces* yeasts (*Candida*, *Cryptococcus*, *Hanseniaspora* and *Zygosaccharomyces*) more sensitive to SO_2, and also less tolerant to ethanol (Romano and Suzzi, 1993; Constantí *et al.*, 1998; Henick-Kling *et al.*, 1998; Albertin *et al.*, 2014; Takahashi *et al.*, 2014). Bokulich *et al.* (2015) studied the effect of different concentrations of SO_2 (between 0 and 150 mg l^{-1}) on bacterial and fungal populations during the alcoholic fermentation of Chardonnay grape musts. Fermentations were slower and extended with low SO_2 concentrations (< 25 mg l^{-1}) due to the growth of bacteria or fungi competing with yeasts, but the development of bacterial and fungal species was greatly reduced with the addition of 25 mg l^{-1} SO_2. However, higher concentrations up to 100 mg l^{-1} had no additional effect on populations or on the progress of fermentation. Beyond this concentration, fermentation was slower than those conducted with concentrations between 25 and 100 mg l^{-1}. The effects of four pre-fermentation oenological practices (clarification degree to 90 NTU and 250 NTU), temperature during pre-fermentation maceration (10 and 15°C), the use of SO_2 (0 and 25 mg l^{-1}) and starter yeast addition on yeast dynamics (*C. zemplinina*, *Hanseniaspora* spp., *Saccharomyces* spp. and *Torulaspora delbrueckii*) were evaluated in a Chardonnay grape must (Albertin *et al.*, 2014). The population dynamics of the four species were impacted differently by oenological practices. For example, the use of SO_2 seemed to favour the genus *Saccharomyces* independently of other practices. Significant interaction effects between practices were revealed. Thus, a low degree of clarification seemed to favour the development of *C. zemplinina* mainly at a pre-fermentation temperature of 10°C. The inhibition of the genus *Hanseniaspora* was observed at a pre-fermentation temperature of 10°C in the presence of SO_2.

Although the dynamics of yeast populations in the composition and organoleptic qualities of wine is an important parameter (Lambrechts and Pretorius, 2000; Swiegers *et al.*, 2005), must composition is also a parameter that cannot be neglected. Does the composition of musts and wines differ according to the phytosanitary protection used? To our knowledge, few studies have been performed on this subject. Existing studies have focused their analyses on compounds related to natural plant defences as they are considered as the compounds directly affected by plant protection (Adrian *et al.*, 1997; Jeandet *et al.*, 2000). Thus, the concentrations of polyphenols and antioxidant activity were higher for berries sampled from an organic plot 30 days before harvesting. However, these differences disappeared at harvest (Mulero *et al.*, 2010). On the contrary, Bunea *et al.* (2012) found a small difference in total polyphenol contents between berry skins for nine different grape varieties from conventional and organic vineyards (respectively 148–1231 and 163–1341 mg kg^{-1} as gallic acid equivalents). Dani *et al.* (2007) also showed that the concentration of total polyphenols, particularly resveratrol, is higher for musts elaborated with grape varieties from *Vitis labrusca* (Bordo and Niagara) from organic viticulture. Levite *et al.* (2000) compared the concentration of resveratrol for conventional and organic wines made from five grape varieties and from six geographical localizations. In most cases, the resveratrol concentration was higher for organic wines. Similarly, Vrček *et al.* (2011) observed a higher concentration of polyphenols for wines from organic viticulture compared with those from conventional viticulture.

The effects of plant protection applied to vineyards on wine composition have been studied mainly from a sensory point of view and are not clearly defined (Moyano *et al.*, 2009; Pagliarini *et al.*, 2013). In addition, most of these studies have not taken into account practices used during winemaking, which can reduce or increase differences due to the protection applied to the vineyard. Finally, to our knowledge no study has determined how the modifications of microbial populations due to plant protection may influence wine composition.

The purpose of this work was to study the combined impacts of three different phytosanitary protections: Organic, Conventional and Ecophyto protections (corresponding to dose reduction compared with conventional treatment); and of oenological practices (pressing-settling and sulfiting) on the biodiversity of fungal populations and on the chemical composition of musts and wines. For the first time, the effects of phytosanitary protection on microbiological and chemical characteristics were evaluated from grape berries to wine in a single study. The use of non-target methods for microbiological (pyrosequencing) and metabolomic (FT-ICR-MS) analyses allowed examining the global effects of three different phytosanitary protections. However, only the non-volatile fraction of the wine composition has been considered in our approach.

Results and discussion

Effect of the phytosanitary protection on fungal populations present on grape berries for the 2012, 2013 and 2014 vintages

The fungal populations present on mature Chardonnay berries (after pressing aseptically) and identified by pyrosequencing for the three blocks treated with different phytosanitary protections and for the three vintages are presented in Fig. 1. Moulds and yeasts were identified by pyrosequencing. As observed by Nisiotou *et al.* (2007) with a culture-dependent method for identifying fungal populations, the highest fungal diversity was observed when moulds were present in quantity. This was also confirmed by the Shannon biodiversity index which was higher in 2012 compared with the other two vintages, regardless of the protection applied. The proportion of moulds seemed primarily related to vintage: between 42% and 70% of the population for the 2012 vintage, between 18% and 19% for the 2013 vintage

and < 2% for the 2014 vintage. This difference in the proportion of mould according to vintage is probably related to differences in temperature and rainfall between vintages during the flowering-harvest period (Sall, 1980; Lalancette *et al.*, 1988; Broome *et al.*, 1995). Indeed for the 2012 vintage, the average temperature for this period was 20.1°C with a rainfall of 306 mm, whereas the average temperature was 19.4°C in 2013 and 2014, with rainfalls of 271 and 244 mm for 2013 and 2014 respectively. Some mould genera appeared to be specific to one vintage. For example, *Botryotinia, Cladosporium* and *Phoma* genera were present only for the 2012 vintage (for the three different protections), while the genus *Monilinia* (for the three types of protection) was present only for the 2013 vintage. Moreover, the number of yeast genera identified was much higher for the 2012 vintage compared with the other two modalities which had a lower proportion of mould. Some yeast genera seem to be linked to the presence of certain moulds. For example, the genus *Candida,* known to be very present on botrytized grapes (Mills *et al.*, 2002; Sipiczki, 2003), was only identified in our study for the 2012 vintage, the only vintage in which the genus *Botrytis* was present. Moreover, the proportions of the genus *Hanseniaspora*, known for its antagonism with *Botrytis cinerea* (Rabosto *et al.*, 2006; Liu *et al.*, 2010), and of the genus *Saccharomyces*, sensitive to glucan produced by *B. cinerea* (Hidalgo, 1978; Donèche, 1993), were much lower in 2012 compared with the other two vintages. It is interesting to note that the genus *Saccharomyces* (2012 vintage) may have been present in amounts too small to be detected by standard methods (0.2% and lower than 0.1% for the Ecophyto and Conventional modalities respectively) or even completely absent (0% in the Organic modality). It has long been known that this genus is difficult to isolate from grapes (Combina *et al.*, 2005; Raspor *et al.*, 2006). These three yeast genera, *Saccharomyces, Hanseniaspora* and *Candida*, have been described as having a strong influence on the quality and organoleptic profile of wines (Zironi *et al.*, 1993; Ciani and Maccarelli, 1998; Andorrà *et al.*, 2010; Moreira *et al.*, 2011). Thus, the modulation of the yeast populations on grape berries by the presence of some phytopathogenic genera could have effects on wine products.

The Shannon index highlighted a plant protection effect: the Organic modality had a lower biodiversity index compared with the other two modalities regardless of the vintage. For the 2014 vintage, the Shannon index was 0.30 for the Organic modality and 0.80 and 2.08 for the Ecophyto and Conventional modalities respectively. Milanović *et al.* (2013) also observed that Organic protection could lead to lower biodiversity on berries compared with Conventional protection. However, these

Fig. 1. Repartition on fungal genera on grape berries (T0) for vineyard with Organic, Conventional and Ecophyto phytosanitary protection. Populations are identified by pyrosequencing for 2012, 2013 and 2014 vintage. H' index are calculated on overall population. Genera representing < 0.2% of the total population are collectively called 'minority genera'.

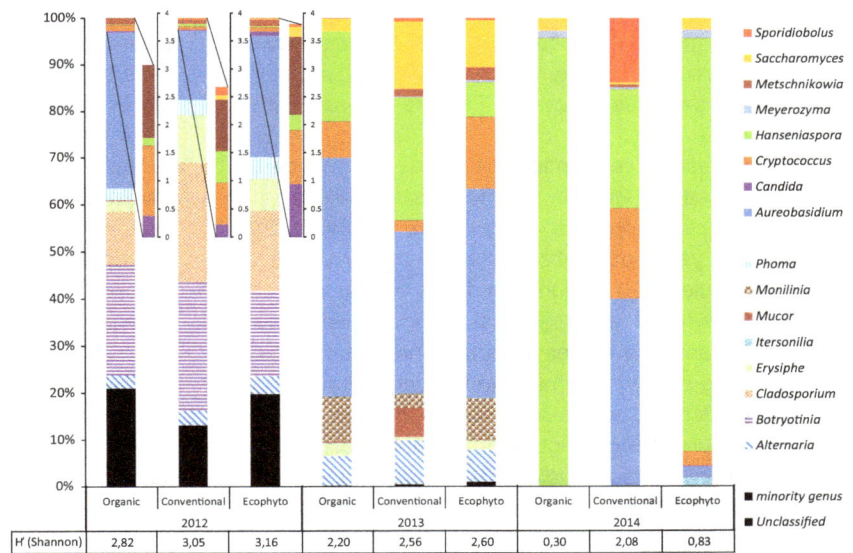

authors studied only species of yeasts. The results obtained in this study suggest that interactions could exist and that the presence of certain fungal genera may promote or inhibit the presence of certain yeast genera. However, this is not enough to explain the lowest biodiversity systematically observed for the Organic modality. The latter could be related to an effect of copper on non-target organisms (e.g. yeast). Copper is a fungicide with a broader spectrum than the synthetic molecules used in the other two protection modalities. Thus, Martins *et al.* (2014) have recently observed a strong correlation between the copper dose used and the decrease of yeast biodiversity observed on berries. In this study, the amounts of copper applied (kg per ha) are: 2.67 for vintage 2012 and 2013, 1.89 for vintage 2014 in the Organic modality, 0.5 for vintage 2012 and 2013, 0.39 for vintage 2013 in the Conventional modality, and 0.25 for vintage 2012, 0 for vintage 2013 and 0.1 for vintage 2014 in the Ecophyto modality.

Furthermore, it is interesting to note that the genus *Sporidiobolus,* representing 13.7% of the population on berries for the Conventional modality, was not detected for the other two modalities. This can be explained partly by the high resistance of this genus to synthetic fungicides (Sláviková and Vadkertiová, 2003) and its sensitivity to copper (Vadkertiová and Sláviková, 2006).

The fungal populations present on berries appeared to result from both the protection and the vintage. The amount of phytosanitary products varied each year (Table S1) regardless of the type of protection. This quantity depended on disease pressure and the risk of leaching related to annual climatic conditions. This difference in disease pressure was very marked between the three vintages studied and caused significant effects on

the overall fungal populations present on the berries, and not only on the proportion of the different plant pathogens. Nevertheless, more than the dose used, it was the type of molecules used that seemed to have a very significant effect on the diversity of fungal genera present in berries. Thus, the grape berries of the Organic modality using copper and sulfur fungicides always presented the lowest biodiversity compared with the other two types of protection using synthetic molecules.

Impact of pre-fermentation steps and vineyard protection on grape musts for the 2013 vintage

Fungal populations. Fungal populations present in the must after pressing-settling (T1) were compared with those present on berries (T0). The results are shown in Table 1. The population of moulds decreased significantly in the grape must after the pressing-settling step. Thus, the genus *Alternaria,* which represented 6.4%, 7% and 9.4% of the total fungi population on the berries of the Organic, Ecophyto and Conventional modalities, respectively, was not found in the musts. The same decrease was observed for the genera *Monilinia* and *Erysiphe.* On the contrary, the genus *Penicillium,* not identified on berries, was represented in must as 2.36% of the total fungi population for the Ecophyto modality and 0.08% for the Organic modality, probably due to the implantation of strains present in the cellar environment. Ocón *et al.* (2011) had already highlighted that this genus was mostly detected in the cellar environment. The Conventional modality differed from the other two modalities by the total absence of mould.

The number of yeasts (CFU ml $^{-1}$) was lower in the three modalities in T0 samples than for T1 samples

(Table 2). This difference could be related to sampling and pressing methods: the quantities of berries were lower at T0 compared with T1; berries at T0 were pressed manually and those of the harvest at T1 were pressed with a vertical press.

The decrease of yeast populations was also observed during the settling step for all the musts whatever the type of protection applied to the vineyard (Table 2). This reduction of yeast populations confirmed the results of Mora and Mulet (1991), who reported a decrease in cell number for several yeast genera during this step. Not only a decrease of total yeast populations was observed after pressing-settling compared to the proportion on berries but a difference was also observed in the proportion mainly for *Aureobasidium* genera. Moreover, some genera present in musts (*Candida, Debaryomyces, Kazachstania* and *Malassezia*) were not present on berries. This could be due to the implantation of the resident flora from the cellar during the pressing-settling steps

(Grangeteau *et al.*, 2015). Nevertheless, the genus *Candida* represented < 0.3% of the population, while the genera *Debaryomyces, Kazachstania* and *Malassezia* represented < 0.1%. The implantation of resident flora from the cellar in the must could also explain the increase in the proportions of the genera *Cryptococcus, Metschnikowia, Meyerozyma* and *Saccharomyces* between berries and grape musts. Otherwise, some genera were affected differently by this step as a function of the protection modality applied in the vineyard. For example, the genus *Aureobasidium* which represented 50.8%, 34.4% and 44.6% of the population on berries for the Organic, Conventional and Ecophyto modalities, respectively, was found in musts at 18.6%, 5.66% and 54.9% for the Organic, Conventional and Ecophyto modalities respectively (Table 1).

A least discriminant analysis effect size (LDA) taxonomic cladogram comparing all the grape musts categorized by the different vineyard protections was performed (Fig. 2) to determine whether, despite these population reshuffles, a specific population was present depending on plant protection. Basidiomycota [especially *Cryptococcus* (48.1%)] were mainly associated with the Organic protection, while Ascomycota, including *Saccharomycotina* [especially *Saccharomyces* (25.7%), *Metschnikowia* (25%) and *Hanseniaspora* (27%)] were mainly associated with the Conventional protection. Among the Ascomycota, *Pezizomycotina* [especially *Aureobasidium* (54.9%)] were associated with the Ecophyto protection. *Fusarium* and *Mucor* (only 0.03% and 0.06% respectively) were associated with the Organic protection because they were not detected in the other modalities. *Penicillium* was associated with the Ecophyto protection even though the source of its presence was probably the cellar. Note that some differences observed for grape must can be related to those observed on grape berries. Indeed, *Hanseniaspora* and *Saccharomyces* represented a larger proportion of the population on berries for the Conventional modality (26.4% for *Hanseniaspora* and 14.6% for *Saccharomyces*) compared with the other modalities (*Hanseniaspora* 19.4% and 7.5% for the Organic and Ecophyto modalities, respectively, and *Saccharomyces* 2.7% and 10.2% for the Organic and Ecophyto modalities respectively). The genera *Cryptococcus*

Table 1. Repartition (%) of fungal genera identified by pyrosequencing on berries (T0) and after pressing-settling (T1) for three phytosanitary vineyard protections for 2013 vintage.

	O		C		E	
	T0	T1	T0	T1	T0	T1
Mould						
Alternaria	6.37	–	9.45	–	7.01	–
Erysiphe	2.90	0.02	0.85	–	1.91	0.04
Fusarium	0.10	0.03	0.24	–	0.74	–
Itersonilia	0.04	0.02	0.02	–	–	–
Monilinia	9.56	0.05	3.00	–	9.11	0.68
Mucor	0.27	0.06	6.15	–	–	–
Penicillium	–	0.08	–	–	–	2.36
Yeast						
Aureobasidium	50.77	18.59	34.39	5.59	44.56	54.91
Candida	–	0.10	–	0.32	–	0.02
Cryptococcus	7.86	48.10	2.47	11.19	15.37	20.89
Debaryomyces	–	0.03	–	0.11	–	0.02
Hanseniaspora	19.39	17.96	26.36	27.02	7.46	2.73
Kazachstania	–	0.03	–	0.08	–	0.04
Malassezia	–	0.02	–	–	–	–
Metschnikowia	–	0.02	1.54	24.97	2.69	3.49
Meyerozyma	–	5.26	0.17	5.03	0.49	1.08
Saccharomyces	2.71	9.64	14.57	25.71	10.17	6.70
Sporidiobolus	0.02	–	0.63	–	0.39	6.54
Unclassified	–	–	0.15	–	0.12	0.47

Table 2. Yeast count on YPD medium for berries (T0) and grape must after pressing and after settling for three phytosanitary vineyard protections for 2013 vintage. Values followed by different letters are significantly different (*P* < 0.01).

	O			C			E		
	T0	After pressing	After settling	T0	After pressing	After settling	T0	After pressing	After settling
Yeast log CFU ml^{-1} (standard deviation)	4.01 (2.48)	5.49[a] (3.92)	4.93[b] (3.68)	3.59[c] (2.42)	5.27[a] (3.48)	5.17[b] (4.33)	3.82[c] (2.18)	5.59[a] (3.30)	5.10[b] (4.06)

and *Aureobasidium* represented a lower proportion of the population on berries for the Conventional modality (34.4% for *Aureobasidium* and 2.5% for *Cryptococcus*) compared with the other modalities (*Aureobasidium* represented 50.8% and 44.6% for the Organic and Ecophyto modalities, respectively, and *Cryptococcus* represented 7.9% and 15.4% for the Organic and Ecophyto modalities respectively).

Chemical composition. The chemical compositions of the three different musts from the 2013 vintage were analysed (Table S3). No significant difference between the musts concerning sugar concentration or acidity level (pH or total acidity) was observed. The available nitrogenous compound content differed slightly depending on the must but was higher than 140 mg N l^{-1} for all them, so there was *a priori* no deficiency preventing the progress of alcoholic fermentation (Agenbach, 1977). Taking the study still further, the grape musts were analysed by FT-ICR-MS. Distributions of CHONSP containing elemental compositions were extremely close for all the musts (Fig. 3B–D), explained by the fact that the grape berries of each must had the same origin. From these results, we can conclude that the direct influence of

plant protection on the composition of the musts was quite low. Nevertheless, PLS-DA (Fig. 3A) allowed partial discrimination of the musts as a function of protection. The grape musts of the Organic modality were separated from the musts of the other two modalities. Associated with the differences between the populations already shown, they could lead to differences in the dynamics of the alcoholic fermentation and chemical composition of wines depending on the phytosanitary protection applied in the vineyard.

Impact of sulfur dioxide use and plant protection on wine during alcoholic fermentation for the 2013 vintage

Fungal populations. The evolution of the number and proportion of yeast populations during AF are presented for the Organic, Conventional and Ecophyto modalities in Figs 4–6 respectively. For the Organic and Conventional modalities, fermentations in the absence of SO$_2$ languished. The maximal population was reached after 7 days for the Conventional modality, with the same kinetics in the presence and absence of SO$_2$. For the Organic modality, the maximal population was also reached after 7 days, but the kinetics and the maximum

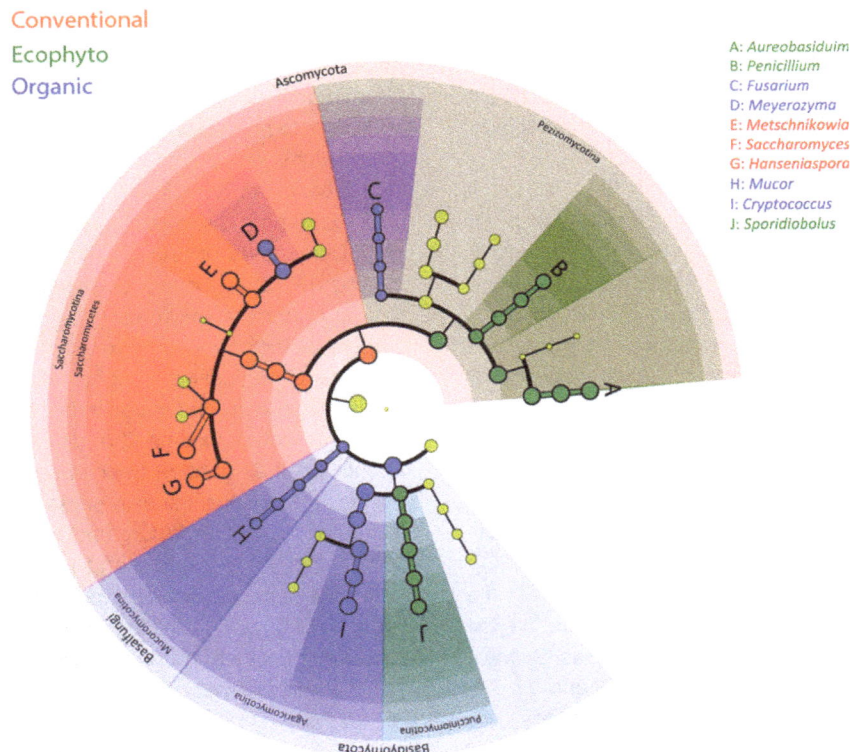

Fig. 2. Least discriminant analysis effect size taxonomic cladogram comparing all grape musts categorized by vineyard protection mode. Significantly discriminant taxon nodes are coloured and branch areas are shaded according to the highest-ranked variety for that taxon. For each taxon detected, the corresponding node in the taxonomic cladogram is coloured according to the highest-ranked group for that taxon. If the taxon is not significantly differentially represented between sample groups, the corresponding node is coloured yellow.

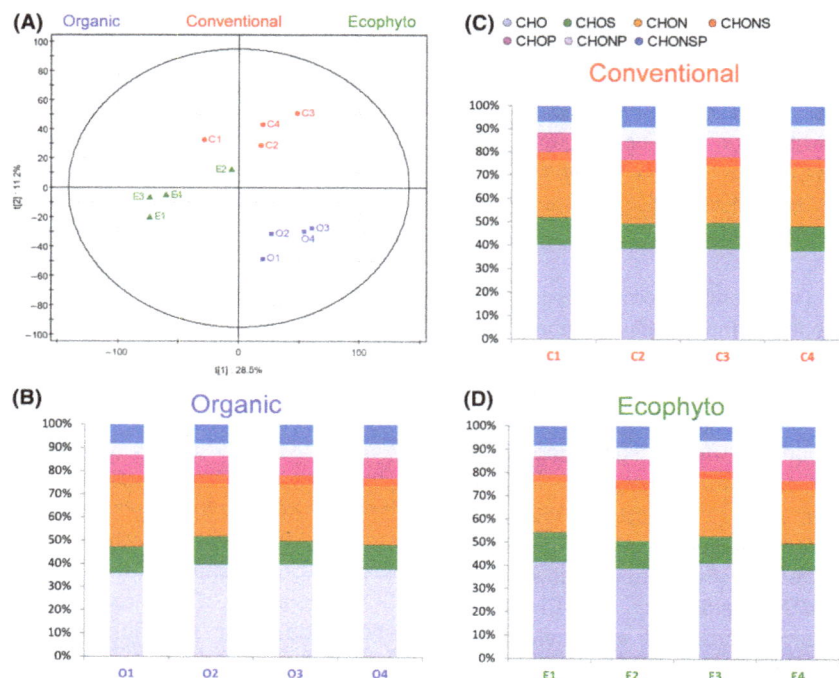

Fig. 3. Analysis of the FT-ICR–MS data for grape musts of 2013 vintage. (A) Scores plot of the PLS-DA depending on the phytosanitary protection mode, the first two components retained 39.7% of the variation. Histograms of elementary composition of Organic (B), (C) Conventional and (D) Ecophyto grape musts.

population level differed as a function of the presence or absence of sulfites. Indeed, in the absence of SO_2, the population increased more quickly but the maximum population was statistically lower: 3. 10^8 CFU ml^{-1} without SO_2 versus 8.10^8 CFU ml^{-1} with SO_2 (Student's test with P-value = 0.0007). These differences in behaviour may be related to the difference in composition of the yeast population present in the must. For the Ecophyto modality, no difference was observed concerning fermentation in the absence of SO_2. The end of alcoholic fermentations occurred simultaneously for the four wines from this modality.

The proportion of *Saccharomyces* increased rapidly in the presence of SO_2 and for the three modalities. It accounted for 80–98% of the population after 3 days of alcoholic fermentation. Thus, the differences in the population dynamics were limited in the presence of SO_2. However differences in the initial population, even in the cases where they disappear quickly, can influence the finished wine (Romano *et al.*, 2003). The genus *Saccharomyces* prevailed in all the wines at the end of AF. Without SO_2, the implantation of *Saccharomyces* was delayed, so differences of populations could persist longer. Additionally, the population of non-*Saccharomyces* remained high at the end of AF for the Organic (30%) and Ecophyto (15%) modalities. For the Conventional modality, *Saccharomyces* represented more than

95% of the total population after 9 days. Alcoholic fermentation was sluggish for the Organic modality. This sluggish fermentation could be due to dead or inactive populations or an external contamination because the *Aureobasidium*, *Cryptococcus*, *Meyerozyma* and *Sporidiobolus* genera were not usually identified at the end of AF. Differences in the *Saccharomyces* strains present during AF could explain differences in alcoholic fermentation dynamics. Furthermore, population size is an important factor for AF dynamics (Albertin *et al.*, 2011). For the Organic and Ecophyto fermentations, a higher percentage and longer persistence of the genus *Hanseniaspora* could explain the difficulties of the development of the genus *Saccharomyces* (Medina *et al.*, 2012). This genus never represented more than 85% of the total population in Organic and Ecophyto modalities. Otherwise, the presence and persistence of the *Hanseniaspora* and *Metschnikowia* genera for fermentations without SO_2 could strongly influence the chemical composition of wines produced. Indeed species belonging to the *Hanseniaspora* genus and the *Metschnikowia* genus are known to produce secondary metabolites which can influence the organoleptic profiles of the wines produced negatively (Ciani and Picciotti, 1995) or positively (Zironi *et al.*, 1993; Rojas, 2003; Moreira *et al.*, 2011; Sadoudi *et al.*, 2012; Medina *et al.*, 2013; Martin *et al.*, 2016).

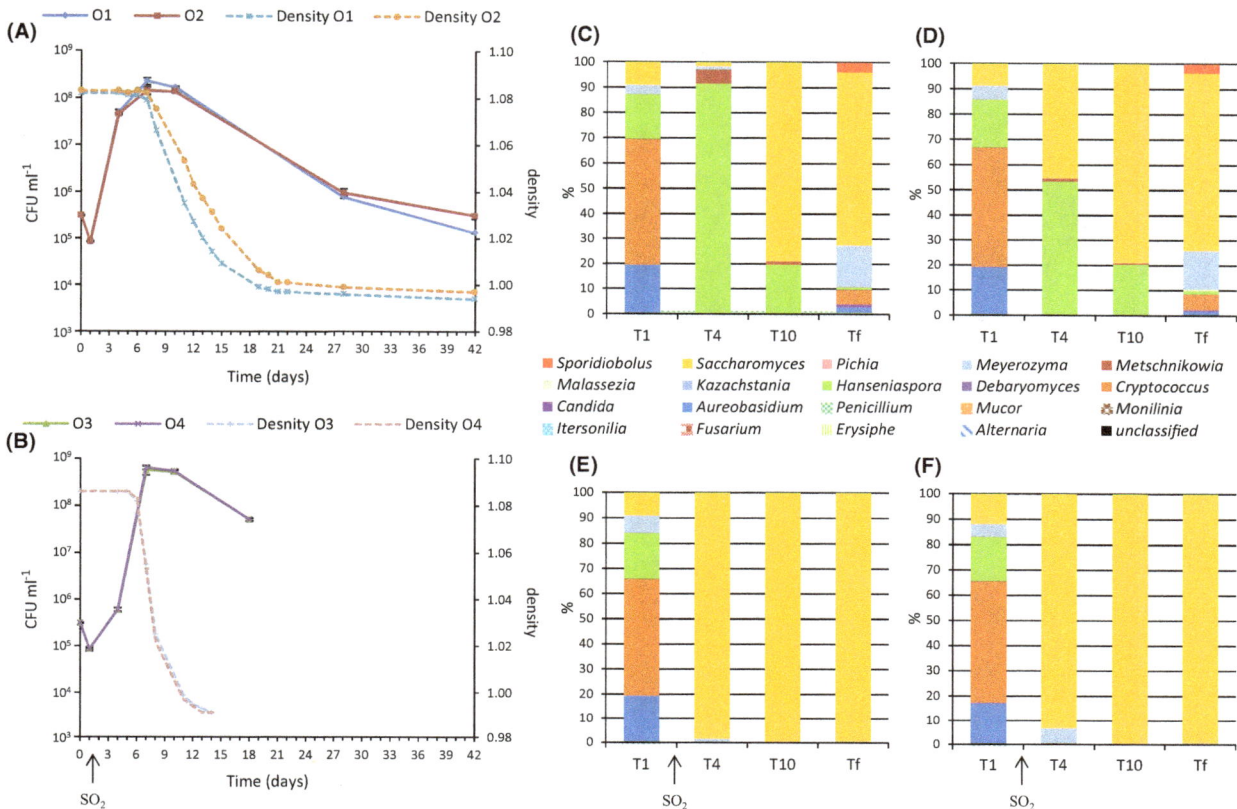

Fig. 4. Monitoring of grape must fermentation of Organic modality for 2013 vintage. Monitoring of yeast population size (CFU ml^{-1}) and fermentation progress (density) without SO2 (A) and with SO2 (B). Repartition of fungal genera during alcoholic fermentations without SO2 for must O1 (C) and must O2 (D) and with SO2 for must O3 (E) and must O4 (F). Populations are identified by pyrosequencing 1 day (T1), 4 days (T4), 10 days (T10) and at end of AF.

Chemical composition. The question is whether these differences in population dynamics (in both number and composition) had an impact on the chemical composition of wine products and whether the differences observed in grape must between the Organic modality and the two other modalities persisted. Wine composition is reported in Table 3. The higher concentrations of ethanol found in the Ecophyto modality can be explained by the complete fermentation observed for the four batches. For the Organic modality, the difference between batches with and without SO$_2$ can also be explained by the presence or absence of residual sugars. However, differences can be noted between the wines of the three modalities and so depending on the protection of the vineyard. The sugar concentrations of all the musts were very close, leading to the assumption that the presence of fermenting sugars at the end of AF was linked to the lower fermentative capacity of some populations related to differences between the strains of *S. cerevisiae* present or to the presence of non-*Saccharomyces* yeast, particularly for wine fermented without SO$_2$ (Charoenchai *et al.*, 1998; Bisson, 1999; Zohre and Erten, 2002; Ferreira *et al.*, 2006).

For all the modalities, volatile acidity was higher for the non-sulfited wines than for the sulfited wines: +0.16–0.18 g acetic acid l^{-1} for Organic modality, +0.35–0.36 g acetic acid l^{-1} for the Conventional modality and +0.12–0.16 g l^{-1} for the Ecophyto modality. The highest values were obtained for the sluggish fermentations, in which the persistence of some non-*Saccharomyces* species is higher.

The 12 wines were analysed by FT-ICR-MS. PLS-DA was used to group the wines depending on the use of SO$_2$ (Fig. 7A). Wines with SO$_2$ could be separated from wines without SO$_2$ along the first component (18.7% of variability). Use of SO$_2$ indeed had a particular impact with a slight increase in sulfur-containing compounds (CHOS) for the wines elaborated with sulfites (Fig. 7E). Although the overall elemental composition remained close between the different wines (Fig. 7B–D), our results show that the effects of adding sulfur dioxide to must were still detectable in wines at the end of vinification.

However, wines can also be clearly separated by PLS-DA (Fig. 8) according to plant protection. Thus, differences linked to plant protection were not masked by

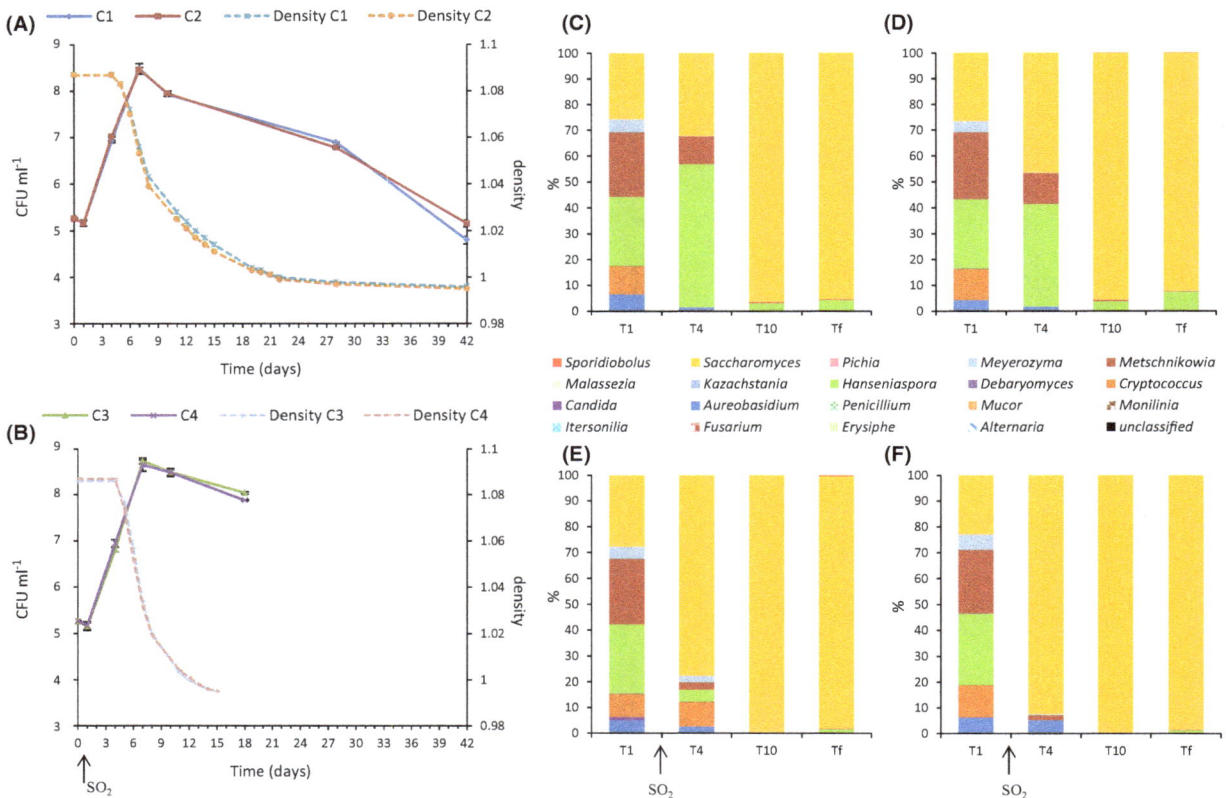

Fig. 5. Monitoring of grape must fermentation of Conventional modality for 2013 vintage. Monitoring of yeast population size (CFU ml⁻¹) and fermentation progress (density) without SO2 (A) and with SO2 (B). Repartition of fungal genera during alcoholic fermentations without SO2 for must C1 (C) and must C2 (D) and with SO2 for C3 (E) and C4 (F) musts. Populations are identified by pyrosequencing at 1 day (T1), 4 days (T4), 10 days (T10) and at the end of AF (Tf).

the use of SO$_2$. Moreover, as already observed for the 'terroir' effect (Roullier-Gall *et al.*, 2014a,b), the differences related to plant protection are more visible after AF and could partly result from microbiological processes. Projecting the masses as filtered from the PLS–DA analysis on van Krevelen diagrams (Fig. S2) reveals specific chemical fingerprints for the Organic, Conventional and Ecophyto wines. It is noteworthy that almost no CHOP- and CHONP-containing compounds are specific to a plant protection type. The Organic wines appear to be characterized by CHONS-, CHONSP- and CHO-containing compounds located in particular in areas of amino acids and carbohydrates according to the area of the van Krevelen diagram. The Conventional wines appear to be specifically richer in CHO-containing compounds with some located in the carbohydrate area and by CHONS- and CHOS-containing compounds. The Ecophyto wines appear to be characterized by CHONS-, CHON- and CHO-containing compounds. Thus, the existence in wines of chemical and microbiological signatures associated with plant protection is highlighted.

Conclusion

In this study, we were able to confirm the strong influence of vintage on fungal populations on grape berries. Moreover, many interactions seemed to exist between the yeasts and mould on grape berries. Therefore, it is essential to study fungal populations of the grape as a whole to better understand the interactions involved. Furthermore, the study solely of yeasts could lead to misinterpretations: that of attributing the influence of some plant pathogenic genera to other parameters. Despite these two important factors: interactions and vintage, a significant effect of plant protection on grape populations has been highlighted with a systematic reduction of biodiversity for grapes treated in Organic modality. The use of broad spectrum fungicides based on copper in the Organic modality could be the cause of this reduction of biodiversity. However, *in vitro* studies are still required to determine the sensitivity of different fungal genera to these products as well as studies in the vineyard to determine the direct effect of the products used at the time of application.

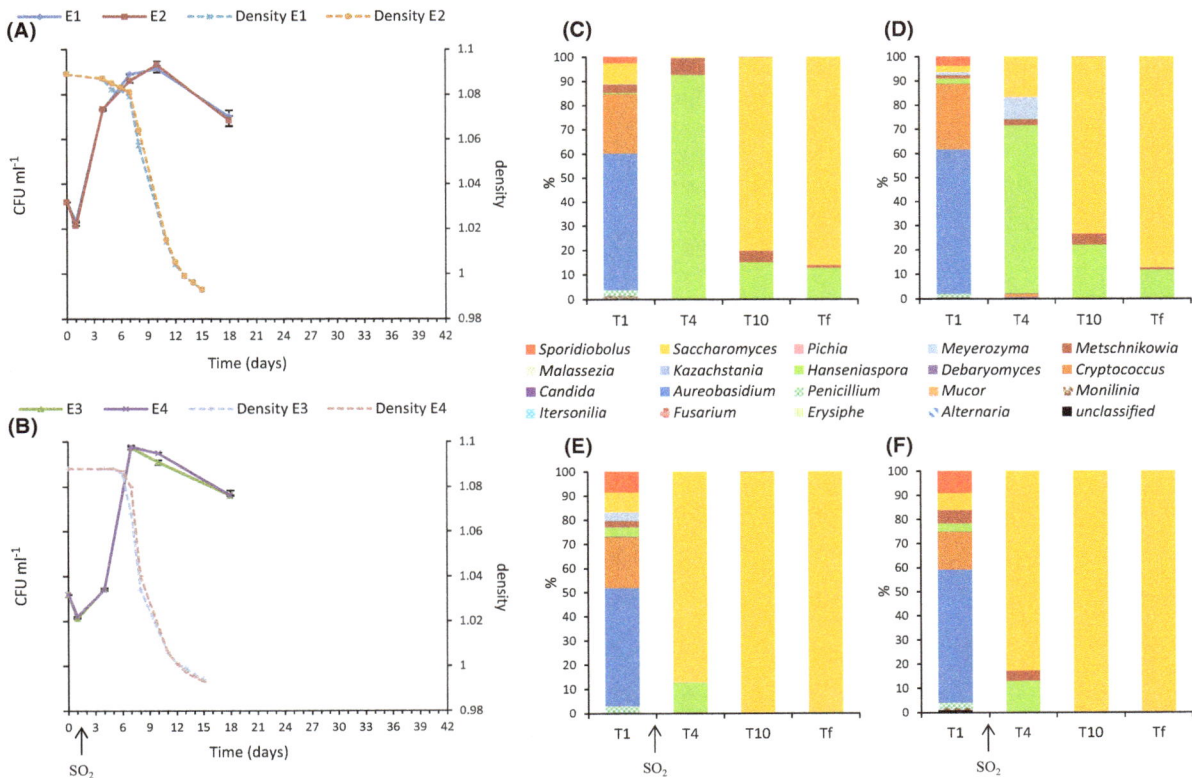

Fig. 6. Monitoring of grape must fermentation of Ecophyto modality for 2013 vintage. Monitoring of yeast population size (CFU ml^{-1}) and fermentation progress (density) without SO2 (A) and with SO2 (B). Repartition of fungal genera during alcoholic fermentations without SO2 for must E1 (C) and must E2 (D) and with SO2 for must E3 (E) and must E4 (F). Populations are identified by pyrosequencing 1 day (T1), 4 days (T4), 10 days (T10) and at end of AF (Tf).

Table 3. Analytical characteristics of the wines elaborated from grape berries harvested in three phytosanitary vineyard protections and fermented with or without SO$_2$ for 2013 vintage.

	Wines	Alcoholic degree (% v/v)	Residual sugars (g l^{-1})	L-malic acid (g l^{-1})	Volatile acidity (g acetic acid l^{-1})	Total SO$_2$ (mg l^{-1})
−SO$_2$	O1	12.55	2.1	2.5	0.68	4
	O2	12.45	2.0	2.3	0.68	5
+SO$_2$	O3	12.80	< 1	2.6	0.50	10
	O4	12.75	< 1	2.6	0.52	10
−SO$_2$	C1	12.80	3.2	2.2	0.78	3
	C2	12.70	4.0	2.3	0.77	5
+SO$_2$	C3	12.90	3.0	1.9	0.42	9
	C4	12.90	3.3	2.6	0.42	11
−SO$_2$	E1	13.00	< 1	2.1	0.36	5
	E2	13.1	< 1	2.1	0.34	3
+SO$_2$	E3	13.15	< 1	2.1	0.22	10
	E4	13.15	< 1	2.1	0.20	9

This study also showed that fungal populations were heavily revamped during the pre-fermentation step with a sharp reduction of mould and the presence of yeast genera not detected on berries. In spite of these reshuffles, the protection applied to the vineyard has a strong influence on fungal populations present in musts. In connection with these differences observed on grape must, populations evolve differently during AF depending on the protection applied. In addition, our results confirm the strong influence of SO2 on populations present during fermentation, and especially the early implantation and domination of the genus *Saccharomyces*. However, despite this selection of *Saccharomyces* yeast by the use of SO$_2$, some differences in

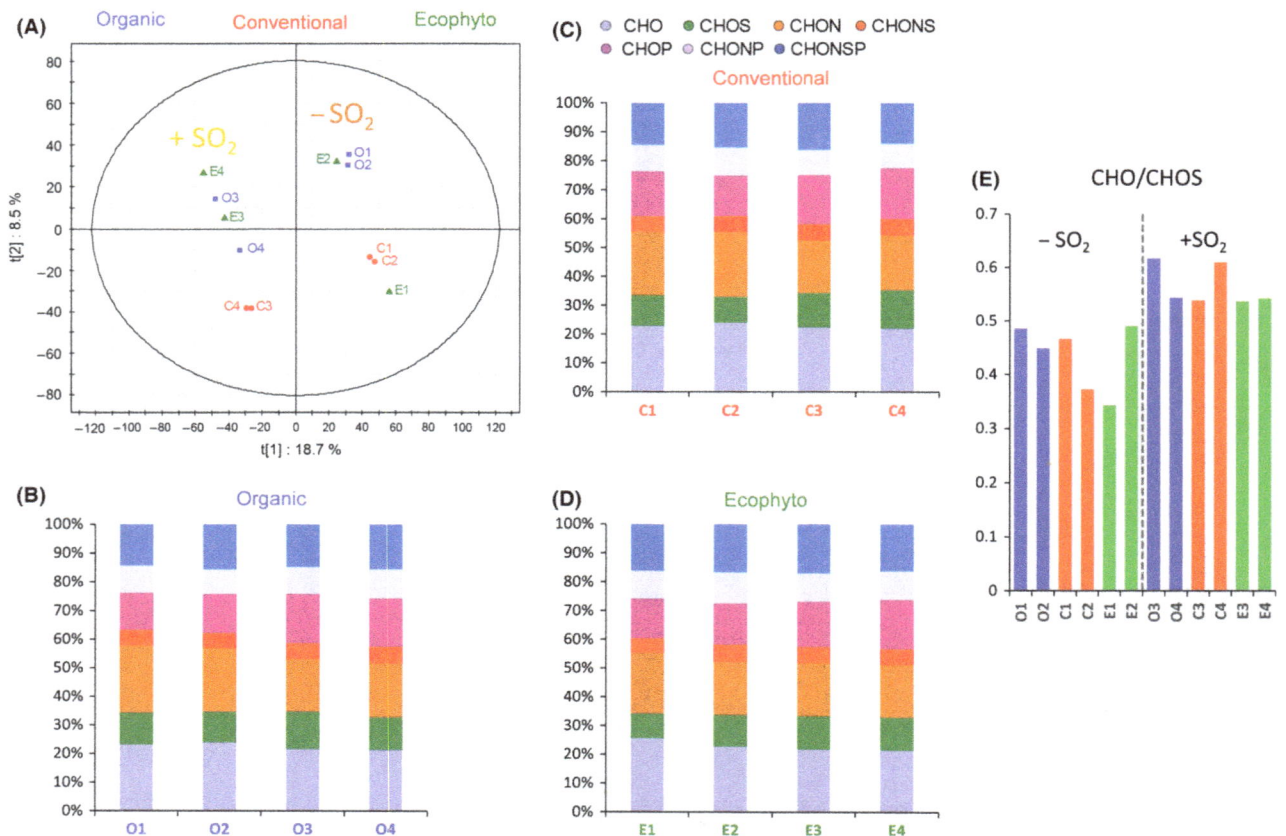

Fig. 7. Analysis of the FT-ICR-MS data for wines of 2013 vintage. (A) Scores plot of the PLS-DA depending on the use of SO2, the first two components retained 27.2% of the variation. Histograms of elementary composition of Organic (B), (C) Conventional and (D) Ecophyto wines. (E) Ratio of CHOS/CHO masses for each analysed wine.

populations persisted for several days. The characterization of differences between populations at species level or intraspecies level is necessary. This could help to highlight an even greater impact on the population than that which was demonstrated in this study. Additionally, the realization of physiological testing for yeasts isolated from each must could help to better understand the mechanisms behind the influence of the protection applied.

Our results showed a significant influence of anthropogenic practices such as the use of sulfur dioxide and plant protection on the composition of wine although the compositions of the grape must from these different protections were quite similar. The wines produced could be clearly distinguished on the one hand, based on the use or not of SO_2 and on the other hand, depending on the plant protection. It is now necessary to identify the discriminating compounds in wines elaborated for each protection to determine whether these compounds are primarily of plant or microbial origin. However, this effect is probably largely indirect and related to the modification of yeast populations during alcoholic fermentation,

as the chemical differences are much more pronounced for wine than for grape must.

Experimental procedures

The vineyard studied

All the grapes were sampled from a plot of Chardonnay planted in 1986 and located in Burgundy, France (46°18′32.2″N, 4°44′17.9″E, 258 m altitude). Since 2007, the plot has been divided into three blocks of eight rows each: one block was managed using phytosanitary products according to conventional viticulture and noted C [pyrethroids, organophosphates, anthranilic diamides, benzamides, pyridinyl-ethyl-benzamides, pyridine-carboxamides, oximino-acetates, cyano-imidazole, triazolo-pyrimidylamine, triazoles, spiroketal-amines, cinnamic acid amides, mandelic acid amides, cyanoacetamide-oxime, phosphonates, benzophenone, dithiocarbamates, phthalimides, quinones and inorganic (sulfur and copper) fungicides]. The second block, Ecophyto (E), was managed with the same products used for the Conventional block, but with dose reduction and/or with a reduced number of

Fig. 8. Analysis of the FT-ICR-MS data for wine of 2013 vintage. Scores plot of the PLS-DA depending on phytosanitary protection mode, the first two components retained 29.4% of the variation.

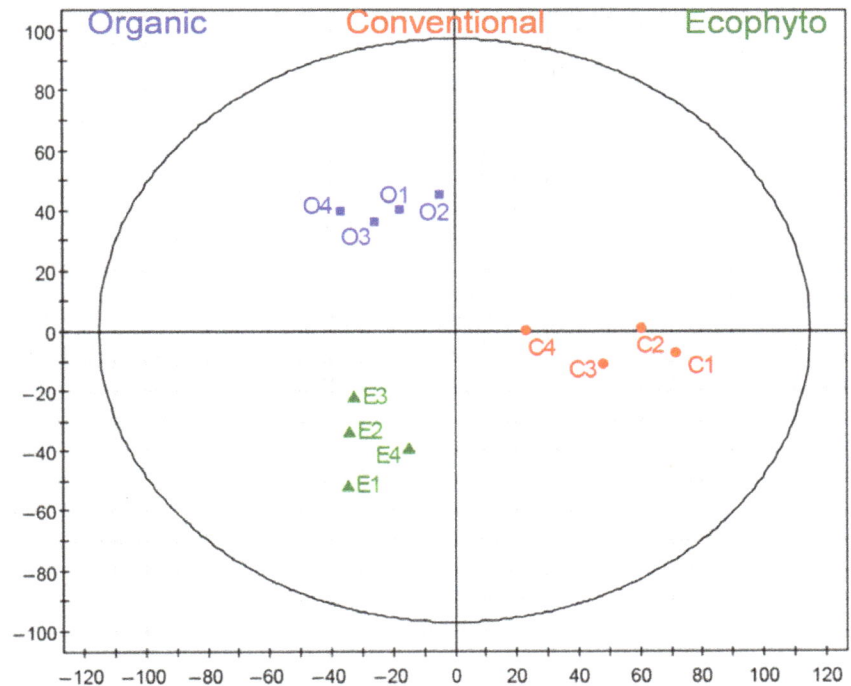

treatments. Block O was managed according to organic viticulture practices for which only pyrethrins, copper and sulfur are allowed. Details on the management procedures can be found in Table S1.

Sampling

The sampling of grape berries, bunches of grapes and total harvest were carried out in the central rows (3 and 5) (Fig. S1) of each block to overcome edge effects related to treatment. For each modality, 6 kg of ripe bunches of grapes were collected aseptically from 20 different vine plants distributed along the rows (one cluster per vine plant) for the 2012 vintage. Ten berries from each vine plant of the two rows (3 and 5) were collected for the 2013 and 2014 vintages (860 berries, 950 berries and 980 berries for the Organic, Conventionnal and Ecophyto modalities respectively). Grape berries were placed in a sterile bag and pressed manually directly in the bag. Immediately after pressing, a sample (50 ml) was taken to analyse the fungal biodiversity of the grape berries.

Moreover for the 2013 vintage, the total harvest of the two rows (3 and 5) of each modality was collected manually and placed in 20 kg crates. The harvests (102, 117 and 78 kg for modalities O, C and E respectively) were then transported to the experimental winery (Beaune, France). The harvests were pressed directly on arrival at the winery using a vertical FASER-PLAST AG press (Rickenbach, Switzerland). The grape must thus

obtained was placed overnight at 10°C for lees sedimentation (settling) and then separated into four replicates for each modality in four 20 l stainless steel vats. About 50 ml of must was sampled (T0) for each vat. The addition of sulfur dioxide (30 mg l^{-1}) was carried out in two of four vats for each modality. During alcoholic fermentation, 50 ml samples were collected from each vat at various times: after 3 days (T3), 6 days (T6), 9 days (T9) and at the end of alcoholic fermentation (when density no longer decreases) (Tf). The wines obtained were then bottled without sulfiting to be used for non-targeted chemical analysis.

Enumeration of yeast

For each sample, serial dilutions were performed and 3 × 100 µl of each dilution were spread on the YPD medium (0.5% w/v yeast extract, 1% w/v peptone, 2% w/v glucose and 2% w/v agar supplemented with chloramphenicol at 200 ppm to inhibit the development of bacteria) and incubated at 28°C. The yeast populations were then estimated by counting the colonies developed and the result was obtained by averaging the three repetitions.

DNA extraction

For each sample, 5 ml of must or wine was collected and centrifuged for 5 min at 4°C at 3000 **g**. The pellet was suspended in 5 ml milliQ water and filtered through glass wool to separate cells from must debris. The filtered suspension was centrifuged again (5 min at 4°C, at 3000 *g*). The pellet was resuspended in 200 µl of lysis buffer (2% Triton X-100, 1% SDS, 100 mM NaCl, 10 mM Tris pH 8.1 mM EDTA pH 8), and the cells were homogenized in a bead beater (Precellys 24, France) with 0.3 g of glass beads (0.5 mm in diameter) in the presence of 200 µL phenol/chloroform/isoamyl alcohol (50:48:2). The mixture was vortexed for 1 min and placed on ice for 1 min. This step was repeated three times. Then 200 µl TE (10 mM Tris, 1 mM EDTA pH 8) was added and the bead/cell mixture was centrifuged for 10 min at 16 000 *g* at 4°C, after which the aqueous phase was collected. The DNA was precipitated from this aqueous phase with 2.5 volumes of 100% ethanol and centrifuged at 16 000 *g* at 4°C for 10 min. Then the pellet was washed with 70% ethanol, dried and suspended in 50 µl of DEPC-treated water (Thermo Fisher Scientific, Waltham, MA, USA). The DNA concentrations of the samples were then standardized (50 ng µl^{-1}) by measuring optical density at 260 nm, and adding DEPC-treated water as appropriate before storage at −20°C.

Pyrosequencing of 18S rRNA gene sequences

Fungal diversity was determined for each sample by using 454 pyrosequencing of ribosomal genes. A 18S rRNA gene fragment with sequence variability and appropriate size (about 350 bases) for 454 pyrosequencing was amplified using the primers FR1 (5′-ANCCATT-CAATCGGTANT-3′) and FF390 (5′-CGATAACGA ACGAGACCT-3′) (Chemidlin Prévost-Bouré *et al.*, 2011). For each sample, 5 ng of DNA was used for a 25 µl PCR conducted under the following conditions: 94°C for 3 min, 35 cycles of 1 min at 94°C, 52°C for 1 min and 72°C for 1 min, followed by 5 min at 72°C. A second PCR of nine cycles was then conducted under similar PCR conditions with purified PCR products and 10 base pair multiplex identifiers were added to the primers at position 5′ to specifically identify each sample and avoid PCR bias. Finally, the PCR products were purified using a MinElute gel extraction kit (Qiagen, Courtaboeuf, France) and quantified using the Pico-Green staining Kit (Molecular Probes, Paris, France). Pyrosequencing was carried out on a GS Junior apparatus (Roche 454 Sequencing System) by the GenoSol platform (INRA, Dijon, France, http://www2.dijon. inra.fr/plateforme_genosol/) and on GS FLX Titanium (Roche 454 Sequencing System) by Genoscreen (Lille, France, http://www.genoscreen.com/).

Analysis of pyrosequencing data

Bioinformatics analyses of reads obtained by pyrosequencing were performed using the GnS-PIPE pipeline

developed by the GenoSol platform (INRA, Dijon, France) (Terrat *et al.*, 2012), or the Qiime pipeline developed by scikit-bio (Caporaso *et al.*, 2010). The parameters chosen for each step were the same for the two pipelines and can be found in supplementary material (Table S2). First, all the 18S raw reads were sorted according to the multiplex identifier sequences. The raw reads were then filtered and deleted based on: (i) their length, (ii) their number of ambiguities (Ns) and (iii) their primer(s) sequence(s). A PERL program was then applied for rigorous dereplication (i.e. clustering of strictly identical sequences). The dereplicated reads were then aligned using Infernal alignment (Cole *et al.*, 2009), and clustered into operational taxonomic units (OTU) using a PERL program that groups rare reads with abundant ones without counting differences in homopolymer lengths. A filtering step was then carried out to check all single-singletons (reads detected only once and not clustered, which might be artefacts such as PCR chimeras) based on the quality of their taxonomic assignments.

The high-quality reads retained were then taxonomically assigned using similarity approaches against dedicated reference databases from *SILVA* (Quast *et al.*, 2013) (see supplementary material) (Table S2). The raw data sets are available on the EBI database system under project accession number PRJEB12990 (awaiting attribution).

Linear discriminant analysis effect size was used to determine significant taxonomic differences between the grape must sample of each phytosanitary protection (Segata *et al.*, 2011). This method employs the factorial Kruskal–Wallis sum-rank test ($\alpha = 0.05$) to identify taxa with significant differential abundances between modalities (using one-against-all comparisons), followed by LDA to estimate the effect size of each differentially abundant feature. Significant taxa were used to generate taxonomic cladograms illustrating differences between phytosanitary protection modalities.

Shannon index (H′) was used to assess the fungal biodiversity identified by pyrosequencing in populations present on grape berries using the number of sequences to calculate *Pi*, where *i* is a genus, S the total number of genera, n_i the number of reads corresponding to genus *i*, *N* the total number of reads and *Pi* the proportion of genus *i* with $Pi = n_i/N$:

$$H' = \sum_{i=1}^{S} Pi(\log_2 Pi)$$

Oenological analysis

The dosage of reducing sugars, available nitrogenous compounds (ammonium and amino acid except proline), L-malic acid, ethanol and acetic acid were determined by

FT-IR spectroscopy (FOSS France). pH was measured using a pH meter.

Non-targeted chemical analyses

Grape musts after settling and wines of the 2013 vintage were analysed by Fourier transformed ion cyclotron resonance mass spectrometry (FT-ICR-MS). The sample preparation consisted of a dilution of the wine in ultrapure methanol in proportions of 50 μl per 950 μl. Centrifuged grape musts were acidified by formic acid (2% v/v) to pH 2 and pre-filtered using C18-SPE cartridges (100 mg ml^{-1} Backerbond SPE columns) to remove sugars. C18 cartridges were conditioned by successive passages of 1 ml methanol and 1 ml of ultrapure water acidified with formic acid (1.25%). One millilitre of acidified must was then passed through the C18 cartridge by gravity, followed by 1 ml of dilute formic acid (1.25%). Finally, the acidified must was eluded with 500 μL of methanol and stored in amber vials at −20°C for analysis. Mass spectra were obtained with an FT-ICR-MS Solarix (Bruker Daltonics, Bremen, Germany) equipped with a 12 Tesla superconducting magnet (Magnex, UK). The instrument was equipped with an electrospray ionization source Apolo II. The samples were injected directly into a micro electrospray source at a rate of 120 μl h^{-1}. The MS was externally calibrated using a 5 ppm solution of arginine (0.2 ppm tolerance). Spectra were recorded in negative ionization mode and for a mass range between *m/z* 100 and 1000. For each sample, 300 scans per sample were accumulated with a time domain of a 4 MW (megaword) per second. Spectra were then internally calibrated using a mass list of ubiquitous wine compounds with a mass error below 50 ppb. Peaks with a signal to noise ratio (S/N) of 4 and higher were used for further data processing.

Partial least square discriminative analysis (PLS–DA) models were used to provide enhanced representations of the sample category discriminations and extract the most discriminative metabolites, which were also checked manually within the spectra. Discriminative masses with a variable importance in projection (VIP) value > 2 and *P* values < 0.05 were considered as relevant. PLS-DA was performed with the SIMCA 9 software (http://www.umetrics.com/). Two dimensional van Krevelen diagrams of discriminative metabolites were obtained using compositional networks (based on elemental compositions) and functional networks, based on selected functional group equivalents that enable improved assignment options of elemental compositions and better classification of organic complexities with tunable validation windows (Tziotis *et al.*, 2011).

Conflict of Interest

The authors declare no conflict of interest.

References

Adrian, M., Jeandet, P., Veneau, J., Weston, L.A., and Bessis, R. (1997) Biological activity of resveratrol, a stilbenic compound from grapevines, against *Botrytis cinerea*, the causal agent for gray mold. *J Chem Ecol* **23**: 1689–1702.

Agenbach, W.A. (1977) A study of must nitrogen content in relation to incomplete fermentations, yeast production and fermentation activity. In Beukman, E.F., (ed.). Proceedings of the South African Society for Enology and Viticulture; (Cape Town, South Africa, Nov., 1977). Stellenbosch, South Africa: South African Society for Enology and Viticulture. pp 66–88.

Albertin, W., Marullo, P., Aigle, M., Dillmann, C., de Vienne, D., Bely, M., and Sicard, D. (2011) Population size drives industrial *Saccharomyces cerevisiae* alcoholic fermentation and is under genetic control. *Appl Environ Microb* **77**: 2772–2784.

Albertin, W., Miot-Sertier, C., Bely, M., Marullo, P., Coulon, J., Moine, V., *et al.* (2014) Oenological prefermentation practices strongly impact yeast population dynamics and alcoholic fermentation kinetics in Chardonnay grape must. *Int J Food Microbiol* **178**: 87–97.

Andorrà, I., Esteve-Zarzoso, B., Guillamón, J.M., and Mas, A. (2010) Determination of viable wine yeast using DNA binding dyes and quantitative PCR. *Int J Food Microbiol* **144**: 257–262.

Barata, A., Seborro, F., Belloch, C., Malfeito-Ferreira, M., and Loureiro, V. (2008) Ascomycetous yeast species recovered from grapes damaged by honeydew and sour rot. *J Appl Microbiol* **104**: 1182–1191.

Barata, A., Malfeito-Ferreira, M., and Loureiro, V. (2012) The microbial ecology of wine grape berries. *Int J Food Microbiol* **153**: 243–259.

Bisson, L.F. (1999) Stuck and sluggish fermentations. *Am J Enol Viticult* **50**: 107–119.

Bokulich, N.A., Swadener, M., Sakamoto, K., Mills, D.A., and Bisson, L.F. (2015) Sulfur dioxide treatment alters wine microbial diversity and fermentation progression in a dose-dependent fashion. *Am J Enol Viticult* **66**: 73–79.

Broome, J.C., English, J.T., Marois, J.J., Latorre, B.A., and Aviles, J.C. (1995) Development of an infection model for *Botrytis* bunch rot of grapes based on wetness duration and temperature. *Phytopathology* **85**: 97–102.

Bunea, C.I., Pop, N., Babeş, A., Matea, C., Dulf, F.V., and Bunea, A. (2012) Carotenoids, total polyphenols and antioxidant activity of grapes (*Vitis vinifera*) cultivated in organic and conventional systems. *Chem Cent J* **6**: 66–77.

Cadez, N., Zupan, J., and Raspor, P. (2010) The efffects of fungicides on yeast communities associated with grape berries. *Yeast Res.* **10**: 619–630.

Caporaso, J.G., Kuczynski, J., Stombaugh, J., Bittinger, K., Bushman, F.D., Costello, E.K., *et al.* (2010) QIIME allows analysis of high-throughput community sequencing data. *Nat Methods* **7**: 335–336.

Charoenchai, C., Fleet, G.H., and Henschke, P.A. (1998) Effects of temperature, pH, and sugar concentration on the growth rates and cell biomass of wine yeasts. *Am J Enol Viticult* **49**: 283–288.

Chemidlin Prévost-Bouré, N., Christen, R., Dequiedt, S., Mougel, C., Lelièvre, M., Jolivet, C., *et al.* (2011) Validation and application of a PCR primer set to quantify fungal communities in the soil environment by real-time quantitative PCR. *PLoS ONE* **6**: e24166.

Ciani, M., and Maccarelli, F. (1998) Oenological properties of non-*Saccharomyces* yeasts associated with wine-making. *World J Microb Biot* **14**: 199–203.

Ciani, M., and Picciotti, G. (1995) The growth kinetics and fermentation behaviour of some non-*Saccharomyces* yeasts associated with wine-making. *Biotechnol Lett* **17**: 1247–1250.

Cole, J.R., Wang, Q., Cardenas, E., Fish, J., Chai, B., Farris, R.J., *et al.* (2009) The Ribosomal Database Project: improved alignments and new tools for rRNA analysis. *Nucleic Acids Res* **37**: 141–145.

Combina, M., Elía, A., Mercado, L., Catania, C., Ganga, A., Martinez, C., and Eli, A. (2005) Dynamics of indigenous yeast populations during spontaneous fermentation of wines from Mendoza, Argentina. *Int J Food Microbiol* **99**: 237–243.

Comitini, F., and Ciani, M. (2008) Influence of fungicide treatments on the occurrence of yeast flora associated with wine grapes. *Ann Microbiol* **58**: 489–493.

Constantí, M., Reguant, C., Poblet, M., Zamora, F., Mas, A., and Guillamón, J.M. (1998) Molecular analysis of yeast population dynamics: effect of sulphur dioxide and inoculum on must fermentation. *Int J Food Microbiol* **41**: 169–175.

Cordero-Bueso, G., Arroyo, T., Serrano, A., Tello, J., Aporta, I., Vélez, M.D., and Valero, E. (2011) Influence of the farming system and vine variety on yeast communities associated with grape berries. *Int J Food Microbiol* **145**: 132–139.

Dani, C., Oliboni, L.S., Vanderlinde, R., Bonatto, D., Salvador, M., and Henriques, J.A.P. (2007) Phenolic content and antioxidant activities of white and purple juices manufactured with organically- or conventionally-produced grapes. *Food Chem Toxicol* **45**: 2574–2580.

Donèche, B.J. (1993) Botrytized wines. In Fleet G. H. (ed.), Wine microbiology and biotechnology. Harwood Academic Publishers, Philadelphia, Pa. pp. 327–352.

Ferreira, J., Du Toit, M., and du Toit, W.J. (2006) The effects of copper and high sugar concentrations on growth, fermentation efficiency and volatile acidity production of different commercial wine yeast strains. *Aust J Grape Wine R* **12**: 50–56.

Grangeteau, C., Gerhards, D., Rousseaux, S., von Wallbrunn, C., Alexandre, H., and Guilloux-Benatier, M. (2015) Diversity of yeast strains of the genus *Hanseniaspora* in the winery environment: What is their involvement in grape must fermentation? *Food Microbiol* **50**: 70–77.

Guerra, E., Sordi, G., Mannazzu, I., Clementi, F., and Fatichenti, F. (1999) Occurrence of wine yeasts on grapes subjected to different pesticide treatments. *Ital J Food Sci* **11**: 221–230.

Henick-Kling, T., Edinger, W., Daniel, P., and Monk, P. (1998) Selective effects of sulfur dioxide and yeast starter culture addition on indigenous yeast populations and sensory characteristics of wine. *J Appl Microbiol* **84**: 865–876.

Hidalgo, L. (1978) Grape rot, methods for its control, and effects on wine quality. *Ann Techn Agric* **27**: 127.

Hierro, N., González, A., Mas, A., and Guillamón, J.M. (2006) Diversity and evolution of non-*Saccharomyces* yeast populations during wine fermentation: effect of grape ripeness and cold maceration. *FEMS Yeast Res* **6**: 102–111.

Jeandet, P., Adrian, M., Breuil, A., Sbaghi, M., Debord, S., Bessis, R., *et al.* (2000) Chemical induction of phytoalexin synthesis in grapevines: application to the control of grey mould in the vineyard. *Acta Hortic* **528**: 591–596.

Lalancette, N., Ellis, M.A., and Madden, L.V. (1988) Development of an infection efficiency model for *Plasmopara viticola* on American grape based on temperature and duration of leaf wetness. *Phytopathology* **78**: 794–800.

Lambrechts, M.G., and Pretorius, I.S. (2000) Yeast and its importance to wine aroma - a review. *S Afr J Enol Vitic* **21**: 97–129.

Levite, D., Adrian, M. and Tamm, L. (2000) Preliminary results of resveratrol in wine of organic and conventional vineyards. Proceedings of the 6th International Congress on organic Viticulture, Basel (Suisse), pp. 256–257.

Liu, H.M., Guo, J.H., Cheng, Y.J., Luo, L., Liu, P., Wang, B.Q., and Long, C.A. (2010) Control of gray mold of grape by *Hanseniaspora uvarum* and its effects on postharvest quality parameters. *Ann Microbiol* **60**: 31–35.

Martin, V., Boido, E., Giorello, F., Mas, A., Dellacassa, E., and Carrau, F. (2016) Effect of yeast assimilable nitrogen on the synthesis of phenolic aroma compounds by *Hanseniaspora vineae* strains. *Yeast* **33**: 323–328.

Martins, G., Vallance, J., Mercier, A., Albertin, W., Stamatopoulos, P., Rey, P., *et al.* (2014) Influence of the farming system on the epiphytic yeasts and yeast-like fungi colonizing grape berries during the ripening process. *Int J Food Microbiol* **177**: 21–28.

Maturano, Y.P., Mestre, M.V., Esteve-Zarzoso, B., Nally, M.C., Lerena, M.C., Toro, M.E., *et al.* (2015) Yeast population dynamics during prefermentative cold soak of Cabernet Sauvignon and Malbec wines. *Int J Food Microbiol* **199**: 23–32.

McGovern, P.E., Glusker, D.L., Exner, L.J., and Voigt, M.M. (1996) Neolithic resinated wine. *Nature* **381**: 480–481.

Medina, K., Boido, E., Dellacassa, E., and Carrau, F. (2012) Growth of non-*Saccharomyces* yeasts affects nutrient availability for *Saccharomyces cerevisiae* during wine fermentation. *Int J Food Microbiol* **157**: 245–250.

Medina, K., Boido, E., Fariña, L., Gioia, O., Gomez, M.E., Barquet, M., *et al.* (2013) Increased flavour diversity of Chardonnay wines by spontaneous fermentation and co-fermentation with *Hanseniaspora vineae*. *Food Chem* **140**: 2513–2521.

Milanović, V., Comitini, F., and Ciani, M. (2013) Grape berry yeast communities: influence of fungicide treatments. *Int J Food Microbiol* **161**: 240–246.

Mills, D.A., Johannsen, E.A., and Cocolin, L. (2002) Yeast diversity and persistence in *botrytis*-affected wine fermentations. *Appl Environ Microb* **68**: 4884–4893.

Mora, J., and Mulet, A. (1991) Effects of some treatments of grape juice on the population and growth of yeast species during fermentation. *Am J Enol Viticult* **42**: 133–136.

Moreira, N., Pina, C., Mendes, F., Couto, J.A., Hogg, T., and Vasconcelos, I. (2011) Volatile compounds contribution of *Hanseniaspora guilliermondii* and *Hanseni-*

aspora uvarum during red wine vinifications. *Food Control* **22:** 662–667.

Moyano, L., Zea, L., Villafuerte, L., and Medina, M. (2009) Comparison of odor-active compounds in sherry wines processed from ecologically and conventionally grown Pedro Ximenez grapes. *J Agr Food Chem* **57:** 968–973.

Mulero, J., Pardo, F., and Zafrilla, P. (2010) Antioxidant activity and phenolic composition of organic and conventional grapes and wines. *J Food Compos Anal* **23:** 569–574.

Nisiotou, A.A., Spiropoulos, A.E., and Nychas, G.-J.E. (2007) Yeast community structures and dynamics in healthy and *Botrytis*-affected grape must fermentations. *Appl Environ Microb* **73:** 6705–6713.

Ocón, E., Gutiérrez, A.R., Garijo, P., Santamaría, P., López, R., Olarte, C., and Sanz, S. (2011) Factors of influence in the distribution of mold in the air in a wine cellar. *J Food Sci* **76:** 169–174.

Pagliarini, E., Laureati, M., and Gaeta, D. (2013) Sensory descriptors, hedonic perception and consumer's attitudes to Sangiovese red wine deriving from organically and conventionally grown grapes. *Front Psycho* **4:** 896.

Quast, C., Pruesse, E., Yilmaz, P., Gerken, J., Schweer, T., Yarza, P., *et al.* (2013) The SILVA ribosomal RNA gene database project: improved data processing and web-based tools. *Nucleic Acids Res* **41:** 590–596.

Rabosto, X., Carrau, M., Paz, A., Boido, E., Dellacassa, E., and Carrau, F.M. (2006) Grapes and vineyard soils as sources of microorganisms for biological control of *Botrytis cinerea*. *Am J Enol Viticult* **57:** 332–338.

Raspor, P., Milek, D.M., Polanc, J., Mozina, S.S., and Cadez, N. (2006) Yeasts isolated from three varieties of grapes cultivated in different locations of the Dolenjska vine-growing region, Slovenia. *Int J Food Microbiol* **109:** 97–102.

Rojas, V. (2003) Acetate ester formation in wine by mixed cultures in laboratory fermentations. *Int J Food Microbiol* **86:** 181–188.

Romano, P. and Suzzi, G. (1993) Sulphur dioxide and wine microorganisms. In Fleet G.H. (ed.). Wine Microbiology and Biotechnology, Switzerland. Harwood Academic Publishers, pp. 373–393.

Romano, P., Fiore, C., and Paraggio, M. (2003) Function of yeast species and strains in wine flavour. *Int J Food Microbiol* **86:** 169–180.

Roullier-Gall, C., Boutegrabet, L., Gougeon, R.D., and Schmitt-Kopplin, P. (2014a) A grape and wine chemodiversity comparison of different appellations in Burgundy: vintage vs terroir effects. *Food Chem* **152:** 100–107.

Roullier-Gall, C., Lucio, M., Noret, L., Schmitt-Kopplin, P., and Gougeon, R.D. (2014b) How subtle is the 'terroir' effect? Chemistry-related signatures of two 'climats de Bourgogne'. *PLoS ONE* **9:** 1–11.

Sadoudi, M., Tourdot-Maréchal, R., Rousseaux, S., Steyer, D., Gallardo-Chacón, J.-J., Ballester, J., and Alexandre, H. (2012) Yeast-yeast interactions revealed by aromatic profile analysis of Sauvignon Blanc wine fermented by single or co-culture of non-*Saccharomyces* and *Saccharomyces* yeasts. *Food Microbiol* **32:** 243–253.

Sall, M.A. (1980) Epidemiology of grape powdery mildew: a model. *Phytopathology* **70:** 338–342.

Segata, N., Izard, J., Waldron, L., Gevers, D., Miropolsky, L., Garrett, W.S., and Huttenhower, C. (2011) Metagenomic biomarker discovery and explanation. *Genome Biol* **12:** R60.

Setati, M.E., Jacobson, D., Andong, U.-C., and Bauer, F. (2012) The vineyard yeast microbiome, a mixed model microbial map. *PLoS ONE* **7:** e52609.

Setati, M.E., Jacobson, D. and Bauer, F.F. (2015) Sequence-based analysis of the *Vitis vinifera* L. cv Cabernet Sauvignon grape must mycobiome in three South African vineyards employing distinct agronomic systems. *Front Microbiol* **6:** 1358.

Sipiczki, M. (2003) *Candida zemplinina* sp. nov., an osmotolerant and psychrotolerant yeast that ferments sweet botrytized wines. *Int J Syst Evol Micr* **53:** 2079–2083.

Sláviková, E., and Vadkertiová, R. (2003) Effects of pesticides on yeasts isolated from agricultural soil. *Z Naturforsch C* **58:** 855–859.

Sturm, J., Grossmann, M., and Schnell, S. (2006) Influence of grape treatment on the wine yeast populations isolated from spontaneous fermentations. *J Appl Microbiol* **101:** 1241–1248.

Swiegers, J.H., Bartowsky, E.J., Henschke, P.A., and Pretorius, I.S. (2005) Yeast and bacterial modulation of wine aroma and flavour. *Aust J Grape Wine R* **11:** 139–173.

Takahashi, M., Ohta, T., Masaki, K., Mizuno, A., and Goto-Yamamoto, N. (2014) Evaluation of microbial diversity in sulfite-added and sulfite-free wine by culture-dependent and -independent methods. *J Biosci Bioeng* **117:** 569–575.

Terrat, S., Christen, R., Dequiedt, S., Lelièvre, M., Nowak, V., Regnier, T., *et al.* (2012) Molecular biomass and MetaTaxogenomic assessment of soil microbial communities as influenced by soil DNA extraction procedure. *Microb Biotechnol* **5:** 135–141.

Tziotis, D., Hertkorn, N., and Schmitt-Kopplin, P. (2011) Kendrick-analogous network visualisation of ion cyclotron resonance Fourier transform mass spectra: improved options for the assignment of elemental compositions and the classification of organic molecular complexity. *Eur J Mass Spectrom* **17:** 415–421.

Vadkertiová, R., and Sláviková, E. (2006) Metal tolerance of yeasts isolated from water, soil and plant environments. *J. Basic Microb* **46:** 145–152.

Vrček, I.V., Bojić, M., Žuntar, I., Mendaš, G., and Medić-Šarić, M. (2011) Phenol content, antioxidant activity and metal composition of Croatian wines deriving from organically and conventionally grown grapes. *Food Chem* **124:** 354–361.

Xufre, A., Albergaria, H., Inácio, J., Spencer-Martins, I., and Gírio, F. (2006) Application of fluorescence in situ hybridisation (FISH) to the analysis of yeast population dynamics in winery and laboratory grape must fermentations. *Int J Food Microbiol* **108:** 376–384.

Zafrilla, P., Morillas, J., Mulero, J., Cayuela, J.M., Martínez-Cachá, A., Pardo, F., and Lopez Nicolas, J.M. (2003) Changes during storage in conventional and ecological wine: phenolic content and antioxidant activity. *J Agr Food Chem* **51:** 4694–4700.

Zironi, R., Romano, P., Suzzi, G., Battistutta, F., and Comi, G. (1993) Volatile metabolites produced in wine by mixed and sequential cultures of *Hanseniaspora guilliermondii* or *Kloeckera apiculata* and *Saccharomyces cerevisiae*. *Biotechnol Lett* **15:** 235–238.

Zohre, D.E., and Erten, H. (2002) The influence of *Kloeckera apiculata* and *Candida pulcherrima* yeasts on wine fermentation. *Process Biochem* **38:** 319–324.

Mechanism and regulation of sorbicillin biosynthesis by *Penicillium chrysogenum*

Fernando Guzmán-Chávez,[1] Oleksandr Salo,[1]
Yvonne Nygård,[1,†] Peter P. Lankhorst,[3] Roel A. L.
Bovenberg[2,3] and Arnold J. M. Driessen[1,*]
[1]*Molecular Microbiology, Groningen Biomolecular
Sciences and Biotechnology Institute, University of
Groningen, Nijenborgh 7, 9747 AG Groningen, The
Netherlands*
[2]*Synthetic Biology and Cell Engineering, Groningen
Biomolecular Sciences and Biotechnology Institute,
University of Groningen, Nijenborgh 7, 9747 AG
Groningen, The Netherlands.*
[3]*DSM Biotechnology Center, Alexander Fleminglaan 1,
2613 AX Delft, The Netherlands.*

Summary

***Penicillium chrysogenum* is a filamentous fungus
that is used to produce β-lactams at an industrial
scale. At an early stage of classical strain improvement, the ability to produce the yellow-coloured
sorbicillinoids was lost through mutation. Sorbicillinoids are highly bioactive of great pharmaceutical
interest. By repair of a critical mutation in one of the
two polyketide synthases in an industrial *P. chrysogenum* strain, sorbicillinoid production was restored
at high levels. Using this strain, the sorbicillin
biosynthesis pathway was elucidated through gene
deletion, overexpression and metabolite profiling.
The polyketide synthase enzymes SorA and SorB
are required to generate the key intermediates sorbicillin and dihydrosorbicillin, which are subsequently converted to (dihydro)sorbillinol by the FAD-
dependent monooxygenase SorC and into the final
product oxosorbicillinol by the oxidoreductase SorD.
Deletion of either of the two *pks* genes not only
impacted the overall production but also strongly
reduce the expression of the pathway genes. Expression is regulated through the interplay of two transcriptional regulators: SorR1 and SorR2. SorR1 acts
as a transcriptional activator, while SorR2 controls
the expression of *sorR1*. Furthermore, the sorbicillinoid pathway is regulated through a novel autoinduction mechanism where sorbicillinoids activate
transcription.**

*For correspondence. E-mail a.j.m.driessen@rug.nl

†Present address: Biology and Biological Engineering, Industrial
Biotechnology, Chalmers University of Technology, Kemigarden 4
Göteborg, Sweden.

Funding information
FGC was supported by Consejo Nacional de Ciencia y Tecnología
(CONACyT, México) and Becas Complemento SEP (México). YN
was supported by funding from the European Union's Seventh
Framework Programme FP7/207-2013, under grant agreement no
607332. OS was funded by NWO|Stichting voor de Technische
Wetenschappen (STW).

Introduction

Sorbicillinoids are a large family of hexaketide metabolites that include more than 90 highly oxygenated molecules. These compounds can be structurally classified
into four groups: monomeric sorbicillinoids, bisorbicillinoids, trisorbicillinoids and hybrid sorbicillinoids (Meng
et al., 2016). Sorbicillinoids were originally isolated from
Penicillium notatum in 1948, but found later also in the
culture broths of marine and terrestrial ascomycetes
(Harned and Volp, 2011). In particular, *P. chrysogenum*
strain NRRL1951 has been reported to be a natural
source of more than 10 sorbicillinoids (Meng *et al.*,
2016). This fungus was the progenitor for the high-β-lac-
tam-yielding strains that are currently used in industry.
These strains were obtained by several decades of classical strain improvement, where an early goal was to
eliminate the production of yellow pigments as contaminants of β-lactams. This resulted in the loss of sorbicillinoid production through mutagenesis of a key polyketide
synthase gene (Salo *et al.*, 2015). Recently, the interest
in sorbicillinoids was revived because of the wide bioac-
tivity spectrum associated with these molecules and their
potential pharmaceutical value. For instance, sor-
bicathecols A/B inhibits the cytopathic effect induced by
HIV-1 and influenza virus A (H1N1) in MDCK cells (Nicoletti and Trincone, 2016), whereas isobisvertinol inhibits
lipid droplet accumulation in macrophages, an event
associated with the initiation of atherosclerosis (Koyama
et al., 2007; Xu *et al.*, 2016). Moreover, the oxidized
form of bisvertinol, bisvertinolone, displays a potent

cytotoxic effect against HL-60 cells and is an antifungal via inhibition of β(1,6)-glucan biosynthesis (Nicolaou et al., 2000; Du et al., 2009). Other sorbicillinoids, such as oxosorbicillinol and dihydrosorbicillinol, were shown to exhibit antimicrobial activity against Staphylococcus aureus and Bacillus subtilis (Maskey et al., 2005).

Despite the wide spectrum of bioactive properties reported for sorbicillinoids, the biosynthetic pathway of these polyketides has not yet been elucidated. Isotope labelling studies suggested that the hexaketide structure of sorbicillinol is assembled by a Claisen-type reaction involved in carbon–carbon bond formation (Sugaya et al., 2008; Harned and Volp, 2011), whereas Diels–Alder- and Michael-type reactions have been proposed as the most probable mechanism for the formation of sorbicillinoid dimers (Maskey et al., 2005; Du et al., 2009). Recently, two polyketide synthases (PKS) have been implicated in the biosynthesis of sorbicillactone A/B in P. chrysogenum E01-10/3. The presumed PKS genes belong to a gene cluster that comprises five additional open reading frames (ORFs) (Avramović, 2011). Commonly, PKS enzymes form the scaffold structure of a molecule that is then further modified by tailoring enzymes, often encoded by genes localized in the vicinity of the key PKS genes (Lim et al., 2012). Indeed, a FAD-dependent monooxygenase has been identified as part of the putative sorbicillin cluster, and this enzyme was shown to convert (2′,3′-dihydro)sorbicillin into (2′,3′-dihydro)sorbicillinol (Fahad et al., 2014).

The putative sorbicillinoid gene cluster of industrial P. chrysogenum strains includes a highly reducing PKS (sorA, Pc21 g05080) and a non-reducing PKS (sorB, Pc21 g05070) (Salo et al., 2016). The sorA gene was shown to be essential for sorbicillinoid biosynthesis, as its deletion abolishes the production of all related compounds (Salo et al., 2015, 2016). In addition, this cluster harbours five further genes, two genes encoding putative transcription factors (sorR1 and sorR2, Pc21 g05050 and Pc21 g05090, respectively), a transporter protein (sorT, Pc21 g05100), a monooxygenase (sorC, Pc21 g05060) and an oxidase (sorD, Pc21 g05110). A recent study in Trichoderma reesei indicates that homologous transcriptional factors are involved in the regulation of sorbicillinoid biosynthesis in this fungus (Derntl et al., 2016), but the exact mechanism of regulation remained obscure. Here, we have resolved the biosynthetic pathway of sorbicillinoid biosynthesis and its regulation by metabolic and expression profiling of individual gene knockout mutants. The data show that SorR1 is a transcriptional activator, whose expression is controlled by the second regulator SorR2. Furthermore, transcription is regulated through an autoinduction mechanism by sorbicillinoids, the products of the pathway.

Results

Metabolic profiling of strains with individual deletions of the sorbicillinoid biosynthesis genes

Penicillium chrysogenum strain DS68530Res13 produces high levels of sorbicillinoids causing yellow pigmentation of the culture broth. This strain is derived from strain DS68530 as described previously (Salo et al., 2016). The proposed sorbicillin biosynthetic gene cluster (Fig. 1A) includes the previously characterized polyketide synthase gene sorA (Pc21 g05080), a second polyketide synthase (Pc21 g05070, sorB), two transcriptional factors (Pc21 g05050, sorR1; Pc21 g05090, sorR2), a transporter protein (Pc21 g05100, sorT), a monooxygenase (Pc21 g05060, sorC) and an oxidoreductase (Pc21 g05110, sorD). These genes were individually deleted, and metabolic profiling was performed on the supernatant fractions of cultures of the respective strains. The metabolic profiling searched for the previously identified sorbicillinoid-related compounds as well as potential new molecules. Furthermore, the expression of the aforementioned genes was analysed by qPCR to exclude possible polar effects of the gene deletions and to assess the impact of the deletion of the two putative regulators on the expression of the sorbicillinoid gene cluster.

In the ΔsorA mutant which lacks the highly reducing polyketide synthase, no sorbicillinoids could be detected in the culture supernatant (Fig. 2B) confirming our earlier observations (Salo et al., 2016). Also in the ΔsorB mutant, which lacks the non-reducing polyketide synthase, sorbicillinoid production was completely abolished (Fig. 2B). These data are consistent with the notion that SorA and SorB are responsible for the formation of the core (dihydro-)sorbicillin structure. The unknown compound [13] was present at elevated levels in both the ΔsorA and ΔsorB mutants as compared to the parental strain and thus is most likely not related to sorbicillins. This compound has a retention time (RT) of 14.21 min, m/z [H]+ of 304.1652 and a calculated empirical formula $C_{16}H_{21}O_3N_3$ (Fig. 2B). All the other unknown compounds listed in Fig. 2 appear to be associated with the sorbicillinoid biosynthetic pathway. Importantly, in both the ΔsorA and in ΔsorB mutants, none of the cluster genes except the sorR1 gene were expressed (Fig. 1B).

Next, the role of the individual genes encoding the enzymes of the pathway was analysed. Deletion of the monooxygenase gene sorC resulted in a 1.3 times increase in dihydrosorbicillinol [4]. The ΔsorC mutant showed lower levels of sorbicillin and sorbicillinol, which is consistent with the proposed role of SorC protein in the oxidative dearomatization of sorbicillin [1] into

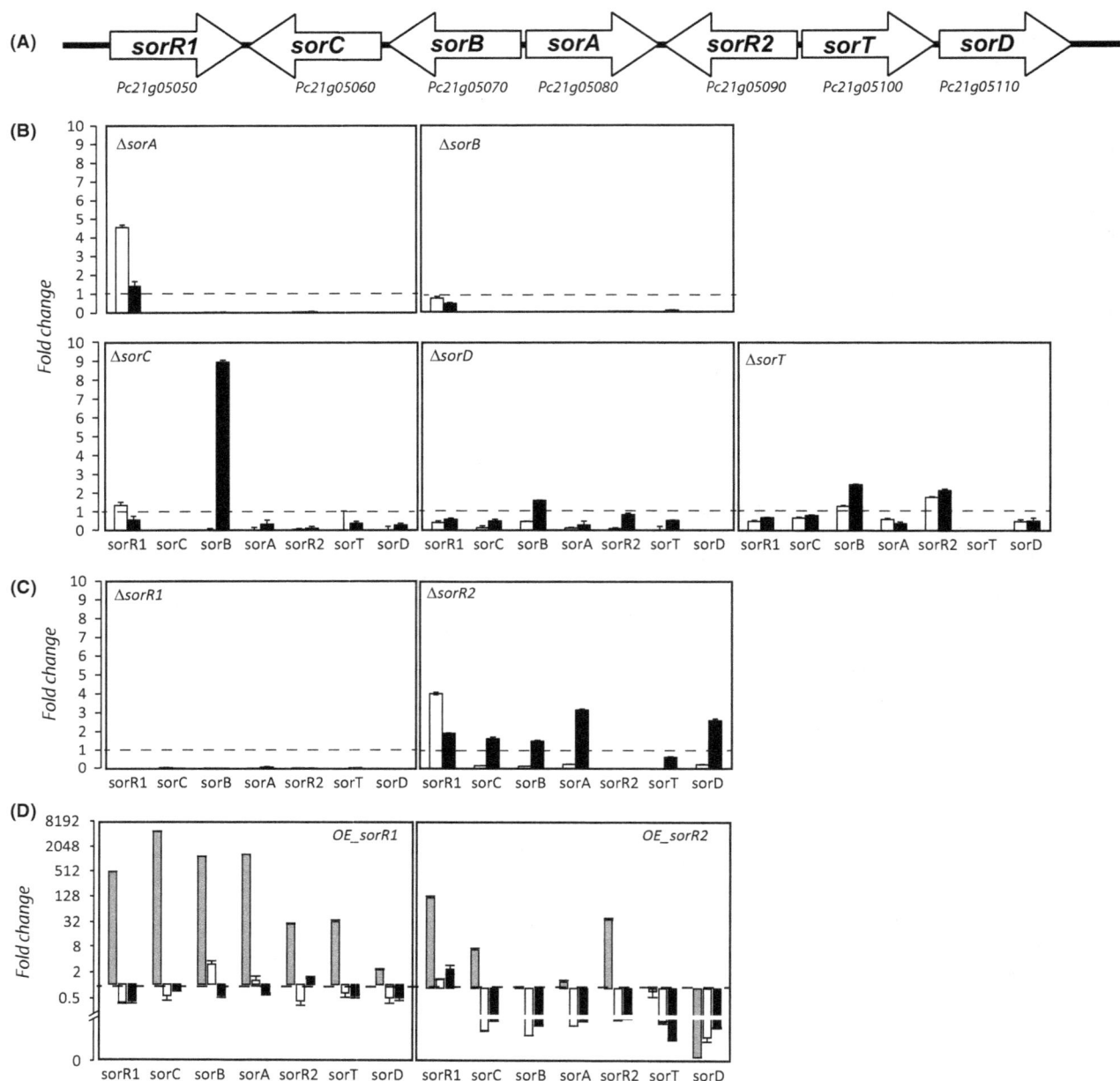

Fig. 1. Relative expression of the genes of the sorbicillinoid biosynthesis gene cluster.
A. Schematic representation of the gene cluster: Pc21 g05050 (sorR1; transcriptional factor), Pc21 g05060 (sorC; monooxygenase), Pc21 g05070 (sorB; non-reduced polyketide synthase), Pc21 g05080 (sorA; highly reduced polyketide synthase), Pc21 g05090 (sorR2; transcriptional factor), Pc21 g05100 (sorT; MFS transporter) and Pc21 g05110 (sorD; oxidase).
B. Quantitative real-time PCR analysis in sorbicillinoid gene cluster expression in strains with individual deleted sorA, sorB, sorC, sorD and sorT.
C. sorR1 and sorR2 genes.
D. qPCR analysis in strains and overexpressed sorR1 and sorR2. Samples were taken at day 1 (grey bars), 3 (white bars) and 5 (black bars). Data shown as fold change relative to P. chrysogenum DS68530Res13 strain.

sorbicillinol [3*] (Fahad et al., 2014). In this strain, we also noted an upregulation of the sorB gene and the partial downregulation of the rest of the cluster (Fig. 1B). The main compound produced after 3 days in the ΔsorD strain, which lacks the putative oxidoreductase, was

sorbicillinol [3;3*]. After 5 days, elevated levels of compound [15] with the m/z [H]+ of 293.1493 were also noted. Oxosorbicillinol [5*] with an empirical formula of $C_{14}H_{16}O_5$ and m/z [H]+ of 265.1069 was not detected in this strain, which suggests that SorD is involved in the

(A)

(B)

No	Compound Name	Formula	Acquired [M+H]+	RT (min)	DS68530Res13 3	5	ΔsorR1 3	5	ΔsorC 3	5	ΔsorB 3	5	ΔsorA 3	5	ΔsorR2 3	5	ΔsorT 3	5	ΔsorD 3	5	OEsorR1 3	5	OEsorR2 3	5
1	Sorbicillin	C14H16O3	233.1174	30.84	0.12	0.37	0.00	0.00	0.00	0.36	0.00	0.00	0.00	0.00	0.00	0.00	0.09	0.19	0.23	0.28	1.03	0.95	0.00	0.00
2	Dihydrosorbicillin	C14H18O3	235.1329	32.01	0.00	0.00	0.00	0.00	0.00	0.14	0.00	0.00	0.00	0.00	0.00	0.00	0.00	0.00	0.00	0.00	0.00	0.04	0.00	0.00
3	Sorbicillinol	C14H16O4	249.1119	21.01	1.82	11.13	0.00	0.00	0.12	0.14	0.00	0.00	0.00	0.00	0.00	0.00	0.86	0.85	4.42	8.91	23.83	14.12	0.00	0.00
4	Dihydrosorbicillinol	C14H18O4	251.1274	22.66	0.04	8.95	0.00	0.00	0.04	11.87	0.00	0.00	0.00	0.00	0.01	0.25	2.00	13.65	0.05	14.98	10.80	38.43	0.00	0.00
5	Oxosorbicillinol	C14H16O5	265.1069	19.34	0.02	0.29	0.00	0.00	0.00	0.00	0.00	0.00	0.00	0.00	0.00	0.00	0.00	0.35	0.00	0.05	0.91	0.38	0.00	0.00
2*	Dihydrosorbicillin*	C14H18O3	235.1329	47.08	0.00	0.00	0.00	0.00	0.00	1.82	0.00	0.00	0.00	0.00	0.00	0.00	0.00	0.00	0.00	0.00	0.00	0.00	0.00	0.00
3*	Sorbicillinol*	C14H16O4	249.1119	22.33	14.47	19.51	0.00	0.00	0.00	0.00	0.00	0.00	0.00	0.00	0.05	0.00	26.72	26.77	83.66	27.72	0.00	0.00	0.00	0.00
4*	Dihydrosorbicillinol*	C14H18O4	251.1274	24.95	0.15	9.88	0.00	0.00	0.00	0.00	0.00	0.00	0.00	0.00	0.00	0.04	7.12	7.81	0.32	13.32	23.29	23.48	0.00	0.00
5*	Oxosorbicillinol*	C14H16O5	265.1069	27.43	0.08	2.45	0.00	0.00	0.00	0.11	0.00	0.00	0.00	0.00	0.00	0.00	0.00	0.00	0.00	0.30	1.84	1.40	0.00	0.00
6	Bisorbicillinol	C28H32O8	497.2164	29.08	0.00	0.17	0.00	0.00	0.00	0.00	0.00	0.00	0.00	0.00	0.00	0.00	0.00	0.00	0.00	0.30	1.53	0.81	0.00	0.00
7	Bisvertinolon	C28H32O9	513.2110	32.97	0.02	2.12	0.00	0.00	0.00	0.00	0.00	0.00	0.00	0.00	0.00	0.00	0.00	0.00	0.00	0.00	35.37	15.12	0.00	0.00
8	Dihydrobisvertinolone	C28H34O9	515.2269	33.50	0.00	1.84	0.00	0.00	0.00	0.00	0.00	0.00	0.00	0.00	0.00	0.00	0.00	0.03	0.00	0.00	3.78	9.98	0.00	0.00
9	Tetrahydrobisvertinolone	C28H36O9	517.2424	33.72	0.00	0.24	0.00	0.00	0.00	0.00	0.00	0.00	0.00	0.00	0.00	0.00	0.19	17.34	0.00	0.00	1.24	11.86	0.00	0.00
10	Unknown	C13H14O3	207.1015	23.98	0.87	0.19	0.00	0.00	0.03	0.48	0.00	0.00	0.00	0.00	0.00	0.00	0.18	0.00	0.22	0.02	3.39	2.27	0.00	0.00
11	Unknown	C11H12O3	193.0860	21.30	7.67	1.53	0.00	0.00	0.33	2.86	0.00	0.00	0.00	0.00	0.00	0.02	8.16	0.00	14.46	0.06	12.03	7.57	0.00	0.00
12	Unknown	C12H17ON	192.1382	14.04	0.22	0.63	0.00	0.00	0.08	0.06	0.00	0.00	0.00	0.00	0.00	0.00	1.16	2.58	0.81	1.41	3.64	3.02	0.00	0.00
13	Unknown	C16H21O3N3	304.1652	14.21	0.00	0.86	0.00	4.21	0.00	5.58	0.29	4.27	0.71	1.12	0.10	0.38	0.01	0.00	0.03	0.42	0.00	1.28	0.12	1.13
14	Unknown	C18H20O3	309.1441	15.86	0.07	3.84	0.00	0.00	0.00	0.05	0.00	0.00	0.00	0.00	0.00	0.00	0.00	0.18	2.86	7.83	12.55	0.00	0.00	
15	Unknown	C15H20O3N2	293.1493	17.55	0.01	12.01	0.00	0.00	0.01	1.25	0.00	0.00	0.00	0.00	0.00	0.38	1.86	26.00	0.04	33.65	20.32	36.83	0.00	0.00
16	Unknown	C32H40O13N6	657.2678	35.89	0.30	7.49	0.00	0.00	0.00	0.00	0.00	0.00	0.00	0.00	0.00	0.00	0.00	0.00	0.57	4.87	32.66	25.34	0.00	0.00
17	Unknown	C11H10O3	223.0601	15.44	0.00	0.28	0.00	0.00	0.00	0.22	0.00	0.00	0.00	0.00	0.00	0.02	0.01	0.88	0.02	5.00	2.34	8.25	0.00	0.00
18	Unknown	C14H18O5	267.1241	17.06	0.03	0.87	0.00	0.00	0.00	0.00	0.00	0.00	0.00	0.00	0.00	0.00	0.00	0.03	0.17	1.09	2.90	4.07	0.00	0.00
19	Unknown	C23H32O13N6	529.2060	28.21	0.00	9.22	0.00	0.00	0.00	0.00	0.00	0.00	0.00	0.00	0.00	0.00	0.00	0.00	0.00	0.00	17.43	29.03	0.00	0.00

Response ratio — Max / Min

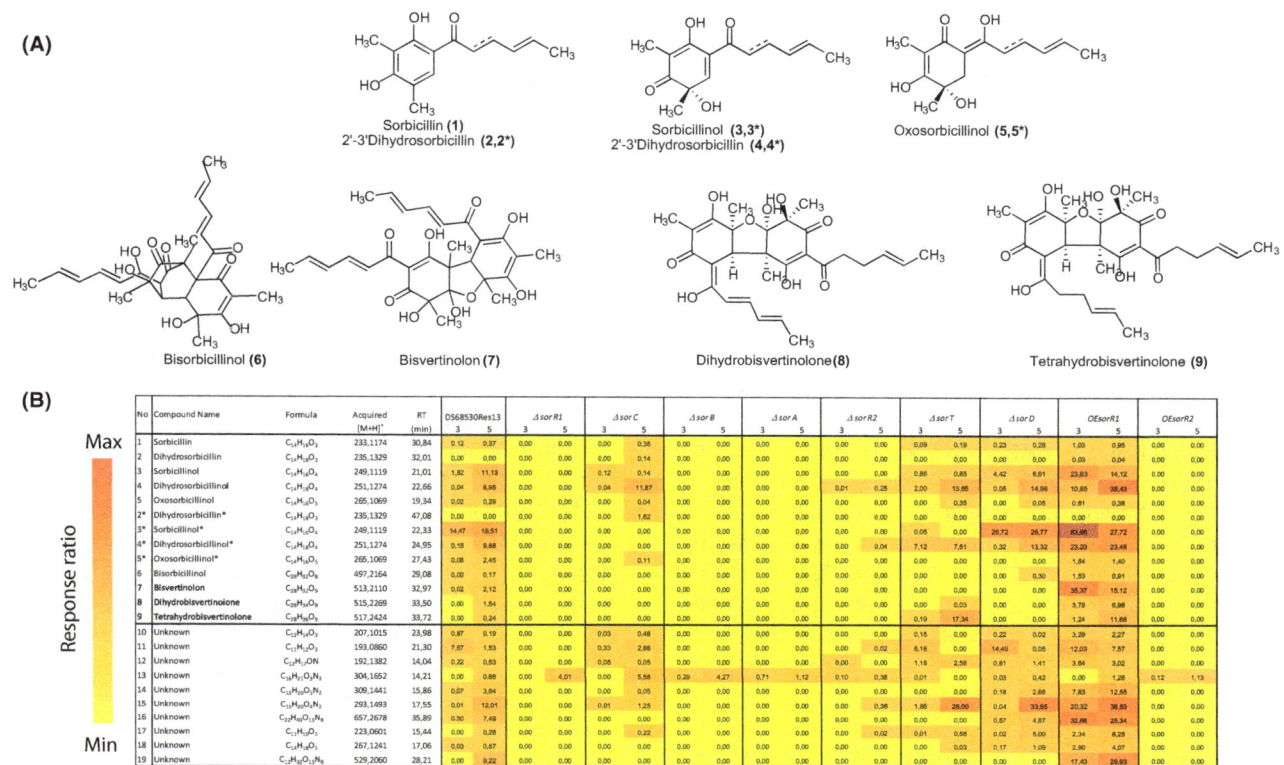

Fig. 2. A. Sorbicillinoid-related compounds detected in this study.
B. Response ratio of the sorbicillinoid concentrations in the culture broth of indicated sorbicillin-producing *P. chrysogenum* strains. Reserpine was used as internal standard for normalization. Compounds were detected after 3 and 5 days of growth. The mass-to-charge ratio (*m/z*) of the protonated metabolites, their empirical formulas and retention time (RT) are indicated. Structures of sorbicillin-related compounds detected in this study. (*) Indicates an isomer of the known sorbicillinoids.

conversion of sorbicillinol into oxosorbicillinol. The *ΔsorD* strain showed a slight overexpression (0.6 times higher) of the *sorB* gene while the other genes of the pathway were about twofold downregulated (Fig. 1B). The *ΔsorT* mutant which lacks the putative transporter showed a similar gene expression as the parental strain with only minor changes in *sorB* and *sorR2* expression. In this strain, the production of sorbicillinoids shifted mostly towards tetrahydrobisvertinolone [9] and the compound with an empirical formula $C_{15}H_{20}O_4N_2$ [15].

Summarizing, our data suggest that the monooxygenase SorC is involved in the oxidative dearomatization of sorbicillin [1] into sorbicillinol [3*] and that the oxidoreductase SorD converts sorbicillinol [3;3*] into oxosorbicillinol [5*]. No clear role can be attributed to the transporter SorT. Furthermore, the individual deletion of the biosynthesis genes also impacts the regulation of the pathway.

Deletion and overexpression of the regulatory genes sorR1 and sorR2

The sorbicillinoid biosynthetic gene cluster contains two genes encoding putative regulators, i.e. *sorR1* and *sorR2*. The deletion of *sorR1* abolished the expression of the entire sorbicillin biosynthesis gene cluster, and consequently, all sorbicillinoid-related compounds were absent in this strain (Figs 1C and 2B). The deletion of *sorR2* impacted the expression after 3 days (Fig. 1C), while after 5 days, the expression profiles were equal or even higher than in the parental strain (Fig. 1C). Intriguingly, despite the biosynthetic genes being expressed, hardly any sorbicillinoids were present in the culture broth of the *ΔsorR2* strain, except for very low levels of dihydrosorbicillinol [4] (Fig. 2B). These data suggested that SorR1 is essential for the regulation of sorbicillinoid biosynthesis, whereas the absence of SorR2 results in a delayed expression of the pathway genes.

Overexpression of *sorR1* (*OEsorR1*) or *sorR2* (*OEsorR2*) resulted in elevated levels of these regulators in the early stages of the cultivation (Fig. 1D). In the *OEsorR1* strain, this also substantially elevated the expression of the other pathway genes which suggests that SorR1 acts as a transcriptional activator. Concomitantly, the overexpression of *sorR1* massively increased the sorbicillinoid production (Fig. 2B). In contrast, in the *OEsorR2* mutant, the expression of nearly all the genes of the sorbicillinoid cluster was strongly reduced, except for *sorR1* expression which was increased. Consequently, production of all sorbicillinoid-related compounds was reduced. These data suggest that SorR2 is involved in a complex mechanism of regulation and likely acts in concert with SorR1 which is a transcriptional activator of the pathway.

Sorbicillinoids activate gene expression

The observation that deletion of the PKS enzymes SorA and SorB, and consequently a loss in sorbicillinoid production, causes a marked reduction in the expression levels of the biosynthesis genes suggests that sorbicillinoids influence the expression of the pathway through an autoinduction regulatory process. To test this hypothesis, a culture of strain DS68530, which itself does not produce sorbicillinoids because of the mutation in SorA, was fed with a sorbicillinoid containing spent medium derived from the DS68530Res13 strain. This resulted in highly increased expression of all sorbicillinoid biosynthetic genes (Fig. 3A), except for the two regulatory genes, the expression of which remained unchanged. As a control, the cells were fed with supernatant derived from the non-sorbicillin-producing strain DS68530, and this had no impact on the expression of the sorbicillinoid gene cluster. These data suggest that the sorbicillinoid biosynthetic gene cluster is regulated through an autoinduction mechanism by which the products of the pathway, the sorbicillinoids, stimulate the expression of the pathway genes.

To examine the autoinduction mechanism in greater detail, the effect of sorbicillinoid addition was also tested for the *ΔsorR1* and *ΔsorR2* strains, and strains overproducing *sorR1* (*OEsorR1*) or *sorR2* (*OEsorR2*) in the genetic background of the non-sorbicillin-producing strain DS68530. Expression of the cluster genes remained unaffected in the *ΔsorR1* and *ΔsorR2* strains when cells were grown in the presence of sorbicillinoids (data not shown, Fig. 3B). Overproduction of *sorR1* resulted in the elevated expression of the pathway genes which was further stimulated by the presence of sorbicillinoids in the culture medium (Fig. 3C). A similar result was obtained with the overexpression of *sorR2*, albeit the effect of sorbicillinoids was at least two orders of magnitude lower (Fig. 3D). It should be noted that *sorT* in the *OEsorR2* strain was highly overexpressed. This gene lies downstream of *sorR2,* and due to strain construction, it is no longer expressed from its endogenous promoter but controlled by the strong *gndA* promoter (Polli *et al.*, 2016). Taken together, these data suggest that sorbicillinoids autoinduce the sorbicillinoid biosynthetic pathway in a process that requires the combined activity of the transcriptional regulators SorR1 and SorR2.

Discussion

Penicillium chrysogenum produces large amounts of sorbicillinoids. In a previous study, we have identified one of the polyketide synthases (SorA) involved in this process (Salo *et al.*, 2016). To resolve the biosynthetic mechanism of sorbicillinoid production, each of the genes of the putative cluster was individually deleted and analysed by metabolic profiling. Our data indicate that the two polyketide synthase genes, *sorA* and *sorB,* are both required for sorbicillinoid production. Our metabolic profile analysis did not reveal possible intermediate products of the polyketide synthases. However, it has previously been suggested that these two proteins are responsible for the synthesis of the basic hexaketide scaffold (Fig. 4A) (Harned and Volp, 2011; Fahad, 2014). Biosynthesis of sorbicillin or dihydrosorbicillin depends on the functionality of the enoylreductase (ER) domain of SorA, while the methylation of the hexaketide derived from SorA, prior to the cyclization, is catalysed by SorB. Our deletion analysis further suggests that SorC, a monooxygenase, is needed for the conversion of dihydrosorbicillin [2*] and sorbicillin [1] into dihydrosorbicillinol [4*] and sorbicillinol [3*], respectively, which confirms previous observations on the biochemical characterization of this enzyme (Fahad *et al.*, 2014). Nevertheless, SorC is apparently not the only enzyme or mechanism involved in this conversion step as low amounts of likely tautomer forms of (dihydro)sorbicillinol [4;3] (Harned and Volp, 2011) were still detected in the supernatant of the *ΔsorC* mutant. In the *ΔsorD* strain, sorbicillinol [3;3*] accumulated while oxosorbicillinol [5*] could not be detected. This suggested that SorD is an oxidase that converts sorbicillinol into oxosorbicillinol [5*], which is also a stable compound. Although low amounts of the potential tautomer [5] have been previously detected (Maskey *et al.*, 2005), this molecule might be the result of the spontaneous oxidation of sorbicillin (Bringmann *et al.*, 2005). Furthermore, we could not detect dihydrobisvertinolone [8] and tetrahydrobisvertinolone [9] in the supernatant, which is in line with the proposed function of SorD as the product oxosorbicillinol is a precursor for bisvertinolone synthesis (Abe *et al.*, 2002). Deletion of the gene specifying the transporter SorT only marginally affected sorbicillinoid production, and thus, no clear transport function could be assigned to this protein. Summarizing, the functional assignment of the various gene products resulted in a biosynthetic scheme depicted in Fig. 4. A similar pathway was recently constructed for *Trichoderma reesei* (Astrid R. Mach-Aigner, pers. comm.).

To understand the mechanism of regulation of the pathway, we analysed the effect of the deletion and overexpression of the two putative transcriptional

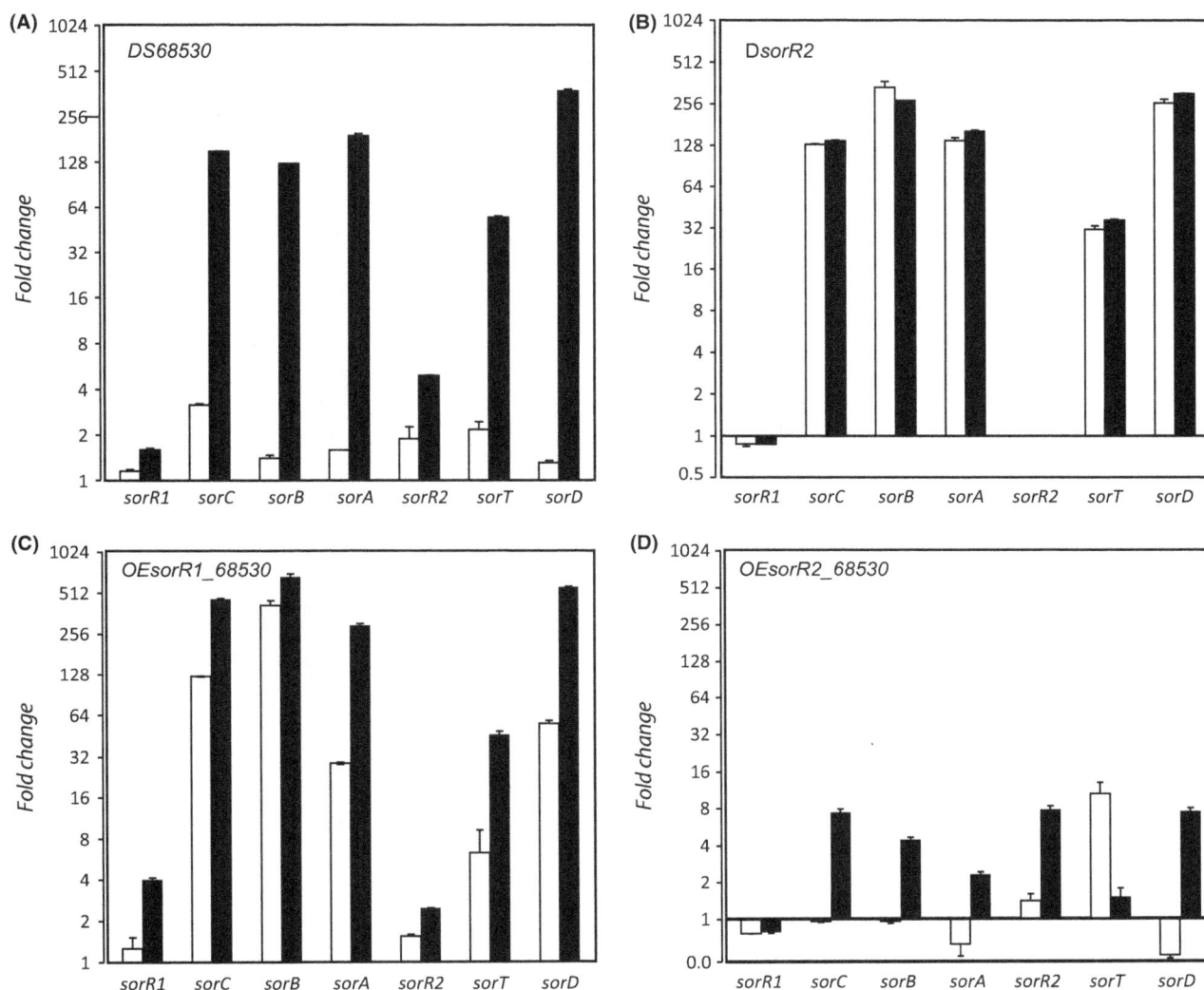

Fig. 3. Relative expression of the sorbicillinoid cluster genes in the presence (black bars) and absence (white bars) of sorbicillinoids in the growth medium. Strains: A. DS68530, B. *ΔsorR2*, C. *OEsorR1*_68530 and D. *OEsorR2*_68530. Samples were taken after 3 days of growth. Data shown as fold change relative to *P. chrysogenum* DS68530 strain.

regulators, *sorR1* and *sorR2,* that are part of the gene cluster. The data indicate that SorR1 is needed for the transcriptional activation of sorbicillinoid gene cluster. In the *sorR1* deletion strain, the expression of all biosynthetic genes was completely abolished and consequently, sorbicillinoid production was eliminated. In the overexpression strain, cluster genes were upregulated causing an earlier onset of sorbicillinoid biosynthesis. SorR2 appears to fulfil a more complex role. Deletion of *sorR2* caused a later onset of the expression of the sorbicillinoid genes, which explains the low amounts of sorbicillinoids that are still detected in that strain. In the *sorR2* overexpression strain, the transcriptional levels of *sorR1* were strongly enhanced, while the expression of the pathway genes was strongly reduced at later stages of growth. Moreover, *sorD* was not expressed and

overall sorbicillinoid production was abolished. This observation suggests a complex mechanism of regulation in which SorR1 and SorR2 cooperate at the protein level. While the data are consistent with a model in which SorR1 acts as a transcriptional activator of the pathway, SorR2 appears to act as an inhibitor of the SorR1 activity. This would also explain why there are still low levels of sorbicillinoids detected in the *sorR2* deletion strain while these are completely absent in the *sorR2* overexpression strain in which the pathway is suppressed. This phenotype resembles that of the *aflJ* deletion strain of *Aspergillus*. AflJ and AflR are transcriptional factors that regulate the aflatoxin and sterigmatocystin cluster in *Aspergillus parasiticus* (Chang, 2003; Yu and Keller, 2005). Deletion of the individual genes abolished the production of these compounds. AflR is a

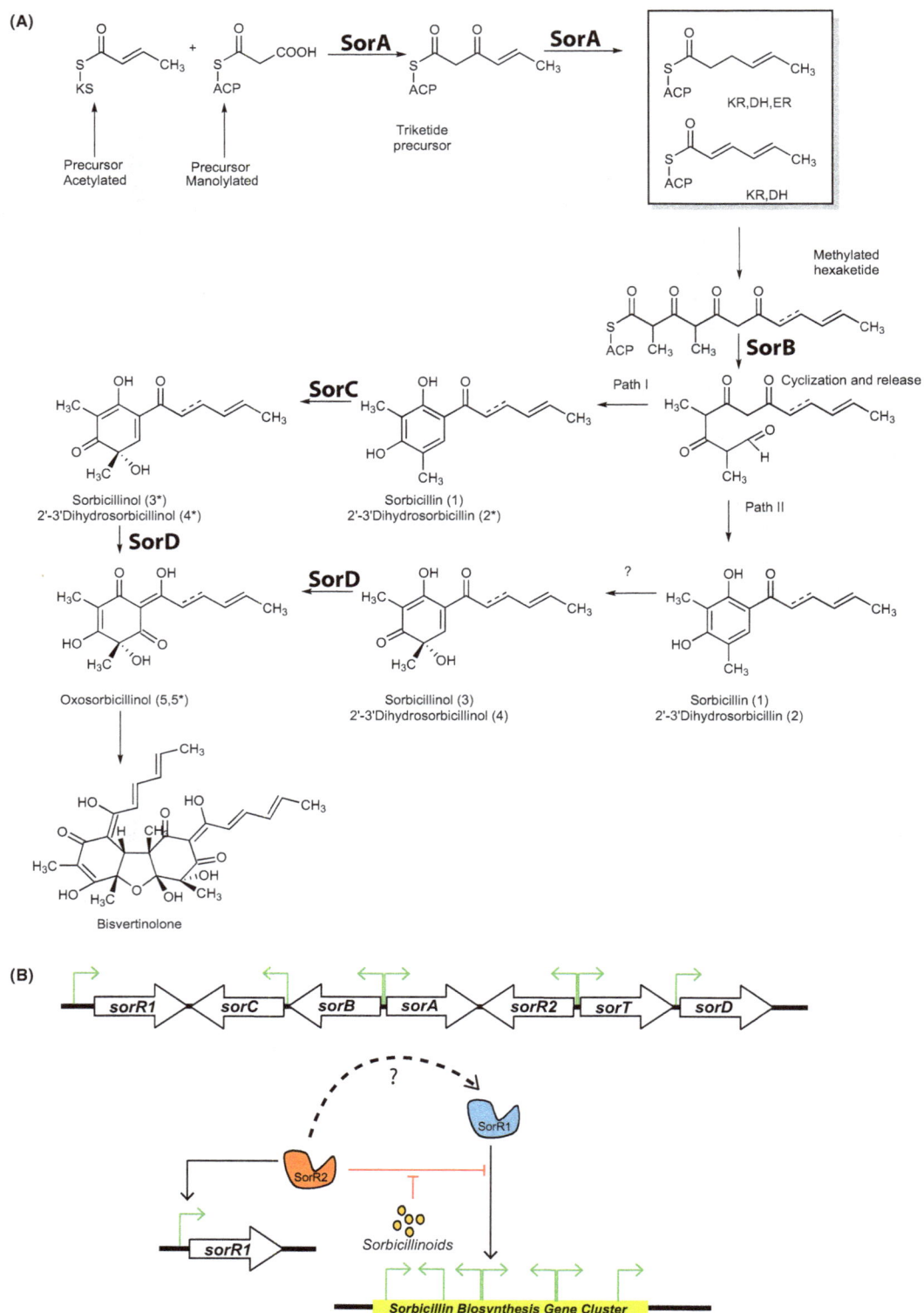

Fig. 4. Proposed model of the sorbicillin biosynthetic pathway and its regulation.
A. PKS domains in SorA and SorB are abbreviated as KS (ketosynthase), ACP (acyl carrier protein), KR (ketoreductase), DH (dehydratase), MT (methyltransferase), ER (enoylreductase). Adapted from Ref. (Avramović, 2011; Fahad *et al.*, 2014; Derntl *et al.*, 2016; Salo *et al.*, 2016).
B. The autoinduction mechanism of pathway gene expression by sorbicillinoids which involves the two transcriptional factors SorR1 and SorR2. On top, a schematic representation of the gene cluster. Black solid arrows indicate a positive stimulation and the red arrows show a negative effect. Green arrows represent the promoters of the indicated genes.

transcriptional activator. Like the ∆sorR2 strain in our study, the aflatoxin structural genes in the aflJ deletion strain also remain expressed at low levels, and it has been suggested that AlfJ forms an active complex with AlfR, the main regulator of the pathway (Georgianna and Payne, 2009). Summarizing our results suggests that both transcriptional factors SorR1 and SorR2 orchestrate the biosynthesis of sorbicillinoids, with SorR1 as main transcriptional activator and SorR2 as a repressor of this biosynthetic pathway (Fig. 4B). A similar regulatory mechanism involving two transcriptional factors has been reported for the homologous cluster in T. reesei (Derntl et al., 2016).

A further observation is that mutational loss of sorbicillinoid production is accompanied by a dramatic reduction in the expression of the pathway genes. A possible explanation of this observation is that sorbicillinoids function as autoinducers. Indeed, when the strain deficient in sorbicillinoid production was fed with filtered medium of a sorbicillinoid-producing strain, a major upregulation of the sorbicillin gene cluster was noted (Fig. 3). Neither the ∆sorR1 nor the ∆sorR2 mutants did show this sorbicillinoid-dependent transcriptional response. Interestingly, the transcriptional response of the core sorbicillin genes (sorA, sorB, sorC) in the sorR2 deletion strain (Fig. 3B) is similar to what is observed when strain DS68530 is fed with sorbicillinoids (Fig. 3A). A possible scenario that explains these observations is that sorbicillinoids act on SorR2, thereby relieving the inhibitory action of SorR2 on the transcriptional activator SorR1. In this respect, the expression of the sorbicillin cluster expression was partially rescued when the OEsorR2 strain was fed with sorbicillinoids (Fig. 3D). Also, overexpression of SorR1 partially restored transcription, and according to our model, the higher levels of SorR1 overcome the inhibitory effect of SorR2 on transcription. We propose that SorR2 interacts with SorR1 to reduce its transcriptional activating activity, while sorbicillinoids relieve this inhibition by acting on SorR1 (Fig. 4B). This is one of the rare reported examples wherein the product of the synthesis pathway acts as an autoinducer of the expression of the pathway genes. The zearalenone (ZEA) biosynthetic cluster gene of Fusarium graminearum, whose regulator isoforms (ZEB2S and ZEB2L) are induced by its own toxin, is another example of this phenomenon (Park et al., 2015).

In silico analysis indicates that the intergenic DNA region between sorB and sorA comprises three nucleotide-binding motifs $5'CGGN_{(9)}CGG$, which may act as binding sites for SorR1 to regulate the cluster. SorR1 belongs to the family of sequence-specific DNA-binding Zn_2-Cys_6 proteins. Such regulators appear to be present in approximately 90% of the PKS-encoding gene cluster in fungi (Brakhage, 2012). In Aspergillus flavus, the deletion of the aflR gene that encodes for a Zn_2-Cys_6-type protein abolished the expression of the aflatoxin and sterigmatocystin cluster, while the overexpression of the same gene increased the expression and production of these secondary metabolites (Yin and Keller, 2011; Brakhage, 2012). SorR1 appears to function in a similar manner. Additionally, our results suggest that there is possible crosstalk between the sorbicillinoid gene cluster regulators (SorR1 and SorR2) and other biosynthetic pathways. When the sorR1 gene was deleted, enhanced production of compound [13] was observed, whereas in the SorR1 overexpression strain, production of this compound was reduced. Remarkably, this secondary metabolite is not related to sorbicillinoids and was also detected in the ∆sorA and ∆sorB strains in which sorbicillinoid biosynthesis is eliminated. Possible crosstalk has been reported before in Aspergillus nidulans where the induction of the silent asperfuranone gene cluster was achieved through expression of the scpR gene that encodes a transcriptional regulator of the inp gene cluster (Bergmann et al., 2010; Fischer et al., 2016). However, we cannot exclude the possibility that the elevated levels of compound [13] are due to a greater availability of precursor molecules not used for sorbicillinoid biosynthesis.

In conclusion, our results support a model for sorbicillinoid production that includes the functions of the various gene products that are part of the sorbicillinoid gene cluster and an alternative branched path independently of SorC. Additionally, it was demonstrated that the regulation of this pathway involves two transcriptional regulators while sorbicillinoids act as autoinducers of this pathway. This work opens possibilities to engineer the sorbicillinoid pathway for the efficient production of novel derivatives of pharmaceutical value.

Experimental procedures

Strains, media and growth conditions

Penicillium chrysogenum DS68530 was kindly provided by DSM Sinochem Pharmaceuticals (Delft, the Netherlands). All gene deletion and overexpression strains were derived from DS68530Res13 (Sorb407) described by Salo et al. (2016); which is a derivative of DS68530. The overexpression strains used in the feed experiments were derived of DS68530 (Table S1). Conidiospores immobilized on rice were inoculated in YGG medium for 48 h to produce fungal protoplasts or for gDNA extraction and for 24 h to produce young mycelium used as pre-culture inoculum for producing the fermentations. After pre-culture, the inoculum was diluted seven times in SMP medium (secondary metabolite production medium (Ali et al., 2013)) and cells were grown for up to 5 days in shaken flasks at 25 °C and 200 rpm. After 3 and 5 days, samples of the culture medium were collected for RNA extraction and metabolite profile analysis. When indicated, phleomycin agar medium (Snoek et al., 2009) supplemented with 60 µg ml^{-1} phleomycin was used for selection and strain purification. Selected transformants were placed on R-agar for sporulation during 5 days, whereupon the conidiospores were used to prepare rice batches for long-term storage (Kovalchuk et al., 2012).

Construction of gene deletion and overexpression strains

Gene deletion and overexpression mutants were built using the Gateway Technology (Invitrogen, USA). For creation of deletion strains, 5′ and 3′ regions of each target genes (*Pc21 g05050* (*sorR1*), *Pc21 g05060* (*sorC*), *Pc21 g05070* (*sorB*), *Pc21 g05080* (*sorA*), *Pc21 g05090* (*sorR2*), *Pc21 g05100* (*sorT*) and *Pc21 g05110* (*sorD*)) were amplified from gDNA of strain DS68530. All primers used in this study are listed in Table S2. The resistance marker gene (*ble*) for phleomycin was amplified from pJAK-109 (Pohl et al., 2016). Phusion HF polymerase (Thermo Fisher Scientific, USA) was used to amplify all the DNA parts used. The *ble* gene was placed under control of pcbC promoter of *P. chrysogenum*. All the fragments generated were cloned in the respective donor vectors PDONR P4-P1R, pDONR2R-P3 and pDONR 221 using BP Clonase II enzyme mix (Invitrogen) and used to transform *E. coli* DH5α, where plasmids were selected for with kanamycin. Next, the constructs were used in an *in vitro* recombination reaction with the pDEST R4-R3 vector employing LR Clonase II Plus enzyme mix (Invitrogen). Following transformation to *E. coli* DH5α, correct constructs were selected for with ampicillin.

The donor vectors containing the 5′ flank were used to generate the 5′ flanks in the overexpression cassettes. The 3′ flank was generated from amplified homologous regions that were located before and after the start codon of each gene (*sorR1; sorR2*). The *pcbC* (isopenicillin N synthase) gene promoter of *P. chrysogenum* was used to induce expression and was inserted between the two flanks selected. To build the DNA fragment that contains the phleomycin-resistant cassette, the *ble* gene was amplified from plasmid pFP-phleo-122 (Polli et al., unpublished) F. Polli, R.A.L. Bovenberg, and A.J.M. Driessen, unpublished data and the *pcbC* promoter which was ordered as a synthetic gene (gBlock) (IDT, USA). The *ble* gene in the phleomycin-resistant cassette (promoter, gene and terminator) is under control of the *gndA* (6-phospho-gluconate dehydrogenase) promoter of *Aspergillus nidulans* (Polli et al., 2016). Next, the two fragments were fused by overlap PCR, as described by Nelson and Fitch (2011).

Fungal transformation

Protoplasts were isolated from *P. chrysogenum* as described previously (Kovalchuk et al., 2012). For all transformations, 5 µg of plasmid DNA was linearized with a suitable restriction enzyme, whereupon transformation was performed as described by Weber et al. (Weber et al., 2012). Screening of transformants was performed by colony PCR using the Phire Plant Direct PCR Kit (Life Technologies, USA). Selected transformants were purified through three rounds of sporulation on R-agar medium. Transformants were further validated by sequencing integration regions amplified from gDNA.

Southern blot analysis

A DNA fragment between 0.7 and 0.1 kb from the upstream or downstream region of every gene was amplified and used as a probe for Southern blot analysis (Fig. S1). The probes were labelled with the HighPrime Kit (Roche Applied Sciences, Almere, the Netherlands). About 15 µg of gDNA, previously digested with suitable restriction enzymes, was separated by electrophoresis on an 0.8% agarose gel. The gel was equilibrated in 20× saline–sodium citrate (SSC) buffer (3 M NaCl; 0.3 M $C_6H_5Na_3O_7$; pH 7) and the DNA was transferred overnight to a positively charged nylon membrane (Zeta-Probe; Bio-Rad, Munchen, Germany). Subsequently, the membrane was incubated overnight with the labelled probe(s). For detection, the membrane was treated with anti-DIG Fab fragment alkaline phosphatase and the CDP-Star chemiluminescent substrate (Roche Applied Sciences). The signal was measured using a Lumi-Imager (Fujifilm LAS-4000, Fujifilm Co. Ltd, Tokio, Japan).

qPCR analysis

Mycelium of strains grown for 3 and 5 days in SMP medium was harvested and disrupted in a FastPrep FP120 system (Qbiogene, Cedex, France) to isolate total RNA. The extraction was performed with the TRIzol (Invitrogen) method, and the total RNA obtained was purified using the Turbo DNA-free kit (Ambion, Carlsbad, CA, USA). RNA integrity was checked on a 2% agarose gel, and the RNA concentration was measured using a NanoDrop ND-1000 device (ISOGEN, Utrecht, the Netherlands). To synthesize cDNA, 500 ng of RNA was used per reaction using iScript cDNA synthesis kit (Bio-Rad). The primers used were described previously (Salo et al., 2016). The γ-actin gene (*Pc20 g11630*) was used as a control for normalization (Nijland et al., 2010). The SensiMIx SYBR Hi-ROX (Bioline, Australia) was used as master mix for the qPCR in a MiniOpticon system (Bio-Rad). The following thermocycler conditions were employed: 95 °C for 10 min, followed by 40 cycles of 95 °C for 15 s, 60 °C for 30 s and 72 °C for 30 s. Measurements were analysed using the Bio-Rad CFX manager program in which the Ct (threshold cycles) values were determined by regression. To determine the specificity of the qPCRs, melting curves were generated. The analysis of the relative gene expression was performed with $2^{-\Delta\Delta C_T}$ method (Livak and Schmittgen, 2001). The expression analysis was performed for two biological samples with at least two technical replicates.

Metabolite analysis

Strains were grown in SMP medium, and supernatant was collected after 3 and 5 days. Samples were centrifuged for 5 min at 14 000 rpm to remove the mycelium, whereupon 1 mL of the supernatant fraction was filtered with a 2 μm pore polytetrafluoroethylene (PTFE) syringe filter and stored at −80 °C. LC-MS analysis was performed as described previously (Salo *et al.*, 2016). Metabolite analysis was performed with two biological samples with two technical duplicates.

Other methods

For the feeding experiments with sorbicillinoids, the parental strain DS68530 and its derivatives *ΔsorR1*, *ΔsorR2*, OEsorR1_68530 and OEsorR2_68530 were grown in YGG medium. After 24 h, the inoculum of 3 mL was transferred into a 100 mL shake flask, supplemented with 20 mL of fresh SMP and 2 mL of filtered supernatant that was obtained from a 3-day-old culture of the sorbicillinoid-producing strain DS68530Res13, also grown in SMP. Controls received supernatant of the DS68530 strain, a non-sorbicillinoid producer. Samples for expression and metabolite analysis were taken at days 1, 3 and 5 of growth.

Acknowledgements

The authors acknowledge DSM Sinochem Pharmaceuticals (Delft, the Netherlands) for kindly providing the DS68530 strain.

Conflict of interest

None declared.

Author contributions

FGC designed the study, performed the experiments, wrote the manuscript and carried out the data analysis. AJMD conceived the study, supervised and coordinated the design, interpreted the data and corrected the manuscript. YN participated in data analysis and helped to draft the manuscript. PPL and OS supported the mass spectrometry and structural analysis and helped to draft the manuscript. RALB contributed to the coordination of the project and the revision of the manuscript.

References

Abe, N., Arakawa, T., and Hirota, A. (2002) The biosynthesis of bisvertinolone: evidence for oxosorbicillinol as a direct precursor. *Chem Commun* **7**: 204–205.

Ali, H., Ries, M.I., Nijland, J.G., Lankhorst, P.P., Hankemeier, T., Bovenberg, R.A.L., *et al.* (2013) A branched biosynthetic pathway is involved in production of roquefortine and related compounds in *Penicillium chrysogenum*. *PLoS ONE* **8**: 1–12.

Avramović, M. (2011) Analysis of the genetic potential of the sponge- derived fungus Penicillium chrysogenum E01- 10/3 for polyketide production. Thesis (Ph.D.) University of Bonn.

Bergmann, S., Funk, A.N., Scherlach, K., Schroeckh, V., Shelest, E., Horn, U., *et al.* (2010) Activation of a silent fungal polyketide biosynthesis pathway through regulatory cross talk with a cryptic nonribosomal peptide. *Appl Environ Microbiol* **76**: 8143–8149.

Brakhage, A.A. (2012) Regulation of fungal secondary metabolism. *Nat Rev Microbiol* **11**: 21–32.

Bringmann, G., Lang, G., Gulder, T.A.M., Tsuruta, H., Wiese, J., Imhoff, J.F., *et al.* (2005) The first sorbicillinoid alkaloids, the antileukemic sorbicillactones A and B, from a sponge-derived *Penicillium chrysogenum* strain. *Tetrahedron* **61**: 7252–7265.

Chang, P. (2003) The *Aspergillus parasiticus* protein AFLJ interacts with the aflatoxin pathway-specific regulator AFLR. *Mol Genet Genomics* **268**: 711–719.

Derntl, C., Rassinger, A., Srebotnik, E., Mach, R.L., and Mach-Aigner, A.R. (2016) Identification of the main regulator responsible for the synthesis of the typical yellow pigment by *Trichoderma reesei*. *Appl Environ Microbiol* **20**: 6247–6257.

Du, L., Zhu, T., Li, L., Cai, S., Zhao, B., and Gu, Q. (2009) Cytotoxic sorbicillinoids and bisorbicillinoids from a marine-derived *Fungus Trichoderma* sp. *Chem Pharm Bull (Tokyo)* **57**: 220–223.

Fahad, A.Al. (2014) Tropolone and Sorbicillactone Biosynthesis in Fungi. Thesis (Ph.D.). University of Bristol.

Fahad, A.al., Abood, A., Fisch, K.M., Osipow, A., Davison, J., Avramovi, M., *et al.* (2014) Oxidative dearomatisation: the key step of sorbicillinoid biosynthesis. *Chem Sci* **5,** 523–527.

Fischer, J., Schroeckh, V., and Brakhage, A.A. (2016) Awakening of Fungal Secondary Metabolite Gene Clusters. In *Gene Expression Systems in Fungi: Advancements and Applications*. Schmoll, M., Dattenbock, C. (eds). Vienna: Springer International Publishing, pp. 253–273.

Georgianna, D.R., and Payne, G.A. (2009) Genetic regulation of aflatoxin biosynthesis: from gene to genome. *Fungal Genet Biol* **46**: 113–125.

Harned, A.M., and Volp, K.A. (2011) The sorbicillinoid family of natural products: isolation, biosynthesis, and synthetic studies. *Nat Prod Rep* **28**: 1790–1810.

Kovalchuk, A., Weber, S.S., Nijland, J.G., Bovenberg, R.A.L. and Driessen, A.J.M. (2012) Chapter 1 fungal ABC transporter deletion and localization analysis. In *Plant Fungal Pathogens: Methods and Protocols.* Bolton, M.D., B.P.H.J., T. (eds), USA: Humana Press, pp. 1–16.

Koyama, N., Ohshiro, T., Tomoda, H., and Ōmura, S. (2007) Fungal isobisvertinol, a new inhibitor of lipid droplet accumulation in mouse macrophages. *Org Lett* **9**: 425–428.

Lim, F.Y., Sanchez, J.F., Wang, C.C., and Keller, N.P. (2012) Toward awakening cryptic secondary metabolite gene clusters in filamentous fungi. *Methods Enzymol* **517**: 303–324.

Livak, K.J. and Schmittgen, T.D. (2001) Analysis of relative gene expression data using real-time quantitative PCR and. *Methods* **25**, 402–408.

Maskey, R.P., Grün-Wollny, I., and Laatsch, H. (2005) Sorbicillin analogues and related dimeric compounds from *Penicillium notatum. J Nat Prod* **68**: 865–870.

Meng, J., Wang, X., Xu, D., Fu, X., Zhang, X., Lai, D., and Zhou, L. (2016) Sorbicillinoids from fungi and their bioactivities. *Molecules* **21**: 1–19.

Nelson, M.D. and Fitch, D.H.A. (2011) Molecular methods for evolutionary genetics. In *Molecular Methods for Evolutionary Genetics.* Orgogozo, V., Rockman, M.V. (eds). USA: Humana Press, pp. 459–470.

Nicolaou, K.C., Vassilikogiannakis, G., Simonsen, K.B., Baran, P.S., Zhong, Y., Vidali, V.P., et al. (2000) Biomimetic total synthesis of bisorbicillinol, bisorbibutenolide, trichodimerol, and designed analogues of the bisorbicillinoids. *J Am Chem Soc* **122**: 3071–3079.

Nicoletti, R., and Trincone, A. (2016) Bioactive compounds produced by strains of penicillium and talaromyces of marine origin. *Mar Drugs* **14**: 1–35.

Nijland, J.G., Ebbendorf, B., Woszczynska, M., Boer, R., Bovenberg, R.A.L., and Driessen, A.J.M. (2010) Nonlinear biosynthetic gene cluster dose effect on penicillin production by *Penicillium chrysogenum. Appl Environ Microbiol* **76**: 7109–7115.

Park, A.R., Son, H., Min, K., Park, J., Goo, J.H., Rhee, S., et al. (2015) Autoregulation of ZEB2 expression for zearalenone production in *Fusarium graminearum. Mol Microbiol* **97**: 942–956.

Pohl, C., Kiel, J.A.K.W., Driessen, A.J.M., Bovenberg, R.A.L., and Nygard, Y. (2016) CRISPR/Cas9 based genome editing of *Penicillium chrysogenum. ACS Synth Biol* **5**: 754–764.

Polli, F., Meijrink, B., Bovenberg, R.A.L., and Driessen, A.J.M. (2016) New promoters for strain engineering of *Penicillium chrysogenum. Fungal Genet Biol* **89**: 62–71.

Salo, O.V., Ries, M., Medema, M.H., Lankhorst, P.P., Vreeken, R.J., Bovenberg, R.A.L., and Driessen, A.J.M. (2015) Genomic mutational analysis of the impact of the classical strain improvement program on β − lactam producing *Penicillium chrysogenum. BMC Genom* **16**: 1–15.

Salo, O., Guzmán-Chávez, F., Ries, M.I., Lankhorst, P.P., Bovenberg, R.A.L., Vreeken, R.J., and Driessen, A.J.M. (2016) Identification of a polyketide synthase involved in sorbicillin biosynthesis by *Penicillium chrysogenum. Appl Environ Microbiol* **82**: 3971–3978.

Snoek, I.S., van der Krogt, Z.A., Touw, H., Kerkman, R., Pronk, J.T., Bovenberg, R.A., et al. (2009) Construction of an hdfA *Penicillium chrysogenum* strain impaired in nonhomologous end-joining and analysis of its potential for functional analysis studies. *Fungal Genet Biol* **46**: 418–426.

Sugaya, K., Koshino, H., Hongo, Y., and Yasunaga, K. (2008) The biosynthesis of sorbicillinoids in Trichoderma sp. USF-2690: prospect for the existence of a common precursor to sorbicillinol and 5-epihydroxyvertinolide, a new sorbicillinoid member. *Tetrahedron Lett* **49**: 654–657.

Weber, S.S., Kovalchuk, A., Bovenberg, R.A.L., and Driessen, A.J.M. (2012) The ABC transporter ABC40 encodes a phenylacetic acid export system in *Penicillium chrysogenum. Fungal Genet Biol* **49**: 915–921.

Xu, L., Wang, Y.-R., Li, P.-C., and Feng, B. (2016) Advanced glycation end products increase lipids accumulation in macrophages through upregulation of receptor of advanced glycation end products: increasing uptake, esterification and decreasing efflux of cholesterol. *Lipids Health Dis* **15**: 1–13.

Yin, W., and Keller, N.P. (2011) Transcriptional regulatory elements in fungal secondary metabolism. *J Microbiol* **49**: 329–339.

Yu, J.-H., and Keller, N. (2005) Regulation of secondary metabolism in filamentous fungi. *Annu Rev Phytopathol* **43**: 437–458.

Engineering *Ashbya gossypii* strains for *de novo* lipid production using industrial by-products

Patricia Lozano-Martínez, Rubén M. Buey, Rodrigo Ledesma-Amaro,[†] Alberto Jiménez and José Luis Revuelta*
Metabolic Engineering Group, Departamento de Microbiología y Genética, Universidad de Salamanca, Edificio Departamental, Campus Miguel de Unamuno, 37007 Salamanca, Spain.

Summary

***Ashbya gossypii* is a filamentous fungus that naturally overproduces riboflavin, and it is currently exploited for the industrial production of this vitamin. The utilization of *A. gossypii* for biotechnological applications presents important advantages such as the utilization of low-cost culture media, inexpensive downstream processing and a wide range of molecular tools for genetic manipulation, thus making *A. gossypii* a valuable biotechnological chassis for metabolic engineering. *A. gossypii* has been shown to accumulate high levels of lipids in oil-based culture media; however, the lipid biosynthesis capacity is rather limited when grown in sugar-based culture media. In this study, by altering the fatty acyl-CoA pool and manipulating the regulation of the main Δ9 desaturase gene, we have obtained *A. gossypii* strains with significantly increased (up to fourfold) *de novo* lipid biosynthesis using glucose as the only carbon source in the fermentation broth. Moreover, these strains were efficient biocatalysts for the conversion of carbohydrates from sugarcane molasses to biolipids, able to accumulate lipids up to 25% of its cell dry weight. Our results represent a proof of principle showing the promising potential of *A. gossypii* as a competitive microorganism for industrial biolipid production using cost-effective feed stocks.**

*For correspondence. E-mail revuelta@usal.es

[†]Present address: Micalis Institute, INRA UMR1319, AgroParisTech, Université Paris-Saclay, 78350 Jouy-en-Josas, France.

Funding Information
This work was supported in part by BASF and by grant BIO2014-56930-P to José Luis Revuelta. Rubén M Buey is supported by a 'Ramón y Cajal' contract from the Spanish Ministerio de Economía y Competitividad.

Introduction

During the last years, studies unravelling the lipid metabolic pathways in microorganisms have boosted the application of systems metabolic engineering to produce biolipids that could be used to produce biofuels and oleochemicals (Beopoulos *et al.*, 2011). Biolipid-derived fuels from renewable or even waste feedstocks and advanced biofuels avoid the severe inconveniences of first- and second-generation biofuels such as competition with food industry, dependence on climate season and longer processing cycle (Stephanopoulos, 2007; Beopoulos *et al.*, 2011; Peralta-Yahya *et al.*, 2012). Biolipids provide a sustainable alternative for fossil fuels that might help to reduce the carbon footprint. Additionally, bio-based oleochemicals could substitute petroleum-based chemically synthesized compounds with interest in pharma, food and polymer industries (Ledesma-Amaro *et al.*, 2014b). Thus, the combination of an engineered microorganism host and a cost-effective feedstock is nowadays a major challenge to produce biofuels and oleochemicals in an environmentally and economically feasible manner.

The model microorganism *Saccharomyces cerevisiae* and the oleaginous yeast *Yarrowia lipolytica* have been extensively manipulated by means of rational metabolic engineering approaches to optimize lipid production (Vorapreeda *et al.*, 2012; Blazeck *et al.*, 2014; Runguphan and Keasling, 2014; Kavšček *et al.*, 2015). Other oleaginous microorganisms such as *Rhodosporidium toruloides* have also been studied for fatty acid-derived products because of its natural ability to accumulate triacylglycerol (Fillet *et al.*, 2015).

Ashbya gossypii is a filamentous fungus first identified for its natural capacity to overproduce riboflavin (vitamin B₂) and, currently, more than half of the worldwide riboflavin industrial production relies on *A. gossypii* fermentation (Stahmann *et al.*, 2000; Schwechheimer *et al.*, 2016). *A. gossypii* is a very convenient fungus for industrial use because it can be readily grown in industrial waste-based culture media. These media include low-cost oils (Schwechheimer *et al.*, 2016), glycerol (Ribeiro *et al.*, 2012) or sucrose (Pridham and Raper, 1950), the main carbon source of sugarcane molasses (Hashizume *et al.*, 1966). This and other advantages have stimulated the use of *A. gossypii* not only for industrial scale riboflavin production, but also for nucleoside production

(Ledesma-Amaro *et al.*, 2015a; Ledesma-Amaro *et al.* 2015b) and recombinant protein production (Magalhes *et al.*, 2014).

We have previously reported *A. gossypii* strains with compromised lipid β-oxidation which are able to accumulate lipids up to 70% of the cell dry weight when grown in culture media supplemented with 2% oleic acid. Nevertheless, when grown in glucose-based media without lipid supplementation, these strains only accumulated up to 10% of their cell dry weight (Ledesma-Amaro *et al.*, 2014a). The development of efficient biocatalyst for the production of biolipids requires not only the conversion of low-cost oily feedstocks into high-value oils, but also a high-yield conversion of carbohydrates to biolipids. In this regard, major bottlenecks exist in the biosynthesis of lipids due to feedback inhibition of lipidogenic enzymes (Fig. 1): acyl-CoA esters regulate the activity of the fatty acid synthase (FAS), the acetyl-CoA carboxylase (*ACC1*) and the Δ9 desaturase (*OLE1*; Chen *et al.*, 2014; Neess *et al.*, 2015). Furthermore, saturated fatty acids also exert a negative effect over the *ACC1* enzyme, thus regulating their own synthesis (Qiao *et al.*, 2015).

Here, we aimed at developing of *A. gossypii* strains that significantly increased lipid production using sugar-based culture media through the manipulation of two major bottlenecks in lipid metabolism: (i) altering the fatty acyl-CoA pool and the subsequent feedback inhibition of lipidogenic genes and (ii) manipulating the regulation of the main Δ9 desaturase encoded by the *OLE1* gene (Fig. 1). We show that rewiring the regulation of lipogenesis can increase significantly the conversion of carbohydrates to lipids in *A. gossypii*. In addition, we demonstrate that engineered strains of *A. gossypii* are able to produce biolipids when grown on a very simple culture medium consisting of sugarcane molasses and tap water. Our results represent a proof of principle showing the promising potential of *A. gossypii* as a competitive microorganism for industrial biolipid production using cost-effective feedstocks.

Results

Aiming at generating *A. gossypii* strains with improved lipid *de novo* biosynthesis, using glucose as the only carbon source, we have manipulated two known bottlenecks for lipid metabolism (Fig. 1): (i) insertion of the first double bond in palmitic and stearic acid by Δ9 desaturases and (ii) alteration of the intracellular concentration of the fatty acyl-CoA pool.

OLE1 endogenous regulation is a bottleneck for the de novo *lipid biosynthesis in* A. gossypii

Ashbya gossypii has two identified and characterized Δ9 desaturases codified by the genes *AgOLE1* and

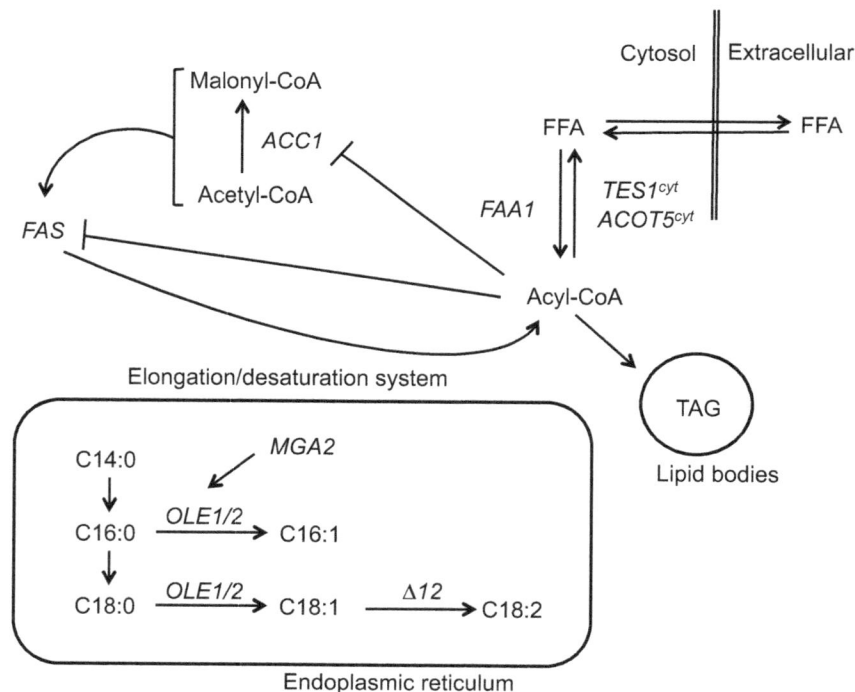

Fig. 1. Schematic – simplified – representation of the lipid metabolism in *A. gossypii*. FFA stands for free fatty acids; TAG for triacylglycerol; FAS for fatty acid synthase.

AgOLE2 that are responsible for the insertion of the first desaturation in stearic and palmitic acid (Fig. 1; Lozano-Martínez et al., 2016). The simultaneous overexpression of these two genes in A. gossypii only slightly increased total fatty acid accumulation in glucose-based medium (up to 1.2-fold with respect to wild type; Lozano-Martínez et al., 2016). Thereby, contrary to what has been reported for Y. lipolytica (Qiao et al., 2015), the overexpression of AgOLE1/OLE2 is not enough to significantly improve the lipid de novo biosynthesis in A. gossypii. We then decided to study AgOLE1 regulation.

In S. cerevisiae, MGA2 – and its homologous SPT23 – are the main regulators of OLE1: Mga2p has been shown not only to activate OLE1 transcription (Zhang et al., 1999; Jiang et al., 2001, 2002; Auld et al., 2006) but also to stabilize OLE1 mRNA transcript when the cells are grown in fatty acid free medium and destabilize it when the cells are exposed to unsaturated fatty acids (Kandasamy et al., 2004). MGA2 codifies for an endoplasmic reticulum membrane protein (Mga2p), but the C-terminal proteolytic cleavage converts Mga2p into a cytoplasmic protein that can be transported to the nucleus (Martin et al., 2007; Liu et al., 2015). The deletion of MGA2 in S. cerevisiae has modest effect on cell fitness, but the double knockout of MGA2 and its homologous SPT23 results in an inviable mutant in the absence of unsaturated fatty acids in the culture medium (Zhang et al., 1999). The overexpression of the cleaved version of MGA2 in S. cerevisiae led to a 1.2-fold increase in triacylglycerides with respect to wild type (Kaliszewski et al., 2008), most probably due to OLE1 overexpression (Chellappa et al., 2001).

Prompted by these results, we decided to investigate the effect on lipid accumulation of the manipulation of the gene ACR165W (AgMGA2), the Ashbya's homologue of S. cerevisiae MGA2/SPT23 (Dietrich et al., 2004). In glucose-based media, AgMGA2 disruption (mga2Δ) caused a slight increase in total lipid accumulation, in contrast to its overexpression (P_GPD-MGA2) that showed a significant increase with respect to the wild-type strain (Table 1). These results agree with previous reports in S. cerevisiae, where the MGA2 orthologues stabilize OLE1 mRNA transcript when the cells are grown in fatty acid free medium (Kandasamy et al., 2004). On the other hand, MGA2 might destabilize OLE1 mRNA transcript when the cells are exposed to unsaturated fatty acids, as reported for S. cerevisiae (Kandasamy et al., 2004), which would result in decreased lipid accumulation (Table 1).

Interestingly, AgMGA2 disruption does not confer auxotrophy for unsaturated fatty acids (not shown), in contrast to what has been described for S. cerevisiae (Kandasamy et al., 2004). This might indicate that MGA2 might be required for maximal transcriptional activation

Table 1. Total fatty acids (TFA) in the engineered A. gossypii strains expressed as the percentage of lipids with respect to dry cell weight. Cultures were grown in MA2 media supplemented with either 8% (w/v) Glucose (MA2-8G) or 1% (w/v) Glucose + 2% (w/v) Oleic Acid (MA2-1G-2O) at 28°C for 7 days in an orbital shaker (150 r.p.m.). Numbers are the mean ± SD of two independent experiments with two replicates each. Total biomass showed no large differences among the different strains tested in this study (8.9 ± 2 mg ml^{-1} in MA2-8G media).

Strain	TFA (%), MA2-8G	TFA (%), MA2-1G-2O
Wild type	5.30 ± 0.4	23.53 ± 0.4
P_{GPD}-MGA2	8.62 ± 0.3	15.61 ± 2.1
ΔMGA2	6.63 ± 0.1	24.08 ± 1.5
MGA2-ΔC-term	10.47 ± 0.1	24.63 ± 0.9
P_{GPD}-FAA1	5.90 ± 0.9	11.55 ± 1.9
ΔFAA1	2.17 ± 0.4	12.88 ± 2.1
P_{GPD}-TES1^{cyt}	7.89 ± 0.4	n.d.
P_{GPD}-ACOT5^{cyt}	9.85 ± 0.1	n.d.
MGA2-ΔC-term/P_{GPD}-ACOT5^{cyt}	20.01 ± 0.2	n.d.

of OLE1/2, but it is possible that OLE1/2 might be expressed at basal levels in the absence of MGA1. In this context, MGA2 deletion would not confer auxotrophy for unsaturated fatty acids. At this point, we do not know if there are additional genes implied in OLE1/2 transcriptional activation, but no additional paralogues of MGA2 exist in the genome of A. gossypii.

The increase in lipid accumulation was significantly higher when a C-terminal truncated version of AgMGA2 was expressed (mga2-ΔC-term), reaching more than 10% of the cell dry weight (Table 1), similarly to what has been reported for S. cerevisiae (Kaliszewski et al., 2008). Remarkably, in media supplemented with oleic acid, P_GPD-MGA2 strains showed a strong decrease in total lipid accumulation with respect to the wild type, in contrast to both the mga2Δ and mga2-ΔC-term strains that showed no significant differences with the wild type (Table 1).

The fatty acyl-CoA pool is a bottleneck for the de novo lipid biosynthesis in A. gossypii

We next decided to study how the alteration of the fatty acyl-CoA pool could influence the de novo lipid biosynthesis in A. gossypii. The intracellular acyl-CoA pool is extensively regulated by the counteraction of acyl-CoA synthetases and acyl-CoA thioesterases (Black and DiRusso, 2007; Chen et al., 2014). Acyl-CoA-thioesterases catalyse the conversion of activated fatty acids into free fatty acids, which is the reverse reaction catalysed by fatty acyl-CoA synthetases (Fig. 1).

We first disrupted the main fatty acyl-CoA synthetase in A. gossypii: AgFAA1, which is the homolog of FAA1 and FAA4 in S. cerevisiae. Our results showed that AgFAA1 is essential when the only carbon source in the culture medium is oleic acid (data not shown) and its disruption (*faa1Δ* strain) significantly decreased lipid accumulation as compared to the wild type when grown on glucose-containing media (Table 1). This finding further supports that *AgFAA1* is the main acyl-CoA synthetase in *A. gossypii*, in agreement with the phenotype of the *S. cerevisiae faa1Δ/faa4Δ* strain, which is also unable to grow in fatty acid-based media (Black and DiRusso, 2007). However, the marked decrease in lipid accumulation in the strain *faa1Δ* in glucose-based media differs from previous results reported for *S. cerevisiae*, where lipid accumulation was not significantly changed with respect to wild type (Færgeman *et al.*, 2001; Black and DiRusso, 2007; Chen *et al.*, 2014). On the other hand, our results agree with the sharp decrease observed when the strain *faa1Δ* of *Y. lipolytica* is grown on oleic acid-containing media (Dulermo *et al.*, 2015). This might indicate that *AgFAA1* has additional functions, apart from the one shown in Fig. 1, in the lipid metabolism of *A. gossypii* that remain unknown. Interestingly, *AgFAA1* overexpression (*P_{GPD}-FAA1*) also decreased lipid accumulation in *A. gossypii* when the strain is grown in media containing 2% of oleic acid, despite it has no significant effects in glucose-based media (Table 1). This might happen because in media-containing oleic acid *FAA1* overexpression might greatly increase the cytoplasmic levels of fatty acyl-CoA that would result in the inhibition of lipidogenic genes and, therefore, the observed decrease in lipid accumulation. Moreover, this result agrees with previous reports describing that the alteration of the intracellular acyl-CoA pool has an inhibitory effect on lipidogenic genes (Færgeman and Knudsen, 1997) and also induced the expression of the genes involved in lipid degradation (Færgeman *et al.*, 2001).

We then intended to alter the fatty acyl-CoA pool by the overexpression of acyl-CoA-thioesterase genes. Acyl-CoA esters are known to repress fatty acid synthesis (Fig. 1) by inhibiting several lipidogenic enzymes, such as *FAS*, *ACC1* and *OLE1* (Bortz and Lynen, 1963; Sumper and Träuble, 1973; Choi *et al.*, 1996; Færgeman and Knudsen, 1997). Thereby, we hypothesized that the decrease in the fatty acyl-CoA intracellular pool (by conversion to free fatty acids) could upregulate lipid accumulation. To test this hypothesis, we constructed two *A. gossypii* strains that ectopically overexpress in their cytoplasm two peroxisomal acyl-CoA thioesterase enzymes: (i) *P_{GPD}-TES1^{cyt}* that overexpress *A. gossypii* *TES1* (the homologue of *TES1* in *S. cerevisiae*) and (ii) *P_{GPD}-ACOT5^{cyt}* that overexpress *Mus musculus ACOT5*, which has been previously shown to increase the accumulation of free fatty acids in *S. cerevisiae* (Chen *et al.*, 2014). Both genes encode fatty acyl-CoA thioesterases involved in fatty acid degradation in the peroxisome (Jones and Gould, 2000; Westin *et al.*, 2004; Maeda

et al., 2006; Chen *et al.*, 2014) and contain the 'SKL' prototypical C-terminal peroxisomal targeting signal (PTS) that it is both necessary and sufficient for directing cytosolic proteins to peroxisomes (Gould *et al.*, 1987). Interestingly, total fatty acid quantification of both strains grown in glucose-based media showed a significant increase (up to twofold) with respect to the wild type (Table 1), suggesting that an excess of cytoplasmic fatty acyl-CoA thioesterase activity results in a decreased acyl-CoA pool that relieves the feedback inhibition of lipidogenic genes (Fig. 1). The excess of free fatty acids in the cytoplasm can be readily excreted to the culture media and, accordingly, the two engineered strains excreted approximately fivefold more fatty acids (118.25 ± 0.4 and 122.38 ± 0.4 mg l^{-1} for *P_{GPD}-TES1-^{cyt}* and *P_{GPD}-ACOT5^{cyt}* respectively) than the wild type (22.00 ± 0.6 mg l^{-1}).

We next combined the two most favourable modifications (Table 1) into a single strain (*MGA2-ΔC-term/P_{GPD}-ACOT5^{cyt}*) and observed an additive effect of both modifications (Table 1). Remarkably, this *A. gossypii* strain accumulates more than 20% of its dry cell weight.

Sugarcane molasses are a very convenient carbon source for biolipid production in A. gossypii

The results obtained with the strain *MGA2-ΔC-term/P_{GPD}-ACOT5^{cyt}* together with the advantages for industrial use convert *A. gossypii* into a very promising microorganism for biolipid production using sugar-based culture media formulations. We therefore studied the use of convenient culture media for industrial use. To this end, we tested sugarcane molasses as the unique carbon source for the culture media of *A. gossypii*.

Molasses, from sugarcane or beet, mainly contains fructose, sucrose and glucose. This sugar is an industrial by-product from sugar manufacturing, and it is considered an ideal raw material for cheap medium culture formulations (Chao *et al.*, 2013). Indeed, sugarcane molasses have been previously proved to be acceptable carbon sources for lipid production in *Y. lipolytica* (Gadjos *et al.*, 2015), as well as for ethanol and butanol production in *S. cerevisiae* (Ni *et al.*, 2012; Arshad *et al.*, 2014).

Remarkably, in contrast to the wild-type strain which slightly decreased lipid accumulation in molasses with respect to glucose-based medium, the *MGA2-ΔC-term/P_{GPD}-ACOT5^{cyt}* *A. gossypii* engineered strain increased lipid accumulation up to 25% of its dry cell weight (Fig. 2). We then quantitatively characterized the lipid profile of the engineered strain when grown in glucose and molasses-based culture media. As can be observed in Fig. 3, there are no significant changes between both media. The strain *MGA2-ΔC-term/P_{GPD}-ACOT5^{cyt}*

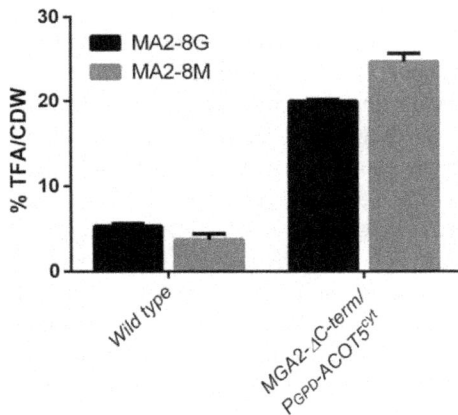

Fig. 2. Comparison of total fatty acid (TFA) per cell dry weight (CDW) in the wild-type and the MGA2-ΔC-term/P$_{GPD}$-ACOT5^{cyt} strains. MGA2-8G and MGA2-8M stand for MA2 medium supplemented with 8% (w/v) of either glucose or sugarcane molasses respectively. The data shown represent the mean of three independent experiments with standard errors.

showed a slight increase in saturated C16:1 and C18:1, in detriment of C16:0 and C18:0 fatty acids (Fig. 3), as an expected consequence of the upregulation of the Δ9 desaturase AgOLE1.

Altogether, by manipulating two genes, we have obtained an A. gossypii strain able to accumulate up to 25% of its dry cell weight in a very convenient culture media composed of sugarcane molasses and tap water. Furthermore, we envisage that this number can be readily increased through a systematic optimization of the culture medium composition as well as the fermentation conditions. Eventually, further genetic manipulations of this strain by means of random and/or rational modifications could also increase this number.

Discussion

Ashbya gossypii has a large capacity to accumulate lipids when grown in media-containing oleic acid (Ledesma-Amaro et al., 2014a). Encouraged by this

result, we aimed at further manipulating A. gossypii to optimize the de novo lipid biosynthesis and have obtained strains with increased de novo lipid biosynthesis, rather than lipid accumulation. These manipulations enable the utilization of A. gossypii as an efficient biocatalyst for the production of biolipids from sugar-based byproducts such as molasses. The most productive strain contained only two modifications that resulted in additive effects on lipid accumulation: a truncated version of MGA2, a main regulator of OLE1 (the main Δ9 desaturase in A. gossypii) and a heterologously expressed murine thioesterase gene, MmACOT5. These modifications are expected to increase the expression of OLE1 and promote the lipid biosynthetic process. In addition, the expression of a heterologous thioesterase is expected to decrease the cytoplasmic pool of fatty acyl-CoA, thus alleviating the feedback inhibition mechanisms of that this metabolite exerts on lipogenesis. Up to our knowledge, this is the first report on the combination of the manipulation of these two genes in microorganisms to enhance de novo lipid biosynthesis. We envisage that further manipulations will readily increase the percentage of lipid accumulation and, indeed, significant efforts are being directed at present in our laboratory towards this aim.

Although the achieved amount of lipid accumulation is not as high as that reported for S. cerevisiae and/or Y. lipolytica (Kamisaka et al., 2013; Blazeck et al., 2014), it must be stressed here that A. gossypii shows important advantages for industrial production of lipids compared to these yeasts that make of it a promising competitive candidate to be taken into account. First, the biomass from a filamentous fungus can be easily separated from culture media by convenient filtration or sedimentation techniques, easy to implement at industrial level (Zheng et al.,2012). Second, large-scale fermentations are nowadays used for riboflavin production, demonstrating the suitability of this fungus for industrial scale-up. Third, A. gossypii hyphae suffer autolysis in the late stationary phase, and triglycerides could be

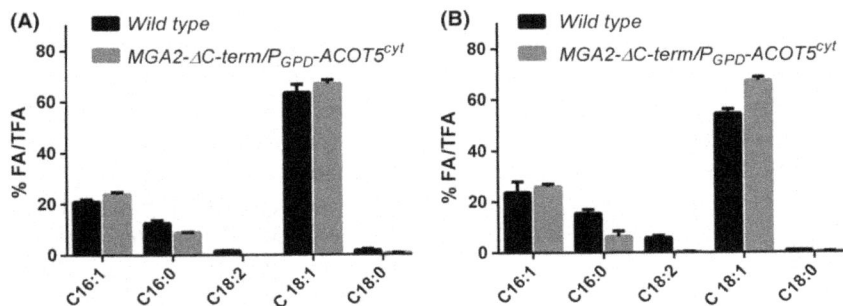

Fig. 3. Lipid profile in the wild-type and the MGA2-ΔC-term/P$_{GPD}$-ACOT5^{cyt} A. gossypii strains in MGA2-8G (A; MA2 medium with 8% (w/v) glucose) and MGA2-8M (B; MA2 medium with 8% (w/v) sugarcane molasses). The data shown represent the mean of three independent experiments with standard errors.

easily recovered by centrifugation, avoiding costly cell-disruption processes. Therefore, our results represent a proof of principle showing that *A. gossypii* is a promising and convenient microorganism that deserves the further investigation of its potential use as a convenient industrial biolipid producer.

On the other hand, one of the disadvantages of using microbial hosts for lipid production is the global cost of the process, which can be notably diminished with the use of alternative feed stocks such as industrial by-products. Thereby, the efficient utilization of alternative sources of carbon will have important economic advantages for the scale-up of lipid production with *A. gossypii* using a cheap and convenient culture media. Sugarcane molasses represent a cheap industrial by-product consisting on sucrose (up to 50%), nitrogen source, proteins, vitamins and amino acids among others. The use of molasses presents important advantages with respect to other waste products such as lignocellulosic biomass, which needs a costly pre-treatment for its consumption by microorganisms (Stephanopoulos, 2007; Taherzadeh and Karimi, 2008). Interestingly, *A. gossypii* can grow on molasses without any further modification, contrary to what happens in *Y. lipolytica* that needs the ectopic overexpression of invertases to degrade sucrose (Gadjos *et al.*, 2015). Thereby, the engineered *MGA2-ΔC-term*/P_{GPD}-*ACOT5*^{cyt} strain is a promising candidate that deserves future attention.

The lipid profile of the engineered oleaginous strain has a composition in monounsaturated, polyunsaturated and saturated methyl esters that correlate with good biodiesel properties, that is neither high levels of polyunsaturated nor long-chain saturated FAs (Ramos *et al.*, 2009). Thereby, this strain accumulates significant amounts of lipids suitable for biodiesel production. Furthermore, we have recently reported that the modification of the lipid profile by manipulating the elongation and desaturation systems enhances biodiesel properties in *A. gossypii* (Ledesma-Amaro *et al.*, 2014b). Thus, future experiments in our laboratory will be focused on the modification of the lipid profile in our engineered oleaginous strain.

Altogether, our results demonstrate that *A. gossypii* is a very promising industrial microorganism that uses a cost-effective feedstock to *de novo* synthetize significant amounts of biolipids that can be used for producing biofuels in an environmentally and economically feasible manner.

Experimental procedures

A. gossypii strains, media and growth conditions

The *A. gossypii* strain ATCC10895 was used and considered wild-type strain. The strains were cultured at 28°C using MA2 rich medium during 7 days (Förster *et al.*, 1999). MA2 is composed of yeast extract, bactopeptone, agar, water and glucose. In this study, for lipid accumulation, the C/N ratio was increased using 8% of glucose (MA2-8G) as carbon source instead of 2%, which is the standard formulation. Alternatively, 1% glucose and 2% oleic acid, previously emulsified by sonication in the presence of 0.02% Tween-40 (MA2-1G-2O), were used. For experiments with molasses, media was prepared with 8% sugarcane molasses (kindly provided by AB Azucarera Iberia S.L.), 0.1% of yeast extract and tap water (MA2-8M). *A. gossypii* transformation, sporulation conditions and spore isolation have been described elsewhere (Santos *et al.*, 2005). Briefly, DNA was introduced into *A. gossypii* by electroporation, and primary transformants were isolated in selective medium. Homokaryon transformant clones were obtained by sporulation of the primary heterokaryon transformants and isolated on antibiotic-containing plates with 250 mg l^{-1} of geneticin (*G418*). Liquid cultures were initiated from spores and were incubated on an orbital shaker at 200 r.p.m at 28°C.

Gene manipulation of A. gossypii

Gene deletion and overexpression were carried out by the construction of recombinant integrative cassettes (Ledesma-Amaro *et al.*, 2014a,b). For gene deletion, a replacement cassette with selection marker (*loxP-KanMX-loxP* module for *G418* resistance) was used. This selection marker is flanked by the repeated inverted sequences *loxP*, which enable the elimination of the selection marker by the expression of a Cre recombinase (Ledesma-Amaro *et al.*, 2014a). The deletion of the C-terminal part of *AgMGA2* was performed by substituting this region by a *G418* antibiotic resistance marker. For gene overexpression, a module based on the *A. gossypii* glycerol 3-phosphate dehydrogenase promoter (P_{GPD}) and phosphoglycerate kinase (T_{PGK1}) terminator sequences, recombinogenic flanks and the antibiotic selectable marker *loxP-KanMX-loxP*, was integrated at the *STE12* locus. DNA constructs were obtained using Golden Gate methodology (Enger *et al.*, 2008). Genome integration of the deletion and overexpression modules was confirmed by analytical PCR and DNA sequencing.

To ectopically express *TES1* and *ACOT5* in the cytosol of *A. gossypii*, we removed the C-terminal prototypical peroxisomal targeting signal (PTS) that it is both necessary and sufficient for directing cytosolic proteins to peroxisomes (Gould *et al.*, 1987). The signal 'SKL' that both *TES1* and *ACOT5* contain is a prototypical PTS and, thereby, its removal avoids peroxisome localization.

Lipid extraction and quantification

Triacylglycerols were extracted and trans-methylated from lyophilized biomass using a modification of the method described by Bligh and Dyer (Bligh and Dyer, 1959). Approximately 200 μg of dried mycelia was mixed with 1 ml of 97.5% methanol/2.5% sulfuric acid and incubated at 80°C for 90 min. The transesterification reaction was stopped by the addition of 1 ml of distilled water. The extraction was performed by mixing the samples with 0.5 ml of hexane and recovery of the upper phase after centrifugation. The hexane-soluble extracted fatty acid methyl esters dissolved were analysed by gas chromatography coupled to mass spectrometry (GC-MS) in an Agilent 7890A gas chromatograph coupled an Agilent MS200 (Agilent, Santa Clara, California, USA) mass spectrometer. A VF50 column (30 m long, 0.25 mm internal diameter and 25 μm film) was used using helium as carrier at 1 ml min^{-1}, with a split ratio of 1:20. The oven programme was as follows: an initial temperature of 90°C for 5 min, a ramp of 12°C min^{-1} up to 190°C and a ramp of 4°C min^{-1} up to 290°C. MS detection was from 50 to 400 Da. Fatty acids were identified by comparison with commercial fatty acid methyl ester standards (FAME32; Supelco), and total quantification of fatty acids, expressed as total fatty acids (TFA), was performed using an internal standard: 50 μg of heptadecanoic acid C17:0 (Sigma-Aldrich, Sigma-Aldrich Quimica SL, Madrid, Spain).

Acknowledgements

This work was supported in part by BASF and by grant BIO2014-56930-P to José Luis Revuelta. Rubén M Buey is supported by a 'Ramón y Cajal' contract from the Spanish Ministerio de Economía y Competitividad. Patricia Lozano-Martínez is supported by a postgraduate fellowship (FPI Program) from the Spanish Ministerio de Economía y Competitividad. We thank María Dolores Sánchez and Silvia Domínguez for excellent technical help.

Conflict of interest

The authors declare that they have no competing interests.

References

Arshad, M., Ahmed, S., Zia, M.A., and Rajoka, M.I. (2014) Kinetics and thermodynamics of ethanol production by *Saccharomyces cerevisiae* MLD10 using molasses. *Appl Biochem Biotechnol* **172:** 2455–2464.

Auld, K.L., Brown, C.R., Casolari, J.M., Komili, S., and Silver, P.A. (2006) Genomic association of the proteasome demonstrates overlapping gene regulatory activity with transcription factor substrates. *Mol Cell* **21:** 861–871.

Beopoulos, A., Nicaud, J.M., and Gaillardin, C. (2011) An overview of lipid metabolism in yeasts and its impact on biotechnological processes. *Appl Microbiol Biotechnol* **90:** 1193–1206.

Black, P.N., and DiRusso, C.C. (2007) Yeast acyl-CoA synthetases at the crossroads of fatty acid metabolism and regulation. *Biochim Biophys Acta* **1771:** 286–298.

Blazeck, J., Hill, A., Liu, L., Knight, R., Miller, J., Pan, A., et al. (2014) Harnessing *Yarrowia lipolytica* lipogenesis to create a platform for lipid and biofuel production. *Nat Commun* **5:** 1–10.

Bligh, E.G., and Dyer, W.J. (1959) A rapid method of total lipid extraction and purification. *Can J Biochem Physiol* **37:** 911–917.

Bortz, W.M., and Lynen, F. (1963) The inhibition of acetyl CoA carboxylase by long chain acyl CoA derivatives. *Biochem Z* **337:** 505–509.

Chao, H., Chen, X., Xiong, L., Chen, X., Ma, L., and Chen, Y. (2013) The possibility and potential of its industrialization. *Biotechnol Adv* **31:** 129–139.

Chellappa, R., Kandasamy, P., Oh, C.S., Jiang, Y., Vemula, M., and Martin, C.E. (2001) The membrane proteins, Spt23p and Mga2p, play distinct roles in the activation of *Saccharomyces cerevisiae* OLE1 gene expression. Fatty acid-mediated regulation of Mga2p activity is independent of its proteolytic processing into a soluble transcription activator. *J Biol Chem* **276:** 43548–43556.

Chen, L., Zhang, J., Lee, J., and Chen, W.N. (2014) Enhancement of free fatty acid production in *Saccharomyces cerevisiae* by control of fatty acyl-CoA metabolism. *Appl Microbiol Biotechnol* **98:** 6739–6750.

Choi, J., Stukey, J.E., Huang, S., and Martin, C.E. (1996) Regulatory elements that control transcription activation and unsaturated fatty acid-mediated repression of the *Saccharomyces cerevisiae* ole1 gene. *J Biol Chem* **271:** 3581–3589.

Dietrich, F.S., Voegeli, S., Brachat, S., Lerch, A., Gates, K., Steiner, S., et al. (2004) The *Ashbya gossypii* genome as a tool for mapping the ancient *Saccharomyces cerevisiae* genome. *Science* **9:** 304–307.

Dulermo, R., Gamboa-Méndez, H., Ledesma-Amaro, R., Thevenieau, F., and Nicaud, J.M. (2015) Unraveling fatty acid transport and activation mechanisms in *Yarrowia lipolytica*. *Biochim Biophys Acta* **1851:** 1202–1217.

Enger, C., Kandzia, R. and Marillonet, S. (2008) A one pot, one step, precision cloning method with high throughput capability. *PLoS ONE* **3:** e3647.

Færgeman, N.J., and Knudsen, J. (1997) Role of long-chain fatty acyl-CoA esters in the regulation of metabolism and in cell signalling. *Biochem J* **323:** 1–12.

Færgeman, N.J., Black, P.N., Zhao, X., Knuædsen, J., and DiRusso, C.C. (2001) The acyl-CoA synthetases encoded within FAA1 and FAA4 in S. cerevisiae function as components of the fatty acid transport system linking import, activation, and intracellular utilization. *J Biol Chem* **276:** 37051–37059.

Fillet, S., Gibert, J., Suárez, B., Lara, A., Ronchel, C., and Adrio, J.L. (2015) Fatty alcohols production by oleaginous yeast. *J Ind Microbiol Biotechnol* **42:** 1463–1472.

Förster, C., Santos, M.A., Ruffert, S., Kramer, R., and Revuelta, J.L. (1999) Physiological consequence of disruption of the VMA1 gene in the riboflavin overproducer *Ashbya gossypii*. *J Biol Chem* **274**: 9442–9448.

Gadjos, P., Nicaud, J.M., Rossignol, T., and Certik, M. (2015) Single cell oil production on molasses by *Yarrowia lipolytica* strains overexpressing DGA2 in multicopy. *Appl Microbiol Biotechnol* **99**: 8065–8074.

Gould, S.J., Keller, G.A., and Subramani, S. (1987) Identification of a peroxisomal targeting signal at the carboxy terminus of firefly luciferase. *J Cell Biol* **105**: 2923–2931.

Hashizume, T., Higa, S., Sasaki, Y., Yamazaki, H., Iwamura, H., and Matsuda, H. (1966) Constituents of cane molasses. Part I. Separation and identification of the nucleic acid derivatives. *Agr Biol Chem* **30**: 319–326.

Jiang, Y., Vasconcelles, M.J., Wretzel, S., Light, A., Martin, C.E., and Godberg, M.A. (2001) MGA2 is involved in the low-oxygen response element-dependent hypoxic induction of genes in *Saccharomyces cerevisiae*. *Mol Cell Biol* **18**: 6161–6169.

Jiang, Y., Vasconcelles, M.J., Wretzel, S., Light, A., Gilooly, L., McDaid, K., *et al.* (2002) Mga2p processing by hypoxia and unsaturated fatty acids in *Saccharomyces cerevisiae*: impact on LORE-dependent gene expression. *Eukariot Cell* **3**: 481–490.

Jones, J.M., and Gould, S.J. (2000) Identification of PTE2, a human peroxisomal long-chain Acyl-CoA thioesterase. *Biochem Biophys Res Commun* **275**: 233–240.

Kaliszewski, P., Szkopińska, A., Ferreira, T., Swiezewska, E., Berges, T., and Zoładek, T. (2008) Rsp5p ubiquitin ligase and the transcriptional activators Spt23p and Mga2p are involved in co-regulation of biosynthesis of end products of the mevalonate pathway and triacylglycerol in yeast *Saccharomyces cerevisiae*. *Biochim Biophys Acta* **1781**: 627–634.

Kamisaka, Y., Kimura, K., Uemura, H., and Yamaoka, M. (2013) Overexpression of the active diacylglycerol acyltransferase variant transforms *Saccharomyces cerevisiae* into an oleaginous yeast. *Appl Microbiol Biotechnol* **97**: 7345–7355.

Kandasamy, P., Vemula, M., Oh, C.S., Chellappa, R., and Martin, C.E. (2004) Regulation of unsaturated fatty acid biosynthesis in *Saccharomyces*. *J Biol Chem* **279**: 36586–36592.

Kavšček, M., Bhutada, G., Madl, T., and Natter, K. (2015) Optimization of lipid production with a genome-scale model of *Yarrowia lipolytica*. *BMC Syst Biol* **9**: 72.

Ledesma-Amaro, R., Santos, M.A., Jimenez, A. and Revuelta, J.L. (2014a) Strain design of *Ashbya gossypii* for single-cell oil production. *Appl Environ Microbiol* **80**, 1237–1244.

Ledesma-Amaro, R., Santos, M.A., Jimenez, A., and Revuelta, J.L. (2014b) Tuning single-cell oil production in *Ashbya gossypii* by engineering the elongation and desaturation systems. *Biotechnol Bioeng* **111**: 1782–1791.

Ledesma-Amaro, R., Buey, R.M., and Revuelta, J.L. (2015a) Increased production of inosine and guanosine by means of metabolic engineering of the purine pathway in *Ashbya gossypii*. *Microb Cell Fact* **14**: 58.

Ledesma-Amaro, R., Lozano-Martínez, P., Jimenez, A., and Revuelta, J.L. (2015b) Engineering *Ashbya gossypii* for efficient biolipid production. *Bioengineered* **6**: 119–123.

Liu, L., Markham, K., Blazeck, J., Zhou, N., Leon, D., Otoupal, P., and Alper, H.S. (2015) Surveying the lipogenesis landscape in *Yarrowia lipolytica* through understanding the function of a Mga2p regulatory protein mutant. *Metab Eng* **31**: 102–111.

Lozano-Martínez, P., Ledesma-Amaro, R., and Revuelta, J.L. (2016) Engineering *Ashbya gossypii* for ricinoleic and linoleic acid production. *Chem Eng Trans* **49**: 253–258.

Maeda, I., Delessert, S., Hasegawa, S., Seto, Y., Zuber, S., and Poirier, Y. (2006) The peroxisomal Acyl-CoA thioesterase Pte1p from *Saccharomyces cerevisiae* is required for efficient degradation of short straight chain and branched chain fatty acids. *J Biol Chem* **281**: 11729–11735.

Magalhes, F., Aguiar, T.Q., Oliveira, C., and Domingues, L. (2014) High-level expression of *Aspergillus niger* β-galactosidase in *Ashbya gossypii*. *Biotechnol Prog* **30**: 261–268.

Martin, C.E., Oh, C.S., and Jiang, Y. (2007) Regulation of long chain unsaturated fatty acid synthesis in yeast. *Biochim Biophys Acta* **1771**: 271–285.

Neess, D., Bek, S., Engelsby, H., Gallego, S.F., and Færgeman, N.J. (2015) Long-chain acyl-CoA esters in metabolism and signaling: role of acyl-CoA binding proteins. *Prog Lipid Res* **59**: 1–25.

Ni, Y., Wang, Y., and Sun, Z. (2012) Butanol production from cane molasses by *Clostridium saccharobutylicum* DSM 13864: batch and semicontinuous fermentation. *Appl Biochem Biotechnol* **166**: 1896–1907.

Peralta-Yahya, P.P., Zhang, F., Cardayre, S.B., and Keasling, J.D. (2012) Microbial engineering for the production of advanced biofuels. *Nature* **488**: 320–326.

Pridham, T.G., and Raper, K.B. (1950) *Ashbya gossypii*: its significance in nature and in the laboratory. *Mycologia* **42**: 603–623.

Qiao, K., Abidi, S.H., Liu, H., Zhang, H., Chakraborty, S., Watson, N., *et al.* (2015) Engineering lipid overproduction in the oleaginous yeast *Yarrowia lipolytica*. *Metab Eng* **29**: 56–65.

Ramos, M.J., Fernández, C.M., Casas, A., Rodríguez, L., and Pérez, A. (2009) Influence of fatty acid composition of raw materials on biodiesel properties. *Bioresour Technol* **100**: 261–268.

Ribeiro, O., Domingues, L., Penttilä, M., and Wiebe, M.G. (2012) Nutritional requirements and strain heterogeneity in *Ashbya gossypii*. *J Basic Microbiol* **51**: 1–8.

Runguphan, W., and Keasling, J.D. (2014) Metabolic engineering of *Saccharomyces cerevisiae* for production of fatty acid-derived biofuels and chemicals. *Metab Eng* **21**: 103–113.

Santos, M.A., Mateos, L., Stahmann, K.P., and Revuelta, J.L. (2005) Insertional mutagenesis in the vitamin B2 producer fungus *Ashbya gossypii*. *Microb Process Prod* **18**: 283–300.

Schwechheimer, S.K., Park, E.Y., Revuelta, J.L., Becker, J., and Wittmann, C. (2016) Biotechnology of riboflavin. *Appl Microbiol Biotechnol* **100**: 2107–2119.

Stahmann, K.P., Revuelta, J.L., and Seulberger, H. (2000) Three biotechnical processes using *Ashbya gossypii*,

Candida famata, or *Bacillus subtilis* compete with chemical riboflavin production. *Appl Microbiol Biotechnol* **53**: 509–516.

Stephanopoulos, G. (2007) Challenges in engineering microbes for biofuels production. *Science* **315**: 801–804.

Sumper, M., and Träuble, H. (1973) Membranes as acceptors for palmitoyl CoA in fatty acid biosynthesis. *FEBS Lett* **30**: 29–34.

Taherzadeh, M.J., and Karimi, K. (2008) Pretreatment of lignocellulosic wastes to improve ethanol and biogas production: a review. *Int J Mol Sci* **9**: 1621–1651.

Vorapreeda, T., Thammarongtham, C., Chhevadhanarak, S., and Laoteng, K. (2012) Alternative routes of acetyl-CoA synthesis identified by comparative genomic analysis: involvement in the lipid production of oleaginous yeast and fungi. *Microbiology* **158**: 217–228.

Westin, M.A., Alexon, S.E., and Hunt, M.C. (2004) Molecular cloning and characterization of two mouse peroxisome proliferator-activated receptor alpha (PPARalpha)-regulated peroxisomal acyl-CoA thioesterases. *J Biol Chem* **279**: 21841–21848.

Zhang, S., Skalsky, Y., and Garfinkel, D.J. (1999) MGA2 or SPT23 is required for transcription of the delta9 fatty acid desaturase gene, OLE1, and nuclear membrane integrity in *Saccharomyces cerevisiae*. *Genetics* **151**: 473–483.

Zheng, Y., Yu, X., Zeng, J., and Chen, S. (2012) Feasibility of filamentous fungi for biofuel production using hydrolysate from dilute sulfuric acid pretreatment of wheat straw. *Biotechnol Biofuels* **5**: 50.

Metallic bionanocatalysts: potential applications as green catalysts and energy materials

Lynne E. Macaskie,[1,]* Iryna P. Mikheenko,[1] Jacob B. Omajai,[1,†] Alan J. Stephen[2] and Joseph Wood[2]
Schools of [1]Biosciences,
[2]Chemical Engineering, University of Birmingham, Edgbaston, Birmingham, B15 2TT, UK.

Summary

Microbially generated or supported nanocatalysts have potential applications in green chemistry and environmental application. However, precious (and base) metals biorefined from wastes may be useful for making cheap, low-grade catalysts for clean energy production. The concept of bionanomaterials for energy applications is reviewed with respect to potential fuel cell applications, bio-catalytic upgrading of oils and manufacturing 'drop-in fuel' precursors. Cheap, effective biomaterials would facilitate progress towards dual development goals of sustainable consumption and production patterns and help to ensure access to affordable, reliable, sustainable and modern energy.

Introduction

In the late 1990s, bacteria were reported to recover soluble palladium (II) via reduction into cell-bound precious metal (PM) nanoparticles (NPs) (Lloyd *et al.*, 1998) with high catalytic activity (Baxter-Plant *et al.*, 2003). Many authors have reiterated the scope for bio-PM catalysts (e.g. reviews by Deplanche *et al.*, 2011; De Corte *et al.*, 2012; Castro *et al.*, 2014; Kulkarni and Maddapur, 2014; Rai *et al.*, 2015; Singh, 2015; Ashok, 2016). A consultancy report (Catalytic Technology Management, unpublished, 2009) concluded that a 'me too' catalyst must be more active than those currently

*For correspondence. E-mail l.e.macaskie@bham.ac.uk

†Present address: Department of Biological Sciences, Faculty of Sciences, Thompson Rivers University, 805 TRU Way, V2C 0C8 Kamloops, British Columbia, Canada.

Funding Information
Natural Environment Research Council (NE/L014076/1)

available, cheaper or both. The paradigm bacterial 'bio-Pd(0)' has significant potential applications in 'green chemistry' and environmental nanotechnology, but the criteria cannot yet be met in full due to high costs of (i) growing dedicated bacteria and (ii) precious metals; (iii) retention of the catalyst for re-use and (iv) potential catalyst poisoning at high reaction temperatures (e.g. by sulfur via degradation of the biomaterial). Waste yeast and bacteria have been successfully reused in 'second life' following primary fermentations (Dimitriadis *et al.*, 2007; Orozco *et al.*, 2010; Zhu *et al.*, 2016) while waste precious metals have been bio-reprocessed into active neo-catalysts (e.g. Mabbett *et al.*, 2006; Deplanche *et al.*, 2007; Yong *et al.*, 2010, 2015; Murray *et al.*, 2017a). Metal attrition from cells was negligible, enabling catalyst re-use [e.g. 6 cycles (Bennett *et al.*, 2013)] and also as an immobilized catalyst (Beauregard *et al.*, 2010). The highest 'green' potential probably lies in 'tandem' one-pot reactions which (e.g.) combine a biotransformation with a bio-Pd-catalysed step [e.g. in an enantioselective deracemization reaction (Foulkes *et al.*, 2011)] although the low Pd loading necessary to permit continued physiological activity may not be optimal for chemical catalysis.

Concerns about nanoparticles in the environment (Valsami-Jones and Lynch, 2015) (biomass will eventually decompose) may restrict pollutant remediation to *ex-situ* applications. Catalyst poisoning is less relevant in a 'dirty' process, but this requires cheap, disposable catalyst. A life cycle analysis is needed to determine where bio-metallic catalysts may outcompete traditional comparators, taking into account socio-environmental as well as economic factors.

Bio-precious metal materials are emerging in energy applications (Table 1). Large-scale oil production may justify a once-through catalyst if a low-grade mixed metal 'dirty' catalyst can be used. On the other hand, in synthesis of fuel precursors from waste CO_2, some bio-nanoparticles [e.g. structured Pd/Au core shells (Deplanche *et al.*, 2012)] could potentially be used for making (e.g.) formic acid electrochemically (Humphrey *et al.*, 2016). Liu *et al.* (2016) reported a bio-Pd/Au alloy with electrocatalytic activity, but the thickness of the Pd-shell is critical for product selectivity [e.g. for formate production a Pd-shell of 10 nm is optimal (Humphrey *et al.*, 2016)]; such fine structure control is probably beyond the reach of biosynthetic capability. Moreover,

Table 1. Microbial precious metal nanoparticles and catalysts in energy applications

Test	Comments	Ref
(a) Fuel cells (anodic and cathodic FC electrocatalysts)		
Anode (PEMFC)	Bio-PM catalyst on *Desulfovibrio desulfuricans*. Required sintering to carbonize Power outputs: commercial Pt- FC, 200 mW; commercial Pt FC catalyst 170 mW; Bio-Pt, 170 mW; bio-Pd, 140 mW*. Metal content was 20 wt%; activated carbon was 80 wt% plus residual biomass component. Loading: 1 mg metal cm^{-2}. *Power density was 9 mW cm^{-2} (electrode area 16 cm^2: ref 3; c.f. ref 6).	1
Anode (alkaline FC)	Bio-Pt catalyst made from waste yeast cells from fermentation, immobilized in polyvinyl alcohol. Activity was ~half that of commercial Pt on carbon catalyst Loading 10 mg Pt cm^{-2}	2
Anode (PEMFC)	Anode as in 1. Power outputs from bio-Pd were: *D. desulfuricans*, 142 mW; *Cupriavidis metallidurans* CH34, 68 mW; *Escherichia coli* MC4100, 29 mW, *E. coli* IC007, 115 mW; *E. coli* IC007 (made from industrial PM waste), 68 mW. Increasing Pd loading onto cells from 5 wt% to 25 wt% doubled power output	3
Anode (PEMFC)	Anode as in 1. Power outputs from bio-Pd were: *Rhodobacter sphaeroides* 001 (biohydrogen-producing), 20 mW and *E. coli* ('second life' cells from biohydrogen process): H-D701, 28 mW; MC4100, 18 mW; IC007, 56 mW.	4
Anode (PEMFC)	Bio-Pd on *Shewanella oneidensis* MR-1. Formate was e-donor in Pd-NP synthesis (NP size 50-10 nm). Pd loading was 20 wt% on cells and 1.28 mg Pd cm^{-2} on anode. Power density was 4.8 mW cm^{-2} for bio-Pd and 5.3 mW cm^{-2} for commercial Pd-catalyst.	5
Catalyst mix pd/activated carbon in EPR	EPR showed more electronic interactions between bio-Pd/C than commercial Pd/C; quenching of free radicals (FR) of activated carbon was higher with sintered bio-Pd$_{D. desulfuricans}$ than with bio-Pd$_{E. coli}$; both bio-Pds gave higher FR quenching than Pd/C catalyst.	6
Native cells in rotating disc electrode	Pd loading 20 wt% (*E. coli*). Use of formate as e^- donor for Pd(0) formation gave small, well separated NPs with no electrochemical activity. Bio-Pd made under H_2 showed proton adsorption/desorption. Similar results obtained using bio-Pd on *Shewanella oneidensis*.	7
Cyclic Votammetry	Palladium NPs on *D. desulfuricans* (native cells). Pd loading not stated Proton-concentration gradients involved in extracellular electron transfer processes. Pd-NPs proposed to augment natural e^- transport chain. Activity increased by adding formate (live cells only).	8
Cathode (PEMFC)	Material as in 1. Bio-Pd$_{E.coli}$ at 25 wt% Pd. Commercial anodic catalyst in FC test rig Paxitech FCT-50s. Cathodic activity of bio-Pd was 25% of that of commercial catalyst. Combining with bio-Pd with Pt (various ratios) did not significantly increase power output	9
O-reduction reaction: cyclic voltammetry	Bio-Pt$_{E. coli}$ cleaned in NaOH gave enhanced cyclic voltammetry response, cf. sintered material. Bio-Pt$_{D. desulfuricans}$ was better when cleaned in phenol–chloroform. Max. current was 8.5 μA and 25 μA, respectively (glassy carbon rotating disc electrode) (see Table 2)	10

Application	Comments	Ref
(b) Upgrading of 5-hydroxymethyl furfural (5-HMF) into 2,5, dimethyl furan (drop-in fuel)		
DMF production	Pd-based catalyst supported on *Bacillus benzeovorans* (Table 3). The conversion was the same as Pd/C catalyst, but selectivity was higher	11
(c) Catalytic upgrading of pyrolysis oil		
Pyrolysis oil algal (*Chlorella*)	Oil was from algae (*Chlorella*) slurry. Bio-Pd$_{D. desulfuricans}$ (5 wt% Pd:biomass; 5 wt% catalyst per reaction) and Pd/C (10% wt% catalyst per reaction: 15 g oil; 4 h; 325°C) had similar activities, O_2 and N_2 contents were reduced by 65% and 35% respectively. Bacterial component of catalyst contributed to bio-oil yield. Nanoparticle size increased in bio-catalyst but not Pd/C. Catalysts were reclaimed but not evaluated in repeated reactions.	12
Pyrolysis oil beechwood	Oil was from beechwood (fast-pyrolyzed 500°C; < 2 h). Catalysts: 5 wt% Pd$_{E. coli}$ 5 wt% Pd/C and commercial Pd/Al_2O_3 (0.6 g catalyst to 20 g oil). Minimum viscosity was ~ 0.035 Pas (250°C, 3 h), increasing thereafter. Bio-Pd gave similar deoxygenation (~20%) at 160°C but performed less well at higher temperatures, attributed to higher coking than with Pd/C at > 259°C (see Fig. S1, Table S1)	13
Pyrolysis oil pine wood	Made from beetle-killed pine trees which were more easily pyrolysed; not tested for upgrading via bio-derived catalysts	14
(d) Catalytic upgrading of heavy oils		
Upgrading of heavy fossil oils from Touchstone (Canada)	Production of bio-magnetite (Fe_3O_4) by *Geobacter sulfurreducens* that reduces Fe(III) to Fe(II) thence to bio-magnetite. Functionalization of bio-magnetite with Pd(0) onto bio-magnetite. Comparable oil upgrading to commercial alumina-supported catalysts. High liquid yield (90%); less viscous than without catalyst. Coking reduced twofold- to fourfold using 9.5 wt% Pd. In situ process proposed by stimulation of natural geomicrobiological system	15
	Bio-PMs made on cells of *Bacillus benzeovorans* and *Desulfovibrio desulfuricans*. Bio-Pd/Pt mix (1:1) was better than Pd alone. Comparable upgrading to commercial catalyst via using 5 wt% and 20 wt% metal loading. Liquid yield was 90%. Coking was reduced by ~20% as compared to using commercial catalyst (Ni-Mo/Al_2O_3). Extensive catalyst characterization.	16

Table 1. (Continued)

Test	Comments	Ref
	Catalyst (Bio-Pd/Pt; 2.5 wt% Pd/2.5 wt% Pt) was made on *B. benzeovorans* and *E. coli*. The former showed ~30% reduced coking. It was discussed that the catalyst may reduce V(V) to a V-species that does not promote free radical reactions leading to coking. The higher inorganic phosphate content of the Gram-positive cell wall (teichoic acids) may sorb Ni^{2+} (although the cells become carbonized to become part of the fuel). Similar results were obtained using metals sourced from simulated road dust waste leaches	17
(e) Other opportunities		
Dry reforming of methane (DRF)	DRF makes syngas from $CH_4 + CO_2$ over (e.g.) Ni-catalyst. Catalyst deactivates due to carbon deposition; noble metals (alone or with Ni) are longer lived. Bio-PM catalysts have not been evaluated but show coking resistance (above)	18
Upgrade waste glycerol	Bio-Au(0) nanocatalyst from jewellery waste was shown to oxidize glycerol to glycerate Application to biodiesel waste glycerol is not yet shown. Glycerate is a substrate for bio-H_2 production	19
Photocatalytic water splitting to make H_2	Noble metals are used to split water to make H_2. (Bio-PMs are not yet tested)	20
	Optically active ZnS was made using mine drainage treatment bioprocess off- gas (H_2S)	21
	CuS has been separated from other metals in biogenic metal recovery from waste	22
	CuS/ZnS hybrid (chemically made) was used in photocatalytic water splitting	23
	CdS quantum dots were made with bacteria and cysteine (Cd is more toxic than Zn; an energy application was not reported).	24
	Biogenic metal selenides were made *ex-situ* (reduction of selenite by *Veilonella atypica*)	25
	CdSe can be made by *Fusarium oxysporum* (intracellularly; metabolic process)	26
	CdSe-hydrogenase hybrids/artificial hydrogenases are reported for water splitting	20

1. Yong *et al.* (2007). 2. Dimitriadis *et al.* (2007). 3. Yong *et al.* (2010). 4. Orozco *et al.* (2010). 5. Ogi *et al.* (2011). 6. Carvalho *et al.* (2009). 7. Courtney *et al.* (2016). 8. Wu *et al.* (2011). 9. Stephen, A.J. unpublished data. 10. Williams (2015). 11. Omajali (2015). 12. Kunwar *et al.* (2017). 13. Deilami *et al.* (unpublished). 14. Luo *et al.* (2017) 15. Brown *et al.* (2016). 16. Omajali *et al.* (2017). 17. Murray *et al.* (2015). 18. Pakhare and Spivey (2014). 19. Deplanche *et al.* (2007). 20. Ran *et al.* (2014). 21. Murray *et al.* (2017b). 22. Nancucheo and Johnson (2012). 23. Zhang *et al.* (2011). 24. Yang *et al.* (2015);. 25. Fellowes *et al.* (2013). 26. Yamaguchi *et al.* (2016).

electrocatalysis does not generate longer chain hydrocarbons and this method may not be readily scalable.

As an alternative, chemical upgrading of CO_2 into hydrocarbon fuels is scalable. Recent advances employ the reverse water gas shift reaction (reduction of CO_2 to CO) in tandem with Fischer–Tropsch chemistry to convert the more reactive CO into hydrocarbons. Consuming CO (a catalyst poison) shifts the equilibrium towards products in the reverse water gas shift reaction, promoting CO_2 consumption. An efficient, abundant, low-cost catalyst(s) must give product selectivity in the desired range (~C5–8 for gasoline; ~C9–16 for diesel fuel). Precious metals (Pd, Pt, Ru, Rh) on SiO_2 have been used in the reverse water gas shift reaction, while catalysts for the Fischer–Tropsch process are usually Fe or Co-based, with recent innovations towards one-pot reactions (see e.g. Mattia *et al.*, 2015; Owen *et al.*, 2016; Prieto, 2017) following early work (Dorner *et al.*, 2010) that showed conversion of CO_2 to hydrocarbons (41% conversion and a C2–C5+ selectivity of 62%) using a doped Fe-based system. The potential for using biogenic catalysts has not been explored although a paradigm hybrid bio-magnetite/Pd(0) catalyst has been reported (Brown *et al.*, 2016). When made from waste, biogenic precious metal catalysts (e.g. Pd/Pt mixtures ~30% Pd) also contained other metals, for example from Degussa processing waste: Al (42%), Ag (6%) and Mg (3%) and from spent automotive catalyst leachate: Fe (14%), Mg (12%)

and Al, 27%), i.e. metals that were present in the original solid material (Mabbett *et al.*, 2006; Macaskie *et al.*, 2010). Potentiation of catalytic activity occurred (e.g. more than 10-fold in the case of reduction of Cr(VI) to Cr (III)) using these waste-derived mixed metal catalysts (Macaskie *et al.*, 2010). Hence, exploration of bio-catalyst made from automotive leachate may be warranted for CO_2 valorization, particularly as waste bacteria for use as catalyst support (Orozco *et al.*, 2010; Zhu *et al.*, 2016; Stephen *et al.*, this volume) are readily available from other scalable biotechnology processes. However, selectivity towards specific hydrocarbon products may not be compatible with such economies.

In contrast to CO_2-valorization, 'proof-of-principle' application of biogenic catalysts has been shown in four key areas of energy and fuels (Table 1).

Fuel cell electrocatalysts

Fuel cells comprise an anode [where fuel, e.g. H_2, is split catalytically to give electrons (current) and protons], a cathode (where protons combine with O_2 in air to give water) (Kraytsberg and Ein-Eli, 2014) and an electrolyte that allows passage of positive ions between them (Kirubakaran *et al.*, 2009). The polymer electrolyte fuel cell (PEM fuel cell) uses purified hydrogen at low temperatures (80°C) with rapid start up (Mehta and Cooper, 2003). PEM fuel cells are applicable for use in (e.g.)

vehicles (Hannan *et al.*, 2014) or larger 'stacks' for domestic power (Staffell *et al.*, 2015). Durability targets (internationally) are 5000 h and 40 000 h of operation for automotive and stationary fuel cells, respectively (Rice *et al.*, 2015). The US Department of Defence installed 5 kW PEM fuel cell systems in ~40 military bases (cost of > $100 000 per system) but, with an operational lifespan of only 500 h, the systems required overhauling annually (Staffell *et al.*, 2015). The cost of precious metal catalysts is restrictive; other catalysts are under development but the power to weight ratio is key, especially for portable and aerospace applications. Substitution of fuel cell Pt-electrocatalysts (0.2–0.8 mg cm^{-2}) would use lighter, equivalently performing (but robust against the high local acidity) metals like Pd (e.g. Meng *et al.*, 2015; Gómez *et al.*, 2016), particularly in the cathodic reduction of O_2 (He *et al.*, 2005) which is rate-limiting. Reducing Pt costs/loadings could also be achieved by optimizing Pt nanoparticles (via size control and increased uniformity), or by developing alloys (Zhu *et al.*, 2015). Developments towards microbially derived fuel cell catalysts are shown in Table 1. The anodic reaction is well reported, and the challenge is now to develop an efficient cathodic oxygen reduction catalyst; electrochemical test data are shown in Table 2.

2,5-dimethyl furan (DMF) production from 5-hydroxymethyl furfural (5-HMF)

Carbohydrates form ~75% of the annual renewable biomass (Schmidt and Dauenhauer, 2007). In thermochemical hydrolysis [e.g. of fermentation feedstocks: Orozco 2011], 5-hydroxymethyl furfural is produced via breakdown of hexoses in cellulose and starch hydrolysates (Román-Leshkov *et al.*, 2007; van Putten *et al.*, 2013). 5-hydroxymethyl furfural is a precursor to 2, 5-dimethyl-furan (DMF), a 'drop-in' fuel for conventional engines (Zhorg *et al.*, 2010). DMF contains comparable energy to gasoline (Davis *et al.*, 2011) (energy contents are 31.5 MJ l^{-1} and 35 MJ l^{-1} respectively). DMF is also

more advantageous than ethanol because of its higher gravimetric energy density (about 40%), higher boiling point and insolubility in water; hence, it is a potential alternative biofuel (Tian *et al.*, 2011).

Various catalytic applications have been developed to achieve good yield and selectivity to 2, 5-dimethyl furan. Most have focused on commercial heterogeneous mono and bimetallic catalysts based on (e.g.) Ru, Pd, Pt, Au, Cu (Tong *et al.*, 2011; Hansen *et al.*, 2012; Nishimura *et al.*, 2014; Zu *et al.*, 2014; Luo *et al.*, 2015) which are costly. A bacterial platform would provide a cheaper, sustainable source of supported precious metal catalyst, possibly using metals biorefined from waste sources. As the first step, conversion of 5-HMF to DMF using a bio-Pd-based catalyst and a Pd/carbon catalyst was compared (Omajali, 2015), with better selectivity to DMF observed using the biomaterial (Table 3). Nishimura *et al.* (2014) showed application of a Pd/Au bimetallic/C catalyst and hence the use of bio-Pd/Au (Deplanche *et al.*, 2012; Hosseinkhani *et al.*, 2012; Liu *et al.*, 2016) is worth evaluating in this application.

Upgrading of pyrolysis oils

Hydrothermal liquefaction (HTL) and fast pyrolysis are thermal treatments of wet organic feedstocks (e.g. agrifood wastes, manure, algae) which produce a highly viscous biofuel (pyrolysis oil). This is unsuitable as an alternative to fossil fuels without further processing. In addition to high viscosity, crude bio-oil contains large quantities of oxygenated molecules which are unsuitable for use directly in vehicle engines and are incompatible for blending with fossil fuels, without further upgrading to remove oxygen (hydrodeoxygenation, HDO).

Hydrodeoxygenation is well studied, but a mechanistic understanding for disassembly of biopolymers and their subsequent deoxygenation is incomplete. New catalytic routes are required to make cost effective drop-in fuels (Huber *et al.*, 2006; Rinaldi and Schuth, 2009). Precious metal-based catalysts feature prominently (e.g. Bouxin

Table 2. Activity of bio-Pt catalyst on *Escherichia coli* and *Desulfovibrio desulfuricans* compared to commercial TKK fuel cell catalyst.

Material/treatment	Specific activity (mA cm^{-2})	Mass activity (mA mg Pt^{-1})	No. electrons transferred per O_2
Bio-Pt$_{E.\ coli}$ (NaOH)	0.68 ± 0.15	75 ± 17	3.78 ± 0.23
Bio-Pt$_{D.\ desulfuricans}$ (phenol–chloroform)	1.43 ± 0.28	304 ± 53	3.84 ± 0.12
TKK catalyst	0.45 ± 0.02	374 ± 4	3.86 ± 0.07

Taken from Williams (2015). Pt loading was 5 wt% of the biomass. Bio-Pt$_{E.\ coli}$ was cleaned using NaOH. Bio-Pt $_{D.\ desulfuricans}$ was cleaned using phenol–chloroform. As with the anodic and cathodic tests in the PEM fuel cell (Table 1), the *E. coli* biomaterial was ~25% as active (mass activity) as that from *D. desulfuricans* and had ~ half the specific activity (mA cm^{-2}). However, growth of *E. coli* is readily scalable and it makes active bio-metal catalyst when used in 'second life' following an independent primary fermentation (Orozco *et al.*, 2010; Zhu *et al.*, 2016). In contrast to *E. coli*, *D. desulfuricans* cells are obligately anaerobic, growth is less readily scalable, and they produce H_2S, a powerful catalyst poison that requires more extensive washing of the cells prior to use. However, a metal bioremediation process that couples excess biogenic H_2S (used for minewater clean-up with respect to heavy metals (Hedrich and Johnson, 2014)) also produces waste biomass of a sulfate-reducing bacterial consortium which may find a 'second life' use as a bio-metallic catalyst for fuel cell application, mitigating waste disposal costs.

Table 3. Pd-catalyst-mediated upgrading of 5-hydroxymethyl furfural into 2,5-dimethyl furan.

Catalyst and H-donor	5-HMF conversion (%)	DMF yield (%)
5 wt% Pd/carbon (formic acid)	97.5	26.5 ± 2.0
Bio-Pd-based (5 wt% metal; formic acid)	96.8	49.8 ± 0.6
5 wt% Pd/carbon (2-propanol)	94.5	32.6 ± 1.8
Bio-Pd-based (5 wt% metal; 2-propanol)	94.5	42.6 ± 1.2

Taken from Omajali (2015). Bio-catalyst was prepared on cells of *Bacillus benzeovorans*. *Bacillus* was selected because this genus is grown at large scale for commercial production of enzymes; a cost-benefit analysis for 'second life' production of catalyst as compared to other current routes for disposal of waste biomass is required.

et al., 2017), and theoretical chemistry can play an important role in understanding the competing reactions on the surface of, for example, Pt (111) faces (Liu *et al.*, 2017). However commercial precious metal catalysts would be uneconomic; moreover, fuel cell electrocatalysts (above) will compete in parallel for limited global resources. The energy demand of winning precious metals from primary ores (e.g. 14 t of CO_2 is emitted per kg of Pt produced: Anon 2008) is a major consideration; carbon-neutral fuel needs, in itself, to produce a low carbon footprint.

Catalytic deoxygenation reactions include dehydration, hydrogenolysis, hydrogenation, decarbonylation and decarboxylation (Fig. S1, Table S1). For fuels in the diesel range, C-C coupling reactions can be achieved through routes such as aldol-condensation, ketonization, oligomerization and hydroxyalkylation, but hydrodeoxygenation is currently considered the most effective method for bio-oil upgrading, improving the effective H/C ratio and leading to hydrocarbons.

To date, several classes of catalysts are reported for hydrodeoxygenation. In addition to precious metal catalysts, other metals (Fe, Ni and Cu) have shown good selectivities in hydrogenation and hydrogenolysis reactions. However, high hydrogen pressures can lead to complete hydrogenation of double bonds (Bykova *et al.*, 2012). Industrial catalysts based on Co-Mo and Ni can provide good hydrodeoxygenation performance, but these deactivate rapidly due to coke formation and water poisoning (Badawi *et al.*, 2011).

The acidic, corrosive biofuels also limit catalyst lifetime and compromise the process economics. Work has focused on upgrading of biocrude oils using various catalysts (e.g. Co-Mo, Ni-Mo Pd/C (e.g. Biller *et al.*, 2015; Si *et al.*, 2017)). Of these, Pd (which is a noble metal and hence is dissolution tolerant) is promising. 5-hydroxymethylfurfural is made within in the product oil (Dang *et al.*, 2016), enabling possible catalytic conversion into 2,5-dimethyl furan into a second fuel stream.

Despite the high costs, precious metal-based catalysts are favoured for bio-oil upgrading (see reviews: Watson, 2014; Pandey *et al.*, 2015; Cheng *et al.*, 2016; Lee *et al.*, 2016; Nam *et al.*, 2017; Fermoso *et al.*, 2017); bio-manufactured catalyst should now be added to the portfolio (Table 1). Bio-reprocessed precious metal waste has not yet been tested as a cheap metal source (c.f. below). The upgrading efficiency, product spectrum and cost savings are being factored into ongoing research via life cycle analysis, towards the dual development goals of sustainable consumption and production patterns for affordable, reliable, sustainable (non-fossil) energy.

Upgrading of fossil oils

The hydrogen economy and carbon-neutral biofuels lag behind the timeline for change articulated by the Stern Review (Stern, 2006). Hence, cleaner production of fossil fuels is vital as these will remain the predominant short-term sources of supply (~80% of global needs; Anon 2014). With globally declining light crude oil reserves, heavy oil and bitumen use will increase from 2 to 7 million barrels day^{-1} by 2030 (Anon 2015). Oil sands production emits more greenhouse gas and uses more water than conventional light oil production (Findlay, 2016) and additional refining is also required (Huc, 2011). Heavy oil exploitation is complicated by the high viscosity, low hydrogen content and high amounts of resins and asphaltenes. (Bio)geochemical processes evolved the hydrocarbons, leading to materials with high contents of heavy molecules rich in sulfur, nitrogen, oxygen and metals (e.g. Ni, V), with high viscosity and acidity (Head *et al.*, 2003; Huc, 2011). Technologies for heavy oil upgrading have been reviewed (Heraud *et al.*, 2011; Castañeda *et al.*, 2014). Upgrading *in situ* gives a cleaner production of less viscous oil, which is more easily transported without the use of diluents (Shah *et al.*, 2010). The THAI-CAPRI (Toe-to-Heel Air Injection coupled with Catalytic Upgrading Process *In situ*) technology combines thermally enhanced oil recovery with down-hole catalytic upgrading of heavy oil into light fractions (Greaves and Xia, 2004; Hart *et al.*, 2014). This catalytic upgrading, using steam, hydrogen and methane, showed significant improvements over non-catalytic thermal processes (Hart *et al.*, 2014). Conventional cracking catalysts such as supported noble metals are prohibitively expensive; one economic option could utilize regenerated catalysts from treated oils but these have lower activity (Hart *et al.*, 2014).

The future is uncertain. Healing (2015) questioned the economics of the THAI process, Findlay (2016) discussed this in view of wider issues (regulation, cost and price uncertainties), and Nduagu *et al.* (2017) discussed the wider issues of oil sands processes in terms of performances and economics including greenhouse gas emissions and other environmental impacts. Meanwhile, emerging researches include modelling of the THAI-CAPRI process (Ado *et al.*, 2017) and the successful use of bio-nanoparticle catalysts sourced from wastes (Table 1). Application of nanoparticles in enhanced oil recovery has been reviewed (Negin *et al.*, 2016; Sun *et al.*, 2017). While bio-nanoparticles are held immobilized on bacterial cells, thermal degradation would evolve them in association with biomass-carbon. Environmental concerns about nanoparticles have been expressed (see earlier), but it is also argued that naturally occurring nanoparticles are ubiquitous in the environment (Montaño *et al.*, 2014) while evidence for natural biogeochemical cycling of platinum has been reported (Reith *et al.*, 2016).

Conclusions and future scope

As far as we are aware, this is the first overview of the potential for biogenic catalysts in various energy applications within the 'whole energy mix picture'. The concept of using microbial technologies to make materials for application to sustainable energy-generating processes (as compared to energy and waste savings via use in 'green chemistry') is a new direction. Palladium occurs in spent nuclear fuels (one ton of spent nuclear fuel contains > 2 kg of Pd or 10% of global requirements: Bourg and Poinssot, 2017). Even though radiation-resistant bacteria are well known, and the ability of microorganisms to discriminate between isotopes of essential metals (Fe, Mo) has been reported (Wasylenki *et al.*, 2007), separation of the active [107]Pd (half-life 6.5 m yrs; 15% of the Pd inventory) from the stable isotopes (85%) is probably beyond the reach of 21st-century biotechnology. Moreover, while biorecovery of precious metals from aggressive solutions (using pre-palladized cells, via chemical catalysis using Pd-bio-nanoparticle 'seeds') has been shown (Murray *et al.*, 2017a), the metal composition of the neo-catalyst reflects the metallic composition of the waste (Macaskie *et al.*, 2010), and hence, selective biorecovery of Pd against higher active radionuclide contaminants would be prohibitively difficult.

Photochemical water splitting to make clean hydrogen is a well-established solar technology, best achieved traditionally using noble metal catalysts (Ran *et al.*, 2014). As potential alternatives, various biogenic, optically active, materials have been made (metal sulfides, selenides; Table 1) but as far as we are aware these have not yet been tested in this application. Biotechnologically, hydrogenase-metal selenide hybrids show potential (Ran *et al.*, 2014). Economic attractiveness is boosted by the potential fabrication of such materials from metallic or H_2S wastes (Nancucheo and Johnson, 2012; Murray *et al.*, 2017b) but, given that the elements are abundant and cheap, the main driver may be waste valorization and mitigation of disposal costs, while incorporation of even a small amount of metal impurity may affect the optical property, and hence, neo-material from metallic waste may not be a useful option.

For biofuels, an unexpected 'biotechnology' has added value towards pine wood-derived pyrolysis oil. Very large areas of pine forest in the USA have been killed by beetles, producing very dry, porous wood. This enables the use of larger wood chips (Luo *et al.*, 2017), reducing comminution costs and energy use. It would be interesting to apply novel biogenic catalysts to pyrolysis oil obtained from this material.

Acknowledgements

This work was funded by NERC (Grant No. NE/L014076/1 to LEM) and EPSRC (studentship to AJS) within the CDT 'Fuel Cells and Their Fuels'.

Conflict of interest

None declared.

References

Ado, M. R., Greaves, M., and Rigby, S. P. (2017) Dynamic simulation of the toe-to-heel air injection heavy oil recovery process. *Energy Fuels* **31:** 1276–1284.

Anon (2008) Resource efficiency KTN, material security – ensuring resource availability for the UK economy.

Anon (2014) *World Energy Outlook*. Paris, France: International Energy Agency (IEA). ISBN: 978 9264 12413 4.

Anon (2015) *Total oil and gas. Reserves for the future*. URL http://www.total.com/en/energies-expertise/oil-gas/explora tion-production/strategic-sectors/heavy-oil/challenges/rese rves-future. Accessed 05/05/2015.

Ashok, B. (2016) Smart bio-palladium nanomaterials synthesised by green method. *J Nanomed Res* **3:** 1–3. https://doi.org/10.15406/jnmj.2016.03.00060

Badawi, M., Paul, J. F., Cristol, S., Payen, E., Romero, Y., Richard, F., *et al.* (2011) Effect of water on the stability of Mo and CoMo hydrodeoxygenation catalysts: a combined experimental and DFT study. *J Catal* **282:** 155–164.

Baxter-Plant, V., Mikheenko, I. P., and Macaskie, L. E. (2003) Sulphate-reducing bacteria, palladium and the reductive dehalogenation of chlorinated aromatic compounds. *Biodegradation* **14:** 83–90.

Beauregard, D. A., Yong, P., Macaskie, L. E., and Johns, M. L. (2010) Using non-invasive magnetic resonance imaging (MRI) to assess the reduction of Cr(VI) using a biofilm–palladium catalyst. *Biotechnol Bioeng* **107:** 11–20.

Bennett, J. A., Mikheenko, I. P., Deplanche, K., Shannon, I. P., Wood, J., and Macaskie, L. E. (2013) Nanoparticles of palladium supported on bacterial biomass; new, recyclable heterogeneous catalyst with comparable activity to homogeneous colloidal Pd in the Heck reaction. *Appl Catal B Environ* **140–141:** 700–707.

Biller, P., Sharma, B. K., Kunwar, B., and Ross, A. B. (2015) Hydroprocessing of bio-crude from continuous hydrothermal liquefaction of microalgae. *Fuel* **159:** 197–205.

Bourg, S., and Poinssot, C. (2017) Could spent nuclear fuel be considered as a non-conventional mine of critical raw materials? *Prog Nucl Energy* **94:** 222–228.

Bouxin, F. P., Zhang, X., Kings, I. N., Lee, A. F., Simmons, M. J. H., Wilson, K., and Jackson, S. D. (2017) Deactivation study of the hydrodeoxygenation of *p*-methylguaiacol over silica supported rhodium and platinum catalysts. *Appl Catal A Gen* **539:** 29–37.

Brown, A. R., Hart, A., Coker, V. S., Lloyd, J. R., and Wood, J. (2016) Upgrading of heavy oil by dispersed biogenic magnetite catalysts. *Fuel* **185:** 442–448.

Bykova, M. V., Ermakov, D. Y., Kaichev, V. V., Bulavchenko, O. A., Sarev, A. A., Lebedev, M. Y., and Yakovlev, V. A. (2012) Ni-based sol-gel catalysts as promising systems for crude bio-oil upgrading: guaiacol hydrodeoxygenation study. *Appl Catal B Environ* **113–114:** 296–307.

Carvalho, R. P., Yong, P., Mikheenko, I. P., Paterson-Beedle, M., and Macaskie, L. E. (2009) Electron paramagnetic resonance analysis of active bio-Pd-based electrode for fuel cells. *Adv Mat Res* **71–73:** 737–740.

Castañeda, L. C., Muñoz, J. A. D., and Ancheyta, J. (2014) Current situation of emerging technologies for upgrading of heavy oils. *Catal Today* **220–222:** 248–273.

Castro, L., Blázquez, M. L., Muñoz, J. A., Gonzalez, F. G., and Ballester, A. (2014) Mechanism and applications of metal nanoparticles prepared by bio-mediated process. *Rev Adv Sci Eng* **3:** 1–18.

Cheng, S., Wei, L., Zhao, X., and Julson, J. (2016) Application, deactivation and regeneration of heterogeneous catalysis in bio-oil upgrading. *Catalysts* **6:** 195–220.

Courtney, J., Deplanche, K., Rees, N. V., and Macaskie, L. E. (2016) Biomanufacture of nano-Pd(0) by *Escherichia coli* and electrochemical activity of bio-Pd(0) made at the expense of hydrogen and formate as electron donors. *Biotechnol Lett* **38:** 1903–1910.

Dang, Q., Hu, W., Rover, M., Brown, R. C., and Wright, M. M. (2016) Economics of biofuels and bioproducts from an integrated pyrolysis biorefinery. *Biofuels Bioprod Biorefin* **10:** 790–804.

Davis, S. E., Houk, L. R., Tamargo, E. E., Datye, A. K., and Davis, R. J. (2011) Oxidation of 5-hydroxymethylfurfural over supported Pt, Pd and Au catalysts. *Catal Today* **160:** 55–60.

De Corte, S., Hennebel, T., De Gusseme, B., Verstraete, W., and Boon, N. (2012) Bio-palladium: from metal recovery to catalytic applications. *Microbial Biotechnol* **5:** 5–17.

Deplanche, K., Attard, G. A., and Macaskie, L. E. (2007) Biorecovery of gold from jewellery wastes by *Escherichia coli* and biomanufacture of active Au-nanomaterial. *Adv Mat Res* **20–21:** 647–650.

Deplanche, K., Murray, A.J., Mennan, C., Taylor, S. and Macaskie, L.E. (2011) Biorecycling of precious metals and rare earth elements. In *Nanomaterials (first edition)*. Rahman, M. (ed.). Croatia: Intech, pp. 279–314, ISBN: 978-953-307-913-4.

Deplanche, K., Merroun, M. L., Casadesus, M., Tran, D. R., Mikheenko, I. P., Bennett, J. A., *et al.* (2012) Microbial synthesis of core/shell gold/palladium nanoparticles for applications in green chemistry. *J Roy Soc Interface* **9:** 1705–1712.

Dimitriadis, S., Nomikou, N., and McHale, A. P. (2007) Pd-based electrocatalytic materials derived from biosorption processes and their exploitation in fuel cell technology. *Biotechnol Lett* **29:** 545–551.

Dorner, R. W., Hardy, D. R., Williams, F. W., and Willauer, H. D. (2010) K and Mn doped iron-based CO_2 hydrogenation catalysts: detection of KAIH 4 as part of the catalyst's active phase. *Appl Catal A Gen* **373:** 112–121.

Fellowes, J. W., Pattrick, R. A. D., Lloyd, J. R., Charnock, J. M., Coker, V. S., Mosselmans, J. F., *et al.* (2013) Ex-situ formation of metal selenide quantum dots using bacterially-derived precursors. *Nanotechnology* **24:** 145603. https://doi.org/10.1088/0957-4484/24/14/15603.

Fermoso, J., Pizarro, P., Coronado, J. M., and Serrano, D. P. (2017) Advanced biofuels production by upgrading of pyrolysis bio-oil. *Wiley Interdiscip Rev Energy Environ* **6:** 1–18.

Findlay, J.P. (2016) *The future of the Canadian oil sands. Growth potential of a unique resource amidst regulation, egress, cost, and price uncertainty.* Oxford, UK: The Oxford Institute for Energy Studies.

Foulkes, J. M., Malone, K. J., Coker, V. S., Turner, N. J., and Lloyd, J. R. (2011) Engineering a biometallic whole cell catalyst for enantioselective deracemization reactions. *ACS Catal* **1:** 1589–1594.

Gómez, J. C. C., Moliner, R., and Maria Jesus Lázaro, M. J. (2016) Palladium-based catalysts as electrodes for direct methanol fuel cells: a last ten years review. *Catalysts* **130:** 130–150.

Greaves, M., and Xia, T. X. (2004) Downhole catalytic process for upgrading heavy oil: produced oil properties and composition. *J Can Petrol Technol* **43:** 25–30.

Hannan, M. A., Azidin, F. A., and Mohamed, A. (2014) Hybrid electric vehicles and their challenges: a review. *Renew Sust Energ Rev* **29:** 135–150.

Hansen, T. S., Barta, K., and Anastas, P. T. (2012) One-pot reduction of 5-hydroxymethylfurfural via hydrogen transfer from supercritical methanol. *Green Chem* **14:** 2457–2461.

Hart, A., Leeke, G., Greaves, M., and Wood, J. (2014) Downhole heavy crude oil upgrading using-CAPRI: effect of steam upon upgrading and coke formation. *Energy Fuels* **28:** 1811–1819.

He, C., Desai, S., Brown, G., and Bollepalli, S. (2005) PEM fuel cell catalysts: cost, performance, and durability. *Electrochem Soc Interface* **14:** 41–46.

Head, I., Jones, D. M., and Larter, S. R. (2003) Biological activity in the deep subsurface and the origin of heavy oil. *Nature* **426:** 344–352.

Healing, D. (2015) Asset sales signal patience short for disappointing THAI heavy oil technology. *The Calgary Herald* July 22nd 2015.

Hedrich, S., and Johnson, D. B. (2014) Remediation and selective recovery of metals from acidic mine waters using novel modular bioreactors. *Environ Sci Technol* **48:** 12206–12212.

Heraud, J.P., Kamp, A. and J.F. Argiller, J.F. (2011) *In-situ* upgrading of heavy oil and bitumen. In *Heavy Crude Oil: From Geology to Upgrading. An Overview.* Alain-Yves, H. (ed.). Paris: Editions Technip, pp. 388–402, ISBN: 978-2-7108-0890-9.

Hosseinkhani, B., Sobjerg, L. S., Rotaru, A. E., Emtiazi, G., Skrydstrup, T., and Meyer, R. L. (2012) Microbially supported synthesis of catalytically active bimetallic Pd-Au nanoparticles. *Biotechnol Bioeng* **109:** 45–52.

Huber, G. W., Iborra, S., and Corma, A. (2006) Synthesis of transportation fuels from biomass: chemistry, catalysis and engineering. *Chem Rev* **106:** 4044–4098.

Huc, Y. (2011) Geological origin of heavy oil. In *Heavy Crude Oil: From Geology to Upgrading. An Overview.* Alain-Yves, H. (ed.). Paris: Editions Technip, pp. 25–31, ISBN: 978-2-7108-0890-9.

Humphrey, J. L., Plana, D., Celorrio, V., Sadasivan, S., Tooze, R., Rodriguez, P., and Fermin, D. J. (2016) Electrochemical reduction of carbon dioxide at gold-palladium core–shell nanoparticles: product distribution versus shell thickness. *ChemCatChem* **8:** 952–960.

Kirubakaran, A., Jain, S., and Nema, R. K. (2009) A review on fuel cell technologies and power electronic interface. *Renew Sust Energ Rev* **13:** 2430–2440.

Kraytsberg, A., and Ein-Eli, Y. (2014) Review of advanced materials for proton exchange membrane fuel cells. *Energy Fuels* **28:** 7303–7330.

Kulkarni, N., and Maddapur, U. (2014) Biosynthesis of metal nanoparticles: a review. *J Nanotechnol* **2014:** 1–8.

Kunwar, B., Deilami, S. D., Macaskie, L. E., Wood, J., Biller, P., and Sharma, B. K. (2017) Nanoparticles of palladium supported on bacterial biomass for hydroprocessing bio-oil from continuous hydrothermal liquefaction (NTL) of algae. *Fuel* [In press].

Lee, H., Kim, C., Yu, M. J., Ko, C. H., Jeon, J.-K., Park, S. H., *et al.* (2016) Catalytic hydrodeoxygenation of bio-oil model compounds over Pt/HY catalyst. *Sci Rep* **6:** 28765. https://doi.org/10.1038/srep28765.

Liu, J., Zheng, Y., Hong, Z., Cai, K., Zhao, F., and Han, H. (2016) Microbial synthesis of highly dispersed PdAu alloy for enhanced electrocatalysis. *Sci Adv* **2:** e1600858. http://advances.sciencemag.org/content/2/9/e1600

Liu, D., Li, G., Yang, F., Wang, H., Han, J., Zhu, X., and Ge, Q. (2017) Competition and cooperation of hydrogenation and deoxygenation reactions during hydrodeoxygenation of phenol on Pt (111). *J Phys Chem C* **121:** 12249–12260.

Lloyd, J. R., Yong, P., and Macaskie, L. E. (1998) Enzymatic recovery of elemental palladium by using sulfate-reducing bacteria. *Appl Environ Microbiol* **64:** 4607–4609.

Luo, J., Arroyo-Ramirez, L., Gorte, R. J., Tzoulaki, D., and Vlachos, D. G. (2015) Hydrodeoxygenation of HMF Over Pt/C in a continuous flow reactor. *AIChE J* **61:** 590–597.

Luo, G., Chandler, D. S., Anjos, L. C. A., Eng, R. J., Jia, P., and Resende, F. L. P. (2017) Pyrolysis of whole wood chips and rods in a novel ablative reactor. *Fuel* **194:** 229–338.

Mabbett, A. N., Sanyahumbi, D., Yong, P., and Macaskie, L. E. (2006) Biorecovered precious metals from industrial wastes. Single step conversion of a mixed metal liquid waste to a bioinorganic catalyst with environmental applications. *Environ Sci Technol* **40:** 1015–1021.

Macaskie, L. E., Mikheenko, I. P., Yong, P., Deplanche, K., Murray, A. J., Paterson-Beedle, M., *et al.* (2010) Today's wastes, tomorrow's materials for environmental protection. *Hydrometallurgy* **104:** 483–487.

Mattia, D., Jones, M. D., O'Byrne, J. P., Griffiths, O. G., Owen, R. E., Sackville, E., *et al.* (2015) Towards carbon-neutral CO_2 conversion to hydrocarbons. *Chemsuschem* **8:** 4064–4072.

Mehta, V., and Cooper, J. S. (2003) Review and analysis of PEM fuel cell design and manufacturing. *J Power Sources* **114:** 32–53.

Meng, H., Zeng, D., and Xie, F. (2015) Recent development of Pd-based electrocatalysts for proton exchange membrane fuel cells. *Catalysts* **5:** 1221–1274.

Montaño, M. D., Gregory, A. F., Lowry, V., von der Kammer, F., Blue, D. J., and Ranvlle, J. F. (2014) Current status and future direction for examining engineered nanoparticles in natural systems. *Environ Chem* **11:** 351–366.

Murray, A. J., Omajali, J. B., Del Mastio, Y., Hart, A., Wood, J., and Macaskie, L. E. (2015) Potential for conversion of waste platinum group metals from road dust into biocatalysts for cracking heavy oil. *Adv Mat Res* **1130:** 623–626.

Murray, A. J., Zhu, J., Wood, J., and Macaskie, L. E. (2017a) A novel biorefinery: biorecovery of precious metals from spent automotive catalyst leachates into new catalysts effective in metal reduction and in the hydrogenation of 2-pentyne. *Mins Eng* [In press].

Murray, A. J., Roussel, J., Rolley, J., Johnson, D. B., and Macaskie, L. E. (2017b) Biosynthesis of zinc sulfide quantum dots using waste off-gas from metal bioremediation process. *RSC Adv* **7:** 21484–21491.

Nam, H., Kim, C., Capareda, S. C., and Adhikan, S. (2017) Catalytic upgrading of fractionated microalgae-bio-oil (*Nannochloropsis oculate*) using a noble metal (Pd/C) catalyst. *Algal Res* **24:** 188–198.

Nancucheo, I., and Johnson, D. B. (2012) Selective removal of transition metals from acidic mine waters by novel consortia of acidophilic sulfidogenic bacteria. *Microbial Biotechnol* **5:** 34–44.

Nduagu, E., Sow, A., Umeozor, E., and Millington, D. (2017) *Economic potentials and efficiencies of oil sands operations: processes and technologies.* Calgary, Alberta, Canada: Canadian Energy Research Institute.

Negin, C., Ali, S., and Xie, Q. (2016) Applications of nanotechnology for enhancing oil recovery: a review. *Petroleum* **2:** 324–333.

Nishimura, S., Ikeda, N., and Ebitani, K. (2014) Selective hydrogenation of biomass-derived 5-hydroxymethylfurfural (HMF) to 2, 5-dimethylfuran (DMF) under atmospheric hydrogen pressure over carbon supported PdAu bimetallic catalyst. *Catal Today* **232:** 89–98.

Ogi, T., Honda, R., Tamaoki, K., Saitoh, N., and Konishi, Y. (2011) Direct room temperature synthesis of a highly dispersed palladium nanoparticle catalyst and its electrical properties in a fuel cell. *Powder Technol* **205:** 143–148.

Omajali, J.B. (2015) Novel bionanocatalysts for green chemistry applications. PhD Thesis. Birmingham, UK: University of Birmingham.

Omajali, J. B., Hart, A., Walker, M., Wood, J., and Macaskie, L. E. (2017) In situ catalytic upgrading of heavy oil using dispersed bio-nanoparticles supported on Gram-positive and Gram-negative bacteria. Appl Catal B Environ 203: 807–819.

Orozco, R. L., Redwood, M. D., Yong, P., Caldelari, I., Sargent, F., and Macaskie, L. E. (2010) Towards an integrated system for bio-energy: hydrogen production by Escherichia coli and use of palladium-coated waste cells for electricity generation in a fuel cell. Biotechnol Lett 32: 1837–1845.

Orozco, R. L. (2011) hydrogen production from Biomass by Integrating Thermochemical and Biological Processes. PhD Thesis. Birmingham, UK: University of Birmingham.

Owen, R. E., Pawel Plucinski, P., Mattia, D., Laura Torrente-Murcianoc, L., Valeska, P., Ting, V. P., and Jones, M. D. (2016) Effect of support of Co-Na-Mo catalysts on the direct conversion of CO2 to hydrocarbons. J CO2 Utiliz 16: 97–103.

Pakhare, D., and Spivey, J. (2014) A review of dry (CO$_2$) reforming of methane over noble metal catalysts. Chem Soc Rev 43: 7813–7837.

Pandey, A., Bhaskar, T., Stöcker, M., and Sukumaran, R. (eds) (2015) Thermochemical Conversion of Biomass. Amsterdam: Elsevier.

Prieto, G. (2017) Carbon dioxide hydrogenates into higher hydrocarbons and oxygenates: thermodynamic and kinetic bounds and progress with homogeneous and heterogeneous catalysts. Chemsuschem 10: 1956–1970.

van Putten, R. J., van der Waal, J. C., de Jong, E., Rasrendra, C. B., Heeres, H. J., and de Vries, J. G. (2013) Hydroxymethylfurfural, a versatile platform chemical made from renewable resources. Chem Rev 113: 1499–1597.

Rai, M., Maliszewska, I., Ingle, A., Gupta, I., and Yadav, A. (2015) Diversity of Microbes in Synthesis of Metal Nanoparticles. In Bio-Nanoparticles: Biosynthesis and Sustainable Biotechnology Applications. Singh, O. V. (ed). Hoboken, NJ: John Wiley & Sons, pp. 1–30.

Ran, J., Zhang, J., Yu, J., Jaroniec, M., and Qiao, S. Z. (2014) Cocatalysts based on noble metals required for reasonable activity in most semiconductor-based photocatalytic systems. Chem Soc Rev 43: 7787–7812.

Reith, F., Zammit, C. M., Shar, S. S., Etschmann, B., Bottrill, R., Southam, G., et al. (2016) Biological role in the transformation of platinum-group mineral grains. Nat Geosci 9: 294–298.

Rice, C. A., Urchaga, P., Pistono, A. O., McFerrin, B. W., McComb, B. T., and Hu, J. (2015) Platinum dissolution in fuel cell electrodes: enhanced degradation from surface area assessment in automotive stress tests. J Electrochem Soc 162: F1175–F1180.

Rinaldi, R., and Schuth, F. (2009) Design of solid acid catalyst for the conversion of biomass. Energy Environ Sci 2: 610–626.

Román-Leshkov, Y., Barrett, C. J., Liu, Z. Y., and Dumesic, J. A. (2007) Production of dimethylfuran for liquid fuels from biomass-derived carbohydrates. Nature 447: 982–985.

Schmidt, L. D., and Dauenhauer, P. J. (2007) Hybrid routes to biofuels. Nature 447: 914–915.

Shah, A., Fishwick, R., Wood, J., Leeke, G., Rigby, S., and Greaves, M. (2010) A review of novel techniques for heavy oil and bitumen extraction and upgrading. Energy Environ Sci 3: 700–714.

Si, Z., Zhang, X., Wang, C., Ma, L., and Ding, R. (2017) An overview on catalytic hydrodeoxygenation of pyrolysis oil and its model compounds. Catalysts 7: 169–191.

Singh, O. V. (ed) (2015) Bio-nanoparticles: Biosynthesis and Sustainable Biotechnological Implications. Hoboken: John Wiley and Sons.

Staffell, I., Brett, D. J. L., Brandon, N. J., and Hawkes, A. D. (2015) Domestic Microgeneration: Renewable and Distributed Energy Technologies. London, UK: Routledge, Taylor & Francis.

Stern, N. (2006) The Economics of Climate Change. Cambridge, UK: Cambridge University Press.

Sun, X., Zhang, Y., Chen, G., and Gai, Z. (2017) Application of nanoparticles in enhanced oil recovery: a critical review of recent progress. Energies 10: 245–378.

Tian, G., Daniel, R. and Xu, H. (2011) DMF – a new biofuel candidate, biofuel production: recent developments and prospects. In Biofuel Production- Recent Developments and Prospects. Dos Santos Bernardes, M.A. (ed.). InTech, pp. 487–520. ISBN: 978-953-307-478-8

Tong, X., Ma, Y., and Li, Y. (2011) Biomass into chemicals: conversion of sugars to furan derivatives by catalytic processes. Appl Catal A Gen 385: 1–13.

Valsami-Jones, E., and Lynch, I. (2015) How safe are nanomaterials? Science 350: 388–389.

Wasylenki, L. E., Anbar, A. D., Liermann, L. J., Mathur, R., Gordon, G. W., and Brantley, S. L. (2007) Isotope fractionation during microbial metal uptake measured by MC-ICP-MS. J Analyt Atom Spectrom 22: 905–910.

Watson, M. J. (2014) Platinum group metal catalysed hydrodexoygenation of model bio-oil compounds. Johnson Matthey Technol Rev 58: 1656–1661.

Williams, A.R. (2015) Biogenic precious metal-based nanocatalysts for enhanced oxygen reduction. PhD Thesis. Birmingham, UK: University of Birmingham

Wu, X., Zhao, F., Rahunen, N., Varcoe, J. R., Avignone-Rossa, C., Thumser, A. E., and Slade, R. C. T. (2011) A role for microbial palladium nanoparticles in extracellular electron transfer. Ang Chemie 123: 447–450.

Yamaguchi, T., Tsuruda, Y., Furukawa, T., Yoshimura, E., and Suzuki, M. (2016) Synthesis of CdSe quantum dots using Fusarium oxysporum. Materials 9: 855–867.

Yang, Z., Lu, L., Berard, V. F., He, Q., Kiely, C. J., Berger, B. W., and McIntosh, S. (2015) Biomanufacturing of CdS quantum dots. Green Chem 17: 3775–3782.

Yong, P., Paterson-Beedle, M., Mikheenko, I. P., and Macaskie, L. E. (2007) From bio-mineralisation to fuel cells: biomanufacture of Pt and Pd nanocrystals for fuel cell electrode catalyst. Biotechnol Lett 29: 539–544.

Yong, P., Mikheenko, I. P., Deplanche, K., Redwood, M. D., and Macaskie, L. E. (2010) Biorefining of precious metals from wastes: an answer to manufacturing of cheap nanocatalysts for fuel cells and power generation via an integrated biorefinery? Biotechnol Lett 32: 1821–1828.

Yong, P., Liu, W., Zhang, Z., Johns, M. L., and Macaskie, L. E. (2015) One step bioconversion of waste precious metals into *Serratia* biofilm-immobilised catalyst for Cr(VI) reduction. *Biotechnol Lett* **37**: 2181–2191.

Zhang, J., Yu, J. G., Zhang, M., Li, Q., and Gong, J. R. (2011) Visible light photocatalytic H_2-production activity of CuS/ZnS porous nanosheets based on photoinduced interfacial charge transfer. *Nano Lett* **11**: 4774–4779.

Zhorg, S., Daniel, R., Xu, H., Zhang, J., Turner, D., Wyszynski, M. L., and Richards, P. (2010) Combustion and emission of 2,5-dimethylfuran in a direct-injection spark-ignition engine. *Energy Fuels* **24**: 2891–2899.

Zhu, F., Kim, J., Tsao, K.-C., Zhang, J., and Yang, H. (2015) Recent development in the preparation of nanoparticles as fuel cell catalysts. *Curr Opin Chem Eng* **8**: 89–97.

Zhu, J., Wood, J., Deplanche, K., Mikheenko, I. P., and Macaskie, L. E. (2016) Selective hydrogenation using palladium bioinorganic catalyst. *Appl Catal B Environ* **199**: 108–122.

Zu, Y. H., Yang, P. P., Wang, J. J., Liu, X. H., Ren, J. W., Lu, G. Z., and Wang, Y. Q. (2014) Efficient production of the liquid fuel 2, 5-dimethylfuran from 5-hydroxymethylfurfural over Ru/Co_3O_4 catalyst. *Appl Catal B Environ* **146**: 244–248.

6

Dynamics of mono-and dual-species biofilm formation and interactions between *Staphylococcus aureus* and Gram-negative bacteria

Jitka Makovcova,[1] Vladimir Babak,[1] Pavel Kulich,[2]
Josef Masek,[3] Michal Slany[1] and Lenka Cincarova[1,*]
[1]*Departments of Food and Feed Safety,* [2]*Chemistry and
Toxicology,* [3]*Pharmacology and Immunotherapy,
Veterinary Research Institute, Brno, Czech Republic.*

Summary

Microorganisms are not commonly found in the planktonic state but predominantly form dual- and multispecies biofilms in almost all natural environments. Bacteria in multispecies biofilms cooperate, compete or have neutral interactions according to the involved species. Here, the development of mono- and dual-species biofilms formed by *Staphylococcus aureus* and other foodborne pathogens such as *Salmonella enterica* subsp. *enterica* serovar *Enteritidis*, potentially pathogenic *Raoultella planticola* and non-pathogenic *Escherichia coli* over the course of 24, 48 and 72 h was studied. Biofilm formation was evaluated by the crystal violet assay (CV), enumeration of colony-forming units (CFU cm^{-2}) and visualization using confocal laser scanning microscopy (CLSM) and scanning electron microscopy (SEM). In general, Gram-negative bacterial species and *S. aureus* interacted in a competitive manner. The tested Gram-negative bacteria grew better in mixed dual-species biofilms than in their mono-species biofilms as determined using the CV assay, CFU ml^{-2} enumeration, and CLSM and SEM visualization. In contrast, the growth of *S. aureus* biofilms was reduced when cultured in dual-species biofilms. CLSM images revealed grape-like clusters of *S. aureus* and monolayers of Gram-negative bacteria in both mono- and dual-species biofilms. *S. aureus* clusters in dual-species biofilms were significantly smaller than clusters in *S. aureus* mono-species biofilms.

*For correspondence. E-mail cincarova@vri.cz

Funding information
European Cooperation in Science and Technology ('COST FA1202', 'COST LD14015').

Introduction

Food-processing environments provide a variety of conditions, which can favour the formation of biofilms, for example the presence of moisture, nutrients and inocula of microorganisms from the raw materials (Bower *et al.*, 1996). Bacterial colonization of food-processing equipment is a source of damage to metal surfaces (pitting and corrosion) and breakdown of plastics (Mittelman, 1998). Biofilms are defined as microbial communities that are adherent to each other and/or to the surface embedded in self-produced extracellular polymeric substances (EPS) composed of polysaccharides, proteins, phospholipids, teichoic and even nucleic acids (Costerton *et al.*, 1995; Hall-Stoodley *et al.*, 2004; Sauer *et al.*, 2007). Microbial cells can adhere to food-contact surfaces within minutes (Hall-Stoodley *et al.*, 2004), and biofilms can form within hours or days (Schlisselberg and Yaron, 2013). Although the majority of bacteria in the food production environment are non-pathogenic (Bagge-Ravn *et al.*, 2003; Schirmer *et al.*, 2013), these bacteria may be involved in reducing the quality of foods and importantly may facilitate colonization and survival of pathogenic bacteria (Nadell *et al.*, 2009; Shi and Zhu, 2009; Van Houdt and Michiels, 2010).

Due to their resistance to disinfectants and sanitizers, biofilms formed by pathogenic microorganisms in food environments can be difficult to completely eliminate from food-processing facilities. For that reason, procedures for the elimination of biofilms must be optimized, usually on an individual basis for different food-processing factories. These procedures are based on a combination of physical factors, chemical products and user conditions (Shi and Zhu, 2009; Jahid and Ha, 2012; Srey *et al.*, 2013).

The dynamic process of biofilm formation is predominantly characterized by initial reversible attachment of planktonic cells, cell aggregation and colonization of surfaces, biofilm maturation and detachment of cells from the biofilm into a planktonic state (Poulsen, 1999; Costerton *et al.*, 2005). Biofilm formation is a general strategy by which microorganisms survive in changing or hostile environments, such as when bacteria are challenged with a limited availability of nutrients, the presence of disinfectants or antibiotics and desiccation or temperature changes (Hall-Stoodley et al., 2004; Bridier et al., 2011). Biofilms can be formed by single, dual and/or multiple species of microorganisms and may constitute a single layer or three-dimensional structures. Mature biofilms represent a highly organized ecosystem with dispersed water channels which ensure the exchange of nutrients, metabolites and waste products (Sauer et al., 2007). The close proximity and complex interactions of species within biofilms underlie both synergistic and antagonistic behaviours (Elias and Banin, 2012).

Polymicrobial growth brings with it interspecies interactions that involve communication, typically via quorum sensing, and metabolic cooperation. The interactions within mixed-species biofilms are suggested to be of a cooperative (synergistic), competitive (antagonistic) or neutral nature based on the genetic background of the involved species (Giaouris et al., 2013). Synergistic interactions within biofilms are based on promotion of biofilm formation by co-aggregation, or metabolic cooperation (one species utilizes a metabolite produced by a neighbouring species), and can also increase resistance to antibiotics or host immune responses compared to mono-species biofilms. Antagonistic interactions are based on competition over nutrients and growth inhibition (Harriott and Noverr, 2009; Schwering et al., 2013). Thus, co-residence of diverse bacteria in biofilms can lead to an increase or decrease in biomass production (Schwering et al., 2013; Ren et al., 2014).

Staphylococcus aureus and Salmonella enterica are two of the most important, globally spread, foodborne pathogens. Staphylococcus aureus is a ubiquitous bacterial species commonly found on the skin and hair, as well as in the noses and throats of people and animals. It is the causative agent of a wide spectrum of human infections (Otto, 2013) and is also often responsible for foodborne intoxications through the production of heat-stable enterotoxins in a variety of food products (Hennekinne et al., 2012). Salmonella enterica is one of the most significant enteric foodborne bacterial pathogens and is classified into more than 2500 serovars of which the serovars Typhimurium and Enteritidis are the most prevalent. Salmonella serovars are responsible for human diseases ranging from gastroenteritis to systemic infections (Ruby et al., 2012; Foley et al., 2013). Escherichia coli is primarily a commensal species which constitutes part of the physiological microflora of the colon and distal ileum. However, E. coli also includes important foodborne pathogenic strains causing intestinal or extraintestinal infections (Kaper et al., 2004). Raoultella planticola (formerly Klebsiella) is generally considered to be an environmental bacterium found in soil and water

(Drancourt et al., 2001) and rarely causes clinical infections. However, this organism was the reported pathogen in several cases of serious infection (Olson et al., 2012; Koukoulaki et al., 2014). R. planticola is an important histamine-producing bacterium in fish, which causes foodborne intoxication due to histamine fish poisoning (Taylor, 1986; Lehane and Olley, 2000).

The above-mentioned bacterial genera and species are able to form biofilms on different surfaces commonly used in the food industry such as glass, plastic or metal (O'Toole et al., 2000; Oliveira et al., 2007). Besides stainless steel, plastic materials are still frequently used in the food industry for the construction of tanks, pipeworks, accessories and cutting surfaces (Pompermayer and Gaylarde, 2000). Various methods, both culture dependent and culture independent, have been developed to study the structure of multispecies biofilms and interactions between different species in various foods and food-contact surfaces (Giaouris et al., 2013; Schwering et al., 2013).

Although various studies have highlighted the importance of multispecies biofilms in foods and food environments, research on this topic is still in an early phase (Manuzon and Blaschek, 2007; Moons et al., 2009). The vast majority of studies have focused on either mixed Gram-negative or Gram-positive biofilms or mixed biofilms consisting of bacteria and fungi. There are only a few detailed studies devoted to dual-species biofilms composed of Gram-positive bacteria, specifically S. aureus and Gram-negative bacteria (Peters et al., 2010; Giaouris et al., 2015). Hence, the aim of our work was to study mono- and dual-species biofilms of S. aureus and three different Gram-negative bacteria, S. enterica, E. coli and R. planticola, and to evaluate interactions between them over the course of 3 days using both quantitative assays (CFU, total biomass) and qualitative methods, namely confocal laser scanning microscopy (CLSM) and Scanning electron microscopy (SEM).

Results

Total biomass quantification

The total biomass of monoculture biofilms was compared to dual-species biofilms formed by S. aureus in the presence of Gram-negative bacteria after 24, 48 and 72 h of incubation at 25 °C (Fig. 1). The biofilm formation capacity of the tested S. aureus and Gram-negative bacteria was determined using the criteria of Štepanović and colleagues (Malone et al., 2009), in TSB at 25 °C. S. aureus and E. coli strains used in this study are moderate biofilm formers, while S. enterica strain is a weak biofilm former and the R. planticola strain forms biofilm so weakly that according to the Štepanović criteria, it should be classified as an isolate that does not form biofilm at

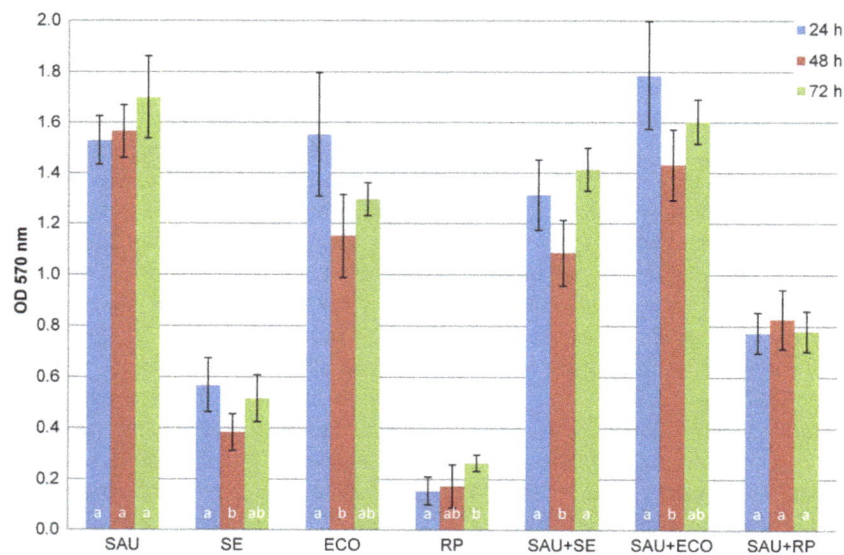

Compared mono- and	Incubation times		
dual-species biofilms	24 h	48 h	72 h
SAU vs SAU+SE	n.s.	$P<0.01$	$P<0.01$
SAU vs SAU+ECO	n.s.	n.s.	n.s.
SAU vs SAU+RP	$P<0.01$	$P<0.01$	$P<0.01$
SE vs SAU+SE	$P<0.01$	$P<0.01$	$P<0.01$
ECO vs SAU+ECO	n.s.	$P<0.01$	$P<0.01$
RP vs SAU+RP	$P<0.01$	$P<0.01$	$P<0.01$

Fig. 1. Crystal violet quantification of mono- and dual-species biofilms of *Staphylococcus aureus* and Gram-negative bacteria. The columns represent mean values of OD 570 nm; the vertical bars denote the 95% confidence intervals of these means. Letters above the x-axis denote statistically significant differences ($P < 0.05$ at least) among incubation times (columns sharing the same letter are not significantly different from each other; columns that have no letter in common are significantly different from each other). The table embedded within the figure shows the significance of differences in OD 570 nm between mono- and dual-species biofilms (ANOVA followed by *post hoc* Bonferroni multiple comparison tests).

all. Mean OD 570 nm values of dual-species biofilms formed by *S. aureus* and Gram-negative bacteria were statistically significantly higher, compared to mono-species biofilms of Gram-negative bacteria in all three incubation periods ($P < 0.01$; ANOVA, Bonferroni tests), except for *S. aureus* co-cultured with *E. coli* compared to mono-species biofilms after the 24 h incubation time ($P > 0.05$). In contrast, when *S. aureus* was co-cultured with *S. enterica* or *R. planticola*, mean OD 570 nm values were statistically significantly lower compared to monoculture of *S. aureus* biofilms in all incubation times, except for a higher mean OD 570 nm value for co-culture of *S. aureus* with *S. enterica* after 24 h of incubation ($P > 0.05$).

Quantification of viable cells in biofilms

Colony-forming unit numbers of *S. aureus* and Gram-negative bacterial strains in single- and dual-species biofilms were enumerated using a plating method after 24, 48 and 72 h of incubation at 25 °C (Fig. 2). *S. aureus*,

S. enterica, *E. coli* and *R. planticola* were able to adhere and to form biofilms both in single as well as in mixed cultures. The number of attached cells of *S. aureus*, *S. enterica*, *E. coli* or *R. planticola* on the polystyrene surface over the time of incubation assayed ranged between 2.40×10^8 and 1.28×10^9 for *S. aureus*, 3.13×10^7 and 2.36×10^8 for *S. enterica*, 1.42×10^8 and 3.74×10^8 for *E. coli* and 1.08×10^8 and 4.41×10^8 CFU cm^{-2} for *R. planticola* respectively. Fig. 2 shows that the CFU cm^{-2} values of all four microorganisms increased with the length of incubation; the slope of covariate (incubation time) was 0.014; that is, CFU cm^{-2} increased on average $2.16\times$ ($P < 0.01$; ANCOVA) in 24 h. Cell numbers of all tested Gram-negative bacteria were not significantly affected when co-cultured with *S. aureus* compared to mono-species biofilms ($P > 0.05$; ANCOVA, contrasts). The data showed that co-culture of *S. aureus* with Gram-negative bacteria in dual-species biofilms resulted in a significant reduction in the counts of *S. aureus* ($P < 0.01$; ANCOVA, contrasts). The reduction in the counts of

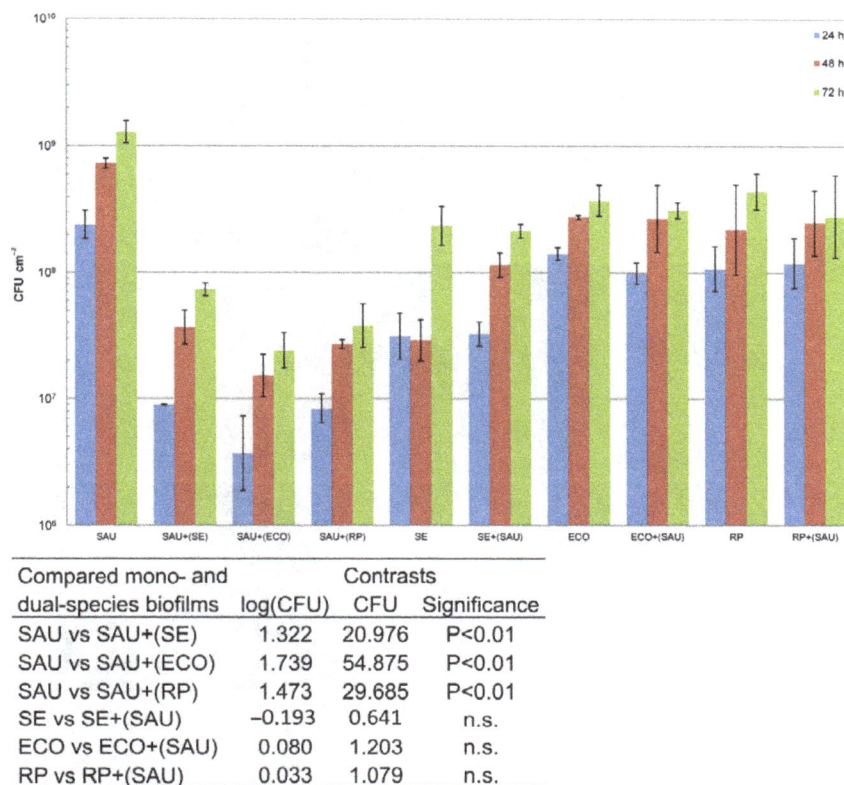

Compared mono- and dual-species biofilms	log(CFU)	Contrasts CFU	Significance
SAU vs SAU+(SE)	1.322	20.976	P<0.01
SAU vs SAU+(ECO)	1.739	54.875	P<0.01
SAU vs SAU+(RP)	1.473	29.685	P<0.01
SE vs SE+(SAU)	–0.193	0.641	n.s.
ECO vs ECO+(SAU)	0.080	1.203	n.s.
RP vs RP+(SAU)	0.033	1.079	n.s.

Fig. 2. Viable cell counts of mono- and dual-species biofilms.
The columns represent values of geometric means of CFU cm^{-2}; vertical bars correspond to geometric standard deviations. The y-axis is scaled logarithmically. Table 1b shows the significance of differences between mono- and dual-species biofilms (ANCOVA followed by a testing of contrasts). The slope of the covariate (incubation time) is 0.014, that is CFU cm^{-2} increased on average 2.16× ($P < 0.01$) in 24 h. SAU, *S. aureus*; SE, *S. enterica*; ECO, *E. coli*; RP, *R. planticola*; SAU+SE, co-culture of *S. aureus* and *S. enterica*, etc.; SAU+(SE), number of viable *S. aureus* cells when co-cultured with *S. enterica*; SE+(SAU). Number of viable *S. enterica* cells when co-cultured with *S. aureus*, etc.; n.s. non-significant.

S. aureus with *S. enterica* was 21x, with *E. coli* 55× and with *R. planticola* 30× respectively (Table in Fig. 2).

Confocal laser scanning microscopy

To confirm the results obtained with the CFU and crystal violet assay, biofilms were visualized using CLSM. Mono-species biofilms of *S. aureus*, *S. enterica*, *E. coli* and *R. planticola* and the influence of Gram-negative bacteria on the development of *S. aureus* cells in mixed-species biofilms were investigated after 24, 48 and 72 h at 25 C. Because different architectures of *S. aureus* mono-species biofilms and dual-species biofilms with Gram-negative bacteria in individual rows were observed, all z-stacks were composed by the transparent snapshot with the cross section and side panels in the positions indicated by the dashed lines. While the majority of fluorescent emission overlap between GFP and mCherry was eliminated by the appropriate setting of the emission filters, a small amount of fluorescent overlap was recorded as yellow colour. Figure 3 shows representative CLSM micrographs of both mono- and dual-species biofilm development after 24, 48 and 72 h.

Confocal laser scanning microscopy confirmed, in general, that Gram-negative bacteria had a suppressive effect on *S. aureus* at every stage of biofilm formation, in comparison with the mono-species counterpart. Gram-negative bacteria, both in mono-species biofilms or in mixed cultures with *S. aureus*, formed monolayers over time. Individual cells of *S. enterica* were evident after 24 h, and after 48 and 72 h of culture, sparsely developed biofilms were visible. Mono-species biofilms of *E. coli* and *R. planticola* were much denser already after 24 h compared to *S. enterica* biofilms. Monolayers were relatively homogenous, with occasional small holes; sometimes small clusters of bacteria were evident. In contrast, *S. aureus* formed three-dimensional structures, and formed biofilms were dense and included holes of different sizes and channels which thickened with the duration of incubation. In the dual-species biofilms, Gram-negative cells formed patchy monolayers on the bottom of the wells. Cells were seen to be more evenly

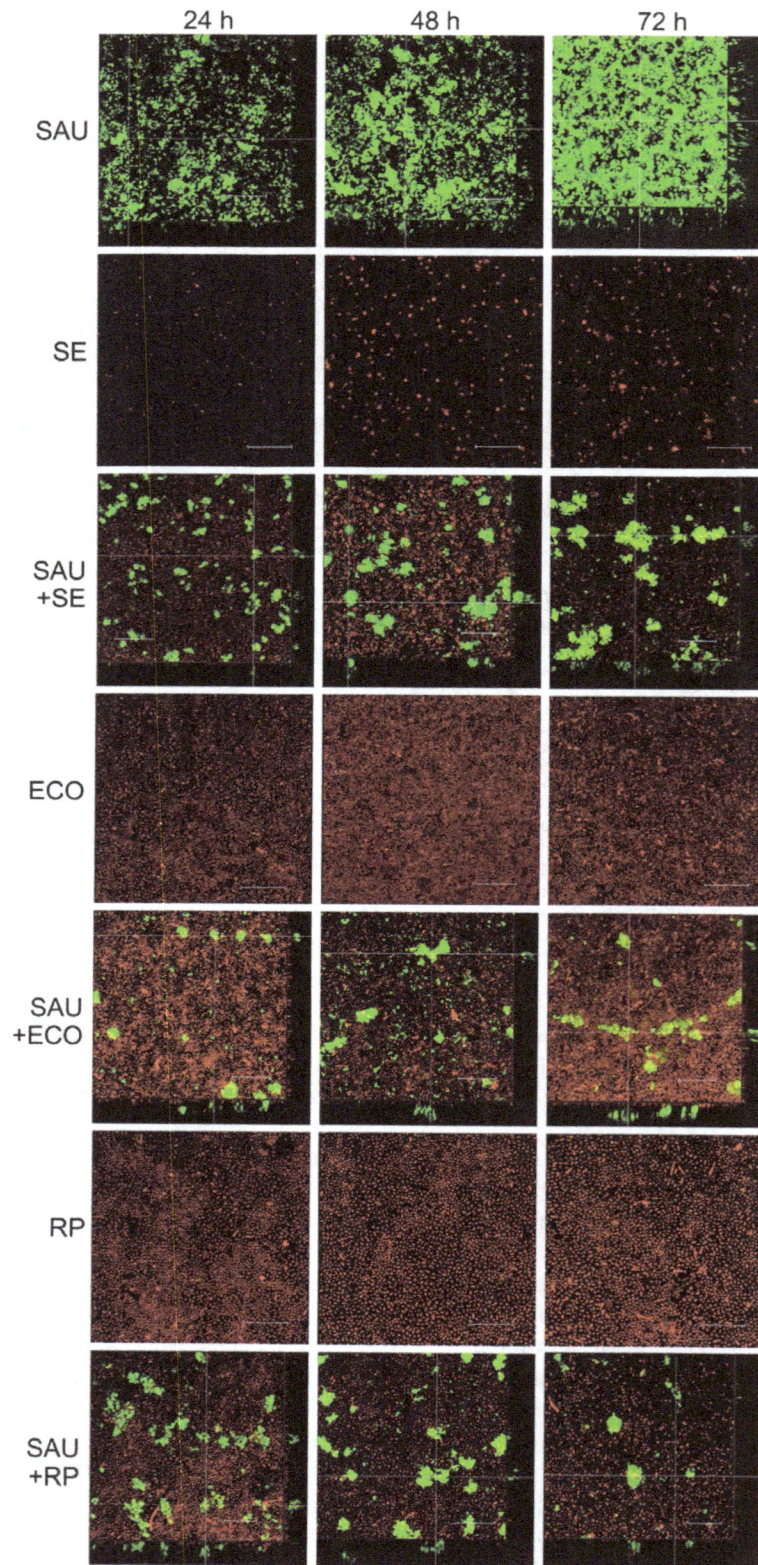

Fig. 3. Representative CLSM images of monocultures and dual-species biofilms formed by *Staphylococcus aureus* and Gram-negative bacteria. SAU, *S. aureus*; SE, *Salmonella enterica*; ECO, *E. coli*; RP, *R. planticola*; SAU+SE, *S. aureus* and *S. enterica* dual-species biofilms, etc. Scale bar represents 25 μm.

attached especially around *S. aureus* clusters, or some of them were found on the surface of grape-like clusters of *S. aureus* (see side panels) and then protruded into the space during incubation. When *S. aureus* was co-cultured with Gram-negative bacteria, clusters were not as robust as compared to mono-species biofilms. The images show that Gram-negative bacteria were attached to the bottom of the well and that *S. aureus* cells attached and formed grape-like structures on Gram-negative bacterial monolayers. The thickness of *S. aureus* biofilms increased both in monocultures and when co-cultured with Gram-negative bacteria over 24, 48 and 72 h of incubation. The thickness of *S. aureus* biofilms was approximately 15, 20 and 27 μm. In dual-species biofilms, *S. aureus* clusters were thinner than their corresponding mono-species variants. When *S. aureus* was co-cultured with *S. enterica*, the thickness of biofilms was 15 and 20 μm after 24 and 48 h, but then fell to 15 μm after 72 h. In dual-species biofilms formed with *E. coli*, *S. aureus* formed biofilms with thicknesses of 10, 13 and 13 μm. *S. aureus* clusters were 10, 15 and 15 μm thick, when co-cultured with *R. planticola*. Gram-negative bacteria formed mono-layered biofilms throughout the incubation time both in mono- and dual-species biofilms. The depth of these slices was not measured as they were smaller than 5 μm.

Confocal laser scanning microscopy images showed that the dual-species biofilms were not composed of both species mixed together in a typical co-aggregation structure but were rather characterized by separate spatial clusters of *S. aureus* and monolayers of Gram-negative bacteria. This confirmed our previous results, where we observed *S. aureus* to have high aggregation propensity (80% aggregation after 24 h) and Gram-negative bacteria to possess moderate aggregation propensity (40–60% aggregation after 24 h; data not shown). However, in mixed culture, the co-aggregation was very low for *S. aureus* and totally absent in Gram-negative bacteria.

Scanning electron microscopy

Scanning electron microscopy was used to examine the structure and interactions in static biofilms formed by *S. aureus* and Gram-negative bacteria after 24, 48 and 72 h of incubation at 25 °C. SEM showed that mixed biofilms consisted of *S. aureus* microcolonies scattered across the surface and Gram-negative bacteria attached to the surface in monolayers. Over the course of co-culture, *S. aureus* grape-like clusters gained in volume, while the number of Gram-negative bacterial cells increased, but remained in monolayers (Fig. 4).

There was evidence that Gram-negative bacteria surrounded *S. aureus* clusters and that some of them had been incorporated into clusters or had attached to the

top of them. Cells could be seen that were embedded in an amorphous matrix (arrow; Fig. 5).

A close-up, side view image of 24 h dual-species biofilms showed an attached *S. enterica* cell on the surface, to which *S. aureus* cell clusters clung. Amorphous matrix, possibly EPS (arrow), was evident on the surface of cocci.

Nascent and fully formed cell-to-cell connections, Gram-negative cells attached to the surface and to other cells using fibril-like structures (tip) and intercellular slime connecting the biofilms (arrow) were seen after 24 h (Fig. 4A, D and G). A few putative outer membrane vesicles (OMVs) were observed on the surfaces of *E. coli* and *R. planticola* cells. Most cells had a normal shape and a smooth cell surface. After 48 h of incubation, the number of fibril-like junctures and cell-to-cell connections between cells had multiplied (tip) and amorphous mass (arrow) had increased (Fig. 4B and E). Buds and putative membrane vesicles (MVs) formed by *S. aureus* and putative OMVs on the surface of *E. coli* and *R. planticolla* cells were evident (Fig. 4E and H). Damaged cells showed a rough and shrunken appearance, and deformation or cellular debris was evident, especially in co-culture of *S. aureus* and *E. coli* (Fig. 4E). Images of 72 h co-culture of *S. aureus* and *S. enterica* and *S. aureus* and *R. planticola* showed biofilms embedded in a large amount of possible EPS (arrow), (Fig. 4C and I). A close-up image of 72 h biofilm formed by *S. aureus* and *S. enterica* revealed an apparent EPS matrix surrounding cells. Damaged cells of *S. aureus* were rough with individual bumps and buds and were deformed on their surfaces. Also, some Gram-negative cells showed signs of damage. Detailed images revealed the first cell-to-cell connections and EPS, but also lysed cells and cell debris (Fig. 6).

Discussion

Biofilms are found in different environments and are usually not formed by single species, but rather by dual or multiple species. Cell-to-cell interactions influence the temporal and spatial organization of biofilm architecture and can be categorized as either cooperative or competitive (Elias and Banin, 2012; Rendueles and Ghigo, 2012).

Attention has been paid to bacterial foodborne pathogens and intra- and interspecies interactions between Gram-positive *Staphylococcus aureus* and the Gram-negative bacteria *Salmonella enterica*, *Raoultella planticola,* or *Escherichia coli*. Because *S. aureus* and *S. enterica* are known to be important foodborne pathogens, we considered it important to study their interspecies interactions. Furthermore, we considered it interesting to include potentially pathogenic *R. planticola*

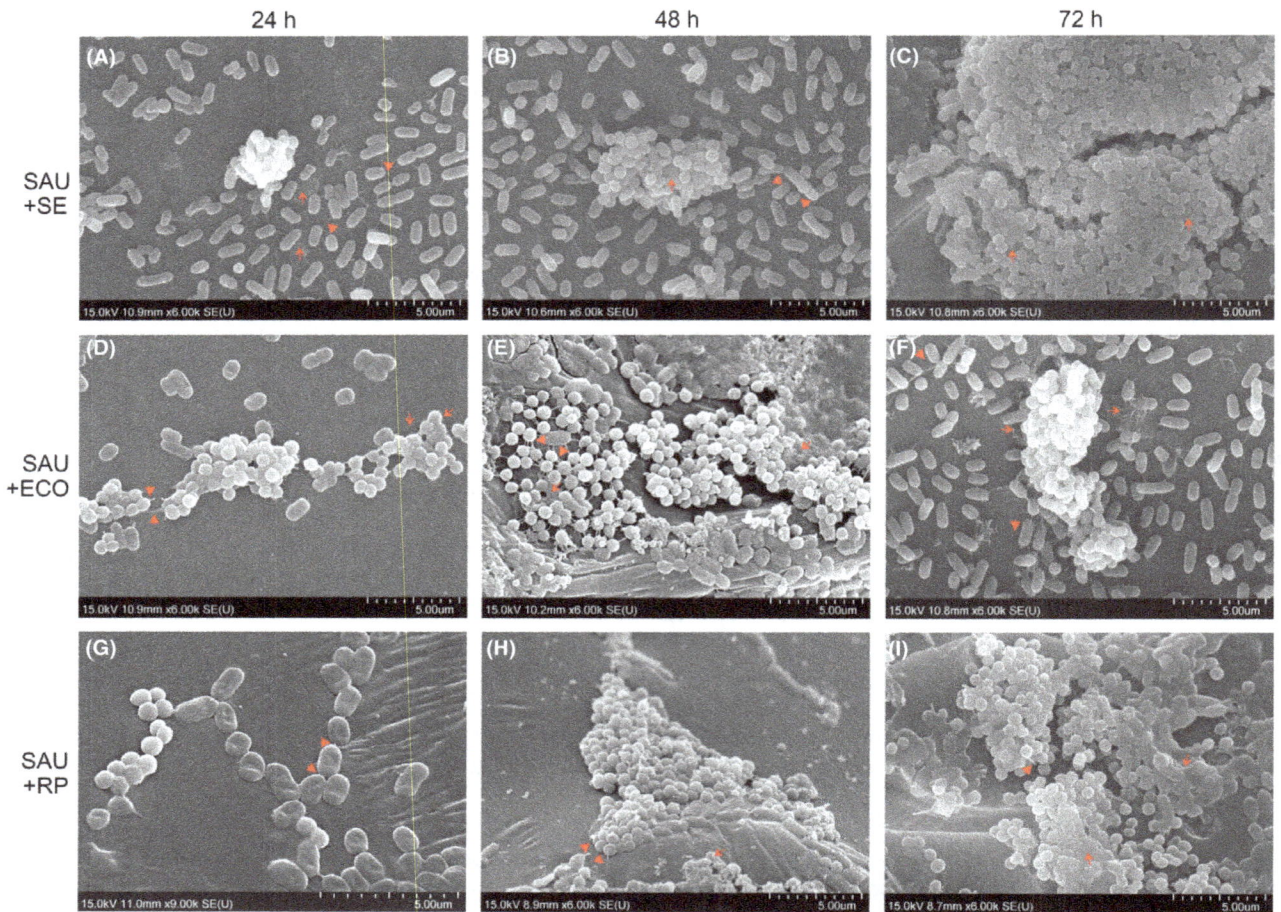

Fig. 4. Representative scanning electron microscopy images of dual-species biofilms formed by *Staphylococcus aureus* and Gram-negative bacteria.
SAU+SE *S. aureus* and *Salmonella enterica*, SAU+ECO *S. aureus* and *E. coli*, SAU+RP *S. aureus* and *R. planticola* dual-species biofilms.
Arrows: amorphous extracellular matrix; tips: adhesive fibres.

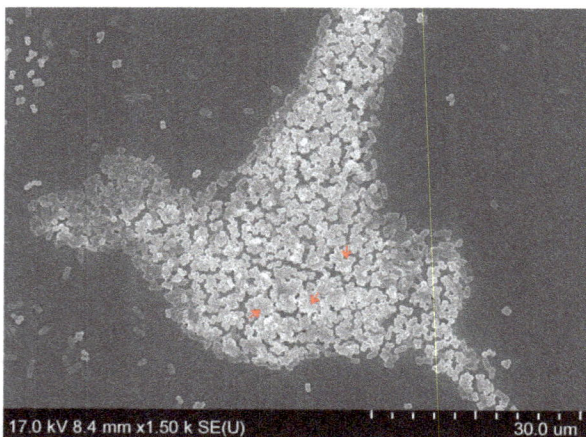

Fig. 5. Scanning electron microscopy image of a *Staphylococcus aureus* microcolony surrounded by *Salmonella enterica* cells, dual-species biofilm after 72 h. Arrows: amorphous extracellular matrix.

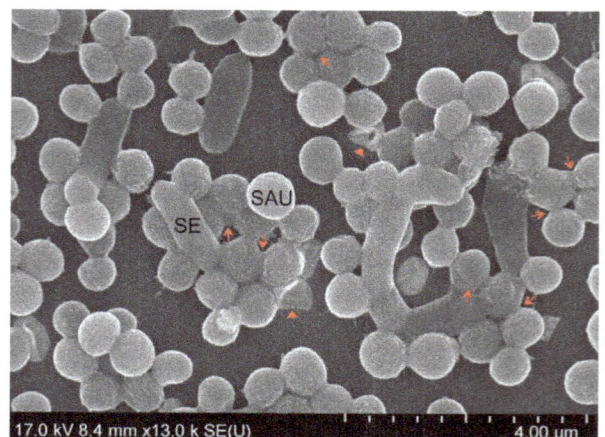

Fig. 6. Scanning electron microscopy close-up image of *Staphylococcus aureus* and *Salmonella enterica* dual-species biofilm after 72 h. Arrows: amorphous extracellular matrix; tips: adhesive fibres.

and non-pathogenic *E. coli*, both obtained from food-processing environments after sanitation, in our study.

The results measured in the crystal violet assay showed that *S. aureus*, *S. enterica* and *E. coli* were able to form mono-species biofilms after 24 h at 25 °C to varying degrees, while *R. planticola* is not a biofilm former. However, *R. planticola* is a species that produces capsules (Shaikh, 2011). Capsules generally are stained very poorly with reagents used in simple crystal violet staining (Breakwell *et al.*, 2009) and thus can be a source of inaccuracy – the total biofilm biomass could be underestimated when simple crystal violet staining is used for determination. Total biomass volume was decreased in dual-species biofilms (except for co-culture of *S. aureus* and *E. coli* after 24 h) compared to *S. aureus* mono-species biofilms over time. These results indicated a competitive relationship between bacteria, an idea which is supported by the obtained CFU data. In mono- and dual-species biofilms, there was a gradual increase in CFU values over time. In dual-species biofilms, the CFU values of Gram-negative bacteria remained almost unchanged, while *S. aureus* CFU values significantly decreased compared to single biofilms of the same species (Fig. 2). This clearly indicates an inhibitory effect of Gram-negative bacteria on *S. aureus* cells and suggests that their overall interactions are competitive rather than cooperative.

There are several studies regarding formation of dual biofilms by *S. aureus* and Gram-negative bacteria. A competitive relationship was observed by Millezi *et al.*, 2012; between *S. aureus* and *E. coli*, in which the number of viable *S. aureus* cells in biofilms was diminished by the presence of *E. coli* (Millezi *et al.*, 2012). Likewise, Pompermayer and Gaylarde, 2000; investigated the adherence of *S. aureus* and *E. coli* and concluded that there is a competition between bacteria with *E. coli* being favoured in dual-species culture (Pompermayer and Gaylarde, 2000). The authors suggested that the adherence of *E. coli* could be greater because of the shorter generation time of this bacterium, which could enable it to develop and maintain dominance. Similarly, this could be the reason for the inhibition of *S. aureus* in our case, because all the Gram-negative bacteria used in our experiments have shorter generation times than *S. aureus*.

It is well recognized that the exopolysaccharide capsule is one of the key bacterial components for biofilm formation (Wang *et al.*, 2015). Capsules are produced by many microorganisms, including *Escherichia, Salmonella, Klebsiella* or *Staphylococcus* strains, and can be either adhesive or anti-adhesive (Hassan and Frank, 2004; Coldren, 2009). Valle *et al.* (2006) demonstrated that *E. coli* expressing group II capsules released a soluble polysaccharide that induces physiochemical surface alterations, which prevent adhesion and biofilm formation by a wide range of both Gram-negative and Gram-positive bacteria, including *S. aureus*. Similarly, an *E. coli* biofilm-associated anti-adhesion polysaccharide which reduces susceptibility to invasion and resulted in rapid exclusion of *S. aureus* from mixed *E. coli* and *S. aureus* biofilms was identified (Rendueles *et al.*, 2011).

Only a few publications have focused on *S. aureus* and *S. enterica* dual-species biofilm formation. In dual-species biofilms formed by *S. aureus* and *S. enterica* serovar Typhimurium in a fermentor, *S. aureus* dominated (99%) over *S.* Typhimurium (Knowles *et al.*, 2005). Given that this experiment is vastly different from the ones described here, the results are not comparable. The development of dual-species biofilms formed by *S. aureus* and *S. enterica* serotype Enteritidis was described by Zhang *et al.* (2014). Mixed biofilms were quantified using colony-forming units and crystal violet assays and *S. aureus* predominated. Biofilm formation was performed at 37 °C, in contrast to our study, in which experiments were performed at 25 °C. It was observed in other studies that biofilm formation of *S. aureus* as well as other species differs depending on the cultivation temperature (Hoštacká *et al.*, 2010; de Souza *et al.*, 2014; Pavlovsky *et al.*, 2015). Moreover, mono-species biofilms of individual species were not evaluated. Therefore, the relationship between bacteria cannot be assessed. Blana and co-authors (Blana *et al.*, 2015) found that *luxS*-positive *Salmonella enterica* serotype Typhimurium culture supernatant significantly increased the *S. aureus* single-cell lag time. We may conclude that the extended lag time of *S. aureus* gives an advantage to *Salmonella* in the covering of surfaces and the formation of biofilms. Unfortunately, to our knowledge, there has been no study regarding dual-species biofilm formation by *S. aureus* and *R. planticola* (previously *K. pneumoniae*).

Other microbes have also been described to show altered characteristics due to interspecies interactions and to exhibit different properties depending on whether they grow in mono- or dual-species biofilms. Varposhti *et al.*, 2014; observed cooperation between *Pseudomonas aeruginosa* and *Acinetobacter baumannii* and *Stenotrophomonas maltophilia*, pathogenic bacteria from the respiratory tract, when cultivated in dual-species biofilms (Varposhti *et al.*, 2014). Giaouris *et al.*, 2013; also observed beneficial cooperation: co-culture with *Listeria monocytogenes* within a dual-species biofilm community strongly increased the resistance of *Pseudomonas putida* to benzalkonium chloride (Giaouris *et al.*, 2013). Peters *et al.*, 2010; in contrast, observed cooperation between *S. aureus* and the fungal species *Candida albicans* based on physical interactions and differential regulation of specific virulence factors (Peters *et al.*, 2010).

They observed an enhanced pathogenesis of *S. aureus* mediated by the association of *S. aureus* with hyphal elements of *Candida albicans* that can penetrate through epithelial layers. Peters *et al.* also observed enhanced virulence of *S. aureus* during co-infection with *Candida albicans*: differential protein expression analysis revealed downregulation of the global transcriptional repressor of virulence factors, CodY, and, as a consequence of this, upregulated expression of *S. aureus* virulence factors.

The competition of microbes can be manifested by the changing of their local environment either directly or as a consequence of their secondary metabolism and physiological by-products. For example, *Lactobacilli* spp. produce lactic acid that lowers environmental pH and thus limits the growth of other species such as *Neisseria gonorrhoeae* (Graver and Wade, 2011). Bacteria also use low-molecular weight compounds (toxic metabolic by-products, bacteriocins or colicins). *Streptococcus pneumoniae* produces hydrogen peroxide as a toxic by-product of aerobic metabolism. In the human nasopharynx, it is an inhibitor of *Neisseria meningitidis* and *Moraxella catarrhalis* (Pericone *et al.*, 2000). Other mechanisms of competition in multispecies biofilms including contact-dependent growth inhibition and predation were characterized by Rendueles *et al.* (2012).

Confocal laser scanning microscopy and SEM were used to study the complexity and structural heterogeneity of mono- and dual-species biofilms. CLSM is a widely used tool for the observation of biofilms because it allows one to obtain a three-dimensional image of the structure of a biofilm and to monitor its development over time without harmful effects on its growth (Canette and Briandet, 2014). CLSM images (Fig. 3) were in agreement with the CFU data (Fig. 2): whereas Gram-negative bacteria are not negatively affected by the presence of *S. aureus*, *S. aureus* is inhibited when co-cultured with Gram-negative bacteria. In CLSM scanning, the projection of captured images on the *z*-axis (data not shown) showed that Gram-negative bacteria were attached to the bottom of the well and that *S. aureus* cells adhered and formed three-dimensional, grape-like structures on Gram-negative bacterial monolayers. CLSM images showed that the dual-species biofilms were not composed of both species mixed together in a typical co-aggregation structure that is typical for mixed biofilms composed of species cooperating or interacting synergistically (Elias and Banin, 2012). Rather, in this case, the observed inability of the species to co-aggregate together with the total biomass data and CFU (Figs 1 and 2) show that *S. aureus* and Gram-negative bacteria compete in mixed-species biofilms.

Using CLSM, different biofilm architectures for mono- and dual-species biofilms composed by *Staphylococcus piscifermentans* and *Salmonella* Agona were observed (Habimana *et al.*, 2010). Habimana *et al.*, 2010, described biofilms composed by *Staphylococcus piscifermentans* and *Salmonella* Agona. While *Staphylococcus* mono-species biofilms were defined as compact, with the presence of holes in the matrix, *S. Agona* mono-species biofilms were found to be composed of more channels. In mixed-species biofilms, *S. Agona* cells were found to partially cover *Staphylococcus* microcolony niches. This architecture is similar to our results, but in our study, the biofilms formed by the *Salmonella* strain were not so dense and compact and *Staphylococcus* did not promote *Salmonella* biofilm formation in mixed-species biofilms. Unfortunately, we could not find any publications that would describe dual-species biofilm formation specifically by *S. aureus* and the Gram-negative bacterial species used in our study using CLSM.

Data obtained using CLSM, CV staining and CFU enumeration were all in conformity with each other, with some exceptions. Differences could arise due to the different principles of these methods. CV staining is a more robust method used for quantification of total biomass (simple CV staining labels living and dead cells as well as the extracellular matrix). In our set-up, CLSM visualizes only cells containing the GFP protein regardless of whether the cells are dead or alive or VBNC (Viable but non-culturable). CLSM does not quantify the extracellular matrix. CV quantifies total biofilm biomass. CFU, in contrast, is a measurement that quantifies changes only in cells that are viable and culturable, and the accuracy heavily depends on the preparation of the sample. CLSM-based visualization of *S. aureus* clearly showed an increase in density and thickness over 24, 48 and 72 h, whereas the CV and CFU data revealed no pronounced increase in total biofilm biomass or in viable and culturable cells at the measured time points. One reason for this inaccuracy may lie in the fact that CFU quantifies only viable and culturable cells; however, a certain proportion of the cells in the biofilms are dead (Bayles, 2007) or non-cultivatable cells, and these are quantified by CLSM but not by CFU enumeration. For CLSM, it is necessary to use special culture plates, which differ slightly in cultivation area and material. This can affect the resulting biofilms that are formed, and therefore, it may not be accurate to compare individual results of different methods.

While CLSM revealed a significant increase in the total cell number of *S. aureus* at 24 h versus 48 h versus 72 h, CV staining showed a relatively unimportant increase in total biofilm quantity. The most significant increase in biofilm biomass would probably occur in the time interval 0–24 h. In this interval, the bacterial cells of the tested isolates mainly adhere to the surface, produce extracellular matrix and multiply. After approximately 24 h, cells that are already part of the biofilm tend to form biofilm mass and multiply less within the biofilm.

Thus, the overall biofilm biomass no longer substantially changes in this phase (this is strongly dependent on conditions – e.g. presence of stress factor).

In the case of *S. enterica*, CLSM and CFU data corresponded with each other. The changes in biofilm quantity revealed in the CV data were not statistically significant for all time points. This could be due to the fact that the tested *Salmonella enterica* isolate is a weak biofilm former and thus is less stable during CV staining. For *Escherichia coli*, CV quantification also did not reveal any significant changes and CFU and CLSM data were in agreement with this result. According to the classification of Stepanović *et al.*, 2004; *Raoultella planticolla* forms biofilm only very weakly. CV data showed no significant increase in quantity between 24 and 72 h. Enumeration of CFU revealed a significant increase in the number of viable cells between 24 and 72 h, an observation that was confirmed using CLSM.

Scanning electron microscopy permits visualization of detailed surface morphologies of microbial biofilms and their structures. SEM images confirmed the CLSM analysis, where Gram-negative bacteria formed monolayers over the surface and *S. aureus* created grape-like structures adhering to Gram-negative bacteria. With increasing incubation times, the number of bacterial cells increased, but damaged or dead cells also appeared. With increasing duration of incubation time, nascent and then established cell-to-cell connections and an increasing amount of amorphous matrix, probably EPS, were clearly visible. In the case of *S. enterica*, roughened cell surfaces, cellular deformation and the formation of depressions in some *S. aureus* cocci were indicative of cellular damage (Fig. 7). Putative membrane vesicles formed by *S. aureus* and *E. coli* or *R. planticola* outer membrane vesicles were observed. A variety of pathogenic Gram-negative and environmental bacteria secrete OMVs during growth. The production of OMVs has among other functions, a role in cell-to-cell communication (Kulp and Kuehn, 2010). OMVs associated with biofilm production have been studied most extensively for *P. aeruginosa*, where they comprise a half of the total lipopolysaccharide content and are associated with the entire biofilm matrix (Schooling and Beveridge, 2006); they have also been described to stimulate biofilm production in *Helicobacter pylori* (Yonezawa *et al.*, 2009). This result suggests that vesicles enable biofilms to form and that their presence in biofilms is not solely the result of their entrapment in the matrix. However, little is known about the MVs produced by Gram-positive bacteria. Lee *et al.* (2009) first demonstrated that *S. aureus* release MVs to the extracellular environment during *in vitro* culture and virulence-associated proteins were identified in *S. aureus* MVs *in vivo* and *in vitro* (Gurung *et al.*, 2011; Thay *et al.*, 2013). Unfortunately, again, we found no

studies of dual-species biofilms formed by *S. aureus* and *S. enterica*, *E. coli* or *R. planticola* supported by SEM analysis.

We believe that a better understanding of the interactions in dual- and multispecies biofilms formed by both foodborne pathogenic and non-pathogenic bacteria occurring in the food industry can lead to important new insights that will facilitate the control of biofilm formation in food-processing environments and thus to an improvement in food safety.

Experimental procedures

Bacterial strains and growth conditions

In this study, *Staphylococcus aureus* subsp. *aureus* RN4220 (Novick, 1990), a kind gift from Julien Deschamps, French National Institute for Agriculture Research, was used. *Salmonella enterica* subsp. *enterica* serovar Enteritidis 147 originated from egg content (Methner *et al.*, 1995); *Escherichia coli* 1685 was isolated from contact surfaces in a meat-processing plant, and *Raoultella planticola* 191 was isolated from contact surfaces in a dairy. Samples originating from food-contact surfaces from meat and dairy-processing plants were collected within 2 h after sanitation by swabbing. Swabs were washed in PBS and subsequently plated onto MacConkey agar. Single colonies were subcultured and identified using ENTEROtest 24 (Erba Lachema s. r. o., Brno, Czech Republic). The identity of all strains was proven by sequencing analysis of two different loci of the *16S rRNA* gene (Harmsen *et al.*, 2003; Slany *et al.*, 2007). Blastn comparison with a publicly available database (http://www.ezbiocloud.net) revealed homology of higher than 99.5% (all analysed strains). The strains were stored in trypticase soy broth (TSB; Oxoid, England) supplemented with 25% glycerol (Lach-ner, s. r. o, Neratovice, Czech Republic) at −80 °C. For strain reactivation

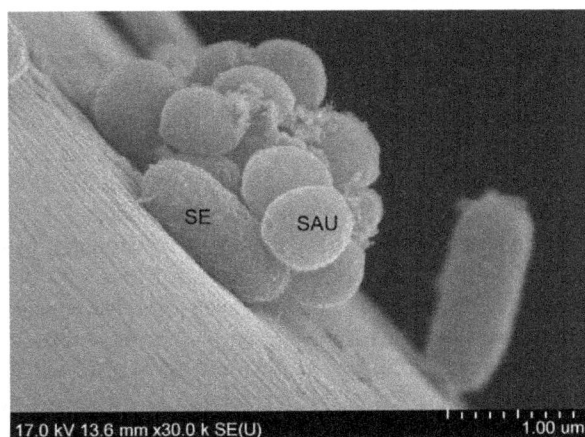

Fig. 7. Scanning electron microscopy detail image of adhesion of *Staphylococcus aureus* and *Salmonella enterica* after 24 h. Arrows: amorphous extracellular matrix.

and use, an aliquot of freezing culture medium was subcultured on trypticase soy agar (TSA; Oxoid., Hampshire, England) and incubated at 37 °C for 24 h. A single colony from a plate was inoculated into 5 ml of TSB and incubated statically for 17 h at 37 °C.

Preparation of mono- and dual-species cultures

Bacterial strains were grown in TSB at 37 °C to the exponential phase of growth (approximately 10^9 CFU ml^{-1}). Bacterial cells were then pelleted at $4000\times g$ for 20 min, and the cell pellet was resuspended in 5 ml of fresh TSB. Optical densities (ODs) of bacterial suspensions were measured using a spectrophotometer (BioPhotometer, Eppendorf, Hamburg, Germany) and adjusted to an absorbance of 1.0 at 600 nm (approximately 10^8 CFU ml^{-1}). Individual bacterial strains were diluted 1:100 in fresh TSB. An equal volume of the two 1:100 diluted mono-species cultures was combined to make the dual-species culture. The cell numbers of individual strains were confirmed using a direct plating method on TSA plates (the final amount of diluted culture was 5×10^6).

Total biomass quantification

Formation of mono- and dual-species biofilms was quantified using the crystal violet (CV) staining method (Stepanović et al., 2007) with slight modifications. Mono- and dual-species cultures, 1000 µl per well, were aseptically dispensed in six replicates to sterile polystyrene, flat-bottom 48-well tissue culture plates (Jet Biofil Jet Bio-Filtration, Guangzhou, China). As a negative control, 1000 µl of TSB only was used. Plates were incubated statically for 24, 48 and 72 h at 25 °C. The culture medium was refreshed every 24 h. After the incubation period, planktonic cells were aspirated carefully and the wells were washed three times with phosphate-buffered saline (PBS; pH 7.2). The plates were inverted and allowed to air-dry at room temperature. The remaining attached bacteria were fixed with 1000 µl of 99% methanol per well, and after 15 min, plates were emptied and air-dried. Subsequently, the biofilms were stained using 1000 µl of 0.25% crystal violet solution per well, followed by 20 min incubation at room temperature. Excess unbound dye was removed by thoroughly washing the plates with distilled water. Finally, after the plates were air-dried, stained biofilms were resolubilized with 750 µl of an ethanol:acetone mixture (80:20) per well and incubated for 20 min at room temperature. The OD of each well was measured at 570 nm, using a microtitre plate reader (spectrophotometer Synergy H1 HYBRID Reader; BioTek, Swindon, UK), and biofilm mass was expressed as OD 570 nm values. The assay was repeated three times.

Quantification of viable cells in biofilm

The quantification of viable cells in single- or dual-species biofilms was determined in four-well polystyrene-bottomed plates with a microwell size of 15 mm (In Vitro Scientific, USA) using the plate counting technique. For removal of cells from the bottom of the well, sterile pipette tip scraping and repeated pipetting of adhered cells was used. The number of viable cells was determined using TSA in mono-species biofilms. In dual-species biofilms, MacConkey agar for S. enterica, E. coli and R. planticola and Columbia blood agar with colistin sulfate (10 mg l^{-1}) and nalidixic acid (10 mg l^{-1}) for S. aureus were used. The dishes were incubated at 37 °C for 24 h. The values were expressed as CFU cm^{-2}. Assays were performed in biological and technical triplicates.

Confocal laser scanning microscopy

To observe the development of mono-species biofilms and interactions of bacteria in dual-species biofilms, fluorescent protein-labelled bacteria were constructed. Briefly, the S. aureus subsp. aureus RN4220 (a kind gift from Julien Deschamps, French National Institute for Agriculture Research) was transformated by expression plasmid pFPV25.1 [gift from Raphael Valdivia (Valdivia and Falkow, 1996); Addgene plasmid # 20668] producing a reporter protein GFP. Tested Gram-negative bacteria were prepared by cloning of a mCherry coding sequence from pLV-mCherry [gift from Pantelis Tsoulfas; Addgene plasmid # 36084, (Dull et al., 1998)] into plasmid pFPV25.1. This plasmid was transformed into Gram-negative bacteria according to a previously published procedure (Gonzales et al., 2013). Commercially available sterilized µ-slide eight-well ibiTreat coverslips (IBIDI, Martinsried, Germany) were used to prepare mono- and dual-species biofilms as described above. Bacterial suspensions of mono- and dual species were added to the plate and incubated for 24, 48 and 72 h at 25 °C. TSB was refreshed every 24 h of incubation. After the incubation time had elapsed, the wells were washed three times in PBS and microscopy was performed. The biofilm architecture was analysed by confocal microscopy using a Leica SP2 (Leica Microsystems, Wetzlar, Germany). Series of images were scanned at 400 Hz using a $63\times$ Leica oil immersion objective (numeric aperture 1.4). The whole well area was inspected to verify the presence of biofilms; then, the most representative location was scanned, providing a stack of horizontal planar images to obtain a three-dimensional view of the biofilms from the substratum to the top of the biofilms. The argon 488 nm laser was used for excitation of GFP, and the 561 nm laser was used for excitation of the mCherry fluorescent protein. The thickness of the biofilms was measured as the maximal distance between the lowest and highest acquired planar images using confocal microscopy. The thickness was determined as the average value from three locations.

Scanning electron microscopy

For SEM, dual-species biofilms were grown on the plastic coverslips as described above. Samples were removed from the dishes at 24, 48 and 72 h, washed three times in PBS and fixed in 3% Millonig phosphate-buffered glutaraldehyde three times for 10 min (Serva, Heidelberg, Germany) and postfixed in 2% Millonig osmium tetroxide-buffered solution for 1 h (Serva, Germany). Samples were washed three times for 10 min in Millonig phosphate buffer. The samples were subsequently dehydrated in increasing concentrations of acetone (50, 70, 90 and 100%), every step for 20 min, and dried in hexamethyldisilazane for 3 h in a hood at room temperature (Sigma-Aldrich, Praha, Czech Republic). Then, the samples were placed on the carbon tabs attached to the aluminium holder and coated with platinum/palladium (Cressington sputter coater 208 HR, UK). The structure and interaction of dual-species biofilms formed by *S. aureus* and Gram-negative bacteria were observed under a Hitachi SU 8010 scanning electron microscope (Hitachi High Technologies, Tokyo, Japan) at a magnification of $1500\times$ (at 17 kV, SE+BSE detector, working distance wd 8.4 mm); $6000\times$ (at 15 kV, wd 10.9 mm); $13\,000\times$ (at 17 kV, wd 8.4 mm); $30\,000\times$ (at 17 kV, wd 13.6).

Statistical analysis

OD 570 nm values (dependent variable) were analysed by ANOVA (biofilm formers, incubation time as categorical predictors), followed by *post hoc* Bonferroni multiple comparison tests. Logarithmically transformed data of CFU cm^{-2} (dependent variable) were analysed using ANCOVA (with biofilm formers as a categorical predictor and incubation time as a covariate), followed by testing of contrasts between mono- and dual-species biofilms. *P*-values lower than 0.05 were considered statistically significant. Data analysis was performed using the statistical software STATISTICA 13.0 (Dell, Tulsa, OK, USA).

Acknowledgements

This work was supported by the MEYS CZ (COST LD 14015 and NPUI LO1218) and EU COST Action CGA FA-1202 BacFoodNet (A European Network for Mitigating Bacterial Colonisation and Persistence on Foods and Food Processing Environments; http://www.bacfoodnet.org/). The authors acknowledge Julien Deschamps, French National Institute for Agriculture Research, for the kind gift of *S. aureus* RN4220. We thank Andrea Durisova (Veterinary Research Institute, Czech Republic) for technical support. Neysan Donnelly (Max Planck Institute of Biochemistry, Munich, Germany) is thanked for the grammatical correction of the manuscript.

Conflict of interest

None declared.

References

Bagge-Ravn, D., Ng, Y., Hjelm, M., Christiansen, J.N., Johansen, C., and Gram, L. (2003) The microbial ecology of processing equipment in different fish industries-analysis of the microflora during processing and following cleaning and disinfection. *Int J Food Microbiol* **87:** 239–250.

Bayles, K.W. (2007) The biological role of death and lysis in biofilm development. *Nat Rev Microbiol* **9:** 721–726.

Blana, V.A., Lianou, A., and Nychas, G.-J.E. (2015) Assessment of the effect of a *Salmonella enterica ser. Typhimurium* culture supernatant on the single-cell lag time of foodborne pathogens. *Int J Food Microbiol* **215:** 143–148.

Bower, C.K., McGuire, J., and Daeschel, M.A. (1996) The adhesion and detachment of bacteria and spores on food-contact surfaces. *Trends Food Sci Technol* **7:** 152–157.

Breakwell, D.P., Moyes, R.B., and Reynolds, J. (2009) Differential staining of bacteria: capsule stain. *Curr Protoc Microbiol* **3:** 1934–8533.

Bridier, A., Briandet, R., Thomas, V., and Dubois-Brissonnet, F. (2011) Resistance of bacterial biofilms to disinfectants: a review. *Biofouling* **27:** 1017–1032.

Canette, A., and Briandet, R. (2014) Microscopy: confocal laser scanning microscopy. In *Encyclopedia of Food Microbiology*. Batt, C., Tortorello, M. (eds). Academic Press, pp. 676–681.

Coldren, P.E. (2009) Encapsulated *Staphylococcus aureus* strains vary in adhesiveness assessed by atomic force microscopy. *J Biomed Mater Res*, Part A **89:** 402–410.

Costerton, J.W., Lewandowski, Z., Caldwell, D.E., Korber, D.R., and Lappin-Scott, H.M. (1995) Microbial biofilms. *Annu Rev Microbiol* **49:** 711–745.

Costerton, J.W., Montanaro, L., and Arciola, C.R. (2005) Biofilm in implant infections: its production and regulation. *Int J Artif Organs* **28:** 1062–1068.

Drancourt, M., Bollet, C., Carta, A., and Rousselier, P. (2001) Phylogenetic analyses of *Klebsiella* species delineate *Klebsiella* and *Raoultella* gen. nov., with description of *Raoultella ornithinolytica* comb. nov., *Raoultella terrigena* comb. nov. and *Raoultella planticola* comb. nov. *Int J Syst Evol Microbiol* **51:** 925–932.

Dull, T., Zufferey, R., Kelly, M., Mandel, R., Nguyen, M., Trondo, D., and Naldini, L. (1998) A third-generation lentivirus vector with a conditional packaging system. *J Virol* **11:** 8463–8471.

Elias, S., and Banin, E. (2012) Multi-species biofilms: living with friendly neighbors. *FEMS Microbiol Rev* **36:** 990–1004.

Foley, S.L., Johnson, T.J., Ricke, S.C., Nayak, R., and Danzeisen, J. (2013) *Salmonella* Pathogenicity and host adaptation in chicken-associated serovars. *Microbiol Mol Biol Rev* **77**: 582–607.

Giaouris, E., Chorianopoulos, N., Doulgeraki, A., and Nychas, G.-J. (2013) Co-culture with *Listeria monocytogenes* within a dual-species biofilm community strongly increases resistance of *Pseudomonas putida* to benzalkonium chloride. *PLoS ONE* **8**: e77276.

Giaouris, E., Heir, E., Desvaux, M., Hébraud, M., Møretrø, T., Langsrud, S., *et al.* (2015) Intra- and inter-species interactions within biofilms of important foodborne bacterial pathogens. *Front Microbiol* **6**: 841.

Gonzales, M.F., Brooks, T., Pukatzki, S.U., and Provenzano, D. (2013) Rapid protocol for preparation of electrocompetent *Escherichia coli* and *Vibrio cholerae*. *J Vis Exp JoVE* **80**: 43–49.

Guðbjörnsdóttir, B., Einarsson, H., and Thorkelsson, G. (2005) Microbial adhesion to processing lines for fish fillets and cooked shrimp: influence of stainless steel surface finish and presence of gram-negative bacteria on the attachment of *Listeria monocytogenes*. *Food Technol Biotechnol* **43**: 55–61.

Graver, M., and Wade, J. (2010) The role of acidification in the inhibition of *Neisseria gonorrhoeae* by vaginal lactobacilli during anaerobic growth. *Ann Clin Microbiol Antimicrob* **10**: 1–5.

Gurung, M., Moon, D.C., Choi, C.W., Lee, J.H., Bae, Y.C., Kim, J., *et al.* (2011) *Staphylococcus aureus* produces membrane-derived vesicles that induce host cell death. *PLoS ONE* **6**: e27958.

Habimana, O., Møretrø, T., Langsrud, S., Vestby, L.K., Nesse, L.L., and Heir, E. (2010) Micro ecosystems from feed industry surfaces: a survival and biofilm study of *Salmonella* versus host resident flora strains. *BMC Vet Res* **6**: 48.

Hall-Stoodley, L., Costerton, J.W., and Stoodley, P. (2004) Bacterial biofilms: from the natural environment to infectious diseases. *Nat Rev Microbiol* **2**: 95–108.

Harmsen, D., Dostal, S., Roth, A., Niemann, S., Rothganger, J., Sammeth, M., *et al.* (2003) RIDOM: comprehensive and public sequence database for identification of Mycobacterium species. *BMC Infect Dis* **11**: 26.

Harriott, M.M., and Noverr, M.C. (2009) *Candida albicans* and *Staphylococcus aureus* form polymicrobial biofilms: effects on antimicrobial resistance. *Antimicrob Agents Chemother* **53**: 3914–3922.

Hassan, A.N., and Frank, J.F. (2004) Attachment of *Escherichia coli* O157:H7 grown in tryptic soy broth and nutrient broth to apple and lettuce surfaces as related to cell hydrophobicity, surface charge, and capsule production. *Int J Food Microbiol* **96**: 103–109.

Hennekinne, J.-A., De Buyser, M.-L., and Dragacci, S. (2012) *Staphylococcus aureus* and its food poisoning toxins: characterization and outbreak investigation. *FEMS Microbiol Rev* **36**: 815–836.

Hoštacká, A., Čižnár, I., and Štefkovičová, M. (2010) Temperature and pH affect the production of bacterial biofilm. *Folia Microbiol (Praha)* **55**: 75–78.

Jahid, I.K., and Ha, S.-D. (2012) A review of microbial biofilms of produce: future challenge to food safety. *Food Sci Biotechnol* **21**: 299–316.

Kaper, J.B., Nataro, J.P., and Mobley, H.L.T. (2004) Pathogenic *Escherichia coli*. *Nat Rev Microbiol* **2**: 123–140.

Knowles, J.R., Roller, S., Murray, D.B., and Naidu, A.S. (2005) Antimicrobial action of carvacrol at different stages of dual-species biofilm development by *Staphylococcus aureus* and *Salmonella enterica* Serovar Typhimurium. *Appl Environ Microbiol* **71**: 797–803.

Koukoulaki, M., Bakalis, A., Kalatzis, V., Belesiotou, E., Papastamopoulos, V., Skoutelis, A., and Drakopoulos, S. (2014) Acute prostatitis caused by *Raoultella planticola* in a renal transplant recipient: a novel case. *Transpl Infect Dis* **16**: 461–464.

Kulp, A., and Kuehn, M.J. (2010) Biological functions and biogenesis of secreted bacterial outer membrane vesicles. *Annu Rev Microbiol* **64**: 163–184.

Lee, P. (2009) Quantitation of microorganisms. In *Practical Handbook of Microbiology*. Goldman, E., and Green, L.H. (eds). USA: CRC Press, Taylor and Francis Group, pp. 11–22.

Lee, E.-Y., Choi, D.-Y., Kim, D.-K., Kim, J.-W., Park, J.O., Kim, S., *et al.* (2009) Gram-positive bacteria produce membrane vesicles: proteomics-based characterization of *Staphylococcus aureus*-derived membrane vesicles. *Proteomics* **9**: 5425–5436.

Lehane, L., and Olley, J. (2000) Histamine fish poisoning revisited. *Int J Food Microbiol* **58**: 1–37.

Malone, C.L., Boles, B.R., Lauderdale, K.J., Thoendel, M., Kavanaugh, J.S., and Horswill, A.R. (2009) Fluorescent reporters for *Staphylococcus aureus*. *J Microbiol Methods* **77**: 251–260.

Manuzon, W.H., and Wang, H.H. (2007) Mixed-species biofilms. In *Biofilms in the Food Environment*. Blaschek, H.P., Wang, H., and Agle, M.E. (eds). IFT Press, pp. 105–125.

Methner, U., Al-Shabibi, S., and Meyer, H. (1995) Experimental oral infection of specific pathogen-free laying hens and cocks with *Salmonella enteritidis* strains. *Zentralbl Veterinärmed B* **42**: 459–469.

Millezi, F.M., Pereira, M.O., Batista, N.N., Camargos, N., Auad, I., Cardoso, M.D.G., and Piccoli, R.H. (2012) Susceptibility of monospecies and dual-species biofilms of *Staphylococcus aureus* and *Escherichia coli* to essential oils: control of biofilms using natural antimicrobial. *J Food Saf* **32**: 351–359.

Mittelman, M.W. (1998) Structure and functional characteristics of bacterial biofilms in fluid processing operations. *J Dairy Sci* **81**: 2760–2764.

Moons, P., Michiels, C.W., and Aertsen, A. (2009) Bacterial interactions in biofilms. *Crit Rev Microbiol* **35**: 157–168.

Nadell, C.D., Xavier, J.B., and Foster, K.R. (2009) The sociobiology of biofilms. *FEMS Microbiol Rev* **33**: 206–224.

Novick, R.P. (1990) The staphylococcus as a molecular genetic system. In *Molecular Biology of the Staphylococci*. Novick, R.P. (eds). VCH Publishers, pp. 1–40.

Oliveira, K., Oliveira, T., Teixeira, P., Azeredo, J., and Oliveira, R. (2007) Adhesion of *Salmonella Enteritidis* to stainless steel surfaces. *Braz J Microbiol* **38**: 318–323.

Olson, D.S., Asare, K., Lyons, M., and Hofinger, D.M. (2012) A novel case of *Raoultella planticola* urinary tract infection. *Infection* **41**: 259–261.

O'Toole, G., Kaplan, H.B., and Kolter, R. (2000) Biofilm Formation as Microbial Development. *Annu Rev Microbiol* **54**: 49–79.

Otto, M. (2013) Staphylococcal infections: mechanisms of biofilm maturation and detachment as critical determinants of pathogenicity. *Annu Rev Med* **64:** 175–188.

Pavlovsky, L., Sturtevant, R.A., Younger, J.G., and Solomon, M.J. (2015) Effects of temperature on the morphological, polymeric, and mechanical properties of Staphylococcus epidermidis bacterial biofilms. *Langmuir ACS J Surf Colloids* **31:** 2036–2042.

Pericone, C., Overweg, K., Hermans, K., and Weiser, J. (2000) Inhibitory and bactericidal effects of hydrogen peroxide production by Streptococcus pneumoniae on other inhabitants of the upper respiratory tract. *Infect Immun* **68:** 3990–3997.

Peters, B.M., Jabra-Rizk, M.A., Scheper, M., Leid, J.G., William Costerton, J.W., and Shirtliff, M.E. (2010) Microbial interactions and differential protein expression in Staphylococcus aureus – Candida albicans dual-species biofilms. *FEMS Immunol Med Microbiol* **59:** 493–503.

Pompermayer, D.M.C., and Gaylarde, C.C. (2000) The influence of temperature on the adhesion of mixed cultures of Staphylococcus aureus and Escherichia coli to polypropylene. *Food Microbiol* **17:** 361–365.

Poulsen, L.V. (1999) Microbial biofilm in food processing. *LWT – Food Sci Technol* **32:** 321–326.

Ren, D., Madsen, J.S., de la Cruz-Perera, C.I., Bergmark, L., Sørensen, S.J., and Burmølle, M. (2014) High-through-put screening of multispecies biofilm formation and quantitative PCR-based assessment of individual species proportions, useful for exploring interspecific bacterial interactions. *Microb Ecol* **68:** 146–154.

Rendueles, O., and Ghigo, J.-M. (2012) Multi-species biofilms: how to avoid unfriendly neighbors. *FEMS Microbiol Rev* **36:** 972–989.

Rendueles, O., Travier, L., Latour-Lambert, P., Fontaine, T., Magnus, J., Denamur, E., and Ghigo, J.-M. (2011) Screening of Escherichia coli species biodiversity reveals new biofilm-associated antiadhesion polysaccharides. *mBio* **2:** e00043-11.

Ruby, T., McLaughlin, L., Gopinath, S., and Monack, D. (2012) Salmonella's long-term relationship with its host. *FEMS Microbiol Rev* **36:** 600–615.

Sauer, K., Rickard, A.H., and Davies, D.G. (2007) Biofilms and biocomplexity. *Microbe* **2:** 347–353.

Schirmer, B.C.T., Heir, E., Møretrø, T., Skaar, I., and Langsrud, S. (2013) Microbial background flora in small-scale cheese production facilities does not inhibit growth and surface attachment of Listeria monocytogenes. *J Dairy Sci* **96:** 6161–6171.

Schlisselberg, D.B., and Yaron, S. (2013) The effects of stainless steel finish on Salmonella Typhimurium attachment, biofilm formation and sensitivity to chlorine. *Food Microbiol* **35:** 65–72.

Schooling, S.R., and Beveridge, T.J. (2006) Membrane vesicles: an overlooked component of the matrices of biofilms. *J Bacteriol* **188:** 5945–5957.

Schwering, M., Song, J., Louie, M., Turner, R.J., and Ceri, H. (2013) Multi-species biofilms defined from drinking water microorganisms provide increased protection against chlorine disinfection. *Biofouling* **29:** 917–928.

Sharma, M., and Anand, S.K. (2002) Biofilms evaluation as an essential component of HACCP for food/dairy processing industry – a case. *Food Control* **13:** 469–477.

Shaikh, M., and Morgan, M. (2011) Sepsis caused by Raoultella terrigena. *JRSM Short Rep* **2:** 49–51.

Shi, X., and Zhu, X. (2009) Biofilm formation and food safety in food industries. *Trends Food Sci Technol* **20:** 407–413.

Slany, M., Freiberger, T., Pavlik, P., and Cerny, J. (2007) Culture-negative infective endocarditis caused by Aerococcus urinae. *J Heart Valve Dis* **16:** 203–205.

de Souza, E.L., Meira, Q.G.S., de Medeiros Barbosa, I., Athayde, A.J.A.A., da Conceição, M.L., and de Siqueira Júnior, J.P. (2014) Biofilm formation by Staphylococcus aureus from food contact surfaces in a meat-based broth and sensitivity to sanitizers. *Braz J Microbiol* **45:** 67–75.

Srey, S., Jahid, I.K., and Ha, S.-D. (2013) Biofilm formation in food industries: a food safety concern. *Food Control* **31:** 572–585.

Stepanović, S., Cirković, I., Ranin, L., and Svabić-Vlahović, M. (2004) Biofilm formation by Salmonella spp. and Listeria monocytogenes on plastic surface. *Lett Appl Microbiol* **38:** 428–432.

Stepanović, S., Vuković, D., Hola, V., Di Bonaventura, G., Djukić, S., Cirković, I., and Ruzicka, F. (2007) Quantification of biofilm in microtiter plates: overview of testing conditions and practical recommendations for assessment of biofilm production by staphylococci. *APMIS* **115:** 891–899.

Taylor, S.L. (1986) Histamine food poisoning: toxicology and clinical aspects. *Crit Rev Toxicol* **17:** 91–128.

Thay, B., Wai, S.N., and Oscarsson, J. (2013) Staphylococcus aureus α-toxin-dependent induction of host cell death by membrane-derived vesicles. *PLoS ONE* **8:** e54661.

Valdivia, R., and Falkow, S. (1996) Bacterial genetics by flow cytometry: rapid isolation of Salmonella typhimurium acid-inducible promoters by differential fluorescence induction. *Mol Microbiol* **22:** 367–378.

Valle, J., Da Re, S., Henry, N., Fontaine, T., Balestrino, D., Latour-Lambert, P., and Ghigo, J.-M. (2006) Broad-spectrum biofilm inhibition by a secreted bacterial polysaccharide. *Proc Natl Acad Sci USA* **103:** 12558–12563.

Van Houdt, R., and Michiels, C.W. (2010) Biofilm formation and the food industry, a focus on the bacterial outer surface. *J Appl Microbiol* **109:** 1117–1131.

Varposhti, M., Entezari, F., Feizabadi, M., Varposhti, M., Entezari, E., and Feizabadi, M. (2014) Synergistic interactions in mixed-species biofilms of pathogenic bacteria from the respiratory tract. *Rev Soc Bras Med Trop* **47:** 649–652.

Wang, H., Wilksch, J.J., Strugnell, R.A., and Gee, M.L. (2015) Role of capsular polysaccharides in biofilm formation: an AFM nanomechanics study. *ACS Appl Mater Interfaces* **7:** 13007–13013.

Yonezawa, H., Osaki, T., Kurata, S., Fukuda, M., Kawakami, H., Ochiai, K., et al. (2009) Outer membrane vesicles of Helicobacter pylori TK1402 are involved in biofilm formation. *BMC Microbiol* **9:** 197.

Zhang, H., Zhou, W., Zhang, W., Yang, A., Liu, Y., Jiang, Y., et al. (2014) inhibitory effects of citral, cinnamaldehyde, and tea polyphenols on mixed biofilm formation by foodborne Staphylococcus aureus and Salmonella enteritidis. *J Food Prot* **77:** 927–933.

Community structure of partial nitritation-anammox biofilms at decreasing substrate concentrations and low temperature

Frank Persson,[1,*] Carolina Suarez,[2]
Malte Hermansson,[2] Elzbieta Plaza,[3]
Razia Sultana[3] and Britt-Marie Wilén[1]

[1]Division of Water Environment Technology, Department of Civil and Environmental Engineering, Chalmers University of Technology, SE-41296 Gothenburg, Sweden.
[2]Department of Chemistry and Molecular Biology, University of Gothenburg, SE-40530 Gothenburg, Sweden.
[3]Department of Sustainable Development, Environmental Science and Engineering (SEED), Royal Institute of Technology (KTH), Teknikringen 76, SE-100 44 Stockholm, Sweden.

Summary

Partial nitritation-anammox (PNA) permits energy effective nitrogen removal. Today PNA is used for treatment of concentrated and warm side streams at wastewater treatment plants, but not the more diluted and colder main stream. To implement PNA in the main stream, better knowledge about microbial communities at the typical environmental conditions is necessary. In order to investigate the response of PNA microbial communities to decreasing substrate availability, we have operated a moving bed biofilm reactor (MBBR) at decreasing reactor concentrations (311–27 mg-N l^{-1} of ammonium) and low temperature (13°C) for 302 days and investigated the biofilm community using high throughput amplicon sequencing; quantitative PCR; and fluorescence *in situ* hybridization. The anammox bacteria (*Ca.* Brocadia) constituted a large fraction of the biomass with fewer aerobic ammonia oxidizing bacteria (AOB) and even less nitrite oxidizing bacteria (NOB; *Nitrotoga*, *Nitrospira* and *Nitrobacter*). Still, NOB had considerable impact on the process performance. The anammox bacteria, AOB and NOB all harboured more than one population, indicating some diversity, and the heterotrophic bacterial community was diverse (seven phyla). Despite the downshifts in substrate availability, changes in the relative abundance and composition of anammox bacteria, AOB and NOB were small and also the heterotrophic community showed little changes in composition. This indicates stability of PNA MBBR communities towards decreasing substrate availability and suggests that even heterotrophic bacteria are integral components of these communities.

Introduction

Autotrophic nitrogen removal from wastewater can be achieved by partial nitritation together with anaerobic ammonium oxidation (anammox). Partial nitritation-anammox (PNA) saves energy due to a reduced need for aeration by > 50% (Siegrist *et al.*, 2008) and enables a higher utilization of organic carbon for production of valuable products, for example, biogas, compared to conventional nitrogen removal with nitrification-denitrification. Together, this makes energy positive wastewater treatment plants (WWTPs) possible (Kartal *et al.*, 2010).

Today, PNA is established for treatment of warm and concentrated wastewater side streams (Lackner *et al.*, 2014), where the conditions for growth of aerobic ammonia oxidizing bacteria (AOB) and anammox bacteria are beneficial and inhibition of unwanted aerobic nitrite oxidation by nitrite oxidizing bacteria (NOB) can be effective. However, the side stream nitrogen removal at WWTPs treats only 15–20% of the total nitrogen. To utilize the benefits of PNA in the wastewater main stream is highly desirable and has recently become a prioritized research area. In the main stream, the conditions for PNA are much more challenging (De Clippeleir *et al.*, 2013; Hu *et al.*, 2013; Laureni *et al.*, 2016). The cold and diluted water causes low activity and slow growth rate of particularly the anammox bacteria (Hendrickx *et al.*, 2014; Lotti *et al.*, 2015), even though some adaptations to low temperatures have been observed (Dosta *et al.*, 2008; Hu *et al.* 2013; Hendrickx *et al.*, 2014). Moreover, the competition between AOB and anammox bacteria with NOB and denitrifying bacteria is challenging at these conditions (see e.g. De Clippeleir *et al.*, 2013; Perez *et al.*, 2014), which necessitates detailed

*For correspondence. E-mail frank.persson@chalmers.se

knowledge about the dynamics of these microorganisms in order to understand process performance.

Despite the challenges, maintenance and activity of AOB and anammox bacteria at low temperatures and low substrate concentrations have been demonstrated in biofilm- and granular sludge reactors (De Clippeleir et al., 2013; Gustavsson et al., 2014; Lotti et al., 2014a; Gilbert et al., 2015; Ma et al., 2015; Laureni et al., 2016) and in a few studies the major population of AOB, anammox bacteria and NOB have been identified (Gilbert et al., 2014; Lotti et al., 2014a). Little is, however, known about the microbial community structure and dynamics at main stream conditions (Gilbert et al., 2014) and how such communities differ from the communities in the PNA reactors treating concentrated wastewater with higher substrate availability. Furthermore, the heterotrophic bacteria in PNA reactors (Gilbert et al., 2014; Pellicer-Nàcher et al., 2014; Chu et al., 2015) most likely affect the nitrogen turnover and process performance, but the composition, diversity and roles of these are little investigated, particularly at main stream conditions.

Here, an experiment was designed to investigate the role of the substrate concentration in shaping the PNA microbial community in a MBBR over 302 days. The influent nitrogen concentration was decreased from 500 to 45 mg-N l^{-1} at low temperature (13°C) to stepwise approach main stream conditions. For investigation of the composition and diversity of the total bacterial community (including heterotrophic bacteria), the abundance of key functional groups and their localization in the biofilms, a multiphase approach of high throughput amplicon sequencing (Illumina MiSeq), quantitative PCR (qPCR) and fluorescence in situ hybridization (FISH) in conjunction with confocal laser scanning microscopy (CLSM) of cryosectioned biofilms was used. Reactor

performance and potential activity of key functional groups was also monitored.

Results

Reactor performance

The concentration of the influent was decreased from target concentrations of 500 to 45 mg-N l^{-1} from period I to period VI (Figure S1) resulting in average ammonium concentrations of 311 to 27 mg-N l^{-1} in the reactor. During periods I to V, the nitrogen removal rate (NRR) was rather similar, with a decrease in period VI, at the lowest influent ammonium concentration (Table 1). The biomass weight on the carriers was stable, with insignificant changes during the study period (Figure S3, ANOVA, $P > 0.05$). The set-up and the performance of the reactor is summarized in Table 1. Time-course displays of nitrogen species are found in Figure S1.

Abundance of autotrophic nitrogen converting bacteria

The anammox bacteria dominated the total bacterial community in all periods, as measured by qPCR (Fig. 1). The gene copy numbers of AOB (amoA) were about two orders of magnitude lower than the anammox bacteria (16S rRNA). The abundances of the NOB, Nitrobacter and Nitrospira, were even lower. There were no major changes in the relative abundances of the anammox bacteria, AOB, Nitrobacter and Nitrospira over the course of the study, as measured by qPCR (ANOVA, $P > 0.05$), and correlations between the relative abundances for each sampling occasion ($n = 17$) and the reactor concentrations of nitrogen species, COD and alkalinity were non-significant ($P > 0.05$). The methods of qPCR, FISH and

Table 1. Study design and operational data of the MBBR.

	Period I	Period II	Period III	Period IV	Period V	Period VI
Operational set-up						
Time (d)	1–55	62–99	106–133	136–189	195–258	262–302
NH$_4$-N $_{infl}$ (mg l^{-1})	496 ± 32	249 ± 14	170 ± 19	129 ± 11	86 ± 8	43 ± 2
COD $_{infl}$ (mg l^{-1})	313 ± 36	174 ± 7	155 ± 26	84 ± 9	56 ± 4	48 ± 8
HRT (d)	3.3 ± 0.2	3.2 ± 0.0	3.2 ± 0.0	2.3 ± 0.1	1.6 ± 0.0	0.8 ± 0.0
NLR (g-N m^{-2} d^{-1})	0.75 ± 0.03	0.39 ± 0.02	0.26 ± 0.03	0.29 ± 0.02	0.28 ± 0.03	0.26 ± 0.02
pH	7.9 ± 0.12	7.7 ± 0.09	7.8 ± 0.08	7.5 ± 0.19	7.1 ± 0.07	7.2 ± 0.12
DO (mg l^{-1})	0.93 ± 0.08	0.67 ± 0.04	0.64 ± 0.07	0.82 ± 0.25	0.49 ± 0.06	0.48 ± 0.10
Reactor performance						
NH$_4$-N $_{effl}$ (mg l^{-1})	311 ± 70	136 ± 21	119 ± 12	63 ± 22	31 ± 10	27 ± 5
NO$_2$-N $_{effl}$ (mg l^{-1})	12.5 ± 5.6	6.0 ± 5.7	2.2 ± 1.0	2.8 ± 1.4	1.6 ± 0.7	1.8 ± 0.3
NO$_3$-N $_{effl}$ (mg l^{-1})	98 ± 41	33 ± 23	8 ± 4	17 ± 13	21 ± 7	9 ± 6
COD $_{effl}$ (mg l^{-1})	281 ± 15	153 ± 19	118 ± 8	67 ± 14	53 ± 8	40 ± 6
ARR (g-N m^{-2} d^{-1})	0.28 ± 0.10	0.18 ± 0.04	0.11 ± 0.07	0.14 ± 0.05	0.18 ± 0.04	0.10 ± 0.04
NRR (g-N m^{-2} d^{-1})	0.11 ± 0.06	0.12 ± 0.03	0.09 ± 0.07	0.10 ± 0.03	0.10 ± 0.03	0.03 ± 0.03
NO$_3$ production (%)	51 ± 19	28 ± 16	15 ± 8	23 ± 11	35 ± 12	58 ± 26

HRT = hydraulic retention time, NLR = nitrogen loading rate, DO = dissolved oxygen, ARR = ammonia removal rate, NRR = nitrogen removal rate. COD and nitrogen species are dissolved (filtered, 0.45 μm).

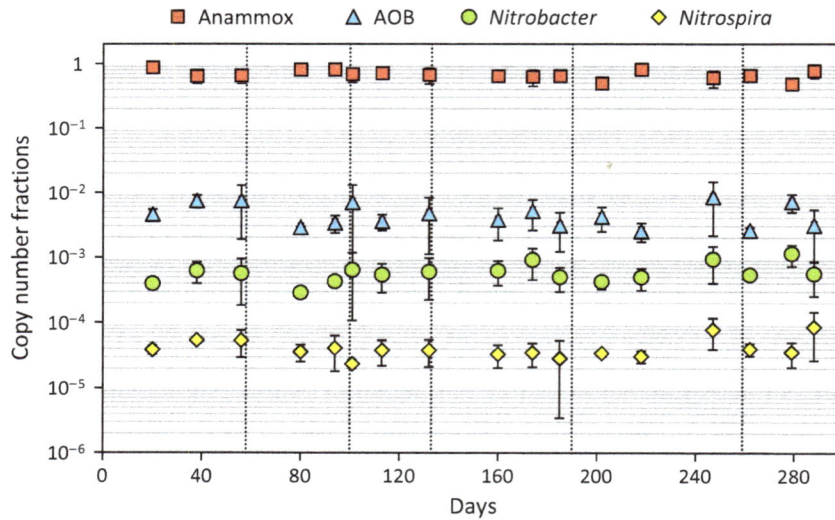

Fig. 1. Relative abundances (copy number fractions) of nitrogen converting bacteria measured by qPCR. The periods in the study are separated by vertical dashed lines. Average values. Error bars show standard deviation.

amplicon sequencing, all showed that anammox bacteria was the largest group followed by AOB and NOB (Fig. 1 and Table S4). FISH detected a higher percentage of AOB than the other methods, but it should be noted that only 20 to 40% of the cells detected by a general DNA stain (SYTO62) were detected by the general FISH EUB probe mix (Figure S4).

Batch activity tests of nitrogen converting groups of microorganisms

Batch tests (Fig. 2) were performed to assess the potential aerobic ammonium oxidation (AOB), aerobic nitrite oxidation (NOB), anammox and nitrate uptake rate (potential denitrification). Although the variations between

samples and sampling occasions were considerable, the data showed a gradual decrease in potential anammox (ANOVA, $P < 0.01$, $F = 5.7$) and an increase in potential nitrite oxidation (ANOVA, $P < 0.01$, $F = 4.6$). For the aerobic ammonium oxidation and the nitrate uptake rate, the changes between the periods were not significant ($P > 0.05$).

Composition and diversity of the biofilm communities

High throughput amplicon sequencing of the 16S rRNA gene (V4 region) showed that the bacterial communities at four periods in the MBBR consisted of similar numbers of OTUs (477 to 523), when resampled at 10 000 sequences (Table S2) and of all the 886 OTUs detected,

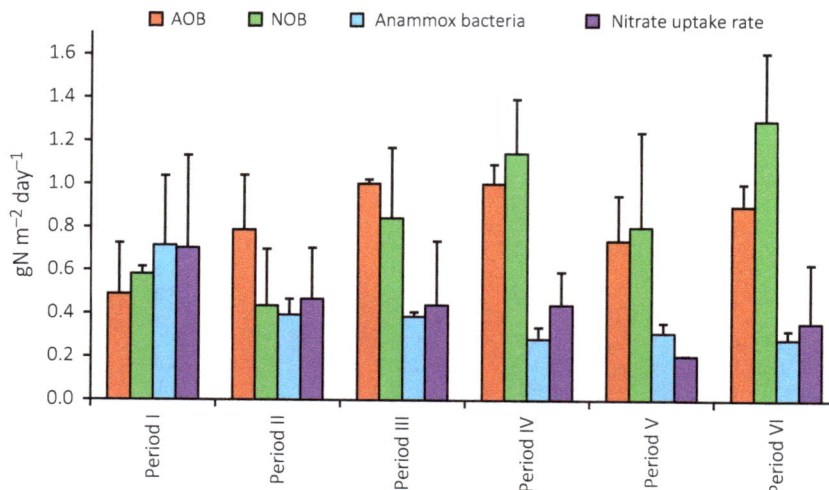

Fig. 2. Potential activity of nitrogen converting bacterial guilds measured in batch tests at period I–VI. Bars show average values, error bars show standard deviation.

236 were shared by all samples. The diversity of the sample from the final period (VI) was somewhat higher, indicative of a slightly more even community (Table S2). Pairwise comparisons of the biofilm communities (Bray-Curtis) in periods I, III and V resulted in coefficients of 0.10–0.11, while period VI differed slightly more (0.15–0.18). However, the dissimilarity coefficients were generally low, indicating similar community structure of all samples.

The composition at the phylum level showed that the majority of the sequences belonged to *Planctomycetes*, with *Chloroflexi* and *Proteobacteria* also having large contributions to the community (Fig. 3). In addition, *Bacteroidetes*, *Acidobacteria*, *Chlorobi*, the candidate phylum BRC1, *Actinobacteria* and *Deinococcus-Thermus* were present in relative sequence abundances > 0.5%.

The anammox bacteria were all affiliated to the genus *Ca.* Brocadia in one single OTU that dominated the bacterial biofilm community in all reactor periods (Table 2). Further subdivision of the sequences within *Ca.* Brocadia, by applying a more stringent criterion of 99% sequence similarity of OTUs, revealed that most anammox bacteria were similar to *Ca.* Brocadia sp. 40, with a smaller population of *Ca.* Brocadia fulgida-like bacteria (Table S3). This population was also detected by FISH (0.5–2.0% of total bacteria).

All identified AOB were affiliated to *Betaproteobacteria* and the major OTU (OTU 85) was affiliated to the *Nitrosomonas europaea/eutropha* cluster (Table 2). Separate populations of *Nitrosomonas europaea* and *Nitrosomonas eutropha* within this cluster could be detected at 99% sequence similarity (Table S3). OTUs similar to *Nitrosomonas* sp. JL21 and *Nitrosospira multiformis* were also observed (Table 2), but at very low relative abundances. The presence of *Nitrosomonas europaea/eutropha*, *Nitrosomonas oligotropha* (JL21) and *Nitrosospira* was confirmed by FISH (Fig. 5) at relative

abundances of 2.4–6.7%, 0–0.2% and 0.06–0.6% respectively.

Among the NOB, one OTU belonged to the order *Nitrospirales* (Table 2) but could not be affiliated to any described bacterium. However, BLAST analysis showed high similarities to unclassified sequences from other nitrogen converting wastewater reactors (OTU1534-1535 in Table S3), suggesting that these non-described *Nitrospirales* converted nitrogen. In addition, the presence of *Nitrospira* was demonstrated by qPCR and FISH (Table S4), but no FISH signal was seen with comammox probes for *Nitrospira nitrosa* and *N. nitrificans* (Table S1). One OTU highly similar to *Nitrotoga* was detected in all biofilm communities (Table 2) and *Nitrotoga* cells were also detected by FISH (Table S4). No OTU was assigned to *Nitrobacter*, although *Nitrobacter* was detected by qPCR and FISH (Table S4). However, OTUs with high similarity to *Nitrobacter* sp., but also to other species within *Bradyrhizobiaceae*, were revealed by BLAST (OTU 0176, 0226 in Table S3). Hence, the sequence information in the V4 region of the 16S rRNA gene was not sufficient for *Nitrobacter* identification.

In the MBBR, 43–50% of the sequences were affiliated to putative heterotrophic bacteria. These bacteria were subdivided into 25 orders with a sequence contribution > 0.5% in any of the samples (Fig. 4).

Localization of key bacterial groups in the biofilm

FISH-CLSM of biofilm cryosections was used to show the localization of key bacterial groups (for FISH probes, see Table S1). Clusters of AOB (*Nitrosomonas europaea/eutropha*, Fig. 5A) and *Nitrosospira*, Fig. 5B), and clusters of NOB (*Nitrospira*, Fig. 5C) were detected near the biofilm–water interface. Anammox bacteria were observed in high numbers deeper in the biofilm (Fig. 5), with two populations present closer to the biofilm–water interface (Fig. 5E). Bacteria within the phylum *Chloroflexi* were also detected both near the biofilm–water interface and deeper in the biofilm (Fig. 5D).

Discussion

Substrate availability is a main factor that determines microbial competitive interactions and thereby shapes the structure of microbial communities (Hibbing *et al.*, 2010; Litchman *et al.*, 2015). PNA is used for treatment of highly concentrated as well as diluted streams of wastewater and the substrate concentrations vary a lot (e.g. Hu *et al.*, 2013; Lackner *et al.*, 2014; Lotti *et al.*, 2014a), but very few systematic studies of the community response to changes in substrate availability have been made. Here, we test the hypothesis that a reduction in substrate availability influences the PNA microbial

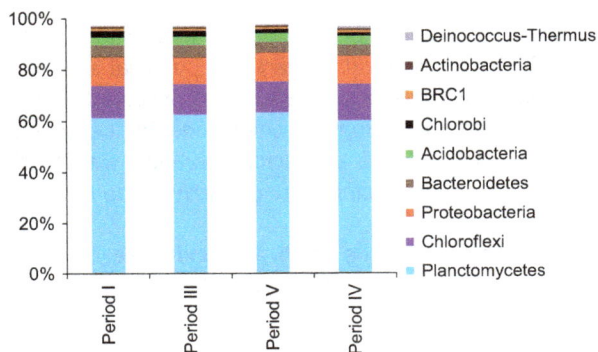

Fig. 3. Major bacterial phyla (> 0.5% relative sequence contribution) in the biofilm communities, as revealed by high throughput amplicon sequencing.

Table 2. Autotrophic nitrogen converting bacteria in the biofilm communities, as revealed by high throughput amplicon sequencing. OTUs clustered at 97% sequence similarity. BLAST analysis was used for classification.

OTU	Classification	Similarity	Period I	Period III	Period V	Period IV
2	*Ca.* Brocadia sp. 40/*Ca. B. caroliniensis* 20b	98	56%	56%	57%	50%
85	*Nitrosomonas europaea/N. eutropha*	99	0.32%	0.27%	0.14%	0.25%
3	*Nitrosomonas* sp. JL21	100	0.038%	0.026%	0.032%	0.014%
2049	*Nitrosospira multiformis*	97	0.017%	0.010%	0.012%	0.028%
930	*Ca.* Nitrotoga sp. clone JS16NT08	100	0.13%	0.051%	0.071%	0.047%
354	*Nitrospirales* 4-29[a]	N.A.	0.11%	0.18%	0.059%	0.079%

a. No described species with > 90% similarity from BLAST analysis. Classification by the Greengenes taxonomy.

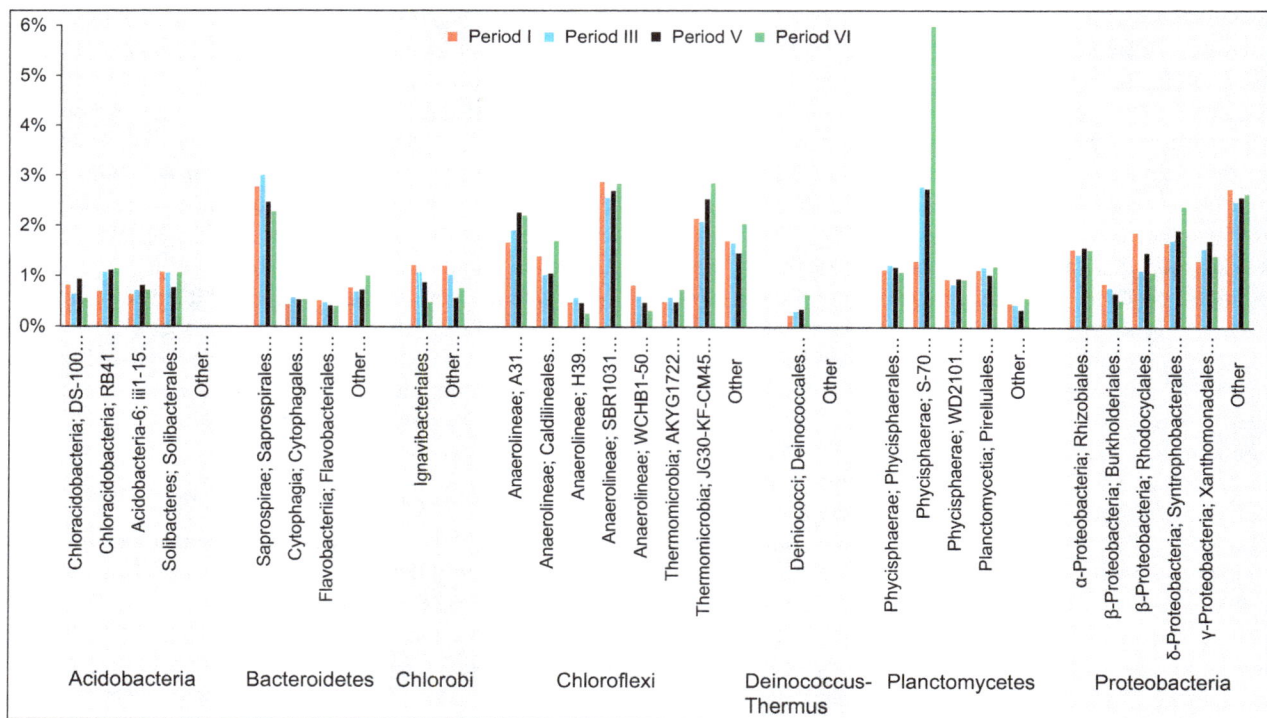

Fig. 4. Major orders (> 0.5% relative sequence contribution) of bacteria not involved in autotrophic nitrogen conversion in the biofilm communities, as revealed by high throughput amplicon sequencing.

community structure and function in a MBBR at low temperature, gradually approaching main stream conditions. The stepwise diluted influent resulted in decreased reactor ammonium concentrations, from 311 to 27 mg-N l^{-1}, as well as decreased concentrations of COD (Table 1). Despite these decreases in substrate concentrations, no major effects on the reactor biomass were observed. The biofilm weight did not change significantly (Table S3). The anammox bacteria dominated the bacterial community with AOB and NOB being considerably fewer (Fig. 1, Table 2, S4), located near the biofilm–water interface (Fig. 5). Changes in the relative abundances of anammox bacteria, AOB and NOB (qPCR) were non-significant between test periods and could not

be related to the reactor concentrations of nitrogen species, COD and alkalinity. Changes in potential activity of the anammox bacteria and NOB were observed (Fig. 2) and reflected the nitrogen conversions in the reactor (Table 1), but in general reactor operation was stable (Table 1), suggesting similar functionality of the microbial community.

The estimation of the relative abundances of anammox bacteria, AOB and NOB generally agreed between the methods (Table S4). The largest deviation was the higher AOB percentage assessed by FISH, but smaller differences among the methods were observed also for anammox bacteria and NOB. Discrepancies between rRNA-based FISH and DNA-based qPCR and

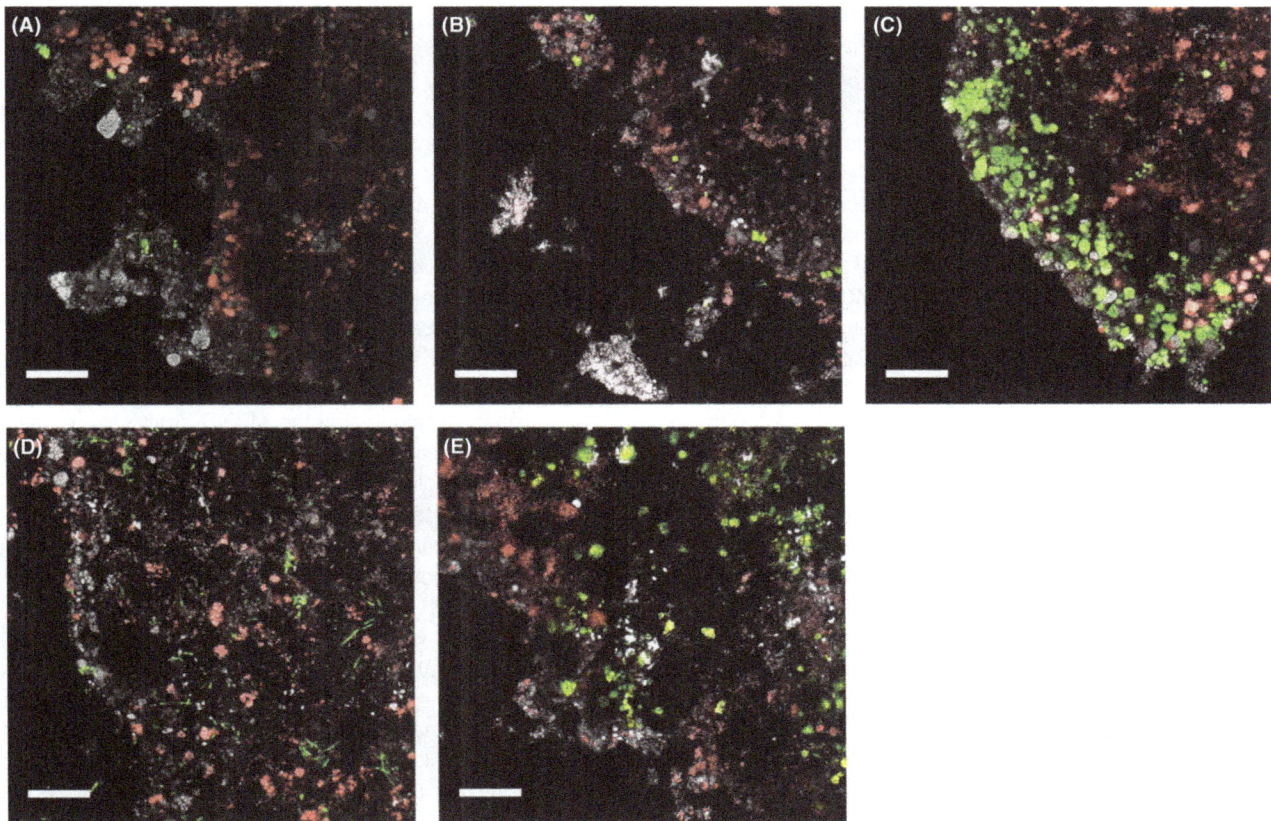

Fig. 5. FISH-CLSM of biofilm cryosections. The water-biofilm interface is oriented to the lower left. In all images, anammox bacteria (Amx820) are in red and nonspecific bacteria (EUB-mix) are in white-grey.
A. In green: AOB within the *Nitrosomonas europaea/eutropha* cluster (Nse1472).
B. In green: AOB within *Nitrosospira* (Nsv443).
C. In green: NOB within *Nitrospira* (Ntspa662).
D. In green: Bacteria within *Chloroflexi* (CFX123 + GNSB941).
E. In green: *Ca.* Brocadia fulgida (Bfu613). Scale bar: 25 µm.

sequencing can be expected. A large fraction of the microbial community had a ribosomal content below the FISH detection limit (Figure S4). However, AOB are known to retain their ribosomes even at challenging conditions such as starvation (Morgenroth *et al.*, 2000), which may help to explain their particularly high relative abundance observed by FISH. Also, the DNA extraction methodology, the choice of PCR primers and the cell copy numbers of target genes influence relative abundances (see e.g. Acinas *et al.*, 2004; Albertsen *et al.*, 2015). Quantification by high throughput amplicon sequencing is furthermore technically challenging (Zhou *et al.*, 2015). Although sequence processing has improved significantly with pipelines, such as Mothur used here (Kozich *et al.*, 2013), relative abundances, especially of rare OTUs (e.g. the NOB), have to be interpreted with caution. Hence, all methods suffer from limitations and multiple methods provide important complementary information. The methods also vary in

response time. Specific populations can be estimated by qPCR and FISH on suspended biofilms within a day or two, which is useful for routine monitoring. High throughput amplicon sequencing and FISH biofilms cryosections provide more detailed information, but takes considerably longer time.

The major anammox population was highly similar to *Ca.* Brocadia sp. 40 (Table 2, Table S3), which has previously been observed in several anammox reactors (van der Star *et al.*, 2008; Park *et al.*, 2010; Costa *et al.*, 2014; Gilbert *et al.*, 2014). Only one population of anammox bacteria is usually observed in PNA reactors (Hu *et al.*, 2013; Gilbert *et al.*, 2014; Laureni *et al.*, 2015). However, using 99% similarity for sequence clustering (Table S3) and a competitor probe to improve the FISH specificity (Table S1, Persson *et al.* (2014)), a second, closely related, *Ca.* Brocadia fulgida-like population was detected (Fig. 5, Table S3). It is likely that the two populations have different niches, just as described for closely related *Nitrospira* strains (Gruber-Dorninger *et al.*, 2014). This was supported by the localization of the smaller *Ca.* Brocadia fulgida-like population near the biofilm–water interface while the *Ca.* Brocadia sp. 40 population was detected throughout the biofilm. Furthermore, despite studies showing that *Ca.* Kuenenia have higher substrate affinity than *Ca.* Brocadia, and hence would be selected for at low substrate concentrations (van der Star *et al.*, 2008; Oshiki *et al.*, 2011), *Ca.* Brocadia fulgida and/or *Ca.* Brocadia sp. 40, rather than *Ca.* Kuenenia have dominated the anammox guild in this and other main stream PNA studies (Gilbert *et al.*, 2014; Lotti *et al.*, 2014a,b).

The AOB community was dominated by two populations within the *Nitrosomonas europaea/eutropha* cluster (Table 2, S3), which are commonly found in PNA reactors (Park *et al.*, 2010; Vlaeminck *et al.*, 2010; Pellicer-Nàcher *et al.*, 2014). Minor OTUs similar to *N. oligotropha* (sp. JL21) and *Nitrosospira multiformis* were also detected. This diverse AOB community was confirmed by FISH.

NOB were constantly present (Fig. 1) and were active during all periods, as seen by the production of nitrate (Table 1, Figure S1) and batch activity tests (Fig. 2). In particular, they had large impact on the process performance in period VI, resulting in low nitrogen removal efficiency (11%). Strategies to abate NOB include careful control of DO- and substrate concentrations (Perez *et al.*, 2014) as well as intermittent periods of anoxic and aerated phases, either at high-or low-DO concentrations (Wett *et al.*, 2013; Ma *et al.*, 2015). Despite maintained ammonium concentrations and careful DO control, NOB could not be repressed in the MBBR biofilms. Unwanted NOB activity in main stream PNA reactors is frequently reported (De Clippeleir *et al.*, 2013; Gilbert *et al.*, 2014; Lotti *et al.*, 2014a) and the strategy for NOB repression depending on the aggregation state of the biomass (suspended, granular, biofilm) and the ecophysiology of AOB and NOB.

The NOB consisted of *Nitrobacter*, *Nitrospira* and *Nitrotoga*. In PNA reactors, the low bulk concentrations of nitrite would select for *Nitrospira*, rather than *Nitrobacter*, due to their higher substrate affinity (Isanta *et al.*, 2015). *Nitrospira* is furthermore hard to outcompete using low DO concentrations due to their high oxygen affinity (Isanta *et al.*, 2015). *Nitrospira* has, in fact, been the only NOB observed in some main stream PNA reactors (De Clippeleir *et al.*, 2013; Gilbert *et al.*, 2014), but as shown here and elsewhere (Liu *et al.*, 2012), even *Nitrobacter* can sustain at such conditions. Interestingly, the long-term operation of the MBBR at 13°C allowed the establishment of a small *Nitrotoga* population (Table 2, S4). *Nitrotoga* has been shown to be important in activated sludge communities at 7–16°C (Lücker

et al., 2015), but so far little is known about their ecophysiology, except for their low temperature requirements. *Nitrotoga* has, to the best of our knowledge, not been previously detected in PNA reactors and this finding may have implications for NOB suppression strategies.

In the MBBR, 43–50% of the sequences were affiliated to putative heterotrophic bacteria. Also in other studies of PNA systems, significant fractions of the microbial communities have been heterotrophs, although little is known about their composition, dynamics and roles (Chu *et al.*, 2015; Gilbert *et al.*, 2014; Pellicer-Nàcher *et al.*, 2014). The relative abundances of the major contributors of the heterotrophic community in the MBBR (from seven phyla) were stable in relative abundance throughout the study (Fig. 4). This implies that the decreasing COD concentrations in the reactor (Table 1) had little impact on the heterotrophic community and suggests that even heterotrophic bacteria were an integral part of the community, possibly with defined roles in the PNA biofilm. Heterotrophic bacteria may contribute to nitrogen removal via denitrification, but their competition with AOB and anammox bacteria for space and electron acceptors can also be detrimental (Kumar and Lin, 2010). Furthermore, they can utilize soluble microbial products (SMP) from the biofilm (Ni *et al.*, 2012) and aid in biofilm formation (Cho *et al.*, 2010). A minor fraction of the influent COD was consistently removed in the MBBR (Table 1) and batch activity tests showed anoxic nitrate uptake (Fig. 2), which suggests some denitrification. We found members of *Rhodocyclales*, *Burkholderiales*, *Rhizobiales* and *Xanthomonadales* (Fig. 4), which all are important contributors to wastewater denitrification (Baytshtok *et al.*, 2009; McIlroy *et al.*, 2016) and have been detected in PNA reactors treating organic-free wastewater (Pellicer-Nàcher *et al.*, 2014; Chu *et al.*, 2015), suggesting SMP utilization. SMP may also have sustained the biofilm population of the non-denitrifying, protein degrading *Saprospirales* (Xia *et al.*, 2008). *Chloroflexi* were abundant here (Figs. 3 and 4), as well as in other PNA communities (Gilbert *et al.*, 2014; Chu *et al.*, 2015). *Chloroflexi* can provide biofilm structural integrity (Cho *et al.*, 2010) and metabolize SMP from autotrophs at both aerobic and anoxic conditions (Okabe *et al.*, 2005; Kindaichi *et al.*, 2012), which would explain their distribution throughout the biofilm (Fig. 5). *Phycisphaerae*, *Ignavibacteriales*, *Deinococcales* (Fig. 4) harbour bacteria with mostly undefined ecophysiologies, but their presence here and in other PNA- and anammox communities (Costa *et al.*, 2014; Chu *et al.*, 2015) suggests defined functions.

There are several possible explanations for the observed stability and maintained diversity of the microbial community at the decreasing substrate concentrations. The main anammox bacteria (*Ca.* Brocadia sp. 40 and *Ca.* Brocadia fulgida) and AOB (*Nitrosomonas* europaea/eutropha) have been detected at high relative abundances at different conditions, including a wide range of substrate concentrations (Park *et al.*, 2010; Vlaeminck *et al.*, 2010; Almstrand *et al.*, 2014; Gilbert *et al.*, 2014; Lotti *et al.*, 2014a; Pellicer-Nàcher *et al.*, 2014), which indicates broad ecophysiologies and a competitiveness at all tested concentrations. Furthermore, the presence of numerous micro-environments in the thick biofilms with gradients of substrate and electron acceptors likely promoted diversity and permitted the coexistence of competing as well as of commensal community members. As mentioned, the community had both active and non-active bacteria; the low ribosomal content of a large fraction of the bacteria indicated inactivity (Figure S4). The protected environment in the biofilm carriers may offer a refuge site for active and inactive cells, which may slow down community changes, as would the low temperature. Although the time between subsequent test periods may have been too short for major community changes to occur, for the entire study, spanning 302 days, time was likely sufficient. In municipal wastewater, the continuous variations over time in influent composition of substrates and suspended bacteria, are factors that may affect the stability of the microbial community, but these were not addressed here. Very few studies have been performed on PNA using real main stream wastewater, and the impact of these factors is yet not valuated. Maintenance and activity of anammox and AOB populations for at least 240 days in PNA MBBRs receiving pre-treated municipal wastewater was recently shown (Laureni *et al.*, 2016), indicating that, at least for biofilm systems, the influence of these factors for the stability of the key functional populations is manageable.

In conclusion, the bacterial community in a PNA MBBR system was stable during decreasing concentrations of substrate, approaching main stream conditions. Within the guilds of AOB, anammox bacteria and NOB, composition and diversity was maintained at all tested concentrations. The composition was largely stable even for the diverse heterotrophic community, suggesting that they were an integral part of the community.

Experimental procedures

The pilot moving bed biofilm reactor

A 200 l MBBR was filled with biofilm carriers (Kaldnes K1) at 40% filling degree. It received reject water after anaerobic digestion from the Himmerfjärden WWTP in Stockholm, diluted with tap water. The study period (302 days) was divided into six periods with stepwise decreased influent concentrations of ammonium from 500 mg-N l^{-1}, representative of reject water, to 45 mg-N l^{-1}, representative of main stream wastewater, with

concomitant decreases in nitrogen loading rate and hydraulic retention time (Table 1). The MBBR was operated at 13°C corresponding to low main stream temperatures in moderate climates.

Redox, pH and DO was measured using online sensors (Cerlic AB, Segeltorp, Sweden). Air supply was provided from the bottom of the reactor and was controlled, via the DO, by a PID controller. The temperature was monitored and controlled by a compact controller (JUMO GmbH & Co. KG, Fulda, Germany) and a cooler (JULABO GmbH, Seelbach, Germany). Mixing of the bulk water and biofilm carriers was achieved by a two-blade stirrer (50 rpm).

For analysis of inorganic nitrogen species and COD in the influent and effluent (filtered 0.45 μm), Dr. LANGE cuvettes were used on a XION 500 Spectrophotometer (HACH LANGE GmbH, Düsseldorf, Germany).

Batch tests for potential activity measurements

Batch tests were performed to measure potential microbial activities at 25°C. The activity measurements of AOB and NOB was based on the oxidation rate of ammonium (present in the test medium) and nitrite (formed by the AOB during the test) by measuring the oxygen uptake rate (OUR) using a DO probe (YSI 5905; YSI Inc. Yellow Springs, OH, USA). The method was adopted from Surmacz-Górska *et al.* (1996). At the start of each measurement, ammonium was available at 100 mg N l^{-1}. First, the total OUR was measured. After 5 minutes, $NaClO_3$ (17 mM) was added for the inhibition of NOB. After 10 minutes, allylthiourea (43 μM) was added for the inhibition of AOB. The activity of NOB and AOB was obtained from the OUR before minus the OUR after the addition of $NaClO_3$ and allylthiourea respectively. The remaining OUR, after addition of both inhibitors, is represented by endogenous respiration and substrate oxidation by heterotrophs and was not reported. The potential anammox activity was measured as the production of nitrogen gas (headspace pressure) according to Dapena-Mora *et al.* (2007) with NH_4^+ and NO_2^- at initial concentrations of 70 mg N l^{-1} each. Nitrate uptake rate measurements were used for measurements of potential heterotrophic denitrifying activity. The tests were performed at anoxic conditions in reject water diluted with distilled water at an initial COD concentration of 200 mg O_2 l^{-1}, with $NaNO_3$ at an initial concentration of 100 mg N l^{-1}. Samples for measurement of NO_3^- were taken every 12 minutes for 4 hours.

DNA extraction

DNA was separately extracted from three carriers at each sampling occasion. From each carrier, 30 mg of biofilm was used for extraction using the FastDNA spin kit for soil (MP Biomedicals, Santa Ana, CA, USA) according to the manufacturers' recommendations. The concentration of the extracted DNA was measured using a NanoDrop ND-1000 spectrophotometer (Thermo Scientific NanoDrop products, Wilmington, DE, USA).

Quantitative PCR

qPCR was used for quantification of autotrophic nitrogen converting bacteria according to Persson *et al.* (2014). In brief, primers for the 16S rRNA gene were used to target all bacteria, anammox bacteria, *Nitrospira* and *Nitrobacter* and primers for the amoA gene were used to target AOB. The qPCR was carried out on an iQ5 (Bio-Rad Laboratories. Inc., Hercules, CA, USA) thermal cycler using the SYBR green chemistry. Plasmid target gene inserts were used as standards. The results are presented as copy number fractions of the nitrogen converging bacteria to all bacteria, to get relative abundances. Differences in the relative abundances between periods I–VI was tested by one-factor analysis of variance (ANOVA). Prior to ANOVA, variance homogeneity was confirmed by Levene's test. To assess whether there was a link between the reactor conditions and the relative abundance of the key microbial groups at each sampling occasion ($n = 17$), the preceding reactor concentrations of nitrogen species, COD and alkalinity was averaged over 10 days. Correlations between these average concentrations and the relative abundances were tested for using a linear model (Pearson's r).

High throughput amplicon sequencing

PCR was carried out using the primers 515F and 806R to amplify partial V4 region sequences of the 16S rRNA gene (Caporaso *et al.*, 2011) with dual indexing of the primers (Kozich *et al.*, 2013). Sequencing was performed on an Illumina MiSeq (Illumina Inc., San Diego, CA, USA) using the MiSeq Reagent Kit v2 with PhiX control library spiked in at 7.5%. For details on PCR, purification, and quality control, see supporting methods. The obtained sequences were processed in Mothur (Schloss *et al.*, 2009) for assembly of contigs, denoising, removal of putative chimera, alignment, classification and construction of operational taxonomic units (OTUs) at 97% taxonomic identity (Kozich *et al.*, 2013). For classification with the Bayesian classifier within Mothur, the Greengenes database v. 13.8.99 (McDonald *et al.*, 2012) was used at 80% confidence threshold. Prior to analysing alpha- and beta-diversity, the OTU dataset was subsampled at 10 000 sequences. Raw sequence reads were deposited at the NCBI Sequence Read Archive, no. SRP059362.

Fluorescence *in situ hybridization and confocal laser scanning microscopy*

For FISH, samples were taken in the periods II, IV and VI. The carriers were fixed in paraformaldehyde (4% w/v) for 8 h at 4°C. For FISH on biofilm suspension, two carriers form each sampling period were used. The fixed biofilm was brushed off the carriers and homogenized in PBS before storage in PBS-ethanol (1:1) at −20°C. The biofilm suspensions (2–4 μl) were spotted on diagnostic microscope slides (8 × 6 mm diameter wells; Menzel GmbH, Braunschweig, Germany) for FISH and images were acquired from 10 random fields of view for each carrier. For FISH on biofilm cryosections, the carriers with fixed biofilm were embedded, frozen and cryosectioned in 20–25 μm thick slices which were captured on microscope slides and subjected to FISH. FISH was carried out at 46°C for 2 h for biofilm suspensions and 4 h for biofilm cryosections according to Almstrand *et al.* (2014). The probes (Table S1) were 5′ labelled with Cy3, Cy5, or Alexa 488. The relative abundances of the anammox bacteria, AOB and NOB was estimated on biofilm suspensions as the ratios of the FISH-targeted biovolumes of the specific populations to the total bacteria (EUB338 I-IV probe mix, Table S1) using daime 2.1 (Daims *et al.*, 2006). See supporting methods for details about embedding, cryosectioning, image acquisition and image analysis.

Acknowledgements

This study was funded by FORMAS (Contract no. 243-2010-2259, 211-2010-140, 2015-1515-30425-28 and 245-2014-1528), SVU (Contract no. 10-105) and the foundations of Carl Trygger (CTS 12:374), Adlerbertska forskningsstiftelsen and Åke & Greta Lissheds. The authors acknowledge the Centre for Cellular Imaging and the Genomics core facility at the University of Gothenburg for support and use of their equipment and the staff at the Hammarby Sjöstadsverk research facility, Stockholm, for maintenance of the pilot plant. The authors also acknowledge the anonymous reviewers for valuable comments.

Conflict of Interest

None declared.

References

Acinas, S. G., Marcelino, L. A., Klepac-Ceraj, V., and Polz, M. F. (2004) Divergence and redundancy of 16S rRNA sequences in genomes with multiple rrn operons. *J Bacteriol* **186:** 2629–2635.

Albertsen, M., Karst, S. M., Ziegler, A. S., Kirkegaard, R. H., and Nielsen, P. H. (2015) Back to basics–the influence of DNA extraction and primer choice on phylogenetic analysis of activated sludge communities. *PLoS ONE* **10(7)**: e0132783.

Almstrand, R., Persson, F., Daims, H., Ekenberg, M., Christensson, M., Wilén, B.-M., *et al.* (2014) Three-dimensional stratification of bacterial biofilm populations in a moving bed biofilm reactor for nitration-anammox. *Int J Mol Sci* **15**: 2191–2206.

Baytshtok, V., Lu, H. J., Park, H., Kim, S., Yu, R., and Chandran, K. (2009) Impact of varying electron donors on the molecular microbial ecology and biokinetics of methylotrophic denitrifying bacteria. *Biotechnol Bioeng* **102**: 1527–1536.

Caporaso, J. G., Lauber, C. L., Walters, W. A., Berg-Lyons, D., Lozupone, C. A., Turnbaugh, P. J., *et al.* (2011) Global patterns of 16S rRNA diversity at a depth of millions of sequences per sample. *Proc Natl Acad Sci USA* **108 (Suppl.** 1): 4516–4522.

Cho, S., Takahashi, Y., Fujii, N., Yamada, Y., Satoh, H., and Okabe, S. (2010) Nitrogen removal performance and microbial community analysis of an anaerobic up-flow granular bed anammox reactor. *Chemosphere* **78**: 1129–1135.

Chu, Z.-R., Wang, K., Li, X.-K., Zhu, M.-T., Yang, L., and Zhang, J. (2015) Microbial characterization of aggregates within a one-stage nitration–anammox system using high-throughput amplicon sequencing. *Chem Eng J* **262**: 41–48.

Costa, M. C., Carvalho, L., Leal, C. D., Dias, M. F., Martins, K. L., Garcia, G. B., *et al.* (2014) Impact of inocula and operating conditions on the microbial community structure of two anammox reactors. *Environ Technol* **35**: 1811–1822.

Daims, H., Lucker, S., and Wagner, M. (2006) Daime, a novel image analysis program for microbial ecology and biofilm research. *Environ Microbiol* **8**: 200–213.

Dapena-Mora, A., Fernandez, I., Campos, J. L., Mosquera-Corral, A., Mendez, R., and Jetten, M. S. M. (2007) Evaluation of activity and inhibition effects on Anammox process by batch tests based on the nitrogen gas production. *Enzyme Microb Tech* **40**: 859–865.

De Clippeleir, H., Vlaeminck, S. E., De Wilde, F., Daeninck, K., Mosquera, M., Boeckx, P., *et al.* (2013) One-stage partial nitration/anammox at 15 degrees C on pretreated sewage: feasibility demonstration at lab-scale. *Appl Microbiol Biotechnol* **97**: 10199–10210.

Dosta, J., Fernandez, I., Vazquez-Padin, J. R., Mosquera-Corral, A., Campos, J. L., Mata-Alvarez, J., and Mendez, R. (2008) Short- and long-term effects of temperature on the Anammox process. *J Hazard Mater* **154**: 688–693.

Gilbert, E. M., Agrawal, S., Karst, S. M., Horn, H., Nielsen, P. H., and Lackner, S. (2014) Low temperature partial nitration/anammox in a moving bed biofilm reactor treating low strength wastewater. *Environ Sci Technol* **48**: 8784–8792.

Gilbert, E. M., Agrawal, S., Schwartz, T., Horn, H., and Lackner, S. (2015) Comparing different reactor configurations for Partial Nitration/Anammox at low temperatures. *Water Res* **81**: 92–100.

Gruber-Dorninger, C., Pester, M., Kitzinger, K., Savio, D. F., Loy, A., Rattei, T., *et al.* (2014) Functionally relevant diversity of closely related Nitrospira in activated sludge. *ISME J* **9**: 643–655.

Gustavsson, D.J., Persson, F. and Jansen, J.L. (2014) *Manammox – mainstream anammox at Sjölunda WWTP.* IWA World Water Congress and Exhibition, September 21-26, 2014. Lisbon, Portugal: IWA

Hendrickx, T. L., Kampman, C., Zeeman, G., Temmink, H., Hu, Z., Kartal, B., and Buisman, C. J. (2014) High specific activity for anammox bacteria enriched from activated sludge at 10 degrees C. *Bioresour Technol* **163**: 214–221.

Hibbing, M. E., Fuqua, C., Parsek, M. R., and Peterson, S. B. (2010) Bacterial competition: surviving and thriving in the microbial jungle. *Nat Rev Microbiol* **8**: 15–25.

Hu, Z. Y., Lotti, T., de Kreuk, M., Kleerebezem, R., van Loosdrecht, M., Kruit, J., *et al.* (2013) Nitrogen removal by a nitration-anammox bioreactor at low temperature. *Appl Environ Microbiol* **79**: 2807–2812.

Isanta, E., Reino, C., Carrera, J., and Perez, J. (2015) Stable partial nitration for low-strength wastewater at low temperature in an aerobic granular reactor. *Water Res* **80**: 149–158.

Kartal, B., Kuenen, J. G., and van Loosdrecht, M. C. M. (2010) Sewage treatment with Anammox. *Science* **328**: 702–703.

Kindaichi, T., Yuri, S., Ozaki, N., and Ohashi, A. (2012) Ecophysiological role and function of uncultured Chloroflexi in an anammox reactor. *Water Sci Technol* **66(12)**: 2556–2561.

Kozich, J. J., Westcott, S. L., Baxter, N. T., Highlander, S. K., and Schloss, P. D. (2013) Development of a dual-index sequencing strategy and curation pipeline for analyzing amplicon sequence data on the MiSeq Illumina sequencing platform. *Appl Environ Microbiol* **79**: 5112–5120.

Kumar, M., and Lin, J. G. (2010) Co-existence of anammox and denitrification for simultaneous nitrogen and carbon removal-strategies and issues. *J Hazard Mater* **178**: 1–9.

Lackner, S., Gilbert, E. M., Vlaeminck, S. E., Joss, A., Horn, H., and van Loosdrecht, M. C. (2014) Full-scale partial nitration/anammox experiences - An application survey. *Water Res* **55**: 292–303.

Laureni, M., Weissbrodt, D.G., Szivak, I., Robin, O., Nielsen, J.L., Morgenroth, E. and Joss, A. (2015) Activity and growth of anammox biomass on aerobically pre-treated municipal wastewater. *Water Res* **80**: 325–336.

Laureni, M., Falås, P., Robin, O., Wick, A., Weissbrodt, D. G., Nielsen, J. L., *et al.* (2016) Mainstream partial nitration and anammox: long-term process stability and effluent quality at low temperatures. *Water Res* **101**: 628–639.

Litchman, E., Edwards, K. F., and Klausmeier, C. A. (2015) Microbial resource utilization traits and trade-offs: implications for community structure, functioning, and biogeochemical impacts at present and in the future. *Front Microbiol* **6**: 1–10.

Liu, T., Li, D., Zeng, H., Li, X., Zeng, T., Chang, X., *et al.* (2012) Biodiversity and quantification of functional bacteria in completely autotrophic nitrogen-removal over nitrite (CANON) process. *Bioresource Technol* **118**: 399–406.

Lotti, T., Kleerebezem, R., Hu, Z., Kartal, B., Jetten, M. S., and van Loosdrecht, M. C. (2014a) Simultaneous partial nitritation and anammox at low temperature with granular sludge. *Water Res* **66C:** 111–121.

Lotti, T., Kleerebezem, R., vanErp Taalman Kip, C., Hendrickx, T.L., Kruit, J., Hoekstra, M. and vanLoosdrecht, M.C. (2014b) Anammox growth on pretreated municipal wastewater. *Environ Sci Technol* **48:** 7874–7880.

Lotti, T., Kleerebezem, R., and van Loosdrecht, M. C. (2015) Effect of temperature change on anammox activity. *Biotechnol Bioeng* **112:** 98–103.

Lücker, S., Schwarz, J., Gruber-Dorninger, C., Spieck, E., Wagner, M., and Daims, H. (2015) Nitrotoga-like bacteria are previously unrecognized key nitrite oxidizers in full-scale wastewater treatment plants. *ISME J* **9:** 708–720.

Ma, B., Bao, P., Wei, Y., Zhu, G., Yuan, Z., and Peng, Y. (2015) Suppressing nitrite-oxidizing bacteria growth to achieve nitrogen removal from domestic wastewater via anammox using intermittent aeration with low dissolved oxygen. *Sci Rep* **5:** 13048.

McDonald, D., Price, M. N., Goodrich, J., Nawrocki, E. P., DeSantis, T. Z., Probst, A., *et al.* (2012) An improved Greengenes taxonomy with explicit ranks for ecological and evolutionary analyses of bacteria and archaea. *ISME J* **6:** 610–618.

McIlroy, S. J., Starnawska, A., Starnawski, P., Saunders, A. M., Nierychlo, M., Nielsen, P. H., and Nielsen, J. L. (2016) Identification of active denitrifiers in full-scale nutrient removal wastewater treatment systems. *Environ Microbiol* **18:** 50–64.

Morgenroth, E., Obermayer, A., Arnold, E., Bruhl, A., Wagner, M., and Wilderer, P. A. (2000) Effect of long-term idle periods on the performance of sequencing batch reactors. *Water Sci Technol* **41(1):** 105–113.

Ni, B. J., Ruscalleda, M., and Smets, B. F. (2012) Evaluation on the microbial interactions of anaerobic ammonium oxidizers and heterotrophs in Anammox biofilm. *Water Res* **46:** 4645–4652.

Okabe, S., Kindaichi, T., and Ito, T. (2005) Fate of 14C-labeled microbial products derived from nitrifying bacteria in autotrophic nitrifying biofilms. *Appl Environ Microbiol* **71:** 3987–3994.

Oshiki, M., Shimokawa, M., Fujii, N., Satoh, H., and Okabe, S. (2011) Physiological characteristics of the anaerobic ammonium-oxidizing bacterium 'Candidatus Brocadia sinica'. *Microbiology* **157:** 1706–1713.

Park, H., Rosenthal, A., Jezek, R., Ramalingam, K., Fillos, J., and Chandran, K. (2010) Impact of inocula and growth mode on the molecular microbial ecology of anaerobic ammonia oxidation (anammox) bioreactor communities. *Water Res* **44:** 5005–5013.

Pellicer-Nàcher, C., Franck, S., Gülay, A., Ruscalleda, M., Terada, A., Al-Soud, W. A., *et al.* (2014) Sequentially aerated membrane biofilm reactors for autotrophic nitrogen removal: microbial community composition and dynamics. *Microb Biotechnol* **7:** 32–43.

Perez, J., Lotti, T., Kleerebezem, R., Picioreanu, C., and van Loosdrecht, M. C. (2014) Outcompeting nitrite-oxidizing bacteria in single-stage nitrogen removal in sewage treatment plants: a model-based study. *Water Res* **66C:** 208–218.

Persson, F., Sultana, R., Suarez, M., Hermansson, M., Plaza, E., and Wilen, B.-M. (2014) Structure and composition of biofilm communities in a moving bed biofilm reactor for nitritation-anammox at low temperatures. *Bioresource Technol* **154:** 267–273.

Schloss, P. D., Westcott, S. L., Ryabin, T., Hall, J. R., Hartmann, M., Hollister, E. B., *et al.* (2009) Introducing mothur: open-source, platform-independent, community-supported software for describing and comparing microbial communities. *Appl Environ Microbiol* **75:** 7537–7541.

Siegrist, H., Salzgeber, D., Eugster, J., and Joss, A. (2008) Anammox brings WWTP closer to energy autarky due to increased biogas production and reduced aeration energy for N-removal. *Water Sci Technol* **57(3):** 383–388.

van der Star, W. R., Miclea, A. I., van Dongen, U. G., Muyzer, G., Picioreanu, C., and van Loosdrecht, M. C. (2008) The membrane bioreactor: a novel tool to grow anammox bacteria as free cells. *Biotechnol Bioeng* **101:** 286–294.

Surmacz-Górska, J., Gernaey, K., Demuynck, C., Vanrolleghem, P. and Verstreate, W. (1996) Nitrification monitoring in activated sludge by oxygen uptake rate (OUR) measurements. *Water Res* **30:** 1228–1236.

Vlaeminck, S. E., Terada, A., Smets, B. F., De Clippeleir, H., Schaubroeck, T., Bolca, S., *et al.* (2010) Aggregate size and architecture determine microbial activity balance for one-stage partial nitritation and anammox. *Appl Environ Microbiol* **76:** 900–909.

Wett, B., Omari, A., Podmirseg, S.M., Han, M., Akintayo, O. and Gómez Brandón, M., *et al.* (2013) Going for mainstream deammonification from bench- to full-scale for maximized resource efficiency. *Water Sci Technol* **68:** 283–289.

Xia, Y., Kong, Y. H., Thomsen, T. R., and Nielsen, P. H. (2008) Identification and ecophysiological characterization of epiphytic protein-hydrolyzing Saprospiraceae ("Candidatus epiflobacter" spp.) in activated sludge. *Appl Environ Microbiol* **74:** 2229–2238.

Zhou, J.Z., He, Z.L., Yang, Y.F., Deng, Y., Tringe, S.G. and Alvarez-Cohen, L. (2015) High-throughput metagenomic technologies for complex microbial community analysis: open and closed formats. *MBio* **6(1):** e02288–14.

Antimicrobial activity of biogenically produced spherical Se-nanomaterials embedded in organic material against *Pseudomonas aeruginosa* and *Staphylococcus aureus* strains on hydroxyapatite-coated surfaces

Elena Piacenza,[1,*] Alessandro Presentato,[1]
Emanuele Zonaro,[2] Joseph A. Lemire,[1]
Marc Demeter,[1] Giovanni Vallini,[2]
Raymond J. Turner[1,**] and Silvia Lampis[2]

[1]*Biofilm Research Group, Department of Biological
Sciences, University of Calgary, 2500 University Dr NW,
Calgary, AB T2N 1N4, Canada.*
[2]*Environmental Microbiology Laboratory, Department of
Biotechnology, University of Verona, Strada Le Grazie
15, 37134 Verona, Italy.*

Summary

In an effort to prevent the formation of pathogenic biofilms on hydroxyapatite (HA)-based clinical devices and surfaces, we present a study evaluating the antimicrobial efficacy of Spherical biogenic Se-Nanostructures Embedded in Organic material (Bio Se-NEMO-S) produced by *Bacillus mycoides* SelTE01 in comparison with two different chemical selenium nanoparticle (SeNP) classes. These nanomaterials have been studied as potential antimicrobials for eradication of established HA-grown biofilms, for preventing biofilm formation on HA-coated surfaces and for inhibition of planktonic cell growth of *Pseudomonas aeruginosa* NCTC 12934 and *Staphylococcus aureus* ATCC 25923. Bio Se-NEMO resulted more efficacious than those chemically produced in all tested scenarios. Bio Se-NEMO produced by *B. mycoides* SelTE01 after 6 or 24 h of Na_2SeO_3 exposure show the same effective antibiofilm activity towards both *P. aeruginosa* and *S. aureus* strains at 0.078 mg ml^{-1} (Bio Se-NEMO$_6$) and 0.3125 mg ml^{-1} (Bio Se-NEMO$_{24}$). Meanwhile, chemically synthesized SeNPs at the highest tested concentration (2.5 mg ml^{-1}) have moderate antimicrobial activity. The confocal laser scanning micrographs demonstrate that the majority of the *P. aeruginosa* and *S. aureus* cells exposed to biogenic SeNPs within the biofilm are killed or eradicated. Bio Se-NEMO therefore displayed good antimicrobial activity towards HA-grown biofilms and planktonic cells, becoming possible candidates as new antimicrobials.

Introduction

In the last 20 years, the potential to use nanoparticles (NPs) as antimicrobial agents has been evaluated (Ankamwar *et al.*, 2005). Primarily, the focus was to synthesize NPs using various chemical methods. However, both the high cost of production and presence of toxic by-products generated a demand for novel methods to synthesize NPs (Ankamwar *et al.*, 2005). Biological systems such as plants, fungi and bacteria have the capacity to convert several toxic metal ions into less toxic forms including metal precipitants or NPs (Suresh *et al.*, 2004; Bhainsa and D'Souza, 2006; Song *et al.*, 2009). Thanks to the potential technological importance of such NPs, research interest has been focused on the use of these organisms to produce NPs with eco-friendly and 'green synthesis' methods (Ingale and Chaudhari, 2013). One of the first classes of biogenic NPs to be evaluated was silver NPs (AgNPs), due to the demonstrated antimicrobial ability of metallic silver (Ag; Dos Santos *et al.*, 2014). Generally, biogenic AgNPs are produced using fungal cultures able to bioaccumulate metals and synthesize NPs, which are excreted outside cells using their filaments (Srivastava and Mukhopadhyay, 2015). Some commonly used fungi for AgNP production are *Verticillium* (Mukherjee *et al.*, 2001), *Aspergillus flavus* (Vigneshwaran *et al.*, 2007) and *Fusarium oxysporum* (Ahmad *et al.*, 2003). AgNP synthesis using bacterial cell extracts from *Lactobacillus acidophilus* (Rajesh *et al.*, 2015) or *Corynebacterium glutamicum* (Gowramma *et al.*, 2015) has also been investigated.

For correspondence. *E-mail elena.piacenza@ucalgary.ca

**E-mail turnerr@ucalgary.ca

Funding Information
Natural Sciences and Engineering Research Council of Canada
(Grant/Award Number: 216887-2010).

Recently, selenium NPs (SeNPs) has emerged as a new class of potential antimicrobial agents. Selenium is an essential micronutrient in biologic systems; it has anticancer and antimicrobial properties, antioxidant effects and modulation functions for the immune system (Sadeghian *et al.*, 2012). Despite these roles, selenium can be found in the environment as toxic forms, such as selenate (SeO_4^{2-}), selenite (SeO_3^{2-}) and selenide (Se^{2-}) (Lampis *et al.*, 2014). Toxicity of these selenium anions depends on their mobilization and availability in soils and water, improving the possibility of exposure for humans and animals (Lampis *et al.*, 2014). In this respect, efforts have been made to identify a non-toxic form of selenium useable in biomedical applications (Wang *et al.*, 2013). Recently, it has been established that SeNPs have higher biocompatibility and lower toxicity in humans compared with bulk selenium (Lampis *et al.*, 2014). Additionally, SeNPs have unique physical and chemical properties due to their large surface–volume ratio, large surface energy, spatial confinement and reduced imperfections (Stroyuk *et al.*, 2008). In this regard, SeNPs possess adsorptive ability, oxidation functions and marked biological reactivity, including antihydroxyl radical efficacy (Lampis *et al.*, 2014). SeNPs are normally synthesized by chemical reduction of selenite or selenous acid by reducing agents such as glutathione (GSH), hydrazine, sodium borohydride ($NaBH_4$), stannous chloride $SnCl_2$, L-cysteine, ascorbic acid, sodium thiosulfate ($Na_2S_2O_3$) and SDS (Stroyuk *et al.*, 2008). These techniques are unfavourable as they produce particles that are subject to photocorrosion (Dobias *et al.*, 2011). Yet, it is now possible to produce SeNPs of various compositions, sizes and morphologies using bacteria (Lampis *et al.*, 2014). A number of bacterial species, residing in diverse terrestrial and aquatic environments, are resistant to selenium oxyanions and possess the ability to reduce selenite and selenate into its less bioavailable elemental form (Se^0; Stolz *et al.*, 2006). This process occurs through both enzymatic and non-enzymatic mechanisms, leading to the formation of SeNPs that are deposited inside the cell periplasm or excreted (Kessi *et al.*, 1999; Oremland *et al.*, 2004; Dhanjal and Cameotra, 2010).

One of the most clinically sought properties of SeNPs is their ability to inhibit pathogenic biofilm formation (Wang and Webster, 2012). Most bacteria have the innate ability to populate, as a biofilm, a vast array of surfaces, including those where sterility is of paramount importance to human health, as food-processing facilities, dental hygiene equipment and medical devices to name a few (Harrison *et al.*, 2004). Two of the most diffuse and harmful bacteria able to grow as biofilms responsible for human pathogenic infections are *Staphylococcus aureus*, which is a Gram-positive bacterium

that causes many serious infections in surgical wounds, bloodstream or in the lungs (Alhede *et al.*,2009), and *Pseudomonas aeruginosa*, which is a Gram-negative bacteria recognized as one of the most important nosocomial opportunistic pathogens, affecting the urinary tract, respiratory system, soft tissue, bone and joint tissue (Alhede *et al.*, 2009). Of particular interest here are biofilms of these species, which are able to grow on artificial orthopaedic implants made of hydroxyapatite (HA) – causing severe infections (Kolmas *et al.*, 2015a). HA is a naturally occurring mineral formed by calcium and phosphate ions – $Ca_5(PO_4)_3(OH)$ – with a recognizable crystalline structure (Elliot, 1994). HA is also the principal component of bones and teeth in humans and animals and is therefore used regularly in orthopaedic surgery and for the replacement of teeth (Gong *et al.*, 2015). To prevent bacterial contamination and subsequent infection, research efforts have focused on developing medical devices and implants out of materials modified with antimicrobial agents such as antibiotics or metallic compounds that are able to kill bacterial cells, or inhibit their growth, without being in general toxic to surrounding tissues (Kolmas *et al.*, 2015a; Lim *et al.*, 2013; Harrison *et al.*, 2004). However, with the broad use and abuse of antibiotics, the emergence of bacterial resistance to these antimicrobials has become one of the greatest health challenges worldwide. Moreover, the use of metallic compounds as coating can cause severe problems, such as ion release in human body and development of several side-effects (Crobb and Schmalzereid, 2006). Indeed, some metal ions can undergo corrosion processes, activating immunological responses and cytotoxic and genotoxic effects (Jacobs *et al.*, 1994; Merritt and Rodrigo, 1996). Furthermore, the ionic species present in HA are normally subject to substitution processes that can dramatically change the structure and properties of HA itself (Kolmas *et al.*, 2015b).

Recently, to overcome both bacterial resistance to antibiotics and the side-effects of metallic compounds, new antimicrobial technologies for HA-based implants have been investigated, such as the use of nano-sized HA conjugated with chemical NPs known for their antimicrobial ability (Kolmas *et al.*, 2015b). In this regard, due to the peculiar chemistry of nanomaterials as compared to one of the bulk ions, the use of NPs as antimicrobial agents could constitute an alternative choice to the use of metallic ions. Specifically, chemically synthesized SeNPs were already evaluated as possible doping for HA-based implants, thanks to either their antimicrobial or anticancer properties (Kolmas *et al.*, 2015b; Wang *et al.*, 2016). Selenium also plays a crucial role in bone growth and proliferation of osteoblasts (Kolmas *et al.*, 2015b), and SeNPs could prove

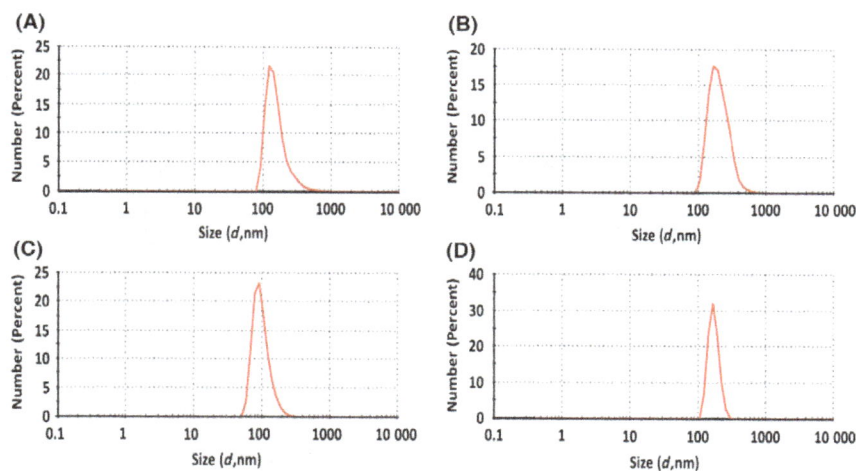

Fig. 1. Dynamic light scattering (DLS) analysis of biogenic SeNPs produced by *Bacillus mycoides* SelTE01 after 6 (A) or 24 h (B) of Na$_2$SeO$_3$ exposure, and chemical SeNPs made using L-cysteine (C) or ascorbic Acid (D).

to be an added value as an antimicrobial coating for HA-based implants.

In this study, we evaluated the antimicrobial and anti-biofilm efficacy of biogenic SeNPs as a useful and alternative approach to chemically synthesized SeNPs, which are already described for their ability to inhibit biofilm proliferation (Wang and Webster, 2012). The antimicrobial properties of both chemical and our biogenic Se-nanomaterials were studied towards *P. aeruginosa* and *S. aureus* biofilms grown onto HA-coated surfaces. Biogenic SeNPs were produced by exposing *Bacillus mycoides* SelTE01 for either 6 or 24 h to Na$_2$SeO$_3$. This strain is a Gram-positive bacterium isolated from the selenium hyperaccumulator plant *Astragalus bisulcatus* described for its ability to reduce selenite oxyanion into its elemental form as SeNPs (Lampis *et al.*, 2014). Chemical SeNPs were synthesized using L-cysteine (Li *et al.*, 2010) or ascorbic acid (Zhang *et al.*, 2004) as reducing agents.

Results

Characterization of SeNPs

The hydrodynamic diameter of chemically and biogenically produced SeNPs has been evaluated by dynamic light scattering (DLS) analysis (Fig. 1). Biogenic SeNPs produced by *B. mycoides* SelTE01 after 6 h of Na$_2$SeO$_3$ exposure are characterized by a sharp peak at 160 ± 58.6 nm (Fig. 1A), while a broad and shifted peak of 209.1 ± 79.1 nm has been detected for those obtained after 24 h of growth in the presence of the selenite precursor (Fig. 1B). DLS number size distributions of SeNPs produced using L-cysteine (L-cys SeNPs) or ascorbic acid (Asc SeNPs) show sharp and defined peaks at 99.8 ± 30.2 nm and 170.5 ± 64.4 nm

respectively (Fig. 1C and D). Furthermore, polydispersity indexes (PDIs) of both chemically and biogenically synthesized SeNPs have been also evaluated to study the stability of NPs in solution. L-cys SeNPs and biogenic SeNPs produced after 6 h of Na$_2$SeO$_3$ exposure are characterized by similar PDI values, namely 0.198 and 0.220. Moreover, comparable PDI values for Asc SeNPs (0.312) and those biogenically synthesized after 24 h of precursor exposure (0.290) have been determined.

Transmission Electron Microscopy (TEM) micrographs of biogenically produced SeNPs highlighted the presence of spherical highly electron-dense NPs different in size and embedded in a light grey and uniform matrix (Fig. 2A and B). Analysis of L-cys SeNPs showed spherical and strongly electron-dense NPs more uniform in size rather than those biogenically produced (Fig. 2C). Considering Asc SeNPs, TEM image showed big NPs with aggregates in the population (Fig. 2D). However, in both chemical SeNP classes, the presence of an embedding matrix has not been detected.

Elemental composition of biogenically and chemically produced SeNPs has been evaluated by performing energy-dispersive X-ray spectroscopy (EDX). Biogenic SeNPs produced by *B. mycoides* SelTE01 after 6 or 24 h of Na$_2$SeO$_3$ exposure revealed the presence of the characteristic Se absorption peaks at 1.37 (SeLα), 11.22 (SeKα) and 12.49 (SeKβ) keV, while only the SeLα peak was detected in both chemically synthesized SeNPs (Fig. 3). Moreover, both biogenic and chemical SeNPs showed similar elemental composition, with the presence of selenium, carbon, oxygen, phosphorus and sulphur, but in different relative percentages (Table 1). Biogenically synthesized SeNPs showed minor

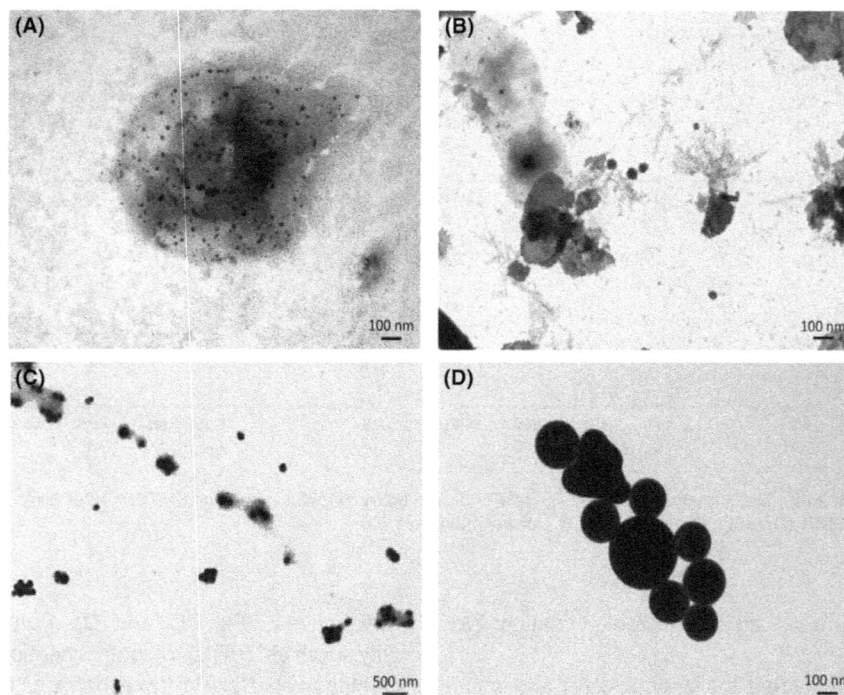

Fig. 2. Transmission Electron Microscopy (TEM) analysis of biogenic SeNPs produced by *Bacillus mycoides* SelTE01 after 6 (A) or 24 h (B) of Na$_2$SeO$_3$ exposure, and chemical SeNPs made using L-cysteine (C) or ascorbic Acid (D).

differences in the detected relative percentage values of P and S, while C, O and Se were present in similar amount in both samples. Overall, the relative percentage values of carbon, oxygen, phosphorous and sulphur for biogenic SeNPs suggested the presence of organic molecules associated with the extracted SeNPs, and TEM observations confirmed the presence of a slightly electron-dense material surrounding the particles. Thus, from here on, we refer to biogenically synthesized SeNPs as Spherical Bio Se-Nanostructures Embedded in an Organic material (Bio Se-NEMO-S). Particularly, Bio Se-NEMO-S$_6$ and Bio Se-NEMO-S$_{24}$ refer to the biogenic SeNPs produced by *B. mycoides* SelTE01 after 6 and 24 h of Na$_2$SeO$_3$ exposure respectively.

For both chemically synthesized SeNPs, the same elements (C, O, Se, P, S) are present yet with different relative percentage values, due to the procedure of the production, using either L-cysteine or ascorbic acid as reducing agent (Table 1).

Zeta potential measurements have been carried out to study the stability of either Bio Se-NEMO-S or chemical SeNPs in solution. In all four cases, a highly negative zeta potential value has been detected (Fig. 4). Particularly, both Bio Se-NEMO-S$_6$ and Bio Se-NEMO-S$_{24}$ generated the same surface charge of −74.2 mV (Fig. 4A and B), while L-cys and Asc SeNPs showed a Z

potential value of −67.9 and −75.9 mV respectively (Fig. 4C and D).

Evaluation of the antimicrobial ability of Bio Se-NEMO and chemical SeNPs

Antimicrobial activity against established biofilms. Eradication of an established biofilm has been evaluated growing both *P. aeruginosa* NCTC 12934 (Figs 5A and 6A) and *S. aureus* ATCC 25923 (Figs 5B and 6B) strains for 24 h prior the treatment with each tested Se-nanomaterial using HA-coated Calgary Biofilm Devices (CBDs) to allow the growth of biofilms (Harrison et al., 2010). Results are shown for both Bio Se-NEMO-S (Bio Se-NEMO-S$_6$ and Bio Se-NEMO-S$_{24}$) and chemical SeNPs (L-cys and Asc SeNPs). Particularly, in the case of *P. aeruginosa* grown biofilm, Bio Se-NEMO$_{24}$ and Asc SeNPs showed a 2 log decrease in the cell viable count at the concentration of 2.5 mg ml^{-1}, while a slight antimicrobial activity (1 log reduction) has been observed for Bio Se-NEMO-S$_6$ and L-cys SeNPs (Figs 5A and 6A). Considering *S. aureus* established biofilm, Bio Se-NEMO-S$_{24}$ and Bio Se-NEMO-S$_6$ affected its fitness with a 2 log reduction in the cellular population at the concentrations of 0.625 and 1.25 mg ml^{-1} respectively (Figs 5B and 6B). Moreover, Bio Se-NEMO-S$_{24}$ exerted their strongest antimicrobial ability (4 log) against *S. aureus* grown

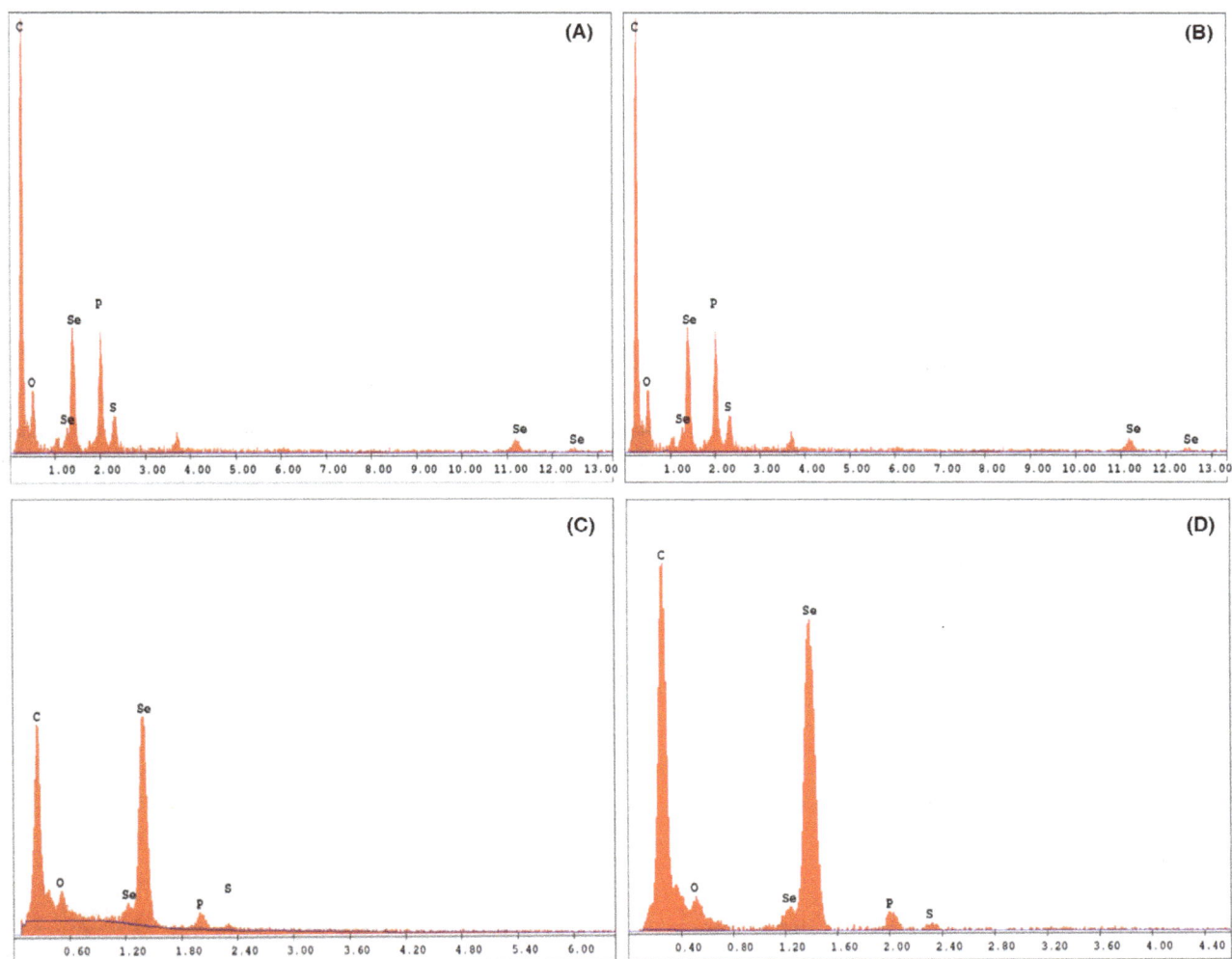

Fig. 3. Energy-dispersive X-ray spectroscopy (EDX) spectra of Bio Se-NEMO-S produced by *Bacillus mycoides* SeITE01 after 6 (A) or 24 h (B) of Na_2SeO_3 exposure, and chemical SeNPs made using L-cysteine (C) or ascorbic Acid (D).

Table 1. Elemental quantification (as weight relative percentage) of Bio Se-NEMO and chemical SeNPs.

Element	Bio Se-NEMO-S		Chemical SeNPs	
	Bio Se-NEMO-S$_6$ Weight (Rel. %)	Bio Se-NEMO-S$_{24}$ Weight (Rel. %)	L-cys SeNPs Weight (Rel. %)	Asc SeNPs Weight (Rel. %)
Selenium (Se)	9.26	9.82	31.61	16.04
Carbon (C)	75.75	80.71	60.91	76.88
Oxygen (O)	10.82	8.45	4.97	6.27
Phosphorous (P)	3.14	0.73	1.88	0.62
Sulphur (S)	1.04	0.29	0.63	0.19

Elemental quantification is expressed as weight relative percentage of the element detected in the SeNPs samples.

biofilm at 2.5 mg ml^{-1}, while the same concentration of L-cys and Asc SeNPs determined a 2 log decrease upon the cells within the biofilm (Figs 5B and 6B). Nevertheless, in the range of tested concentrations, none of the studied Bio Se-NEMO-S or SeNPs classes showed a complete eradication of a pre-formed biofilm. Thus, a Minimal Biofilm Eradication Concentration (MBEC), that is the concentration of either Bio Se-NEMO-S or chemical SeNPs at which there is the eradication of an already established biofilm, could not be determined.

(A)

(B)

(C)

(D)

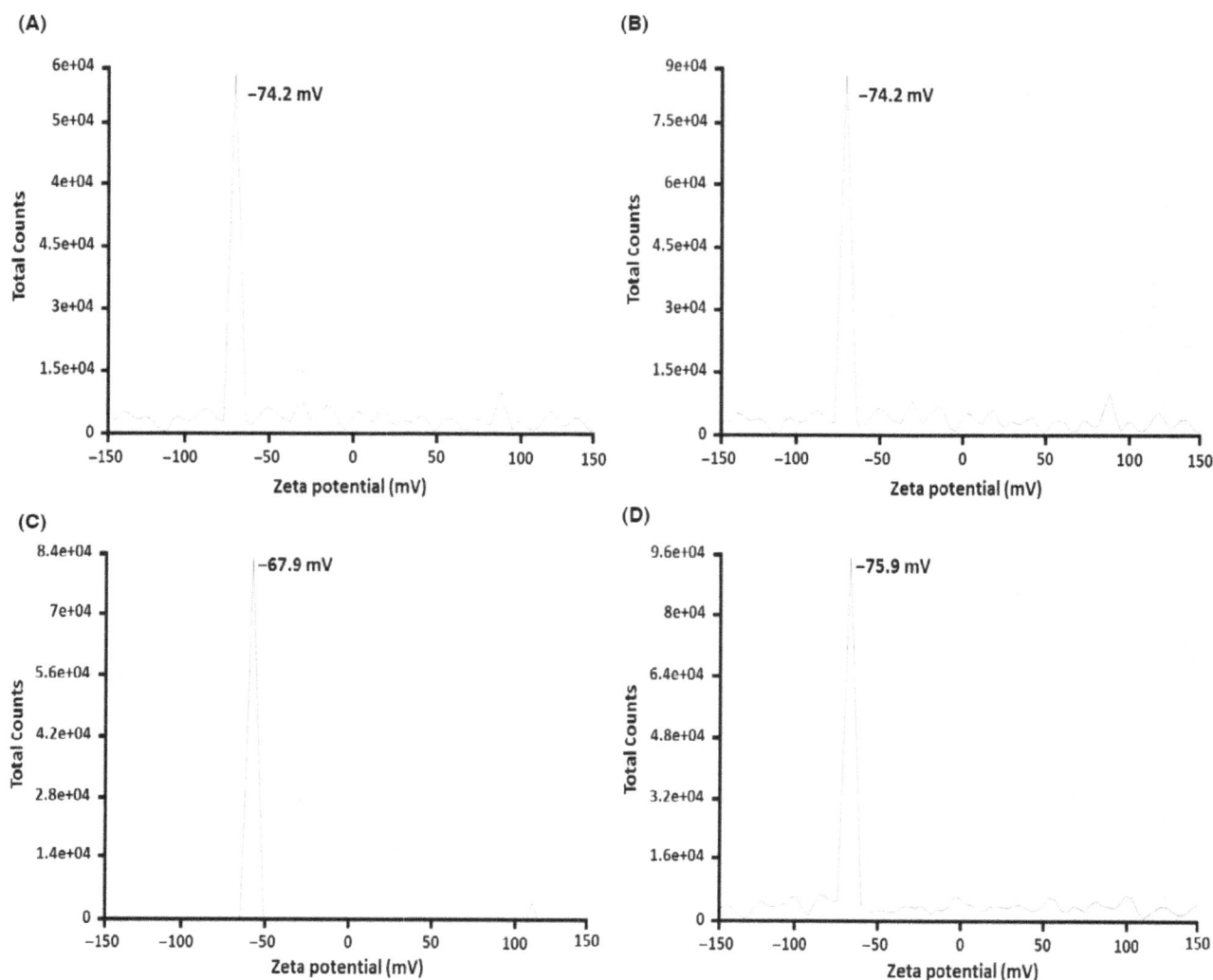

Fig. 4. Zeta potential measurements of Bio Se-NEMO-S produced by *Bacillus mycoides* SeITE01 after 6 (A) or 24 h (B) of Na$_2$SeO$_3$ exposure, and chemical SeNPs made using L-cysteine (C) or ascorbic acid (D).

Inhibition of biofilm formation and growth. The CBDs were also used to evaluate the potential of either Bio Se-NEMO-S or SeNPs to inhibit biofilm formation by adding Se-nanomaterials at the time of inoculation. In this assay, the bacterial cells must survive the stress/challenge of the biocide (Bio Se-NEMO-S or SeNPs) in the planktonic state long enough to initiate attachment and then proliferate as a biofilm. The data are plotted as number of *P. aeruginosa* (Figs 7A and 8A) or *S. aureus* (Figs 7B and 8B) viable cells (24 h) at varying concentrations of biocide.

For either Bio Se-NEMO-S or chemical SeNPs that demonstrated high antimicrobial activity, a Minimal Biofilm Prevention Concentration (MBPC), defined as the concentration after which an antimicrobial agent is able to totally inhibit and prevent cell attachment and biofilm proliferation, was established. Using this approach, an

antibiofilm efficacy with differences in trends between chemical SeNPs and Bio Se-NEMO-S has been observed. Both Bio Se-NEMO-S$_6$ and Bio Se-NEMO-S$_{24}$ showed very good antimicrobial activity with defined MBPC that totally inhibited the establishment and growth of *P. aeruginosa* and *S. aureus* biofilms. In particular, Bio Se-NEMO-S$_6$ has strong antibiofilm efficacy as well as Bio Se-NEMO-S$_{24}$ against both tested strains, with MBPC values of 0.078 and 0.3125 mg ml^{-1} respectively (Fig. 7A and B). These results also showed that both L-cys and Asc SeNPs are not able to completely prevent the formation of the pathogen indicator strain biofilms (Fig. 8A and B). However, L-cys SeNPs still maintained antibiofilm activity, as they caused a decrease in the number of *P. aeruginosa* (3 log) and *S. aureus* (4 log) viable cells. Finally, Asc SeNPs showed a slight antimicrobial ability for the two tested strains. In particular, we

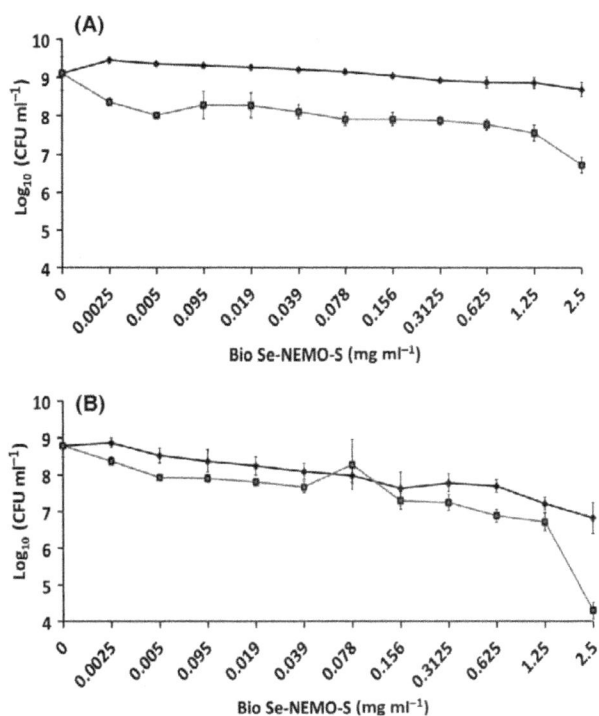

Fig. 5. Minimal Biofilm Eradication Concentration (MBEC) assays of *Pseudomonas aeruginosa* NCTC 12934 (A) and *Staphylococcus aureus* ACTT 25923 (B) established biofilms for 24 h, and subsequently exposed for 24 h to –◆– Bio Se-NEMO-S$_6$ and –■– Bio Se-NEMO-S$_{24}$. Error bars show the standard deviation.

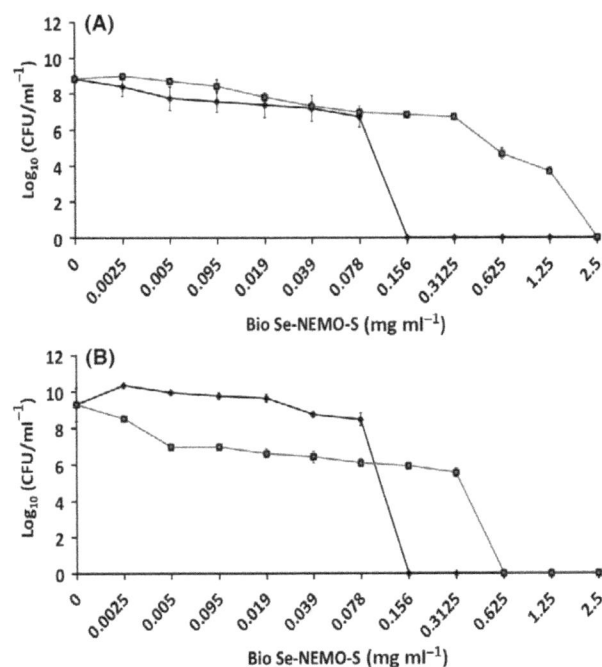

Fig. 6. Minimal Biofilm Eradication Concentration (MBEC) assays of *Pseudomonas aeruginosa* NCTC 12934 (A) and *Staphylococcus aureus* ACTT 25923 (B) established biofilms for 24 h, and subsequently exposed for 24 h to ··▲·· L-cys SeNPs and ✕ Asc SeNPs. Error bars show the standard deviation.

observed a 1 log reduction in the number of biofilm-resident bacteria against *P. aeruginosa* and 2 log in the case of *S. aureus*. As chemically synthesized SeNPs show a moderate antimicrobial ability, MBPC values were determined only for Bio Se-NEMO-S.

Efficacy on planktonic cultures. Efficacy of the both Bio Se-NEMO-S and chemical SeNPs against planktonic cells of *P. aeruginosa* (Figs 9A and 10A) and *S. aureus* (Figs 9B and 10B) was also tested. Results demonstrated a strong antimicrobial effect of Bio Se-NEMO-S on planktonic cultures of the tested strains. In this case, a Minimal Inhibitory Concentration (MIC), that is the concentration of an antimicrobial agent after which there is no planktonic cells growth, has been established for Bio Se-NEMO-S. Particularly, the MIC values of Bio Se-NEMO-S were equal to the MBPC values established for inhibiting biofilm growth, being 0.078 mg ml^{-1} for Bio Se-NEMO-S$_6$ and 0.3125 mg ml^{-1} for Bio Se-NEMO-S$_{24}$ against both *P. aeruginosa* and *S. aureus* cultures (Fig. 9A and B). Among the chemically synthesized SeNPs, L-cys SeNPs reveal a moderate antimicrobial activity at the highest tested concentrations (1.25–2.5 mg ml^{-1}) with a 4 log decrease in the number of *P. aeruginosa* and *S. aureus* growing cells (Fig. 10A and B). Moreover, Asc

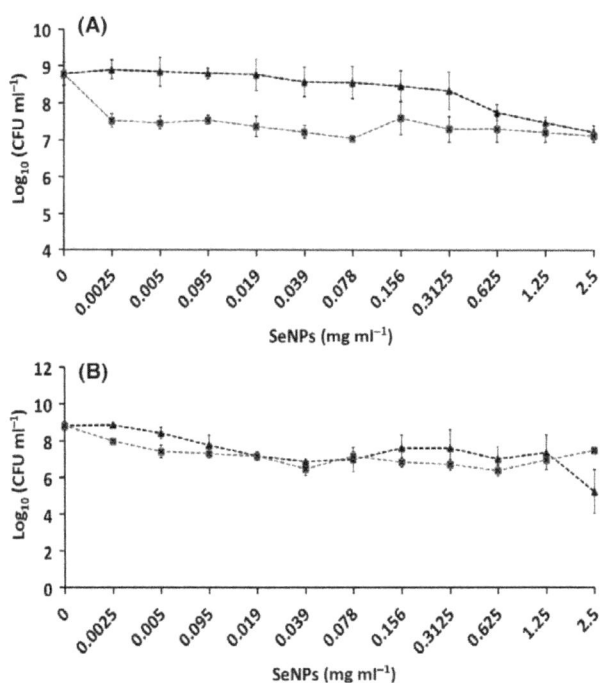

Fig. 7. Minimal Biofilm Prevention Concentration (MBPC) assays of *Pseudomonas aeruginosa* NCTC 12934 (A) and *Staphylococcus aureus* ACTT 25923 (B) growing biofilms exposed for 24 h to –◆– Bio Se-NEMO-S$_6$ and –■– Bio Se-NEMO-S$_{24}$. Error bars show the standard deviation.

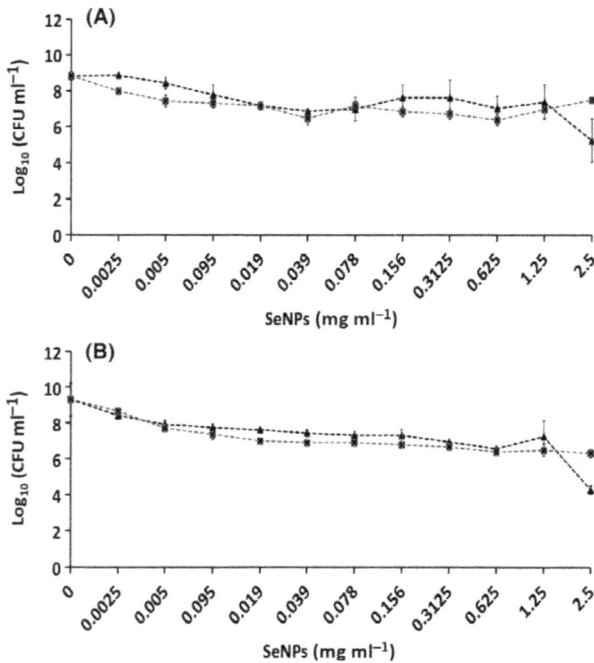

Fig. 8. Minimal Biofilm Prevention Concentration (MBPC) assays of *Pseudomonas aeruginosa* NCTC 12934 (A) and *Staphylococcus aureus* ACTT 25923 (B) growing biofilms exposed for 24 h to ··▲·· L-cys SeNPs and ·×· Asc SeNPs. Error bars show the standard deviation.

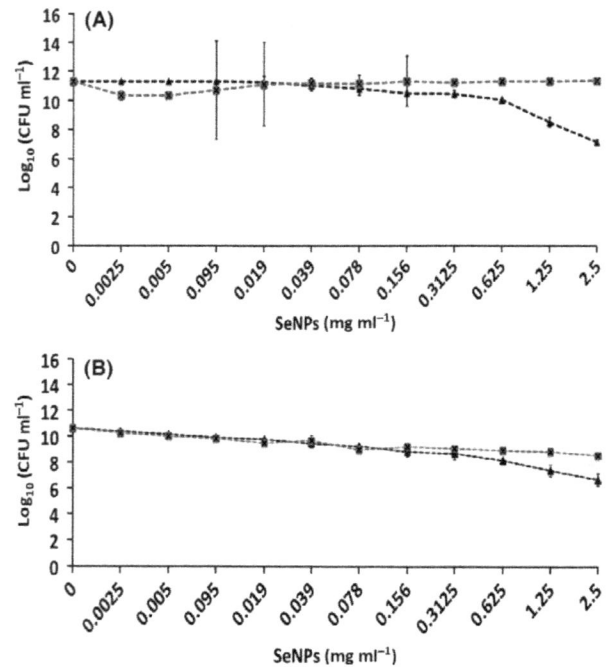

Fig. 9. Minimal Inhibition Concentration (MIC) assays of *Pseudomonas aeruginosa* NCTC 12934 (A) and *Staphylococcus aureus* ACTT 25923 (B) growing planktonic cells exposed for 24 h to ·◆· Bio Se-NEMO-S$_6$ and ·■· Bio Se-NEMO-S$_{24}$. Error bars show the standard deviation.

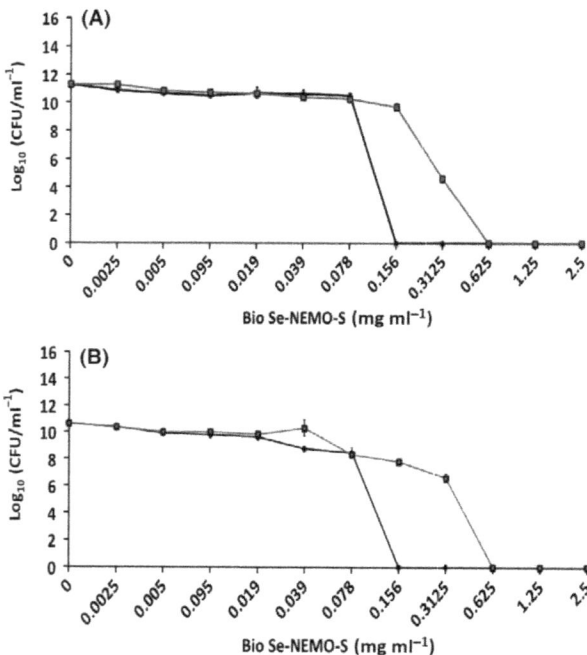

Fig. 10. Minimal Inhibition Concentration (MIC) assays of *Pseudomonas aeruginosa* NCTC 12934 (A) and *Staphylococcus aureus* ACTT 25923 (B) growing planktonic cells exposed for 24 h to ··▲·· L-cys SeNPs and ·×· Asc SeNPs. Error bars show the standard deviation.

SeNPs display a slight efficacy only against *S. aureus* planktonic population (2 log reduction), without any decrease in the number of *P. aeruginosa* cells (Fig. 10A and B). Due to their moderate antimicrobial efficacy, no MIC values have been established for chemical SeNPs.

Confocal Laser Scanning Microscopy (CLSM) analysis

P. aeruginosa (Fig. 11) and *S. aureus* (Fig. 12) biofilms growing on HA-coated pegs in the presence of either Bio Se-NEMO-S or chemical SeNPs have been observed by performing CLSM analyses. We chose to analyse Bio Se-NEMO-S$_6$, as this nanomaterial possessed the strongest antibiofilm activity, making a comparison with L-cys SeNPs, which showed the highest antimicrobial efficacy among the tested chemical SeNPs. Specifically, we analysed three samples of *P. aeruginosa* and *S. aureus* growing biofilms exposed to the different Bio Se-NEMO-S$_6$ SeNP concentrations of 0.039, 0.078 mg ml^{-1} (established MBPC) and 1.25 mg ml^{-1}. Due to the absence of an defined MBPC for L-cys SeNPs, the antimicrobial effect of these NPs has been evaluated exposing *P. aeruginosa* and *S. aureus* growing biofilms to the three highest SeNP concentrations of the tested range: 0.625, 1.25 and 2.5 mg ml^{-1}. Samples were stained using Live/Dead® BacLight™ staining kit, as described in

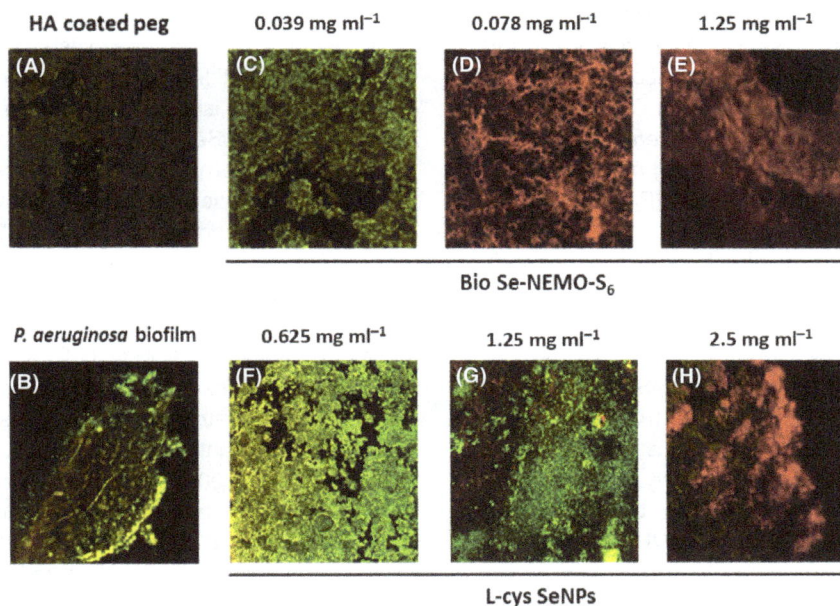

Fig. 11. Confocal Laser Scanning Microscopy (CLSM) of HA-coated peg (A), *Pseudomonas aeruginosa* biofilm grown onto HA-coated peg (B) and in the presence of bio6 SeNPs at the concentration of 0.039 mg ml^{-1} (C), 0.078 mg ml^{-1} (MBBC) (D), 1.25 mg ml^{-1} (E), or in the presence of L-cys SeNPs at the concentration of 0.625 mg ml^{-1} (F), 1.25 mg ml^{-1} (G), 2.5 mg ml^{-1} (H).

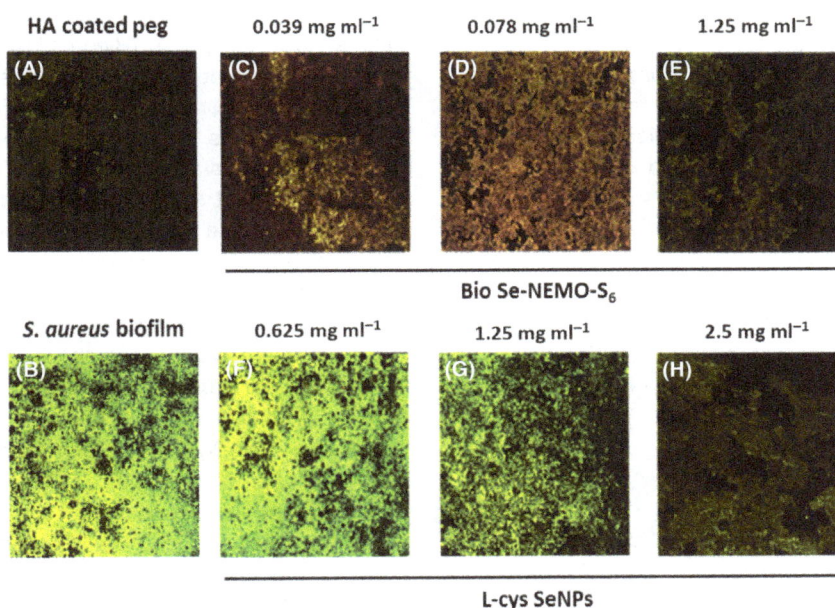

Fig. 12. Confocal Laser Scanning Microscopy (CLSM) of HA-coated peg (A), *Staphylococcus aureus* biofilm grown onto HA-coated peg (B) and in the presence of bio6 SeNPs at the concentration of 0.039 mg ml^{-1} (C), 0.078 mg ml^{-1} (MBBC) (D), 1.25 mg ml^{-1} (E), or in the presence of L-cys SeNPs at the concentration of 0.625 mg ml^{-1} (F), 1.25 mg ml^{-1} (G), 2.5 mg ml^{-1} (H).

the Methods section. Due to the ion exchange chemistry of HA, the cations of the stain were able to bind to HA, resulting in a dark green or red background noise (Figs 11A and 12A), which was different from the brilliant and strong fluorescence corresponding to the stained cells.

P. aeruginosa biofilms grown in the presence of the lowest concentrations of Bio Se-NEMO-S$_6$ (0.039 mg ml^{-1}) or L-cys SeNPs (0.625 mg ml^{-1}; Fig. 11C and F) were characterized by a strong green-fluorescent signal, indicating a mature and vibrant biofilm. However, *P. aeruginosa* biofilm exposed to the action of Bio Se-NEMO-S$_6$ showed a reduced consistency (375 μm), while L-cys SeNPs had a slight effect on biofilm thickness (415 μm), being comparable to the non-treated one (450 μm; Fig. 11B). Observations are drastically different considering *P. aeruginosa* biofilms grown in the presence of 0.078 mg ml^{-1} (MPBC) or 1.25 mg ml^{-1} of Bio Se-NEMO-S$_6$ (Fig. 11D and G), which showed strong and brilliant red colour, indicating not viable cells. Moreover, the thickness of these biofilms dropped to values of 150.4 μm for 0.078 mg ml^{-1} or 75.3 μm in the case of the highest tested concentration (1.25 mg ml^{-1}). By contrast, samples grown with 1.25 mg ml^{-1} or 2.5 mg ml^{-1} L-cys SeNPs had both green- and red-fluorescent signals (Fig. 11E and H), demonstrating that there was a coexistence of live and dead bacterial cells. Furthermore, analysis of these samples showed the presence of thick biofilms for both tested concentrations of chemical SeNPs, being 223 μm (1.25 mg ml^{-1}) and 185,5 μm (2.5 mg ml^{-1}).

S. aureus biofilm grown on HA-coated peg (Fig. 12B) appeared to be thinner (285 μm) as compared with that of *P. aeruginosa* (450 μm). A slight fluorescent signal was observed when *S. aureus* cells growing as a biofilm were exposed to 0.039 or 0.078 mg ml^{-1} (MPBC) of Bio Se-NEMO-S$_6$ (Fig. 12C and D), resulting in an evaluated thickness of 94.2 and 2.5 μm respectively. This value decreased till 0.95 μm when *S. aureus* biofilm was grown in the presence of 1.25 mg ml^{-1} of Bio Se-NEMO-S$_6$, with a fluorescence emission comparable to the HA-coated peg itself (Fig. 12A and E). On the opposite, the lowest concentration of L-cys SeNPs (0.625 mg ml^{-1}) slightly affected *S. aureus* biofilm growth, as highlighted by the strong and green fluorescence due to the presence of a live and thicker biofilm (192.7 μm; Fig. 12F). Moreover, when *S. aureus* cells were incubated with 1.25 or 2.5 mg ml^{-1} of L-cys SeNPs, the growing biofilm displayed a thickness of 114 or 25 μm respectively.

Discussion

In this study, we analyse the potential of either biogenic SeNPs produced by *B. mycoides* SelTE01 (Lampis et al., 2014) indicated as Bio Se-NEMO-S or chemically synthesized SeNPs (L-cys SeNPs and Asc SeNPs) as antimicrobial agents against pathogenic indicator strains *P. aeruginosa* NCTC 12934 and *S. aureus* ATCC 25923 grown on CBDs coated with HA, which is an important bioactive ceramic material used in orthopaedic and dental surgery (Gong et al., 2015). Considering the possible medical application of such nanomaterials, the toxicity of either Bio Se-NEMO-S produced by *B. mycoides* SelTE01 or chemical SeNPs towards human cells has been investigated by Cremonini et al. (2016), showing

no cytotoxicity and biocompatibility with human cell lines (i.e. dendritic cells and fibroblasts).

In the evaluation of Se-nanomaterials' ability to eradicate both established *P. aeruginosa* and *S. aureus* biofilms, either Bio Se-NEMO-S or chemical SeNPs led to a decrease in the number of viable cells. Bio Se-NEMO-S$_{24}$ SeNPs showed the strongest antimicrobial activity at concentration ranging from 0.625 to 2.5 mg ml^{-1} against *S. aureus* grown biofilm. In this respect, the moderate efficacy of antimicrobials towards an already established biofilm is due to the complexity of the biofilm structure itself. In particular, bacterial cells within a biofilm are surrounded by an extracellular polymeric substance (EPS), which provides structural stability, as well as acts like a barrier against antibiotics, metallic ions and bactericides (Harrison et al., 2004). Moreover, a subpopulation of biofilm-resident bacteria behaves as persister cells, which are dormant and metabolically inactive (Hobby et al., 1942). These cells were observed for the first time by Hobby et al. (1942) and Bigger (1944) studying the effect of penicillin against *S. aureus* and *S. pyogenes* strains respectively. In both cases, some cells were able to survive in the presence of the antibiotic without undergoing genetic change. In this regard, persister cells should be considered as a subpopulation of bacteria with tolerance to antibiotics (Bigger, 1944). The presence of both EPS and persister cells inside a biofilm structure can explain the modest antimicrobial activity of tested Se-nanomaterials in the eradication of *P. aeruginosa* and *S. aureus* established biofilms.

In the present study, the antimicrobial efficacy of Bio Se-NEMO-S and chemically synthesized SeNPs has been also evaluated towards *P. aeruginosa* and *S. aureus* cells growing under either planktonic or biofilm conditions. In both scenarios, Bio Se-NEMO-S highlighted a great and equivalent antimicrobial ability against the tested strains. Particularly, growth of planktonic cells and biofilm formation are inhibited at the same concentrations of Bio Se-NEMO-S$_6$ (0.078 mg ml^{-1}) and Bio Se-NEMO$_{24}$ (0.3125 mg ml^{-1}) for both *P. aeruginosa* and *S. aureus* strains respectively. Notably, Bio Se-NEMO-S$_6$ were more efficient than Bio Se-NEMO-S$_{24}$ in preventing the biofilm establishment and growth. Zonaro et al. (2015) have observed a similar behaviour for SeNPs produced by *Stenotrophomonas maltophilia* SelTE02, when it was exposed to Na$_2$SeO$_3$ for 6 or 24 h. In this case, those produced after 6 h showed a stronger antimicrobial activity against *P. aeruginosa* PAO1, *S. aureus* ATCC 25923 and *E. coli* JM109 strains rather than those obtained after 24 h (Zonaro et al., 2015). This diversity in action for biogenic SeNPs is ascribed to the different size of SeNPs themselves (Zonaro et al., 2015). Similar results have been observed by Lu et al. (2013) in the evaluation of AgNPs' action as antimicrobial agents against oral anaerobic pathogenic bacteria. Lu *et al.* (2013) indicated a size-dependent mechanism of AgNPs' action, by which smaller NPs have higher antimicrobial effect. In our study, the same observation is supported by DLS analyses, in which Bio Se-NEMO-S$_6$ showed a smaller hydrodynamic diameter (160 nm) as compared to Bio Se-NEMO-S$_{24}$ (209.1 nm). Furthermore, the Bio Se-NEMO-S sizes suggested the existence of a time-dependent mechanism of biogenic SeNPs' production. In this regard, improving the exposure time of *B. mycoides* SeITE01 to the selenium precursor (Na$_2$SeO$_3$) resulted in the production of bigger Se-nanomaterials, as already described by Lampis *et al.* (2014). As Bio Se-NEMO-S$_6$ showed the highest antimicrobial ability among all tested NPs, we further performed CLSM analyses to evaluate the fitness of *P. aeruginosa* and *S. aureus* cells growing as a biofilm in the presence of this class of biogenic SeNPs. As we observed a reduction in the thickness of both studied biofilms, CLSM observations correlate with our kill curve results. Nevertheless, *P. aeruginosa* and *S. aureus* growing biofilms showed completely different fluorescence emission at the efficient concentration of Bio Se-NEMO-S$_6$, resulting in a strong red-fluorescent signal or in one comparable to the HA-coated peg control respectively.

These data suggest two possible distinct modes of antimicrobial action of Bio Se-NEMO-S$_6$ depending on the pathogenic indicator strain analysed. The presence of large amount of red cells in *P. aeruginosa* biofilm underlines Bio Se-NEMO$_6$ ability to kill bacterial cells or to limit the number of viable cells. At the same time, the absence of thick *S. aureus* biofilm highlights Bio Se-NEMO-S$_6$ action in inhibiting irreversible attachment of bacterial cells to form a biofilm.

Chemically synthesized SeNPs showed less antimicrobial activity towards both *P. aeruginosa* and *S. aureus* planktonic cell growth and biofilm formation compared to those biogenically produced. Tran and Webster (2011) observed similar results in the evaluation of the antimicrobial efficacy of chemical SeNPs against planktonic cells of *S. aureus*. CLSM observations of both *P. aeruginosa* and *S. aureus* growing biofilms in the presence of L-cys SeNPs confirm our kill curve results. Even at the highest L-cys SeNPs tested concentration, *P. aeruginosa* cells exhibited a green-fluorescence emission, indicating the presence of live cells in the biofilm. Moreover, an established and thick biofilm was observed in the case of *S. aureus* cells incubated in the presence of 1.25 mg ml^{-1} L-cys SeNPs, while a slight inhibition is revealed at the highest tested concentration (2.5 mg ml^{-1}).

The higher antimicrobial ability of Bio Se-NEMO-s as compared to those chemically synthesized could be explained by the presence of an embedding organic matrix that surrounded these nanomaterials, as confirmed by TEM analyses, EDX spectra and zeta potential measurements. Specifically, L-cys and Asc SeNPs TEM images underlined evidence of SeNPs in a bright and white field, while both Bio Se-NEMO-S were characterized by a grey slightly electron-dense surrounding material. EDX analyses of all Se-nanomaterials highlighted the presence of carbon, oxygen, phosphorous and sulphur, along with selenium. Considering the synthesis procedure of chemical SeNPs, the detection of these elements in EDX analyses is explained by the use of either L-cysteine, rich in C, O and S, or ascorbic acid, rich in C and O, for the production (Zhang *et al.*, 2004; Li *et al.*, 2010). In the case of Bio Se-NEMO-S, the existence of C, O, P and S suggested the evidence of selenium nanoparticles surrounded by an organic material. These observations were validated by zeta potential measurements, in which both Bio Se-NEMO-S and chemical SeNPs showed strongly negative surface charge values. Considering that elemental selenium (Se^0) does not have a net charge, these results suggested that the stability of the Se-nanomaterials in solution is facilitated by the presence of negatively charged organic chemical groups. The association of SeNPs produced by *B. mycoides* SeITE01 with a surrounding material has been observed also by Lampis *et al.* (2014), which ascribed the nature of this external layer to the adhesion of proteins to SeNPs. The existence of proteins associated with biogenic SeNPs has been established and studied for other bacterial strains able to reduce selenium oxyanions to elemental selenium (Se^0). Debieux *et al.* (2011) have identified some proteins associated with SeNPs produced by *Thauera selenatis* that played an important role in stabilizing the formation of NPs, while Lenz *et al.* (2011) highlighted that SeNPs produced by *Bacillus selenatarsenatis* and *Sulfurospirillum barnesii* were characterized by the association with several high-affinity proteins. Recently, Zonaro *et al.* (2015) have observed the presence of several organic compounds surrounding biogenic SeNPs produced by *S. maltophilia* SeITE02, which have been characterized as carbohydrates, lipids and proteins by Lampis *et al.* (2016) using FT-IR analyses. This evidence complements our characterization analyses regarding the presence of associated organic material surrounding biogenic SeNPs.

In our study, despite DLS results that showed a defined size for all the studied Se-nanomaterials, TEM analyses highlighted the presence of smaller yet monodispersed SeNPs in Bio Se-NEMO-S samples of less than 100 nm in size. Probably, this discrepancy in size between DLS measurements and TEM observations is due to the contribution of the organic surrounding material in which biogenic SeNPs are embedded.

Conclusions

In our study, we evaluated the antimicrobial efficacy of biogenic SeNPs produced by *B. mycoides* SelTE01 (Bio Se-NEMO-S) compared to chemical SeNPs against two pathogenic indicator strains, *P. aeruginosa* NCTC 12934 and *S. aureus* ATCC 25923, grown onto HA-coated pegs. Bio Se-NEMO-S showed higher antimicrobial efficacy as compared to chemically synthesized SeNPs in all three evaluated scenarios for both tested strains. Difference in antimicrobial ability of these Se-nanomaterials suggests either distinct chemistry or the involvement of the organic surrounding material associated with the SeNPs (Bio Se-NEMO-S) originating from *Bacillus mycoides* SelTE01. Finally, this study demonstrates good promise of utilizing Bio Se-NEMO-S as antimicrobial agents to inhibit biofilm formation onto HA-coated pegs.

Experimental procedures

Materials, media and organisms used

Bacillus mycoides SelTE01 has been isolated previously from the rhizosphere of the Se-hyperaccumulating plant *Astragalus bisulcatus* (Lampis *et al.*, 2014). The two defined indicator strains used are *P. aeruginosa* NCTC 12934 and *S. aureus* ATCC 25923. Nutrient broth, agar and sodium chloride (NaCl) were obtained from Oxoid, Basingstoke Hampshire, England. Sodium selenite (Na_2SeO_3), octanol, L-cysteine, ascorbic acid, L-histidine, reduced glutathione were obtained from Sigma Aldrich, St. Louis, MO, USA. Tris–HCl 1.5 mM was made by adding 0.091 g of Trizma base (Sigma Aldrich) to 500 ml of water, and then, using HCl from Sigma Aldrich, pH was adjusted to 7.0. HA-coated Calgary Biofilm Device (CBD) plates or MBEC™was obtained from Innovotech, Edmonton, AB, Canada.

Methods

Chemical synthesis of SeNPs. SeNPs were chemically synthesized by adding 500 μl of 100 mM Na_2SeO_3 to 500 μl of 50 mM L-cysteine as the reducing agent and 1 ml of milli-Q water (L-cys SeNPs) (Li *et al.*, 2010), or using 1 ml of 30 mM ascorbic acid (reducing agent) to 200 μl of 100 mM Na_2SeO_3 and 800 μl of milli-Q water (Asc SeNPs; Zhang *et al.*, 2004).

Biogenic SeNPs (Bio Se-NEMO) synthesis. Following the protocol established by Lampis *et al.* (2014), *B. mycoides* SelTE01 was grown in culture tubes with 4 ml of nutrient broth medium for a minimum 24 h at 130 rpm. These cultures were used to inoculate 1-l flasks containing 400 ml of nutrient broth medium and 8 ml of filter sterilized Na_2SeO_3 (100 mM). The cultures were subsequently incubated at 27°C at 130 rpm, for 24 h. During this incubation, reduction of SeO_3^{2-} ions to Se^0 was observed visually through the formation of a red colour solution, typical of colloidal elemental selenium. Samples were collected after 6 and 24 h.

To extract the Se-nanomaterial from *B. mycoides* cultures, the contents of each flask were divided into eight 50-ml conical tubes and collected by centrifugation at 10 000 × *g* for 10 min. Pellets were washed twice with 0.9% NaCl solution, resuspended in Tris/HCl buffer (pH 8.2) and then disrupted by ultrasonication at 100 W for 5 min. The suspension was then centrifuged at 10 000 × *g* for 30 min to separate disrupted cells (pellet) from nanomaterials (supernatant), which were recovered after centrifugation at 40 000 × *g* for 30 min, washed twice with saline solution (0.9% NaCl) and resuspended in deionized water. This washed suspended material is what we characterize here and is defined as Spherical Bio Se-Nanostructures Embedded in an Organic material (Bio Se-NEMO-S).

Characterization of Bio Se-NEMO-S and SeNPs. Size, polydispersity index (PDI) and surface charge of chemical and biogenic SeNPs (Bio Se-NEMO-S) have been studied by evaluating their dynamic light scattering (DLS) and measuring their zeta potential. DLS measurements for all Se-nanomaterials have been evaluated using Zen 3600 Zetasizer Nano ZS from Malvern Instruments (Worcestershire, UK) (Cremonini *et al.*, 2016). Hydrodynamic diameter and PDI values were obtained using the software provided by the Malvern with the instrument. All the samples were then transferred to a quartz cuvette (10 mm path length), and the zeta potential at pH 7 has been measured at 25°C using the Malvern software. TEM analyses have been carried out to study morphology, size and shape for both the chemical and biogenic (Bio Se-NEMO-S) Se-nanomaterials. For TEM observations, 5 μl of each Se-nanomaterials was mounted on carbon-coated copper grids (CF300-CU; Electron Microscopy Sciences, Hatfield, PA, USA). Then, samples were air-dried and visualized using a Hitachi H7650 TEM. Finally, EDX spectra have been performed using an XL30 ESEM (FEI) equipped with an EDX micro-analytical system. Each sample was fixed, dehydrated through an increasing ethanol concentration series and dried in liquid CO_2. The Se-nanomaterials were then mounted on metallic stubs and directly observed to perform EDX analyses.

Measuring the biocidal activity of Bio Se-NEMO-S and chemical SeNPs. Kill curve assays were performed to evaluate the antimicrobial activity of either Bio Se-NEMO-S or chemical SeNPs. Here, we used the Calgary Biofilm Device (CBD) that is a 96-well microtitre plate lid with pegs protruding into the microtitre plate wells (Ceri et al., 2001). The use of this device to culture the bacteria permits high-throughput challenging with chemically synthesized SeNPs and Bio Se-NEMO-S. In this study, the CBD pegs had been pre-treated with a HA coating, generating a system relevant to dental and orthopaedic medicine. The assay was conducted using P. aeruginosa (NCTC 12934) and S. aureus (ATCC 25923), both established laboratory indicator strains of pathogens.

The CBD was inoculated with P. aeruginosa or S. aureus cells grown in nutrient broth medium along with solutions of either Bio Se-NEMO-S or chemical SeNPs at various concentrations. We tested two biogenic Se-nanomaterials produced by B. mycoides SelTE01 after 6 h (Bio Se-NEMO-S_6) and 24 h (Bio Se-NEMO-S_{24}) of Na_2SeO_3 exposure. Additionally, we tested two chemically generated SeNPs made using L-cysteine (L-cys SeNPs; Li et al., 2010) or ascorbic acid (Asc SeNPs; Zhang et al., 2004) as reducing agents. Inoculum was prepared by resuspending biomass collected from a second subculture on a nutrient broth–agar plate in sterile distilled deionized water. Turbidity of the solution was adjusted to achieve an equivalent optical density of 0.5 McFarland standardized suspension. This suspension was then diluted in nutrient broth medium to achieve an approximate cell density of 10^5 per ml. Overall, we examined three different scenarios:

i　The ability of Bio Se-NEMO-S and SeNPs to eradicate 24-h pre-grown biofilms (minimal biofilm eradication concentration; MBEC).

ii　The ability of Bio Se-NEMO-S and SeNPs to prevent the formation of a biofilm (minimal biofilm prevention concentration; MBPC).

iii　The ability of Bio Se-NEMO-S and SeNPs to inhibit planktonic cells growth (minimal inhibitory concentration; MIC)

For i, the CBD was inoculated with P. aeruginosa or S. aureus strains at a 0.5 McFarland standard concentration in nutrient broth medium and placed in a gyrorotary shaker operating at 150 rpm, 37°C for 24 h. Following the establishment of a biofilm, solutions containing various concentrations of each tested Se-nanomaterials were added to the wells in twofold serial dilutions between 2.5 and 0.0025 mg ml^{-1}. Experiments were performed in triplicate. The inoculated CBDs were placed on a shaker at 150 rpm, 37°C for 24 h.

For ii, CBDs were inoculated with a 0.5 McFarland standard of the bacterial strains into nutrient broth medium and Bio Se-NEMO-S or chemical SeNPs solutions were added as described in scenario i, allowing biofilm growth for 24 h.

To collect biofilm biomass from scenarios i and ii, peg lids were rinsed twice with 0.9% NaCl to remove loosely bound cells and then placed into 96-well recovery plates containing 180 µl each of a mixture 1:100 of universal neutralizer solution and nutrient broth medium. Finally, we sonicated for 30 minutes to remove biofilm biomass from the peg surface.

For iii, 96-well plates containing nutrient medium and Se-nanomaterial solutions described in scenario i were inoculated with P. aeruginosa and S. aureus cells. The inoculated plates were incubated at 37°C with shaking (150 rpm) for 24 h.

Biofilm and planktonic cells exposed to either Bio Se-NEMO-S or chemical SeNPs from all three scenarios were collected and serially diluted into 0.9% saline solution. Viable cell numbers were determined using spot plates count. All assays were conducted using three biological replicates. Data are reported as means with standard deviations.

Confocal laser scanning microscopy (CLSM). A LEICA model DM IRE2 microscope was used to collect confocal images further processed using IMARIS ×64 (Bitplane, Concord, MA, USA) software. Briefly, hydroxyapatite-coated CBD pegs containing biofilms were aseptically removed from the lid using alcohol-flamed pliers. The collected pegs were rinsed twice with 200 µl of 0.9% NaCl solution in a microtiter plate and stained using the Live/Dead® BacLight™ stain (Molecular Probes, Ontario, Canada) for 30 minutes prior to visualization (Harrison et al., 2006; Cerca et al., 2012). This kit contains SYTO 9 green-fluorescent nucleic acid stain and propidium iodide, which is a red-fluorescent dye (Dailey, 2006). These stains have diverse spectral characteristics, and they are able to differentially penetrate bacterial cells. SYTO 9 generally labels all bacterial cells in population, while propidium iodide penetrates only bacterial cells with damaged membranes (Dailey, 2006). The thickness of the biofilms has been estimated through obtaining high-resolution optical images with depth selectivity, allowing 3D images of complex biological samples (Dailey, 2006).

Acknowledgements

This study was supported by the Natural Science and Engineering Research Council of Canada (NSERC) to RT. Innovotech® is gratefully acknowledged for the possibility to use hydroxyapatite-coated Calgary Biofilm Devices (CBDs). We also acknowledge Microscopy Imaging Facility (MIF) of the University of Calgary to allow and provide the use of transmission electronic microscopy (TEM).

Conflict of interest

The authors have no conflict of interest to declare.

References

Ahmad, A., Mukherjee, P., Senapati, S., Mandal, D., Khan, M.S.I., Kumar, R., and Sastry, M. (2003) Extracellular biosynthesis of silver nanoparticles using the fungus *Fusarium oxysporum*. *Colloids Surf B* **28**: 313–318.

Alhede, M., Jensen, P., Givskov, M., and Bjarnshot, T. (2009) Biofilm medical importance. *Biotechnology* **12**: e6581112.

Ankamwar, B., Chaudhary, M., and Sastry, M. (2005) Gold nanoparticles biologically synthesized using Tamarind leaf extract and potential application in vapour sensing. *Synth React Inorg, Met-Org, Nano-Met Chem* **35**: 19–26.

Bhainsa, K.C., and D'Souza, S.F. (2006) Extracellular biosynthesis of silver nanoparticles using the fungus *Aspergillus fumigatus*. *Colloids Surf B Biointerfaces* **47**: 160–164.

Bigger, J.W. (1944) Treatment of Staphylococcal infections with penicillin by intermittent sterilization. *Lancet* **244**: 497–500.

Cerca, N., Gomes, F., Pereira, S., Teixeira, P., and Oliveira, R. (2012) Confocal laser scanning microscopy analysis of *S. epidermidis* biofilms exposed to farnesol, vancomycin and rifampicin. *BMC Res Notes* **5**: 244.

Ceri, H., Olson, M.E., Morck, D.W., Storey, D., Read, R.R., Buret, A.G., and Olson, B. (2001) The MBEC assay system: multiple equivalent biofilms for antibiotic and biocide susceptibility testing. *Methods Enzymol* **337**: 377–384.

Cremonini, E., Zonaro, E., Donini, M., Lampis, S., Boaretti, M., Dusi, S., *et al.* (2016) Biogenic selenium nanoparticles: characterization, antimicrobial activity and effects on human dendritic cells and fibroblasts. *Microb Biotechnol* **9**: 758–771.

Crobb, A.G., and Schmalzereid, T.P. (2006) The clinical significance of metal ion release from cobalt-chromium metal-on-metal hip-joint arthroplasty. *Proc Inst Mech Eng H* **220**: 385–398.

Dailey, M.E. (2006) Chapter 18: confocal microscopy in living cells. In *Handbook of Biological Confocal Microscopy*, 3rd edn. Pawley, J.B. (ed.). New York, USA: Springer, pp. 381–403.

Debieux, C.M., Dridge, E.J., Mueller, C.M., Splatt, P., Paszkiewicz, K., Knight, I., *et al.* (2011) A bacterial process for selenium nanosphere assembly. *Proc Natl Acad Sci USA* **108**: 13480–13485.

Dhanjal, S., and Cameotra, S.S. (2010) Aerobic biogenesis of selenium nanospheres by *Bacillus cereus* isolated from coalmine soil. *Microb Cell Fact* **5**: 52.

Dobias, J., Suvorova, E.I., and Bernier-Latmani, R. (2011) Role of proteins in controlling selenium nanoparticles size. *Nanotechnology* **22**: 195605.

Dos Santos, C.A., Seckler, M.M., Ingle, A.P., Gupta, I., Galdiero, S., Galdiero, M., *et al.* (2014) Silver nanoparticles: therapeutical uses, toxicity, and safety issues. *J Pharm Sci* **103**: 1931–1944.

Elliot, J.C. (1994) Chapter 3: Hydroxyapatite and Nonstoichiometric Apatites. In *Studies in inorganic chemistry 18: Structure and chemistry of the apatites and other calcium*

orthophosphates. Elliot, J.C. (ed.). Amsterdam: Elsevier, pp. 111–189.

Gong, T., Xie, J., Liao, J., Zhang, T., Lin, S. and Lin, Y. (2015) Nanomaterials and bone regeneration. *Bone Res* **3**, 15029.

Gowramma, B., Keerthi, U., Rafi, M., and Muralidhara Rao, D. (2015) Biogenic silver nanoparticles production and characterization from native stain of *Corynebacterium* species and its antimicrobial activity. *Biotech* **5**: 195–201.

Harrison, J.J., Ceri, H., Stremick, C., and Turner, R.J. (2004) Biofilm susceptibility to metal toxicity. *Environ Microbiol* **6**: 1220–1227.

Harrison, J.J., Ceri, H., Yerly, J., Stremick, C.A., Hu, Y., Martinuzzi, R., and Turner, R.J. (2006) The use of microscopy and three-dimensional visualization to evaluate the structure of microbial biofilms cultivated in the Calgary Biofilm Device. *Biol Proced Online* **8**: 194–215.

Harrison, J.J., Stremick, C., Turner, R.J., Allan, N.D., Olson, M.E., and Ceri, H. (2010) Microtiter susceptibility testing of microbes growing on peg lids: a miniaturized biofilm model for high-throughput screening. *Nat Protoc* **5**: 1236–1254.

Hobby, G.L., Meyer, K., and Chaffee, E. (1942) Observations on the mechanism of action of penicillin. *Proc Soc Exp Biol Med* **50**: 281–285.

Ingale, A.G., and Chaudhari, A.N. (2013) Biogenic synthesis of nanoparticles and potential applications: an ecofriendly approach. *J Nanomed Nanotechnol* **4**: 165.

Jacobs, J.J., Skipor, A.K., Urban, R.M., Black, J., Manion, L.M., Starr, A., *et al.* (1994) Systemic distribution of metal degradation products from titanium alloy total hip replacements: an autopsy study. *Trans Orthop Res Soc* **19**: 838.

Kessi, J., Ramuz, M., Wehrli, E., Spycher, M., and Bachofen, R. (1999) Reduction of selenite and detoxification of elemental selenium by the phototrophic bacterium *Rhodospirillum rubrum*. *Appl Environ Microbiol* **65**: 4734–4740.

Kolmas, J., Groszyk, E., and Piotrowska, U. (2015a) Nanocrystalline hydroxyapatite enriched in selenite and manganese ions: physicochemical and antibacterial properties. *Nanoscale Res Lett* **10**: 278–286.

Kolmas, J., Kuras, M., Oledzka, E., and Sobczak, M. (2015b) A solid-state NMR study of selenium substitution into nanocrystalline hydroxyapatite. *Int J Mol Sci* **16**: 11452–11464.

Lampis, S., Zonaro, E., Bertolini, C., Bernardi, P., Butler, C.S., and Vallini, G. (2014) Delayed formation of zerovalent selenium nanoparticles by *Bacillus mycoides* SelTE01 as a consequence of selenite reduction under aerobic conditions. *Microb Cell Fact* **3**: 35.

Lampis, S., Zonaro, E., Bertolini, C., Cecconi, D., Monti, F., Micaroni, M., *et al.* (2016) Selenite biotransformation and detoxification by *Stenotrophomonas maltophilia* SelTE02. Novel clues on the route to bacterial biogenesis of selenium nanoparticles. *J Hazard Mater* **324**: 3–14.

Lenz, M., Kolvenbach, B., Gygax, B., Moes, S., and Corvini, P.F.X. (2011) Shedding light on selenium biomineralization: proteins associated with bionanominerals. *Appl EnvironMicrobiol* **77**: 4676–4680.

Li, Q., Chen, T., Yang, F., Liu, J., and Zheng, W. (2010) Facile and controllable one-step fabrication of selenium nanoparticles assisted by L-cysteine. *Mater Lett* **6**: 614–617.

Lim, P.N., Teo, E.Y., Ho, B., Tay, B.Y., and Thian, E.S. (2013) Effect of silver content on the antibacterial and bioactive properties of silver substituted hydroxyapatite. *J Biomed Mater Res, Part A* **101A:** 2456–2464.

Lu, Z., Rong, K., Li, J., Yang, H., and Chen, R. (2013) Size-dependent antibacterial activities of silver nanoparticles against oral anaerobic pathogenic bacteria. *J Mater Sci Mater Med* **24:** 1465–1471.

Merritt, K., and Rodrigo, J.J. (1996) Immune response to synthetic materials. Sensitization of patients receiving orthopedic implants. *Clin Orthop* **326:** 71–79.

Mukherjee, P., Ahmad, A., Mandal, D., Senapati, S., Sainkar, S.R., Khan, M.I., *et al.* (2001) Fungus-mediated synthesis of silver nanoparticles and their immobilization in the mycelial matrix: a novel biological approach to nanoparticle synthesis. *Nano Lett* **1:** 515–519.

Oremland, R.S., Herbel, M.J., Blum, J.S., Langley, S., Beveridge, T.J., Ajayan, P.M., *et al.* (2004) Structural and spectral features of selenium nanospheres produced by Se-respiring bacteria. *Appl Environ Microbiol* **70:** 52–60.

Rajesh, S., Dharanishanthi, V., and Vinoth Kanna, A. (2015) Antibacterial mechanism of biogenic silver nanoparticles of *Lactobacillus acidophilus*. *J Exp Nanosci* **10:** 1143–1152.

Sadeghian, S., Kojouri, G.A., and Mohebbi, A. (2012) Nanoparticles of selenium as species with stronger physiological effects in sheep in comparison with sodium selenite. *Biol Trace Elem Res* **146:** 302–308.

Song, J.Y., and Kim, B.S. (2009) Rapid biological synthesis of silver nanoparticles using plant leaf extracts. *Bioprocess Biosyst Eng* **32:** 79–84.

Srivastava, N., and Mukhopadhyay, M. (2015) Biosynthesis and structural characterization of selenium nanoparticles using *Gliocladium roseum*. *J Clust Sci* **26:** 1473–1482.

Stolz, J.F., Basu, P., Santini, J.M., and Oremland, R.S. (2006) Arsenic and selenium in microbial metabolism. *Annu Rev Microbiol* **60:** 107–130.

Stroyuk, A.L., Raevskaya, A.E., Kuchmiy, S.Y., Dzhagan, V.M., Sahn, D.R.T., and Schulze, S. (2008) Structural and optical characterization of colloidal Se nanoparticles prepared via the acidic decomposition of sodium selenosulfate. *Colloids Surf A Physicochem Eng Asp* **320:** 169–174.

Suresh, K., Prabagaran, S.R., Sengupta, S., and Shivaji, S. (2004) *Bacillus indicus* sp. nov., an arsenic-resistant bacterium isolated from an aquifer in West Bengal, India. *J Syst Evol Microbiol* **54:** 1369–1375.

Tran, P.A., and Webster, T.J. (2011) Selenium nanoparticles inhibit *Staphylococcus aureus* growth. *Int J Nanomedicine* **6:** 1553–1558.

Vigneshwaran, N., Ashtaputre, N.M., Varadarajan, P.V., Nachane, R.P., Paralikar, K.M., and Balasubramanya, R.H. (2007) Biological synthesis of silver nanoparticles using the fungus *Aspergillus flavus*. *Mater Lett* **66:** 1413–1418.

Wang, Q., and Webster, T.J. (2012) Nanostructured selenium for preventing biofilm formation on polycarbonate medical devices. *J Biomed Mater Res, Part A* **100A:** 3205–3210.

Wang, Y., Yan, X., and Fu, L. (2013) Effect of selenium nanoparticles with different sizes in primary cultured intestinal epithelial cells of crucian carp, *Carassius auratus gibelio*. *Int J Nanomedicine* **8:** 4007–4013.

Wang, Y., Wang, J., Hao, H., Cai, M., Wang, S., Ma, J., *et al.* (2016) In vitro and in vivo mechanism of bone tumor inhibition by selenium-doped bone mineral nanoparticles. *ACS Nano* **10:** 9927–9937.

Zhang, S., Zhang, J., Wang, H., and Chen, H. (2004) Synthesis of selenium nanoparticles in the presence of polysaccharides. *Mater Lett* **58:** 2590–2594.

Zonaro, E., Lampis, S., Turner, R.J., Qazi, S.J.S., and Vallini, G. (2015) Biogenic selenium and tellurium nanoparticles synthesized by environmental microbial isolates efficaciously inhibit bacterial planktonic cultures and biofilms. *Front Microbiol* **6:** 584.

Synthetic extreme environments: overlooked sources of potential biotechnologically relevant microorganisms

Timothy Sibanda,* Ramganesh Selvarajan and
Memory Tekere

*Department of Environmental Sciences, College of
Agriculture and Environmental Science, UNISA Florida
Campus, PO Box X6, Florida 1709, South Africa.*

Summary

**Synthetic extreme environments like carwash efflu-
ent tanks and drains are potential sources of
biotechnologically important microorganisms and
molecules which have, however, remained unexplored. Using culture- and molecular-based methods,
a total of 17 bacterial isolates belonging to the genera *Shewanella, Proteus, Paenibacillus, Enterobacter*
and *Citrobacter, Aeromonas, Pseudomonas* and *Pantoea* were identified. Hydrocarbon utilization and
enzyme production screening assays showed that
Aeromonas sp. CAC11, *Paenibacillus* sp. CAC12 and
Paenibacillus sp. CAC13 and *Citrobacter* sp. PCW7
were able to degrade benzanthracene, naphthalene
and diesel oil, *Paenibacillus* sp. CAC12 and *Paenibacillus* sp. CAC13 could produce cellulase enzyme,
while *Proteus* sp. BPS2, *Pseudomonas* sp. SAS8 and
Proteus sp. CAL3 could produce lipase. GC-MS analysis of bacterial secondary metabolites resulted in
identification of 107 different compounds produced
by *Proteus* sp. BPS2, *Paenibacillus* sp. CAC12,
Pseudomonas sp. SAS8, *Proteus* sp. CAL3 and
Paenibacillus sp. CAC13. Most of the compounds
identified by both GC-MS and LC-MS have previously
been determined to have antibacterial, antifungal
and/or anticancer properties. Further, microbial
metabolites which have previously been known to be
produced only by plants or microorganisms found in
natural extreme environments were also identified in
this study. This research has revealed the immense**

*For correspondence. E-mail timsibanda@gmail.com

Funding Information
The authors wish to acknowledge UNISA for funding this research
through the Women in Science Research Program.

**bioresource potential of microorganisms inhabiting
synthetic extreme environments.**

Introduction

Microbial life has been found to adapt to and thrive in
extreme conditions ranging from extremely low water
availability, intense solar radiation, high salinity and
extreme temperatures, pH and pressure (Rampelotto,
2013; Azua-Bustos and González-Silva, 2014). Natural
extreme environments can be defined as habitats that
experience steady or fluctuating exposure to one or
more environmental factors such as salinity, osmolarity,
desiccation, UV radiation, barometric pressure, pH and
temperature (Seufferheld *et al.*, 2008). Some extreme
environments are, however, synthetic (Maes *et al.*, 2016)
and may be created by a sustained and consistent discharge of pollutants (like acid mine drainage, untreated
municipal sewage effluents and industrial effluent dis-
charges among others) into the environment (Selbmann
et al., 2013), making it very selective and often hostile to
major life forms.

Microorganisms that inhabit extreme environments are
referred to as extremophiles and are broadly classified
into extremophilic organisms (which require one or more
extreme conditions to grow), and extremotolerant organisms (which can tolerate extreme values of one or more
physicochemical parameters though growing optimally at
normal conditions) (Rampelotto, 2013). Extremophiles
from a wide range of natural extreme environments have
been characterized, and indeed, a lot of biotechnological
applications have been made stemming from their
secondary metabolites (Seufferheld *et al.*, 2008; Azua-
Bustos and González-Silva, 2014). Relatively few
researchers (Chronakova *et al.*, 2010; Perks, 2011) have
characterized microorganisms from synthetic extreme
environments with the aim to biotechnologically exploit
them. Therefore, the varieties of microorganisms in synthetic extreme environments, together with the molecular
mechanisms they have evolved to cope with abiotic stres-
ses in their environments, still need to be elucidated.

The current challenge besetting the study of biological
molecules produced by extremophiles is that their potential applications may not be well known, or even
unknown (Azua-Bustos and González-Silva, 2014). This,
however, offers an immense potential for future

development. To this end, evidence from past research suggests biotechnologically significant roles ranging from antimicrobials to anticancer molecules (Perks, 2011). Among the most explored applications of extremophiles are their production of enzymes (also referred to as extremozymes) (Demirjian *et al.*, 2001; Enache and Kamekura, 2010), fatty acids and proteins (Cavicchioli *et al.*, 2002; Reed *et al.*, 2013), antibiotics or biomedicines (Azua-Bustos and González-Silva, 2014), and the direct application of cultures in chemical processes like desulfurication of flue gases and in biohydrometallurgical processes (Huber and Stetter, 1998). Biohydrometallurgical processes include, but are not limited to, microbial mining of precious (and rare) metals, oil recovery, bioleaching and water treatment.

Among microbial groups most commonly found in extreme environments, actinobacteria are arguably the richest source of small molecule diversity, with widespread global and environmental dispersal (Jami *et al.*, 2015; Kuang *et al.*, 2015; Ettoumi *et al.*, 2016). Synthetic extreme environments like carwash effluent tanks and drains remain unexplored, and their potential as a source of biotechnologically important molecules remains unknown. While synthetic extreme environments are (almost all) seriously polluted sites which are in need of remediation, they may conversely be harbouring life-saving agents and other biotechnologically important molecules. As the survival strategies of extremophiles are often novel and unique (Seufferheld *et al.*, 2008), the necessity for microbial cell components to adapt to extreme environments (natural or synthetic) implies that a broad range of cellular products (genes and metabolites) are available for biotechnological applications.

While there exists no previous literature records elucidating the microbial composition of carwash effluents, Tekere *et al.* (2016) characterized the physicochemical properties of carwash effluents which they have shown to have high levels of, among other pollutants, petroleum hydrocarbons, polycyclic aromatic hydrocarbons (PAHs), phenols and heavy metals. Further, these effluents were shown to be toxic not only to animals and plants but were also found to inhibit light production by the bacterium *Vibrio fischeri*. PAHs are the most common environmental pollutants which enter both the terrestrial and aquatic environments in large volumes, posing a great hazard to ecosystems (Bayat *et al.*, 2015). Research has shown that in the short to medium term, contamination of environmental media with heavy metals results in reduced microbial biomass, coupled with reduced enzyme activity (Wang *et al.*, 2007; Akmal, 2009). In the long term, environmental media contaminated with PAHs showed reduced microbial biodiversity and species richness index (Markowicz *et al.*, 2016). However, the same study by Markowicz *et al.* (2016) showed that while soils

contaminated with heavy metals had reduced microbial activity and changes in microbial community structure, a high microbial evenness index was observed. This suggests that some populations of the microbial community are 'enriched' by the presence of the very same pollutants which diminish the population sizes of the less tolerant microorganisms. Assessment of the impact of long-term diesel contamination on soil microbial community structure, for example, showed both a decrease in the relative abundance of certain bacterial phyla and an increase in the relative abundance of other phyla (Sutton *et al.*, 2013). The microbes that undergo natural attenuation to adapt to extreme environments could be sources of biotechnologically indispensable molecules with a wide range of applications ranging from bioremediation of polluted sites to producing industrially relevant enzymes and other products. The aim of this study therefore was to characterize bacterial isolates from carwash effluent samples and to determine their ability to produce some target enzymes as well as other secondary metabolites, previously known or unknown, and create a metabolic fingerprint.

Experimental procedures

Description of study area

Carwash effluent samples were obtained from six carwash stations in the Johannesburg Metropolitan Municipality located in Gauteng Province in South Africa. Three carwash stations were located in Johannesburg City, while the other three were in the affluent business area, Sandton City (Fig. 1).

For confidentiality purposes, the exact locations and/or real names of the carwashes are not mentioned. However, the sampling stations were code-named as BPS, CAL and SAS (Johannesburg), and CAC, PCW and TCWPM (Sandton).

Sample collection, bacterial isolation and characterization

One-litre (1 l) grab samples were obtained from the onsite treatment facility of each carwash using sterile plastic sampling bottles. The samples were immediately chilled by placing them in cooler boxes containing ice and transported to the laboratory at UNISA Florida Campus for analysis within 6 h of collection. Once in the laboratory, the samples were shaken to ensure homogeneity after which 100 µl aliquots of each sample were spread plated in triplicate onto freshly prepared nutrient agar (Sigma Aldrich, Pretoria, South Africa). The plates were incubated at 30°C for 48 h. The resultant (mixed) cultures were separated and purified by subculturing onto nutrient agar plates until axenic cultures were

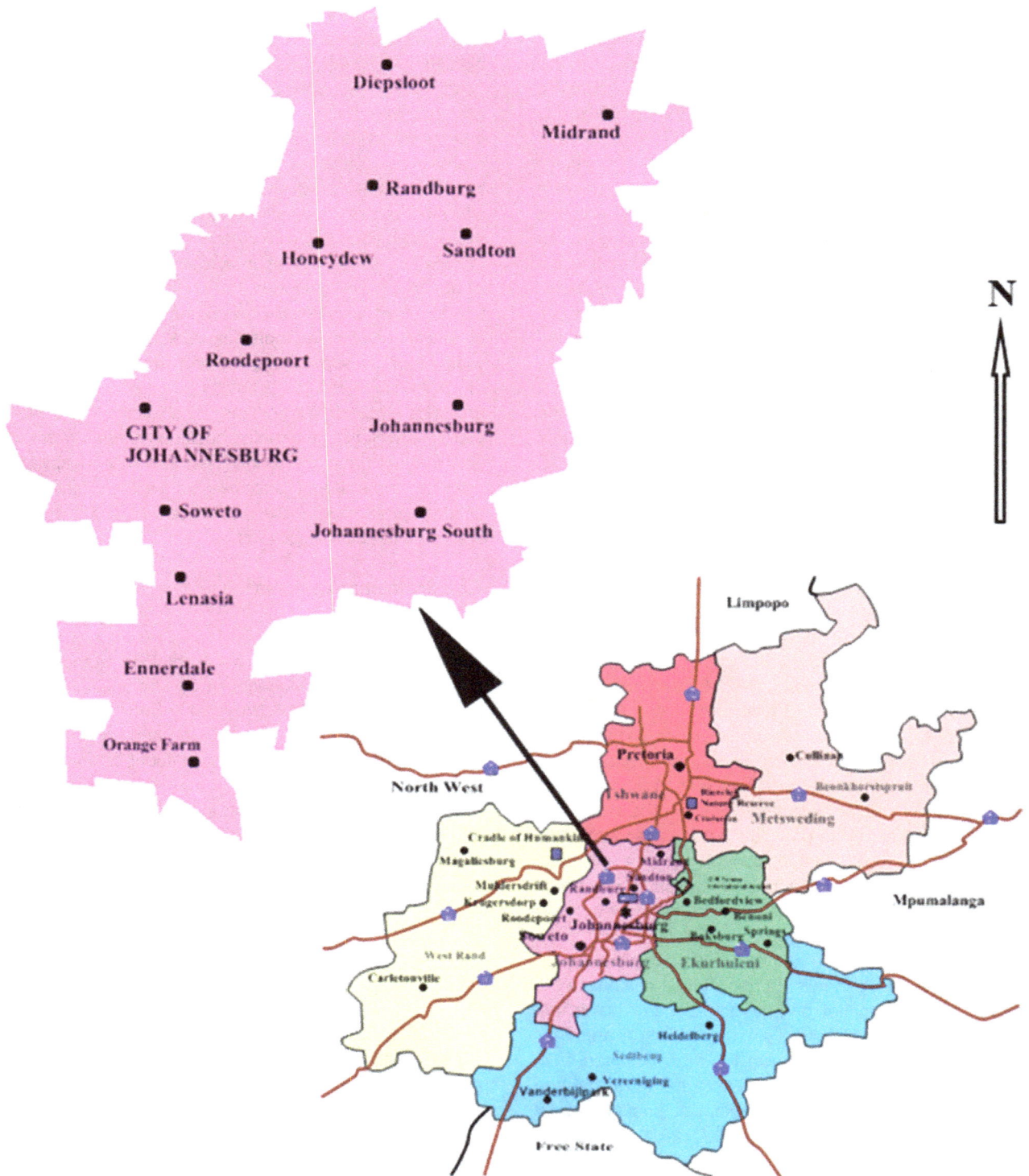

Fig. 1. Map of Gauteng Province in South Africa showing the Johannesburg Metropolitan Municipality demarcation (not drawn to scale).

obtained. DNA was then isolated from each culture using a Quick *g*-DNA extraction kit (Zymo Research, Irvine, CA 92614, U.S.A.) followed by PCR amplification using

16S universal bacterial primers (27 F and 1492 R). The PCR amplicons were confirmed by gel electrophoresis and then sent to Inqaba Biotec (Pretoria, South Africa)

for sequence analysis. The resultant sequence chromatograms were manually edited with Chromas software v2.6.1 (Technelysium Pty Ltd., South Brisbane, Qld, Australia) and subjected to BLAST analysis to compare the identity of the isolates. The sequences were then used for phylogenetic analysis using the MOLECULAR EVOLUTIONARY GENETIC ANALYSIS v6.0 (MEGA6) software (Tamura *et al.*, 2013) using an alignment created with SINA Aligner. A phylogenetic tree was constructed using maximum-likelihood analysis with kimura 2-parameter model and 500 times of bootstrap replications. Finally, the sequences were submitted to Gene bank to obtain the accession numbers.

Screening for hydrocarbon utilization

The isolates were tested for the capacity to degrade hydrocarbons following a modified protocol previously described by Um *et al.* (2010). Minimal salt medium (MSM) containing in g/l: NaCl 5.0, K_2HPO_4 1.0, $NH_4H_2PO_4$ 1.0, $(NH_4)_2SO_4$ 1.0, $MgSO_4 . 7H_2O$ 0.2 and KNO_3 3.0 supplemented with 1.25% (w/v) of agar, was autoclaved for 15 min and poured out onto Petri plates as 'bottom agar'. This bottom agar was then overlaid with 100 μl of a polycyclic aromatic hydrocarbon (PAH) solution prepared by dissolving 10 mg/50 ml naphthalene and 10 mg/10 ml benzanthracene (Sigma, Pretoria, South Africa) in methanol. The PAH solution was evenly spread over the agar surface using sterile disposable spreaders and the plates were left in the lamina flow for the solvent to evaporate, leaving behind a visible thin white layer of PAHs on the surface of the bottom agar. To inoculate the bottom agar, 100 μl of each bacterial isolate was mixed with 900 μl of 35–40°C molten 'top agar' medium containing 0.5% agar and same mineral composition as the bottom agar and immediately poured on Petri plates containing bottom agar. The plates were then swirled gently to spread molten top agar medium over bottom agar, which resulted in the white layer of PAHs to move to the surface of the top agar medium upon solidifying. The inoculated plates were incubated at 25°C and examined daily for the presence of growth and clear zones.

The isolates which were able to degrade PAHs (seen by formation of clear halos colonies) were selected for hydrocarbon-degrading assays following the method of Oliveira *et al.* (2012). The isolates were inoculated into 20-ml tubes containing 10 ml of fermentation broth made up of minimal salt medium supplemented with 10 g/l of glucose and 1.0 g/L of yeast extract and incubated at 25°C for 72 h with shaking at 120 r.p.m. After incubation, the inoculum was centrifuged at 10 000 r.p.m. for 5 min at 10°C. The cell pellet was resuspended in phosphate buffer (pH 7) and again centrifuged at

10 000 r.p.m. for 5 min at 10°C, discarding the supernatant. To remove all culture medium residues, this step was performed twice. After the final centrifugation, the isolates were resuspended in phosphate buffer and the OD adjusted to McFarland 0.5. The assay was carried out in a sterile 96-well microtitre plate by adding 20 μl of the isolate suspension in triplicate, to a row of wells containing 168 μl of minimal salt medium, 12 μl of 2,6-dichlorophenolindophenol (DCPIP) and 2 μl of diesel oil. Each isolate was inoculated into a row of wells containing three different hydrocarbons, the other two being naphthalene and benzanthracene. In addition to the test wells containing isolates and different hydrocarbons, the experiment had a positive control produced by mixing 148 μl of minimal medium with 12 μl of DCPIP solution, 20 μl of 10% glucose solution and 20 μl of the *Pseudomonas* cell suspension; and a negative control consisting of 168 μl of minimal medium, 20 μl of the isolate suspension and 12 μl of DCPIP solution. The plates were incubated at 30°C, and the readings were taken after 24, 48 and 72 h of incubation using a spectrophotometer at 600 nm. The percentage reduction in DCPIP was obtained using the following equation:

$$\%\text{DCPIP reduction} = \frac{0h\,OD_{600} - Yh\,OD_{600}}{0h\,OD_{600}} \times 100\%,$$

where the value of Y was, sequentially, 24, 48 and 72 h.

Screening for enzyme production (lipase and cellulase)

Production of lipase was determined by growing the bacterial isolates in rhodamine–olive oil–agar medium following the method of Kumar *et al.* (2012) with little modifications. The agar medium contained, in g/l, agar–agar 20, $MgSO_4$ 0.2, $CaCl_2$ 0.02, KPO_4 1.0, K_2PO_4 1.0, NH_4NO_3 1.0, $FeCl_3$ 1.0 and yeast extract 5.0. The medium was adjusted to pH 7.0, autoclaved and cooled to about 50°C after which 31.25 ml of olive oil and 10 ml of rhodamine B solution (1.0 mg/ml distilled water and sterilized by filtration) were added with vigorous stirring. It was then poured into Petri plates under aseptic conditions and allowed to solidify. The plates were inoculated by smearing buttons of bacterial cultures on the agar surface. After incubation of the plates for 48 h at 30°C, they were viewed under UV irradiation and lipase-producing strains were identified by formation of orange fluorescent halos around bacterial colonies due to the hydrolysis of substrate.

For cellulase screening, agar medium containing 0.2% (w/v) carboxymethylcellulose sodium salt (CMC), 1% agar and minimal salt medium was prepared and poured into Petri plates. Spot inoculation was then performed using axenic cultures of the test isolates followed by incubation at 30°C for 48 h. Hydrolysis zones were visualized by flooding the plates with 0.1% Congo red stain (Glass World, Johannesburg, South Africa) and allowing to stand for 15 min followed by destaining with 1 M NaCl.

Analysis of bacterial secondary metabolites

Bacterial isolates which were positive for enzyme production screens were grown in fermentation media consisting of minimal salt medium supplemented with olive oil and CMC-Na salt for 7 days in a shaking incubator at 30°C. After incubation, cell debris was separated from the supernatant under high-speed centrifugation at 4°C. The supernatant was subjected to solvent extraction of secondary metabolites. Briefly, the supernatant was a mixture of chloroform and methanol (1:1, v/v) and shaken for 12 h at 120 r.p.m. and 25°C. The ethyl acetate fraction was then separated from the aqueous fraction using a separating funnel. The solvent fraction was evaporated to dryness *in vacuo* at 80°C, and the residue was reconstituted in a mixture of 1:1 (v:v) acetonitrile and hexane. The acetonitrile fraction was used to analyse for the polar secondary metabolites using LC-MS, while the hexane fraction was used to analyse for nonpolar, volatile compounds using GC-MS.

For GC analysis, ionization energy was set at 70 eV using He as the carrier gas. The flow rate was set at 1 ml/min, injection volume at 2 μl, split ratio at 10:1, injection temperature at 250°C, ion source temperature at 200°C and oven temperature at 110°C (isothermal at 2 min) with increase of 10°C/min to 200°C then 5°C/min to 280°C (with 9 min isothermal at 280°C). An HP-5MS fused silica capillary column (30 m, 0.25 mm i.d., 0.25-μm film, cross-linked to 5% phenyl methyl siloxane stationary phase) was used. For MS, ionization energy was set at 70 eV, scan interval at 0.5 s, fragments at 45–450 kD and solvent delay at 0–2 min. The identification of compounds was based on comparison of their mass spectra with the National Institute of Standards and Technology (NIST 2005) library.

LC-MS was performed using an ultra high-performance liquid chromatography–mass spectrophotometer (Compass otofSeries 1.9, Bruker Instrument: ImpactII) system. For LC analysis, the column was Acquity UPLC BEH C18 1.7 μm, diameter 2.1 × 100 mm (Miscrosep Waters, Johannesburg, South Africa), and the solvents were 0.1% formic acid (FA) in water and 0.1% FA in acetonitrile. The column flow was set at 0.3 ml/min, column oven temp at 35°C and draw speed at 2 μl/s with a total injection volume of 2 μl. Mass spectrometer (MS) conditions were set at a mass range of 50–1600 m/z, capillary 4500 v, dry gas 8 l/m, gas temperature 220°C, ion energy 4.0 eV, collision energy 7.0 eV, cycle time

0.5 s. Data analysis was performed using the Bruker Software (Bruker Compass DataAnalysis 4.3; Bruker Daltonik GmbH 2014, Bremen, Germany).

Results and discussion

Isolation and characterization of bacterial isolates in carwash samples

A total of 17 different isolates were obtained from all the six carwash sampling sites. Phylogenetic comparison of PCR-amplified 16S rDNA sequence data of each isolate with the database of the databases of known species using the NCBI server revealed that isolated bacteria belonged to the following genera: *Shewanella* (35.32%), *Proteus* (11.76%), *Paenibacillus* (11.76%), *Enterobacter* (11.76%), *Citrobacter* (11.76%), *Aeromonas* (5.88%), *Pseudomonas* (5.88%) and *Pantoea* (5.88%) (Table 1 and Fig. 2). The 1500-bp-long 16S rRNA gene containing nine hypervariable regions (V1–V9) (Chakravorty *et al.*, 2007) was sequenced in single chain and, in all instances during phylogenetic analysis, the percentage similarity was ≥ 99%. Isolation of bacterial strains in the present work was performed under the conditions and culture medium explored, and the percentages observed only correspond to data observed in this work although other bacteria which could not grow under the explored conditions could exist in the same environment. There were no available data in literature to compare with the data obtained in this study, making this study potentially the first study to isolate and characterize bacterial isolates from carwash effluents.

The phylogenetic tree diagram (Fig. 2) showed that isolates *Pseudomonas* sp. SAS8, *Paenibacillus* sp. CAC12 and *Paenibacillus* sp. CAC13 were distinctly different from the rest of the isolates, while isolates *Shewanella* sp. TCWPM1, *Shewanella* sp. TCWPM2, *Shewanella* sp. PCW5, *Shewanella* sp. BPS1, *Shewanella* sp. SAS10 and *Shewanella* sp. CAL2 formed were closely related.

Screening for enzyme production (lipase and cellulase) and hydrocarbon utilization

Four isolates, three from carwash CAC namely *Aeromonas* sp. CAC11, *Paenibacillus* sp. CAC12 and *Paenibacillus* sp. CAC13 and one from carwash PCW namely *Citrobacter* sp. PCW7, were able to utilize hydrocarbon as a carbon source during preliminary screening (Table 2). Further, *Paenibacillus* sp. CAC12 and *Paenibacillus* sp. CAC13 were the only isolates to utilize the CMC sodium salt as a carbon source, indicating their ability to produce cellulase. Screening for lipase production showed that only isolate *Proteus* sp. BPS2, *Pseudomonas* sp. SAS8 and *Proteus* sp. CAL3 could hydrolyse olive oil.

Table 1. Characterization of isolates by isolate codes, sequence length, percentage similarity to closest matching strains and accession numbers.

Isolate code	Sequence length (nt)	Closest match	% similarity	Accession number
BPS1	916	*Shewanella xiamenensis*	99	KX885451
BPS2	942	*Proteus penneri* strain wf-4	99	KX885439
CAC11	928	*Aeromonas tecta* strain L47	100	KX885441
CAC12	826	*Paenibacillus lautus*	100	KX885440
CAC13	822	*Paenibacillus* sp. HF_07	100	KX885444
CAL2	948	*Shewanella* sp. LH8	100	KX885437
CAL3	974	*Proteus vulgaris* strain NWG20141026	100	KX885445
PCW5	868	*Shewanella* sp. JG353	100	KX885449
PCW6	815	*Citrobacter braakii* strain 425C3	100	KX885442
PCW7	931	*Citrobacter* sp. 29Kp3	100	KX885448
SAS10	935	*Shewanella putrefaciens* strain CZ-BHG004	99	KX885452
SAS8	915	*Pseudomonas protegens*	100	KX885438
SAS9	904	*Pantoea* sp. 82353	99	KX885443
TCWPM1	847	*Shewanella* sp. JG353	100	KX885453
TCWPM2	871	*Shewanella* sp. UIWRF0468	99	KX885450
TCWPM3	894	*Enterobacter xiangfangensis* strain CDDS 11	100	KX885447
TCWPM4	887	*Enterobacter xiangfangensis* strain CDDS 11	99	KX885446

There are environments which, consequent to their substrate profiles, are better suited for the growth and proliferation of microorganisms producing certain kinds of enzymes. For example, a municipal solid waste site is rich in lignocellulosic materials, thus providing a suitable environment for cellulase-producing organisms (Ali *et al.*, 2013), while oil mill soils are likely to be endowed with lipase-producing organisms (Ramesh *et al.*, 2014). However, extreme environments are known to support microbial life forms with highly flexible metabolic pathways (Stathopoulou *et al.*, 2013; Dalmaso *et al.*, 2015). It is this flexibility, which causes extremophilic microorganisms to be the best known producers of biotechnologically important molecules/products (Bisht and Panda, 2011; Bull, 2011; Azua-Bustos and González-Silva, 2014). As in this study, production of cellulase by *Paenibacillus* sp. has previously been confirmed (Pandey *et al.*, 2013). Similarly, production of lipase by *Pseudomonas* and *Proteus* has been confirmed in previous studies (Golani *et al.*, 2016; Kumar *et al.*, 2016). What has not been reported, however, is the occurrence of enzyme-producing microorganisms from carwash effluent samples. In the wake of growing demand for enzymes with improved catalytic performance and tolerance to process-specific parameters (Guazzaroni *et al.*, 2015), it is imperative that every kind of environment be surveyed for enzyme-producing microbes.

Hydrocarbon degradation assays

Isolates *Aeromonas* sp. CAC11, *Paenibacillus* sp. CAC12, *Paenibacillus* sp. CAC13 and *Citrobacter* sp. PCW7 were further applied to hydrocarbon degradation assays using three classes of hydrocarbons namely benzanthracene, naphthalene and diesel fuel using

2,6-dichlorophenolindolphenol (DCPIP) as the indicator. DCPIP is an electron acceptor that becomes reduced (decolourized) when redox reactions occur, in this case when NADH is converted to NAD$^+$ during microbial degradation of hydrocarbons (Kubota *et al.*, 2008; Bidoia *et al.*, 2012). During the assay, the OD_{600} at 24 and 48 h was still almost similar to the OD_{600} at 0 h with, however, a sudden significant reduction in OD_{600} at 72 h. The results are presented in Fig. 3. Ionescu *et al.* (2015) also observed that bacterial cells needed about 4 days for the activation of the transcriptional regulation of metabolic operons involved in producing the necessary enzymes to metabolize an unusual substrate.

Isolate *Citrobacter* sp. PCW7 showed the highest potential of the four test isolates to metabolize all three hydrocarbon classes, with the highest activity against benzanthracene (45% DCPIP reduction). Next was isolate *Aeromonas* sp. CAC11, which showed 25% reduction in DCPIP when metabolizing naphthalene and 15% when metabolizing benzanthracene. Isolates *Paenibacillus* sp. CAC12 and *Paenibacillus* sp. CAC13 did not result in any significant DCPIP reduction. High levels of polycyclic aromatic hydrocarbons in environmental media have generated considerable interest from a toxicological standpoint (Al-Daghri *et al.*, 2014). Benzanthracene, for instance, is an environmentally recalcitrant pollutant, which is classified as a group 2A carcinogen by the International Agency for Research on cancer (Kunihiro *et al.*, 2013). Besides naphthalene and benzanthracene, diesel fuel contains alkylbenzenes, toluene, naphthalenes and polycyclic aromatic hydrocarbons (Irwin *et al.*, 1997), constituents which have already been implicated in asthma and cancer cases (Al-Daghri *et al.*, 2013). As such, it is of public and environmental health interest to find microorganisms which can biodegrade such

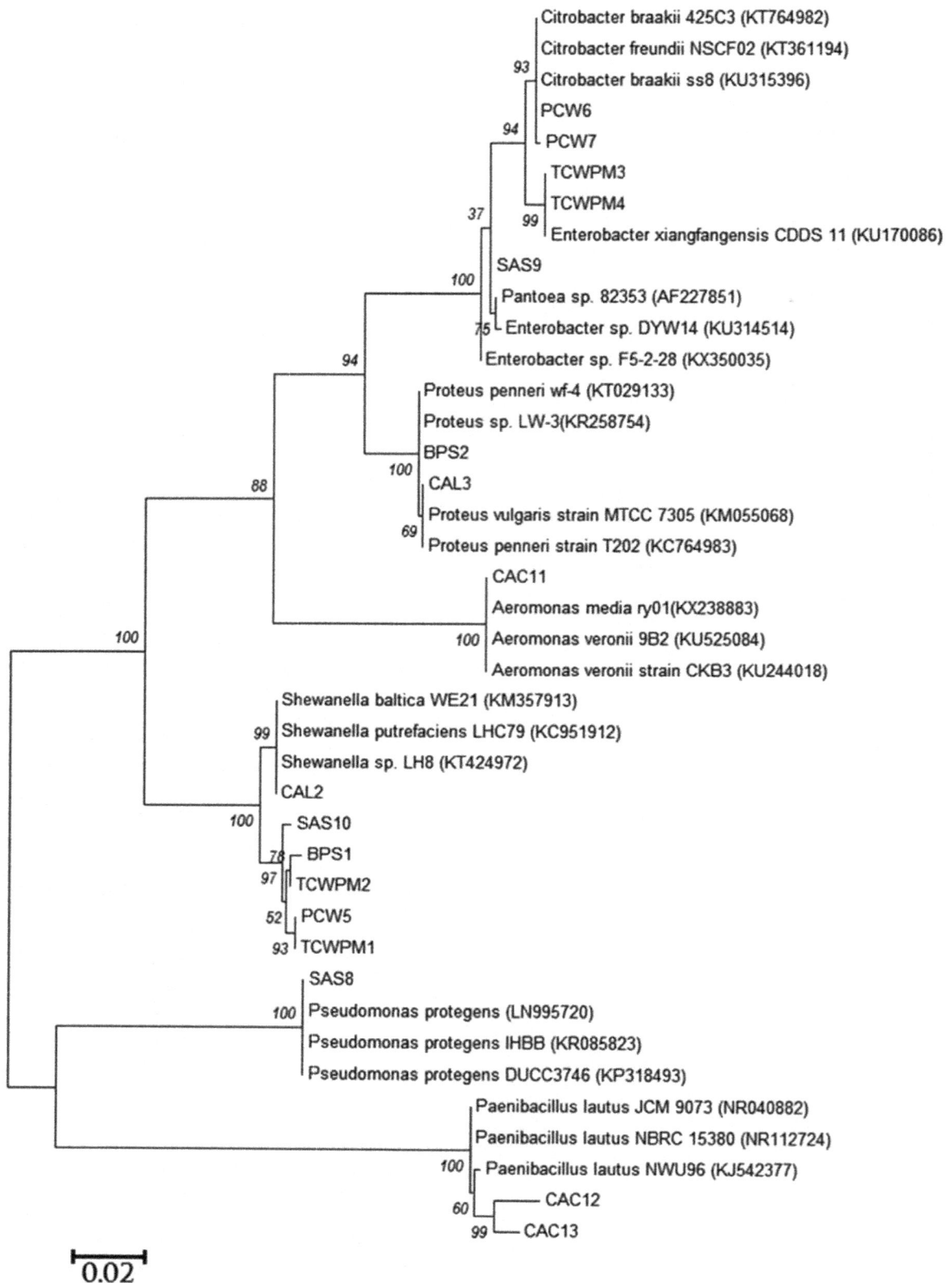

Fig. 2. Phylogenetic tree based on 16S rDNA gene sequences obtained by the maximum-likelihood method showing the phylogenetic relationship among the 17 bacterial isolates of this study (code names) and related bacteria.

Table 2. Screening results for hydrocarbon degradation, cellulase and lipase production.

Isolate	Substrate		
	Hydrocarbon	CMC salt	Olive oil
BPS1	−	−	−
BPS2	−	−	+
CAC11	+	−	−
CAC12	+	+	−
CAC13	+	+	−
CAL2	−	−	−
CAL3	−	−	+
PCW5	−	−	−
PCW6	−	−	−
PCW7	+	−	−
SAS10	−	−	−
SAS8	−	−	+
SAS9	−	−	−
TCWPM1	−	−	−
TCWPM2	−	−	−
TCWPM3	−	−	−
TCWPM4	−	−	−

hydrocarbons to rid or lessen the danger that they pose to public health, more so if such organisms are indigenous to the polluted environments. Isolates *Citrobacter* sp. PCW7 and *Aeromonas* sp. CAC11 showed much promise as potential microorganisms that can be used for the bioremediation of environments polluted by high molecular weight hydrocarbons. Other studies (Mandri and Lin, 2007; Affandi *et al.*, 2014) have also proved the high diesel oil and grease degradation potentials of *Citrobacter* (*C. freundii*) and *Aeromonas* (*A. hydrophila*), which were, however, isolated from hydrocarbon-contaminated environments, as in the present study. Most other studies, which have assessed the hydrocarbon degradation potentials of *Aeromonas* and *Shewanella*,

used test strains that were isolated from hydrocarbon-contaminated environmental media (Martín-Gil *et al.*, 2004; Ben Said *et al.*, 2008; Hamzah *et al.*, 2010). However, the study of Deppe *et al.* (2005) showed that *Shewanella* sp. isolated from a hydrocarbon-free environment were unable to degrade crude oil except when it was applied in consortia with other bacteria. This shows that these bacterial strains have evolved in the explored environments to degrade hydrocarbons. Bacteria of the genera *Alcaligenes*, *Stenotrophomonas*, *Sphingomonas* and *Pseudomonas* have also been documented to biodegrade benzanthracene (Kunihiro *et al.*, 2013), although the metabolites were not documented, as was the case in this study.

Elsewhere, Bonfá *et al.* (2011) described the isolation of aromatic hydrocarbon-degrading haloarchaea in hypersaline environments. As in this particular study, the hypersaline environment from which they isolated the organisms was synthetic, created as a result of the discharge of 'produced' hypersaline water from gas and oil extraction activities. One point comes to the fore; that environmental pollution of whatever nature may indeed eliminate most forms of life (including some microorganisms), but it will always result in selective proliferation of other groups of microorganisms that will utilize the pollutant as a food source or a means to it. It is that particular ability to not only survive but thrive in polluted/adverse environments that makes such microorganisms potential bioresources (or sources thereof) for the ecological recovery of polluted sites. Further to, the existence of polyaromatic hydrocarbon-degrading bacteria like *Citrobacter* sp. PCW7 and *Aeromonas* sp. CAC11 in the natural environment could be exerting a protective effect on higher organisms by continuously removing potentially harmful hydrocarbons from the environment.

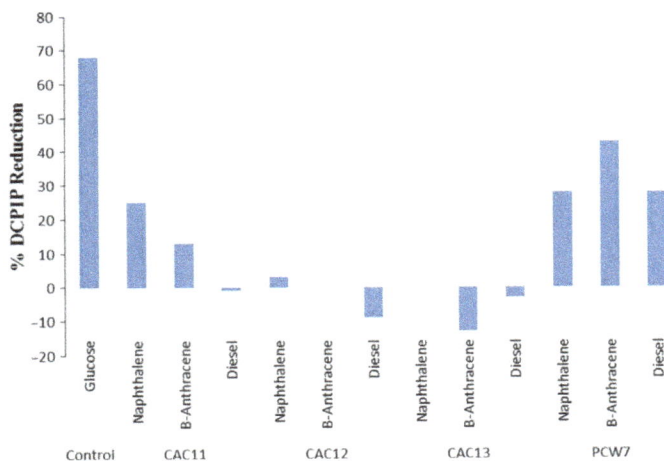

Fig. 3. Percentage reduction in DCPIP during hydrolysis of hydrocarbons by four bacterial isolates.

Secondary metabolite mapping using GC-MS and UHPLC-MS

GC-MS secondary metabolite elucidation was able to identify 107 different compounds produced by five bacterial isolates, which had tested positive when screened for production of either lipase or cellulase. These were *Proteus* sp. BPS2, *Paenibacillus* sp. CAC12, *Pseudomonas* sp. SAS8, *Proteus* sp. CAL3 and *Paenibacillus* sp. CAC13. The compounds were identified by comparison of their mass spectra with the NIST library based on their molecular weight and retention time. To ensure a higher degree of accuracy, the minimum

similarity match cut-off for compound identification was pegged at 700. The compounds were configured into a map (Fig. 4) through multivariate analysis (seriation) using the PALEONTOLOGICAL STATISTICAL software (PAST3.13, Hammer 1999–2016). Looking at the number of compounds that were produced, it would have been virtually impossible to discuss each of them in detail; hence, the data were also subjected to principal component analysis (PCA) using SIMCA (SIMCA 14.0 Ink, UMETRICS Company 1998–2015, Malmö, Sweden) and the compounds which fell off the nucleus were chosen for further analysis and discussion. These are presented in Table 3.

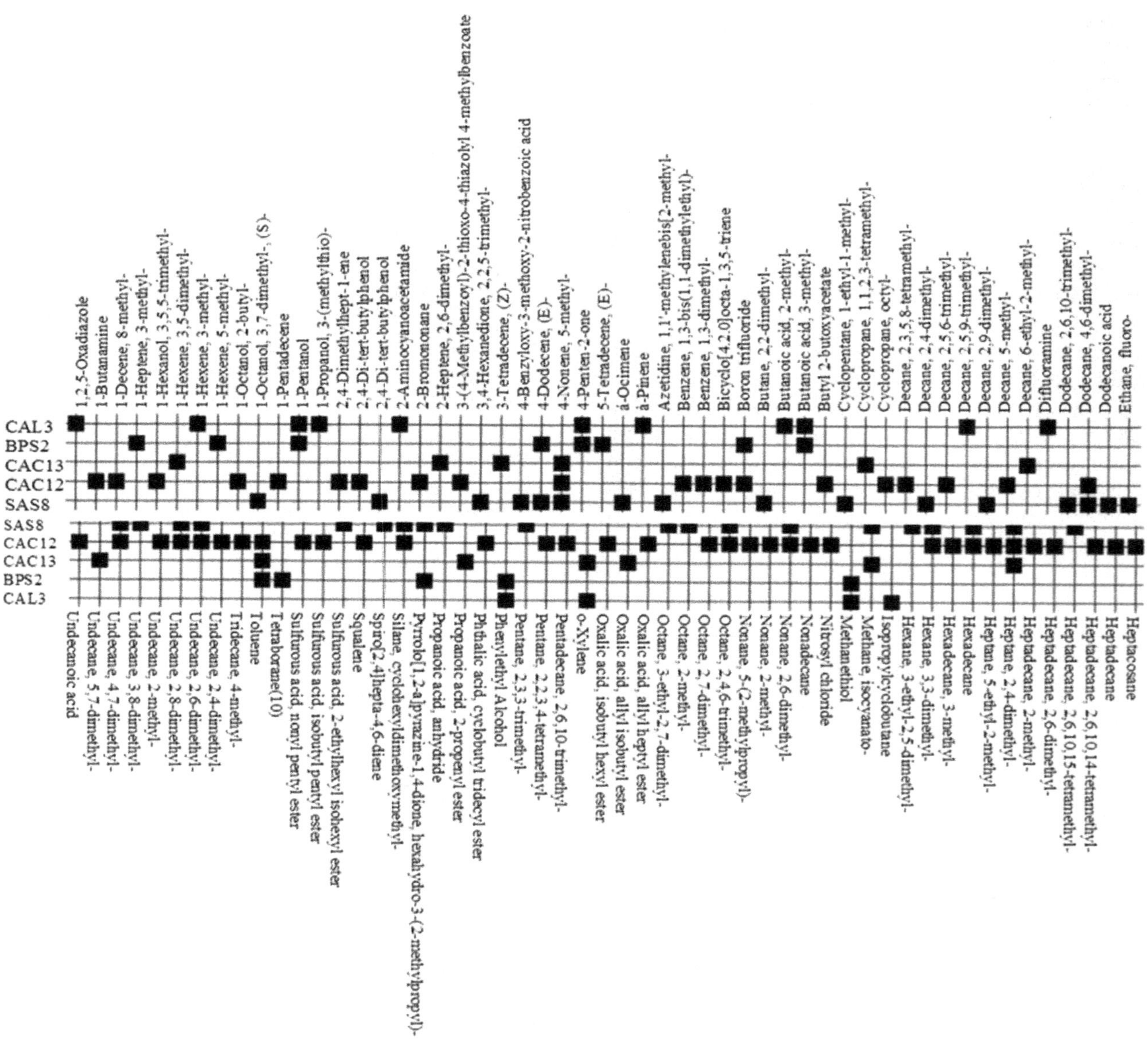

Fig. 4. A map of secondary metabolites produced by bacterial isolates from carwash effluents as detected by GC-MS.

Isolate *Paenibacillus* sp. CAC12 produced the most number of biotechnologically important molecules. These included cyclopentane, silane (cyclohexyldimethoxymethyl-), spiro[2,4]hepta-4,6-diene and 3-(4-methylbenzoyl)-2-thioxo-4-thiazolyl 4-methylbenzoate. With the exception of silane, which is used as an important component of polymeric substances for filling dental cavities (Antonucci *et al.*, 2005), all the other metabolites are known bioactive compounds exhibiting either antifungal or antibacterial activities. While no studies have previously reported production of any of these compounds by any member of the genus *Paenibacillus*, Pohl *et al.* (2011) reported microbial production of cyclopentane in the form of a C17 cyclopentane fatty acid from *Pseudomonas*. Zhang *et al.* (2015) have, however, previously extracted cyclopentane from *Cunninghamia lanceolate* wood extracts known for their rich drug compositions, using GC-MS. Peng *et al.* (2013) also extracted it from *Illicium verum* fruit, a medicinal plant that is known for having many active ingredients. As according to Zhang *et al.* (2015), the output of bioactive ingredients from herbaceous plants is too little to meet market demand, biotechnological approaches targeting enzyme-producing microorganisms or their genes can be employed to amplify production of such biotechnologically important molecules.

Production of the compound 3-(4-methylbenzoyl)-2-thioxo-4-thiazolyl 4-methylbenzoate was of particular interest in this study as no research records could be found of its production by any microorganism. The only information that could be accessed on this compound was about its multiple uses which ranged from inhibition of tau fibril formation to inhibition of *Bacillus subtilis* virulence mechanisms to inhibition of parasitic mechanisms of the malaria parasite *Plasmodium falciparum* (NCBI, 2016). Formation of tau fibrils has been linked to Alzheimer's disease, which is one of the most common causes of dementia in adults (Morozova *et al.*, 2013; Dinkel *et al.*, 2015). Clearly, this is a very important compound with potential for many other applications, and further research on its production and applications is highly recommended.

A glance at Table 3 also shows that most of the compounds produced by the isolates are bioactive ingredients of commercial antimicrobials or plant extracts. The compound pyrrolo[1,2a]pyrazine-1, 4-dione, hexahydro-3-(2-methylpropyl), for instance, has antifungal, antibacterial, nematicidal and anticancer activities. Its antimicrobial profile suggests a need by the producing isolates (*Pseudomonas* sp. SAS8 and *Proteus* sp. BPS2) to competitively suppress the growth of other competing microorganisms, which could be likely due to restricted resources in their environment. Hibbing *et al.* (2010) allude to the fact that in cases of stiff competition

for resources, microbial communities could resort to production of colicins (bacteriocins targeting enteric bacteria) and/or antibiotics among other strategies of limiting the growth of competing species. Microorganisms, which produce anticancer drugs, deserve a much bigger attention given the growing global incidence of this disease, and the discomfort of the current chemotherapeutic methods of treating it.

Although a few selected compounds were given extensive analysis and discussion, a potentially high number of biotechnologically important molecules from the metabolite map (Fig. 4) remain unexplored. This opens up new avenues for further research to fully tap into these microbial bioresources. Also, while compound determination by both GC-MS and LC-MS was only qualitative, the presence of any of the identified compounds points to the presence of the encoding gene(s) within the bacterial isolates. Application of biotechnological techniques like gene cloning can therefore be used to amplify production of the compounds to commercial levels.

Compounds identified by LC-MS

Elucidation of microbial secondary metabolites was performed using the Bruker software. LC-MS results were first subjected to principal component analysis (PCA), which helped in identifying those compounds which were significantly different from the rest. Structural elucidation of the compounds was performed using online libraries including KEGG, Pubchem and Chemspider. Principal component 1 (PC1) showed that isolates *Pseudomonas* sp. SAS8, *Paenibacillus* sp. CAC12 and *Paenibacillus* sp. CAC13 produced metabolites, which were significantly different from those of other isolates. PC2 further showed that while metabolites produced by isolates *Paenibacillus* sp. CAC12 and *Paenibacillus* sp. CAC13 were comparably similar, they were, however, different from metabolites produced by isolate *Pseudomonas* sp. SAS8. PCA loadings showed that all the isolates could produce the compounds cyclo(phenylalanyl-prolyl), 2,2,3-trihydroxybutanoic acid, 5,7-dihydroxyflavone, cyclo(L-leucyl-L-leucyl), vidarabine, 2-(4-hydroxyphenyl)ethyl tricosanoate and 1-(2-hydroxyethyl)-2,5,5,8a-tetramethyldecahydro-2-naphthalenol, while only isolate *Paenibacillus* sp. CAC12 and *Paenibacillus* sp. CAC13 could produce cyclohexyl(2,4-dimethylcyclohexyl)hydroxyacetic acid. Similarly, the compound 4-hydroxy-1-phenylbenzimidazole hydrochloride could only be produced by the isolate *Pseudomonas* sp. SAS8. The molecular structures of the compounds are shown in Fig. 5.

The compound 5,7-dihydroxyflavone (chrysin), which was produced by all isolates, is a known natural flavonoid found in many plant extracts that are commonly used as

Table 3. Description of structure and uses of some bacterial secondary metabolites identified by GC-MS.

Isolate	Compound name and chemical structure	Known/potential applications	References
CAL3	1,2,5-oxadiazole	Major scaffold in the development of potential anticancer agents	(Boiani et al., 2001; Kumar et al., 2014)
CAC12	Spiro[2,4]hepta-4,6-diene	Used in synthesis of biologically active compounds ranging from pesticides to therapeutic drugs to enzymes	(Menchikov and Nefedov, 2016)
SAS8	α-pinene	Major component of some therapeutic plant oils known for antibacterial activity against any bacterial species	(Baik et al., 2008; Hernández et al., 2013)
CAL3, BPS2	Phenylethyl alcohol	Antimicrobial preservative in pharmaceutical products like nasal sprays. Used to make selective agar for growth of Gram-positive microbes	(Brewer and Lilley, 2011; Reza et al., 2015)
CAC12, SAS8	Silane, Cyclohexyldimethoxymethyl-	Used to mediate interfacial bonding in mineral reinforced dental polymeric composites	(Antonucci et al., 2005)
CAC12, CAC13	Cyclopentane	Cyclopentane fatty acids have potential antifungal activity against Candida albicans, C. glabrata and Pythium ultimum	(Pohl et al., 2011)
SAS8, BPS2	Pyrrolo[1,2a]pyrazine-1,4-dione, hexahydro-3-(2-methylpropyl)-	Exhibits antibacterial, antifungal, nematicidal and anticancer properties. Microbial and plant extracts containing this compound are commonly used as broad spectrum antibiotics	(Moniruzzaman et al., 2015; Mohan et al., 2016; Ser et al., 2016; Sharma et al., 2016)

Table 3. (Continued)

Isolate	Compound name and chemical structure	Known/potential applications	References
CAC12	3-(4-Methylbenzoyl)-2-thioxo-4-thiazolyl 4-methylbenzoate	Uses include inhibition of tau fibril formation and thioflavin T binding, inhibition of *Bacillus subtilis* Sfp phosphopantetheinyl transferase (PPTase), and identification of small molecule inhibitors of *Plasmodium falciparum* Glucose-6-phosphate dehydrogenase via a fluorescence intensity assay	(NCBI, 2016)

wound-healing, skin-protective and anticancer medicines (Sathishkumar *et al.*, 2015). Chrysin is also known for its anti-inflammatory and antioxidation properties (Engelmann *et al.*, 2005). The research of Kim *et al.* (2011) also found chrysin to be an effective inhibitor against hyperpigmentation by directly inhibiting the activity of adenylyl cyclase, a key enzyme involved in cAMP-induced melanogenesis. Other researchers (Li *et al.*, 2015) have proven that chrysin is an antihypertensive drug, which acts by inhibiting the proliferation and migration of pulmonary artery smooth muscle cells, together with the associated extracellular matrix components, principally, collagens. Given its important range of biological activities, microbial production of chrysin could be manipulated to significantly increase its production both as a therapeutic agent and also as a dietary supplement.

The compound 2-(4-hydroxyphenyl)ethyl tricosanoate, which was also identified in the crude extracts of all isolates, was previously isolated by Acevedo *et al.* (2000)

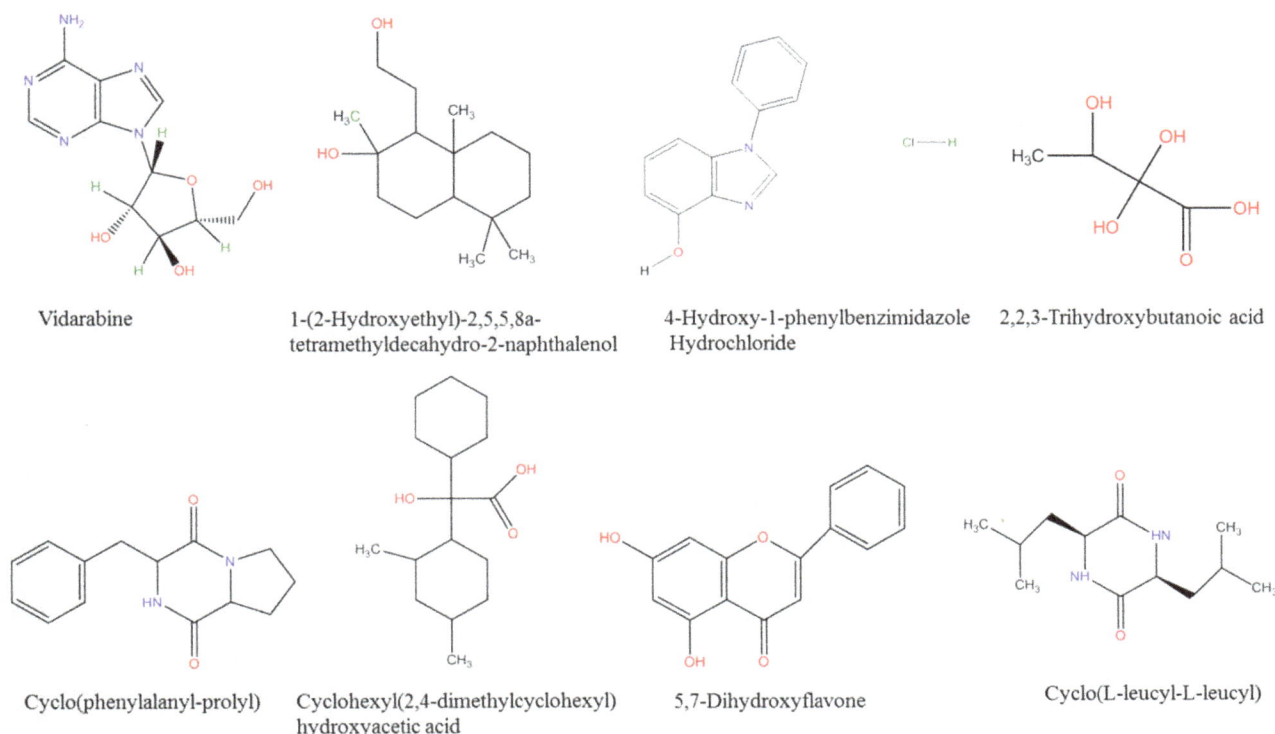

Vidarabine

1-(2-Hydroxyethyl)-2,5,5,8a-tetramethyldecahydro-2-naphthalenol

4-Hydroxy-1-phenylbenzimidazole Hydrochloride

2,2,3-Trihydroxybutanoic acid

Cyclo(phenylalanyl-prolyl)

Cyclohexyl(2,4-dimethylcyclohexyl) hydroxyacetic acid

5,7-Dihydroxyflavone

Cyclo(L-leucyl-L-leucyl)

Fig. 5. Structural elucidation of bacterial secondary metabolites identified using UHPLC-MS.

as one of the components of the crude extracts of *Buddleja cordata* subsp. *cordata* known for antimycobacterial activity. In addition, the same compound is traditionally used in various other applications including treatment of diarrhoea, headaches and kidney ailments (Acevedo *et al.*, 2000). Literature records of microbial production of 2-(4-hydroxyphenyl)ethyl tricosanoate are scarce, suggesting there could be many metabolites which are currently known to be produced by plants only when, in fact, they are produced by microorganisms also.

One of the identified compounds, cyclo(phenylalanyl-prolyl), is a known cyclic dipeptide, which is used as a scaffold for drugs besides its use as an antiviral, antibiotic and antitumour drug (Wickrama Arachchilage *et al.*, 2012). Microbial production of this dipeptide has previously been reported in a *Streptomyces* sp. isolated from marine sediments (Macherla *et al.*, 2005). Vidarabine is yet another compound with antiviral properties that was identified in the crude extracts of all isolates. Vidarabine is a purine nucleotide analog that inhibits viral DNA synthesis and has previously been identified in the extracts of a Caribbean sponge *Cryptotethia crypta* (Suzuki *et al.*, 2006). Literature is awash with evidence of production of vidarabine by marine bacteria and sponges (Sagar *et al.*, 2010; Wang *et al.*, 2013; Nadeem *et al.*, 2016). To our knowledge, this is the first report of the production of vidarabine by bacteria isolated from a non-marine environment. Of late, however, use of vidarabine as an antiviral agent has been declining owing to its neurotoxicity, which has prompted research for more effective and less toxic alternatives (Vajpayee and Malhotra, 2000). Other identified compounds including 1-(2-hydroxyethyl)-2,5,5,8a-tetramethyldecahydro-2-naphthalenol, 4-hydroxy-1-phenylbenzimidazole hydrochloride, 2,2,3-trihydroxybutanoic acid and cyclohexyl(2,4-dimethylcyclohexyl)hydroxyacetic acid have no clear known uses, as does a host of other compounds which neither of the LC-MS linked online libraries could detect. This shows that exploration of novel microbial metabolites still has a long way to go. Dalmaso *et al.* (2015) estimated that between 1 and 10% of prokaryotes have so far been described, and this equates more or less to microbial secondary metabolites.

Most of the metabolites identified in this study have previously been documented to have antibacterial, antifungal and/or anticancer properties, although this was not experimentally proven in the present study. This finding, however, was to be expected since currently, the most marketed antimicrobials and anticancer drugs are natural products of microbial origin (Shaaban *et al.*, 2013). It follows that microbial populations in different ecological settings including natural and synthetic extreme environments still have a lot to offer in terms of bioresources. In a bid to increase the success rate in the identification of novel microbial secondary metabolites,

the use of GC-MS and UHPLC-MS in chemical profiling of crude fermentation extracts can be a very useful tool for assessing the chemical novelty of the crude extracts by comparing the mass spectra to in-built and online compound libraries such as Chemspider, Pubchem and KEGG.

Conclusion

Synthetic extreme environments have been overlooked as potential sources of biotechnologically relevant microbiota. The outcome of this research indicates that some microbial metabolites which have previously been known to be produced only by microorganisms in natural extreme environments like the marine environment can also be produced by microorganisms in synthetic extreme environments like carwashes. Further, this research has revealed the immense bioresource potential of microorganisms inhabiting synthetic extreme environments, which harbour potential as agents of bioremediation and/or producers of biotechnologically relevant molecules including enzymes.

Acknowledgements

The authors wish to acknowledge UNISA for funding this research through the Women in Science Research Program.

Conflict of interest

The authors wish to declare no conflict of interests with respect to submission, review and potential publication of this manuscript.

References

Acevedo, L., Martínez, E., Castañeda, P., Franzblau, S., Timmermann, B.N., Linares, E., *et al.* (2000) New phenylethanoids from *Buddleja cordata* subsp. cordata. *Planta Med* **66:** 257–261.

Affandi, I.E., Suratman, N.H., Abdullah, S., Ahmad, W.A., and Zakaria, Z.A. (2014) Degradation of oil and grease from high-strength industrial effluents using locally isolated aerobic biosurfactant-producing bacteria. *Int Biodeterior Biodegrad* **95:** 33–40.

Akmal, M. (2009) Microbial biomass and bacterial community changes by Pb contamination in acidic soil. *J Agric Biol Sci* **1:** 30–37.

Al-Daghri, N.M., Alokail, M.S., Abd-Alrahman, S.H., Draz, H.M., Yakout, S.M., and Clerici, M. (2013) Polycyclic aromatic hydrocarbon exposure and pediatric asthma in children: a case-control study. *Environ Heal A Glob Access Sci Source* **12:** 2–7.

Al-Daghri, N.M., Alokail, M.S., Abd-Alrahman, S.H., and Draz, H.M. (2014) Polycyclic aromatic hydrocarbon distribution in serum of Saudi children using HPLC-FLD: marker elevations in children with asthma. *Environ Sci Pollut Res Int* **21**: 12085–12090.

Ali, J., Ahmad, B., Nigar, S., Sadaf, S., Shah, A., Bashir, S., et al. (2013) Isolation and identification of cellulose degrading bacteria from municipal waste and their screening for potential antimicrobial activity. *World Appl Sci J* **27**: 1420–1426.

Antonucci, J.M., Dickens, S.H., Fowler, B.O., and Xu, H.H.K. (2005) Chemistry of silanes: interfaces in dental polymers and composites. *J Res Natl Inst Stand Technol* **110**: 541.

Azua-Bustos, A., and González-Silva, C. (2014) Biotechnological applications derived from microorganisms of the Atacama Desert. *Biomed Res Int* **2014**: 1–7.

Baik, J.S., Kim, S.-S., Lee, J.-A., Oh, T.-H., Kim, J.-Y., Lee, N.H., and Hyun, C.-G. (2008) Chemical composition and biological activities of essential oils extracted from Korean endemic citrus species. *J Microbiol Biotechnol* **18**: 74–79.

Bayat, Z., Hassanshahian, M., and Cappello, S. (2015) Immobilization of microbes for bioremediation of crude oil polluted environments : a mini review. *Open Microbiol J* **9**: 48–54.

Ben Said, O., Goñi-Urriza, M.S., El Bour, M., Dellali, M., Aissa, P., and Duran, R. (2008) Characterization of aerobic polycyclic aromatic hydrocarbon-degrading bacteria from Bizerte lagoon sediments, Tunisia. *J Appl Microbiol* **104**: 987–997.

Bidoia, E.D., Montagnolli, R.N., and Lopes, P.R.M. (2012) Microbial biodegradation potential of hydrocarbons evaluated by colorimetric technique: a case study. *Appl Microbiol Microb Biotechnol* **3**: 1277–1288.

Bisht, S.S., and Panda, A.K. (2011) Biochemical characterization and 16S rRNA sequencing of few lipase-producing thermophilic bacteria from Taptapani hot water spring, Orissa, India. *Biotechnol Res Int* **2011**: 452710.

Boiani, M., Cerecetto, H., and González, M. (2001) 1, 2, 5-Oxadiazole N-oxide derivatives as potential anti-cancer agents: synthesis and biological evaluation. Part IV. *Eur J Med Chem* **36**: 771–782.

Bonfá, M.R.L., Grossman, M.J., Mellado, E., and Durrant, L.R. (2011) Biodegradation of aromatic hydrocarbons by Haloarchaea and their use for the reduction of the chemical oxygen demand of hypersaline petroleum produced water. *Chemosphere* **84**: 1671–1676.

Brewer, J.H., and Lilley, B.D. (2011) Phenylethyl Alcohol Agar. *Microbe Libr* **5**: 1–8.

Bull, A.T. (2011) *Bioprospecting Among the Actinobacteria of Extreme Environments*. University of Kent, Canterbury, UK.

Cavicchioli, R., Siddiqui, K.S., Andrews, D., and Sowers, K.R. (2002) Low-temperature extremophiles and their applications. *Curr Opin Biotechnol* **13**: 253–261.

Chakravorty, S., Helb, D., Burday, M., and Connell, N. (2007) A detailed analysis of 16S ribosomal RNA gene segments for the diagnosis of pathogenic bacteria. *J Microbiol Methods* **69**: 330–339.

Chronakova, A., Kristufek, V., Tichy, M., and Elhottova, D. (2010) Biodiversity of streptomycetes isolated from a succession sequence at a post-mining site and their evidence in Miocene lacustrine sediment. *Microbiol Res* **165**: 594–608.

Dalmaso, G.Z.L., Ferreira, D., and Vermelho, A.B. (2015) Marine extremophiles a source of hydrolases for biotechnological applications. *Mar Drugs* **13**: 1925–1965.

Demirjian, D.C., Morís-varas, F., and Cassidy, C.S. (2001) Enzymes from extremophiles. *Curr Opin Chem Biol* **5**: 144–151.

Deppe, U., Richnow, H.H., Michaelis, W., and Antranikian, G. (2005) Degradation of crude oil by an arctic microbial consortium. *Extremophiles* **9**: 461–470.

Dinkel, P.D., Holden, M.R., Matin, N., and Margittai, M. (2015) RNA binds to tau fibrils and sustains template-assisted growth. *Biochemistry* **54**: 4731–4740.

Enache, M., and Kamekura, M. (2010) Hydrolytic enzymes of halophilic microorganisms and their economic values. *Rom J Biochem* **47**: 47–59.

Engelmann, M.D., Hutcheson, R., and Cheng, I.F. (2005) Stability of ferric complexes with 3-hydroxyflavone (flavonol), 5,7-dihydroxyflavone (chrysin), and 3',4'-dihydroxyflavone. *J Agric Food Chem* **53**: 2953–2960.

Ettoumi, B., Chouchane, H., Guesmi, A., Mahjoubi, M., Brusetti, L., Neifar, M., et al. (2016) Diversity, ecological distribution and biotechnological potential of Actinobacteria inhabiting seamounts and non-seamounts in the Tyrrhenian Sea. *Microbiol Res* **186–187**: 71–80.

Golani, M., Hajela, K., and Pandey, G.P. (2016) Screening. Identification, characterization and production of bacterial lipase from oil spilled soil. *Asian J Pharm Clin Res* **5**: 745–763.

Guazzaroni, M.E., Silva-Rocha, R., and Ward, R.J. (2015) Synthetic biology approaches to improve biocatalyst identification in metagenomic library screening. *Microb Biotechnol* **8**: 52–64.

Hamzah, A., Rabu, A., Azmy, R.F.H.R., and Yussoff, N.A. (2010) Isolation and characterization of bacteria degrading Sumandak and South Angsi oils. *Sains Malaysiana* **39**: 161–168.

Hernández, V., Mora, F., Araque, M., De Montijo, S., Rojas, L., Meléndez, P., and De Tommasi, N. (2013) Chemical composition and antibacterial activity of *Astronium graveolens* JACQ essential oil. *Rev Latinoam Quim* **41**: 89–94.

Hibbing, M.E., Fuqua, C., Parsek, M.R., and Peterson, S.B. (2010) Bacterial competition: surviving and thriving in the microbial jungle. *Natl Rev Microbiol* **8**: 15–25.

Huber, H., and Stetter, K.O. (1998) Hyperthermophiles and their possible potential in biotechnology. *J Biotechnol* **64**: 39–52.

Ionescu, R., Măruțescu, L., Tănase, A.-M., Chiciudean, I., Csutak, O., Pelinescu, D., et al. (2015) Flow cytometry based method for evaluation of biodegradative potential of *Pseudomonas fluorescens*. *Agric Agric Sci Procedia* **6**: 567–578.

Irwin, R., Van Mouwerik, M., Stevens, L., Seese, M. and Basham, W. (1997) *Environmental Contaminants Encyclopedia: Diesel Oil Entry*. Fort Collins, CO: National Park Service, Water Resources Divisions, Water Operations Branch.

Jami, M., Ghanbari, M., Kneifel, W., and Domig, K.J. (2015) Phylogenetic diversity and biological activity of culturable Actinobacteria isolated from freshwater fish gut microbiota. *Microbiol Res* **175**: 6–15.

Kim, D.C., Rho, S.H., Shin, J.C., Park, H.H., and Kim, D. (2011) Inhibition of melanogenesis by 5,7-dihydroxyflavone (chrysin) via blocking adenylyl cyclase activity. *Biochem Biophys Res Commun* **411**: 121–125.

Kuang, W., Li, J., Zhang, S., and Long, L. (2015) Diversity and distribution of Actinobacteria associated with reef coral *Porites lutea*. *Front Microbiol* **6**: 1–13.

Kubota, K., Koma, D., Matsumiya, Y., Chung, S.Y., and Kubo, M. (2008) Phylogenetic analysis of long-chain hydrocarbon-degrading bacteria and evaluation of their hydrocarbon-degradation by the 2,6-DCPIP assay. *Biodegradation* **19**: 749–757.

Kumar, D., Kumar, L., Nagar, S., Raina, C., Parshad, R., and Gupta, V.K. (2012) Screening, isolation and production of lipase/esterase producing Bacillus sp. strain DVL2 and its potential evaluation in esterification and resolution reactions. *Arch Appl Sci Res* **4**: 1763–1770.

Kumar, A., Ito, A., Takemoto, M., Yoshida, M., and Zhang, K.Y.J. (2014) Identification of 1,2,5-oxadiazoles as a new class of SENP2 inhibitors using structure based virtual screening. *J Chem Inf Model* **54**: 870–880.

Kumar, A., Dhar, K., Kanwar, S.S., and Arora, P.K. (2016) Lipase catalysis in organic solvents: advantages and applications. *Biol Proced Online* **18**: 2.

Kunihiro, M., Ozeki, Y., Nogi, Y., Hamamura, N., and Kanaly, R.A. (2013) Benz[a]anthracene biotransformation and production of ring fission products by Sphingobium sp. strain KK22. *Appl Environ Microbiol* **79**: 4410–4420.

Li, X.-W., Wang, X.-M., Li, S., and Yang, J.-R. (2015) Effects of chrysin (5,7-dihydroxyflavone) on vascular remodeling in hypoxia-induced pulmonary hypertension in rats. *Chin Med* **10**: 4.

Macherla, V.R., Liu, J., Bellows, C., Teisan, S., Nicholson, B., Lam, K.S., and Potts, B.C.M. (2005) Glaciapyrroles A, B, and C, pyrrolosesquiterpenes from a Streptomyces sp. isolated from an Alaskan marine sediment. *J Nat Prod* **68**: 780–783.

Maes, S., Props, R., Fitts, P., Smet, R.De., Vilchez-vargas, R., Vital, M., *et al.* (2016) Platinum recovery from synthetic extreme environments by halophilic bacteria. *Environ Sci Technol* **50**: 2619–2626.

Mandri, T., and Lin, J. (2007) Isolation and characterization of engine oil degrading indigenous microrganisms in Kwazulu-Natal, South Africa. *Afr J Biotechnol* **6**: 23–27.

Markowicz, A., Cyco, M., and Piotrowska-Seget, Z. (2016) Microbial community structure and diversity in long-term hydrocarbon and heavy metal contaminated soils. *Int J Environ Res* **10**: 321–332.

Martín-Gil, J., Ramos-Sánchez, M.C., and Martín-Gil, F.J. (2004) Shewanella putrefaciens in a fuel-in-water emulsion from the Prestige oil spill. *Antonie van Leeuwenhoek. Int J Gen Mol Microbiol* **86**: 283–285.

Menchikov, L.G., and Nefedov, O.M. (2016) Spiro[2.4]hepta-4,6-dienes: synthesis and application in organic synthesis. *Russ Chem Rev* **85**: 205–225.

Mohan, G., Thangappanpillai, A.K., and Ramasamy, B. (2016) Antimicrobial activities of secondary metabolites and phylogenetic study of sponge endosymbiotic bacteria, Bacillus sp. at Agatti Island, Lakshadweep Archipelago. *Biotechnol Rep* **11**: 44–52.

Moniruzzaman, S., Haque, A., Khatun, R., and Yaakob, Z. (2015) Gas chromatography mass spectrometry analysis and in vitro antibacterial activity of essential oil from

Trigonella foenum-graecum. *Asian Pac J Trop Biomed* **5**: 1033–1036.

Morozova, O.A., March, Z.M., Robinson, A.S., and Colby, D.W. (2013) Conformational features of tau fibrils from Alzheimer's disease brain are faithfully propagated by unmodified recombinant protein. *Biochemistry* **52**: 6960–6967.

Nadeem, F., Oves, M., Qari, H., and Ismail, I.M. (2016) Red sea microbial diversity for antimicrobial and anticancer agents. *J Mol Biomark Diagn* **7**: 1–14.

NCBI (2016) National Center for Biotechnology Information. PubChem Compound Database; CID=576784. Ncbi.

Oliveira, N.C.De., Rodrigues, A.A., Alves, M.I.R., Filho, N.R.A., Sadoyama, G., and Vieira, J.D.G. (2012) Endophytic bacteria with potential for bioremediation of petroleum hydrocarbons and derivatives. *Afr J Biotechnol* **11**: 2977–2984.

Pandey, S., Singh, S., Yadav, A.N., Nain, L., and Saxena, A.K. (2013) Phylogenetic diversity and characterization of novel and efficient cellulase producing bacterial isolates from various extreme environments. *Biosci Biotechnol Biochem* **77**: 1474–1480.

Peng, W., Lin, Z., Chang, J., Gu, F., and Zhu, X. (2013) Biomedical molecular characteristics of YBSJ extractives from *Illicium verum* fruit. *Biotechnol Biotechnol Equip* **27**: 4311–4316.

Perks, B. (2011) Extreme potential. *Chem World* 48–51.

Pohl, C.H., Kock, J.L.F. and Thibane, V.S. (2011) *Antifungal free fatty acids: A Review*. University of the Free State, Bloemfontein, South Africa.

Ramesh, S., Kumar, R., Devi, R.A., and Balakrishnan, K. (2014) Isolation of a lipase producing bacteria for enzyme synthesis in shake flask cultivation. *Int J Curr Microbiol Appl Sci* **3**: 712–719.

Rampelotto, P.H. (2013) Extremophiles and extreme environments. *Life* **3**: 482–485.

Reed, C.J., Lewis, H., Trejo, E., Winston, V., and Evilia, C. (2013) Protein adaptations in archaeal extremophiles. *Archaea* **2013**: 1–14.

Reza, H., Fereshteh, N., and Saman, A.N. (2015) Determination of phenylethyl alcohol by reversed-phase high-performance liquid chromatography (RP-HPLC) in Budesonide nasal spray. *Afr J Pure Appl Chem* **9**: 81–90.

Sagar, S., Kaur, M., and Minneman, K.P. (2010) Antiviral lead compounds from marine sponges. *Mar Drugs* **8**: 2619–2638.

Sathishkumar, G., Bharti, R., Jha, P.K., Selvakumar, M., Dey, G., Jha, R., *et al.* (2015) Dietary flavone chrysin (5,7-dihydroxyflavone ChR) functionalized highly-stable metal nanoformulations for improved anticancer applications. *RSC Adv* **5**: 89869–89878.

Selbmann, L., Egidi, E., Isola, D., Onofri, S., Zucconi, L., de Hoog, G.S., *et al.* (2013) Biodiversity, evolution and adaptation of fungi in extreme environments. *Plant Biosyst – An Int J Deal with all Asp Plant Biol* **147**: 237–246.

Ser, H.-L., Palanisamy, U.D., Yin, W.-F., Chan, K.-G., Goh, B.-H. and Lee, L.-H. (2016) *Streptomyces malaysiense* sp. nov.: a novel Malaysian mangrove soil actinobacterium with antioxidative activity and cytotoxic potential against human cancer cell lines. *Sci Rep* **6,** 24247.

Seufferheld, M.J., Alvarez, M., and Farias, M.E. (2008) Role of polyphosphates in microbial adaptation to extreme environments. *Appl Environ Microbiol* **74**: 5867–5874.

Shaaban, M., Abdel-Razik, A.S., Abdel-Aziz, M.S., Abou Zied, A., and Fadel, M. (2013) Bioactive secondary metabolites from marine *Streptomyces albogriseolus* isolated from Red Sea coast. *J Appl Sci Res* **9**: 996–1003.

Sharma, P., Kalita, M.C., and Thakur, D. (2016) Broad spectrum antimicrobial activity of forest-derived soil actinomycete, Nocardia sp. PB-52. *Front Microbiol* **7**: 1–17.

Stathopoulou, P.M., Savvides, A.L., Karagouni, A.D., and Hatzinikolaou, D.G. (2013) Unraveling the lipolytic activity of thermophilic bacteria isolated from a volcanic environment unraveling the lipolytic activity of thermophilic bacteria isolated from a volcanic environment. *Biomed Res Int* **2013**: 1–13.

Sutton, N.B., Maphosa, F., Morillo, J.A., Al-Soud, W.A., Langenhoff, A.A.M., Grotenhuis, T., *et al.* (2013) Impact of long-term diesel contamination on soil microbial community structure. *Appl Environ Microbiol* **79**: 619–630.

Suzuki, M., Okuda, T., and Shiraki, K. (2006) Synergistic antiviral activity of acyclovir and vidarabine against herpes simplex virus types 1 and 2 and varicella-zoster virus. *Antiviral Res* **72**: 157–161.

Tamura, K., Stecher, G., Peterson, D., Filipski, A., and Kumar, S. (2013) MEGA6: Molecular Evolutionary Genetics Analysis Version 6.0. *Mol Biol Evol* **30**: 2725–2729.

Tekere, M., Sibanda, T., and Maphangwa, K.W. (2016) An assessment of the physicochemical properties and toxicity potential of carwash effluents from professional carwash outlets in Gauteng Province, South Africa. *Environ Sci Pollut Res* **23**: 11876–11884.

Um, Y., Chang, M.W., and Holoman, T.P. (2010) A simple and effective plating method to screen polycyclic aromatic hydrocarbon-degrading bacteria under various redox conditions. *Appl Microbiol Biotechnol* **88**: 291–297.

Vajpayee, M., and Malhotra, N. (2000) Antiviral drugs against herpes infections. *Indian J od Pharmacol* **32**: 330–338.

Wang, Y., Shi, J., Wang, H., Lin, Q., Chen, X., and Chen, Y. (2007) The influence of soil heavy metals pollution on soil microbial biomass, enzyme activity, and community composition near a copper smelter. *Ecotoxicol Environ Saf* **67**: 75–81.

Wang, R., Billone, P.S., and Mullett, W.M. (2013) Nanomedicine in action: an overview of cancer nanomedicine on the market and in clinical trials. *J Nanomater* **2013**: 421–428.

Wickrama Arachchilage, A.P., Wang, F., Feyer, V., Plekan, O., and Prince, K.C. (2012) Photoelectron spectra and structures of three cyclic dipeptides: PhePhe, TyrPro, and HisGly. *J Chem Phys* **136**: 124301–124308.

Zhang, X., Huang, K., Ye, Y., Shi, J., and Zhang, Z. (2015) Biomedical molecular of woody extractives of *Cunninghamia lanceolata* biomass. *Pak J Pharm Sci* **28**: 761–764.

10

Advances and bottlenecks in microbial hydrogen production

Alan J. Stephen,[1] Sophie A. Archer,[1]
Rafael L. Orozco[2] and Lynne E. Macaskie[2,*]
*Schools of [1]Chemical Engineering, [2]Biosciences,
University of Birmingham, Edgbaston, Birmingham, B15
2TT, UK.*

Summary

Biological production of hydrogen is poised to become a significant player in the future energy mix. This review highlights recent advances and bottlenecks in various approaches to biohydrogen processes, often in concert with management of organic wastes or waste CO$_2$. Some key bottlenecks are highlighted in terms of the overall energy balance of the process and highlighting the need for economic and environmental life cycle analyses with regard also to socio-economic and geographical issues.

Introduction

Hydrogen provides a CO$_2$-free sustainable alternative to fossil fuels. A pioneering global initiative, the 'Hydrogen Council', comprising thirteen leading energy, transport and related industries, intends to increase investment in the hydrogen and fuel cell sectors (currently €1.4 Bn year^{-1}) to stimulate hydrogen as a key part of the future energy mix via new policies and schemes (Anon, 2017).

Hydrogen is currently obtained mainly by steam reforming of hydrocarbons, releasing multiple greenhouse gas emissions (DOE, 2013). Hence, new H$_2$ production methods are required such as biological production (bio-H$_2$; Dincer and Acar, 2015). H biotechnologies are maturing towards benchmarking against established clean energy from electrolysis of water, solar photovoltaics and wind farms. Biohydrogen can be made fermentatively from wastes, providing a simultaneous method of organic waste management (Chang *et al.*,

*For correspondence. E-mail L.E.Macaskie@bham.ac.uk

Funding Information
Natural Environment Research Council (NE/L014076/1).

2011). This short review highlights progress and bottlenecks of bio-H$_2$ towards a sustainable development goal to ensure access to affordable, reliable, sustainable and modern energy for all. Biohydrogen has been reviewed in comparison with other hydrogen production processes (Nikolaidis and Poullikkas, 2017).

Biohydrogen embraces any H$_2$ production involving biological material (Mohan and Pandey, 2013). The energy source can be solar or can come from conversion of fixed carbon substrates (or both, in various combinations). An approach to CO$_2$-end of pipe treatment (e.g. from flue gas from fossil fuel combustion or carbon-neutral fermentation of biomass) is to grow algae on waste CO$_2$. Algal biohydrogen production is well-described, but O$_2$ from algal oxygenic photosynthesis inhibits the hydrogenase that makes H$_2$. A key study (Kubas *et al.*, 2017) will open the way to developing O$_2$-resistant hydrogenase. Emerging technology uses cyanobacteria (blue-green algae) that make H$_2$ via hydrogenase and also nitrogenase; their O$_2$-sensitivity is managed by temporal separation of photosynthetic O$_2$ evolution and nitrogenase action, and by compartmentalization into microanaerobic heterocysts (Tiwari and Pandey, 2012). Despite a note that cyanobacterial biohydrogen is probably uneconomic (Singh *et al.*, 2016), an environmental life cycle analysis (LCA) has shown for the first time that cyanobacterial bio-H$_2$ has the potential to be a competitor to desulfurized natural gas; the associated environmental impact of producing and extracting each gas, including use in a solid oxide fuel cell, was calculated and simulated respectively using the LCA software SIMAPRO (Archer *et al.*, 2017). This research used published data from a raceway growth system (James *et al.*, 2009). However, at latitudes above ~40°N, the generally low incident solar energy makes stand-alone photobiological H$_2$ systems seasonal and uneconomic without some form of process intensification. Boosting light delivery (e.g. LEDs, quantum dots) can be effective, but these may risk photopigment saturation and inhibition; this approach may be questionable economically and would be best addressed by a life cycle analysis. In sunny countries, light is plentiful, but in this case, 'delivering cold' is needed to extend crop product and food life; cooling is energy-demanding and a global challenge (Strahan, 2017).

Another challenge is organic materials from agri-food and municipal wastes, which must be managed to avoid landfilling which yields methane, a potent greenhouse gas. Current practices use anaerobic digestion (AD) with biogas – methane used for power. We review some options for combining waste treatments with bio-H_2 technology as possibly the best approach to tackling effectively these dual socio-economic problems; stand-alone biohydrogen is possibly uneconomic, but this awaits a life cycle analysis, currently in progress.

Biohydrogen production from waste: fermentation strategies for sustainable 'waste to hydrogen energy'

Fermentation is the disposal of excess metabolic reductant (NADH) onto organic compounds in the absence of alternative electron acceptors such as O_2 and NO_3^- (Guo et al., 2010). The mixed-acid fermentation ('dark fermentation') pathway of the paradigm *Escherichia coli* (Fig. 1A) is simple, has high rates of H_2 production but has limitations (Saratale et al., 2013; Fig. 1A inset). Hexose sugars can stoichiometrically deliver 12 mol H_2 mol hexose^{-1}.

The mixed-acid fermentation, while irreversible, is thermodynamically limited to 2–4 mol H_2 mol hexose^{-1} (Hallenbeck, 2012). The 'NADH pathway' of some microorganisms (Hallenbeck, 2012, 2017) can deliver a higher H yield, but is reversible under a positive H_2 partial pressure, which is required for with a downstream H fuel cell. Thermophilic bacteria have advantages but require input of heat energy. Hence, the focus has been mainly on mesophilic bacteria (Balachandar et al., 2013).

Most mixed-acid fermentations follow a similar schematic: the cell forms reduced metabolic end-products: organic acids (including toxic formate) and alcohol (Fig. 1A). Up to 2 mol H_2 mol^{-1} hexose (Hallenbeck and Ghosh, 2009) is produced via the activity of formate hydrogen lyase (which splits formate to $H_2 + CO_2$), that is < 20% of the theoretical maximum H_2. Sustained bio-H_2 production is limited by end-product (ethanol) toxicity and acidification of the medium by accumulating organic acids (Redwood, 2007).

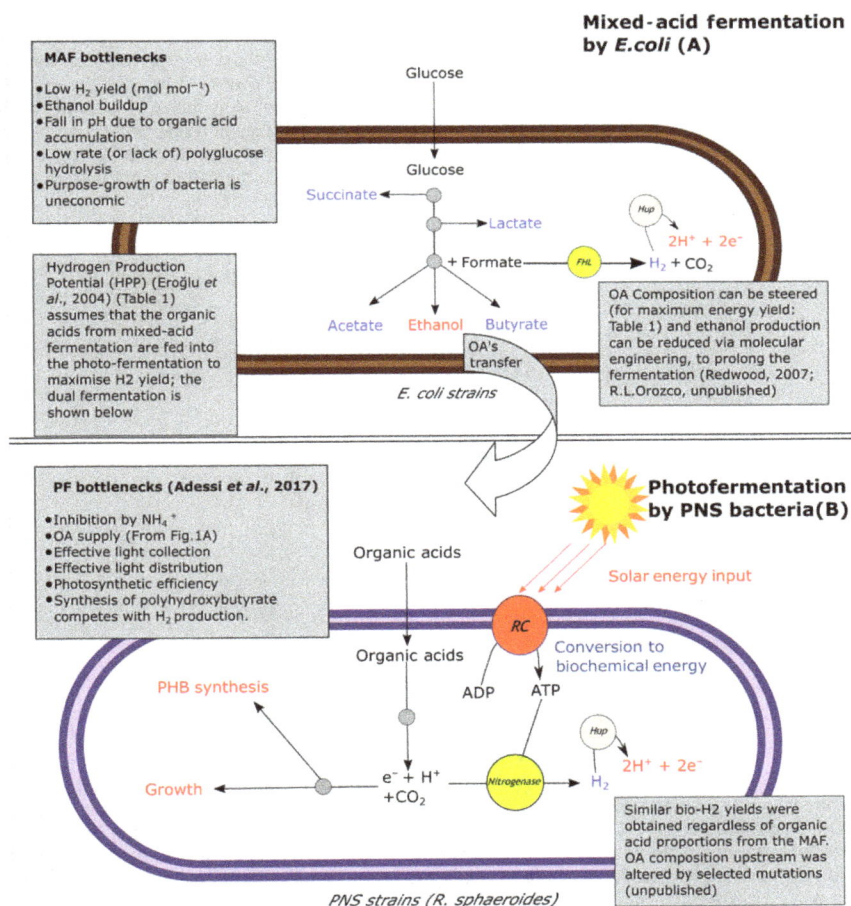

Fig. 1. Mixed-acid fermentation (MAF) of *E. coli* (A) and use of purple non-sulfur bacteria (B) in photofermentation (PF) of organic acids (OAs) into H_2. The organic acids are taken up by (e.g.) *R. sphaeroides,* and reducing power is generated as NADH (not shown). This reducing power can either be used for polyhydroxybutyrate synthesis or growth to maintain cellular redox or alternatively can be used for H_2 production under light when growth is restricted by limitation of N or P source. Italicized bottlenecks are those overcome by use of the dual system (see text).

The organic acids provide a means to overcome the thermodynamic limitation via their use in a coupled photofermentation reactor (Redwood et al., 2012a,b; Hallenbeck, 2013, 2017) via electrodialysis (Fig 2). If organic acid mixtures are fed to purple non-sulfur bacteria (e.g. Rhodobacter sphaeroides), the off-gas (typically > 90% H) is suitable for direct use in fuel cells (Nakada et al., 1999). This anoxygenic photofermentative H_2 process (Fig. 1B) requires input of light energy (to help overcome the thermodynamic barrier in converting organic acids into H_2 (Hallenbeck, 2013)). Nitrogen-deficient conditions are essential; in purple non-sulfur bacteria, H_2 biogenesis is a side reaction of nitrogenase, which normally fixes N_2 and is downregulated in the presence of fixed nitrogen. Utilizable organic acids also feed a competing pathway to make polyhydroxybutyrate which detracts from the H_2 yield (Fig. 1B). Redwood et al. (2012a,b) incorporated an electrodialysis step to concentrate the organic acids (by ~eightfold) and link the mixed-acid and photofermentation steps (Fig. 2). Electrodialysis separates anions (negatively charged organic acids in the dark fermentation medium), removing them and also preventing the transfer of inhibitory NH_4^+ into the photofermentation medium. This continuous dual fermentation process combines high H_2 production rates and yield (Redwood et al., 2012b); the electrical energy demand of electrodialysis is counterbalanced, in part, by a third H_2 stream from electrolysis of water.

Redwood (2007) calculated the break-even current efficiency to quantify the role played by specific organic acids (Table 1). Butyrate is the most attractive organic acid for electrodialysis with the lowest break-even current efficiency at 13% (Table 1). Butyrate is a neglected organic acid product from E. coli which can predominate under some conditions (Redwood, 2007; R.L. Orozco unpublished). Using this example (Figs 1 and 2), the energy balance for bio-H_2 (via fermentation of food

Table 1. Properties of organic acids relevant to their separation from spent medium by electrodialysis.

Organic acid	Carbons	Valence	pK_a	HPP mol^{-1}	BCE (%)
Butyrate	4	1	4.81	10	13
Lactate	3	1	3.86	6	21.6
Formate	1	1	3.75	2	N/A
Acetate	2	1	4.76	7	32.5
Succinate	4	2	4.19, 5.57	7	27.1

The break-even current efficiency (BCE: (energy expended/energy gained) × 100)) was calculated for individual organic acids. The lower the BCE, the less energy required to transport the organic acid. The electrical energy required for organic acid transport via electrodialysis relates to the number of charges and number of carbons; butyrate (4 carbons, 1 charge) is the most favourable and also has the highest proportion of charged butyrate (c.f. butyric acid) according to the pK_a. HPP is hydrogen production potential of the dual system as defined by Eroğlu et al. (2004).

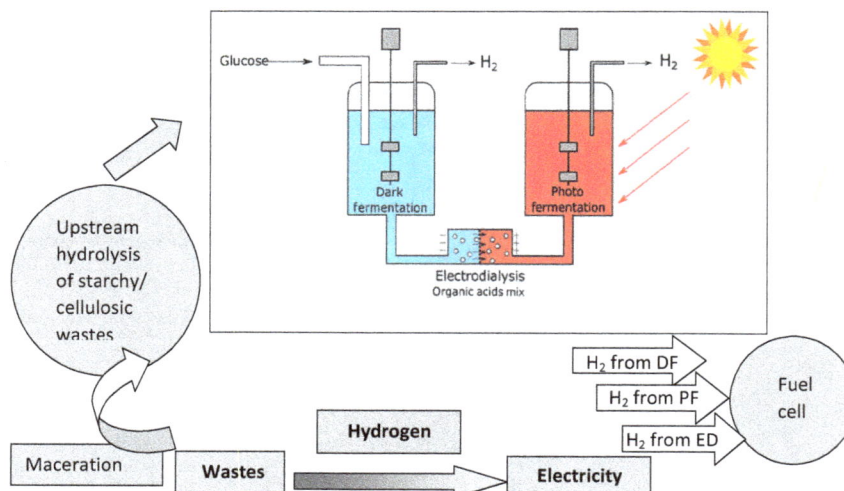

Fig. 2. System for energy delivery from wastes via biohydrogen A fusion of chemical and biochemical engineering for conversion of waste into electricity via integrated biohydrogen technology. Electrodialysis (ED) separates the organic acid (OA) products from the mixed-acid fermentation of (e.g.) E. coli (formate is converted to H_2 + CO_2 via formate hydrogen lyase). OAs pass from the dark fermentation medium to the photofermentation, typically being concentrated by ~eightfold via electrodialysis for dilution into the photofermentation vessel. Alcohol is not removed by ED; this would require a catalytic oxidation stage to give the corresponding organic acid; this has been achieved via using Au(0) nanoparticle catalyst made on E. coli cells (Deplanche et al., 2007). Two bio-H_2 streams are formed from the combined dark- and photofermentations, with a third H_2 stream from electrolysis of water. The maximum H_2 yield from the mixed-acid fermentation is 2 mol sugar^{-1}; hence, the dark fermentation can be viewed as a generator of OAs rather than as the primary H supply. A schematic of upstream waste conversion into sugar feed is shown (see text), and downstream use of hydrogen in a fuel cell for electricity production. Note that bio-H_2 is free of catalyst poisons, which extends fuel cell life. Not all wastes (e.g. sugary fruits, bakery products) require extensive upstream treatment. The main box is the biotechnology; the grey flow sheet is the chemical engineering required to realize the positive energy balance. Both are equally important.

waste) exceeded that from anaerobic digestion, wind and solar power, even without factoring in the additional electrochemically made H_2. (Redwood et al., 2012b). Although ~half of the organic acid is available (anionic) at the pH of the fermentation (according to the pK_a values: Table 1), the electrodialysis chamber itself is alkaline due to OH^- release.

Two key findings are salient. First, the role of the dark fermentation is more important as a supply of organic acids into the photofermentation than for its bio-H_2 per se. Second, recent work (R.L. Orozco and A.J. Stephen, unpublished) showed that the H_2 yield in the photofermentation was largely independent of the actual organic acid proportions in the feed from the mixed-acid fermentation and was optimal at ~40 mM organic acids. Hence, any source of organic acids could be potentially used from a dual system or, indeed, in a stand-alone photofermentation.

Bacterial photofermentation

Purple non-sulfur photosynthetic bacteria produce H_2 from a variety of organic substrates including organic acids (Lazaro et al., 2012), sugars (Keskin and

Table 2. Some approaches to increase photofermentation H productivity (Reviewed by Adessi et al., 2017).

Approach/Rationale	Outcomes/comments	References
'Black box' mathematical relationships between input and output streams Box-behnken statistical design/methods	Permits multivariable analysis: measures cause and effect; hence can be empirical SCE (glycerol) > doubled (R.palustris)	Abo-Hashesh et al. (2013), Show and Lee (2013) and Ghosh et al. (2012a,b,c)
Modelling metabolic fluxes	Guided interventions: success using lactate but not malate or acetate	Golomysova et al. (2010) and Hädicke et al. (2011)
Deletion of polyhydroxybutyrate synthesis pathway	Increased H_2 yield (by 1.5-fold c.f. wild type)	Kim et al. (2011)
Reducing pigment concentration	Allows greater light penetration[a]	Ma et al. (2012)
Use of quantum dots to 'upgrade' light	Doubled photosynthetic efficiency	M.D. Redwood, unpublished[b]

SCE, substrate conversion efficiency.
a. 27% increase in H_2 yield was obtained.
b. Collaborative study with Photon Science Institute, University of Manchester: M.D. Redwood, L.E. Macaskie and D.J. Binks, unpublished work. But note: current commercial quantum dots would be grossly uneconomic at scale.

Table 3. Options for delivery of bio-H_2 into power, all via electro-photofermentation (Figs 1 and 2; M.D. Redwood, R.L.Orozco and L.E. Macaskie, unpublished work)[a].

Feedstock (upstream)	Power (downstream)	Comments
Fermentation of food wastes	Fuel cell electricity[b] or combined heat and power[c]	Food wastes (FW) required (tonnages). Anaerobic digestion (AD) has monopoly on FW. Bio-H_2 can power a fuel cell directly.
Fermentation of cellulosic wastes	Fuel cell electricity or CHP	Comminution/maceration energy demand adversely affects overall energy balance[e]. Upstream hydrolysis is required.
OAs obtained from anaerobic digestion (AD)	'Hythane': mix of CH_4 (AD) + bio-H_2; CHP	AD interrupted at acetogenesis stage; organic acids diverted into a bolt-on photofermentation. Overall AD residence time is reduced. This increases process complexity but gives a higher energy output. Gas is compatible with current infrastructure. Scenario 1: 20% more power[d]. Scenario 2: 70% more power[d]
OAs used directly from wastes (e.g. wastewaters) or CHP	Fuel cell electricity	Organic acid waste streams (tonnage scale) are (e.g.) vinasse (from bioethanol production) and municipal wastewater treatment plants (see text).

a. Calculations were made independently of incentivization schemes as these tend to be ephemeral and skew the longer term picture. Likewise, increasing/decreasing feed-in tariffs would complicate economic assessments.
b. Fuel cell technology is still emergent at large scale, and FCs fail prematurely (see Rabis et al., 2012).
c. Combined heat and power (CHP: well-established technology). In this scenario, the methane stream from anaerobic digestion can be supplemented with photofermentatively derived H_2 to make 'hythane' for CHP.
d. Scenario 1: diversion of 10% of the organic acids into photofermentation and use of hythane in CHP. Scenario 2: diversion of 80% of the organic acids into photofermentation and use of AD-methane in CHP plus use of the photofermentation H_2 in a fuel cell would give 70% more power (R.L. Orozco, unpublished). The proportion of flow diverted from the acetogenesis step of anaerobic digestion (via electroseparation) could be simply ramped in response to incident light intensity to feed the photofermentation; at night the flow would pass to the methanogenic reactor as normal. By combining the two processes, the residence time in the system would also be reduced as compared to traditional anaerobic digestion due to reduced flow entering the methanogenesis reactor daily.
e. Using Miscanthus as an example, the energy demand of comminution to 4 mm particles is 184 kJ kg dry matter^{-1}; energy from H is 10 kJ l^{-1} (at 1 atm and 125°C); that from the dark fermentation was only 110 kJ kg cellulose; hydrolysate; hence the PF (~4 times the H_2 as the dark fermentation) is key to a positive energy balance from complex substrates.

Hallenbeck, 2012) and industrial and agricultural efflu-ents (Saratale *et al.*, 2013), with high H_2 yields from acetic, butyric and lactic acids (Hallenbeck, 2013). Bac-teria used include *Rhodobacter sphaeroides* (Han *et al.*, 2013)*, R. rubrum* (Zürrer and Bachofen, 1979)*, R. palus-tris* (Oh *et al.*, 2004; Xiaobing, 2012) and *R. capsulatus* (Zhang *et al.*, 2016); despite some differences, they all follow a similar general scheme (Fig. 1B), metabolizing organic acids to reduce NAD^+ to the cellular reductant NADH (Oh *et al.*, 2013). Excess reductant must be dissi-pated to reoxidize NADH and maintain cellular redox bal-ance. This is achieved via cellular growth, channelling of carbon into cellular reserves (synthesis of polyhydroxy-butyrate) or via H_2 production under nitrogen-deficient conditions, via nitrogenase, which produces H_2 as an electron sink for excess reducing power (as with cyanobacteria: above). Nitrogenase normally fixes N_2 into NH_3 under light (to supply the large energy demand

of N-fixation, via ATP). Without N, the enzyme uses the reductant and ATP to produce H_2 ($2H^+ + 2e^- + 4$ ATP $\Rightarrow H_2 + 4ADP + Pi$). NADH is not a sufficiently strong reductant for this reaction; it is 'upgraded' to the stronger reductant ferredoxin via the input of energy, which is supplied by light through the action of the photosynthetic apparatus, via reverse electron transport. This apparatus also produces the ATP required for nitrogenase action (Hallenbeck, 2011). Various papers have studied the role of light (e.g. Uyar *et al.*, 2007; Nath, 2009), showing that optimum light conversion efficiency occurs at light inten-sities much lower than light saturation points; e.g. Uyar *et al.* (2007) showed light saturation for *R. sphaeroides* at 270 W m^{-2} but similar substrate conversion efficiency could be achieved at light intensities as low 88 W m^{-2}. Furthermore, optimum light intensities can be species specific; e.g. *R. sphaeroides* and *R. palustris* under simi-lar conditions (Light intensity = 2500 Lux) had substrate

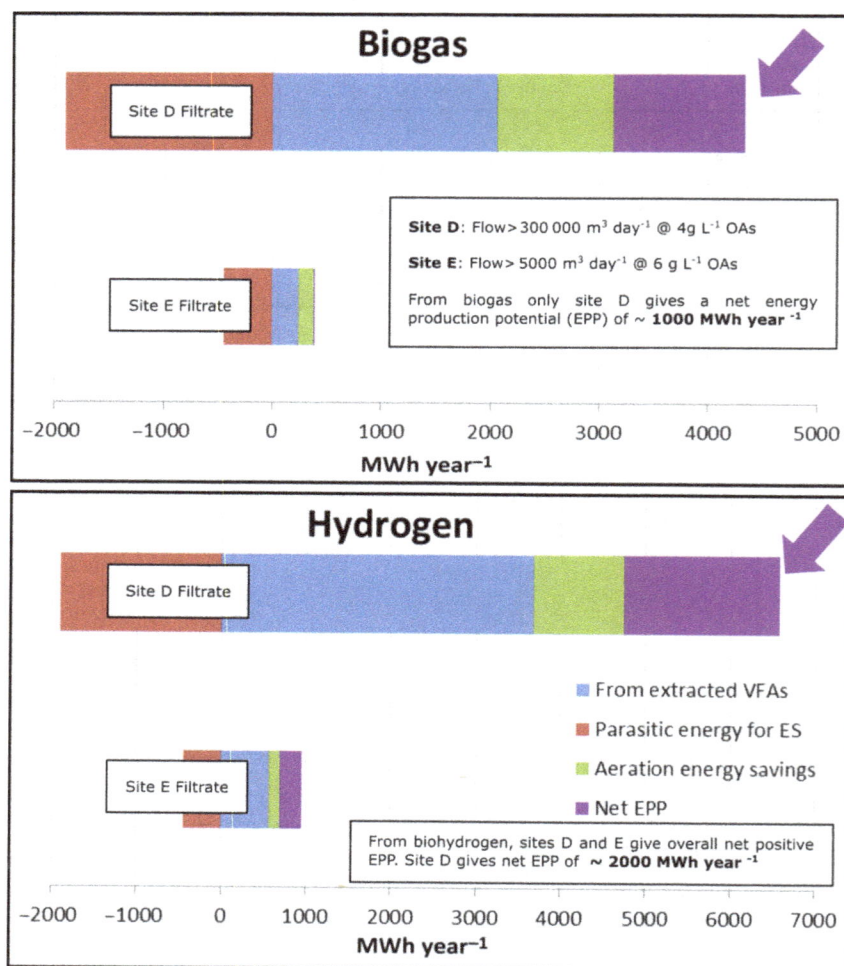

Fig. 3. Energy production potential (EPP) from use of real wastewater organic acids in a stand-alone photofermentation (real test data using *R. sphaeroides*: R.L.Orozco, I. Mikheenko and L.E. Macaskie, unpublished). As an organic acids liquid stream is used directly, the upstream dark fermentation is not required, and there is no sacrificial energy demand for maceration.

conversion efficiencies of 60–70% and 47% respectively (Han *et al.*, 2013; Oh *et al.*, 2013).

Hallenbeck and Liu (2015) reviewed advances in the field, highlighting various approaches to improve substrate conversion efficiency (Table 2), while recent publications provide an up-to-date overview of recent developments for photobiological biohydrogen technologies (Adessi *et al.*, 2017; Hallenbeck, 2017).

Towards an economically competitive biohydrogen process from waste

Table 3 summarizes various options for a biohydrogen process. In the UK, food wastes at scale are generally centralized and 'committed' by agreements into anaerobic digestion and a 'bolt-on' addition into existing anaerobic digestion and combined heat and power (CHP) processes is one option as there is insufficient waste available for a realistic stand-alone bio-H_2 process (unpublished survey; Sustainable Resource Solutions Ltd). Agricultural wastes are currently unattractive due to high energy demands of comminution/maceration and upstream hydrolysis. A survey of wastes has indicated that vinasse (from bioethanol production) and in-process streams from UK Utility companies contain sufficient organic acids to warrant trialling for data into a full life cycle analysis.

The organic acid content of a typical vinasse waste is > 40 g l^{-1} (Ryznar-Luty *et al.*, 2008; Esapaña-Gamboa *et al.*, 2012); the high concentration of betaine (trimethylglycine, a zwitterionic osmoprotectant; 20 g l^{-1}) is not potentially problematic because at the low pH of vinasse (pH 3–4), it would be protonated (i.e. inaccessible to the anion transfer in electroseparation). Moreover, betaine was reported to stimulate nitrogenase activity, but it was not used as a nitrogen source (Igeño *et al.*, 1995).

Selected UK utility company wastewaters were trialled as potential targets for hydrogen bioenergy following filtration to remove debris but with no other modifications (Fig. 3). The energy production potential from biohydrogen via photofermentation was twice that from biogas (Fig. 3). Hence, H_2 energy from organic acid wastes is a viable option for energy production by heavily populated, industrialized countries but may be limited seasonally by available natural sunlight. Stand-alone photobiological hydrogen production has major potential in solar-rich countries with the option to also treat wastes in areas of high population density. An environmental life cycle analysis has been developed for cyanobacterial bio-H_2 (Archer *et al.*, 2017). The next step is to apply a similar LCA for various options with respect to geographical location, other socio-economic factors and the global increase in demand for cooling to safeguard food supplies for expanding populations.

Acknowledgements

AJS and SAA acknowledge with thanks studentships from the EPSRC Doctoral Training Centre '*Fuel Cells and their Fuels*'. The work was supported in part by NERC (grant NE/L014076/1) to LEM.

Conflict of Interest

None declared.

References

Abo-Hashesh, M., Desaunay, N., and Hallenbeck, P.C. (2013) High yield single stage conversion of glucose to hydrogen by photofermentation with continuous cultures of *Rhodobacter capsulatus* JP91. *Biores Technol* **128:** 513–517.

Adessi, A., Corneli, E., and Philippi, S. (2017) Photosynthetic purple nonsulfur bacteria in hydrogen production systems: new approaches in the use of well known and innovative substrates. In *Modern Topics in the Phototrophic Prokaryotes.* Hallenbeck, P.C. (ed.). Cham, Switzerland: Springer, pp. 321–350.

Anon (2017) URL http://www.hydrogeneurope.eu/wp-content/uploads/2017/01/170113-Hydrogen-Council-International-Press-Releases.pdf.

Archer, S.A., Murphy, R.J. and Steinberger-Wilckens, R.. (2017). Systematic review and life cycle analysis of biomass derived fuels for solid oxide fuel cells. In *Proceedings: Fuel Cell and Hydrogen Technical Conference 2017. 31st May – 1st June 2017. Birmingham, UK.*

Balachandar, G., Khanna, N., and Das, D. (2013) Biohydrogen production from organic wastes by dark fermentation. In *Biohydrogen.* Pandey, A., Chang, J.-S., Hallenbeck, P.C., and Larroche, C. (eds). Burlington, MA, USA: Elsevier, pp. 103–144.

Chang, A.C.C., Chang, H.-F., Lin, F.-J., Lin, K.-H., and Chen, C.-H. (2011) Biomass gasification for hydrogen production. *Int J Hydrogen Energy* **36:** 14252–14260.

Deplanche, K., Attard, G.A., and Macaskie, L.E. (2007) Biorecovery of gold from jewellery waste by *Escherichia coli* and biomanufacture of active Au-nanomaterial. *Adv Mater Res* **20–21:** 647–652.

Dincer, I., and Acar, C. (2015) Review and evaluation of hydrogen production methods for better sustainability. *Int J Hydrogen Energy* **40:** 11094–11111.

DOE (2013). *Report of the hydrogen production expert panel: a subcommittee of the hydrogen and fuel cell technical advisory committee.* URL https://www.hydrogen.energy.gov/pdfs/hpep_report_2013.pdf.

Eroğlu, E., Gündüz, U., Yücel, M., Türker, L., and Eroğlu, I. (2004) Photobiological hydrogen production by using olive mill wastewater as a sole substrate source. *Int J Hydrogen Energy* **29:** 163–171.

Esapaña-Gamboa, E., Mijangos-Cortēs, J.O., Hernández-Zárate, G., Maldonado, J.A.D., and Alzate-Gaviria, L.M. (2012) Methane production from hydrous ethanol using a modified UASB. *Biotechnol Biofuels* **5:** 82–92.

Ghosh, D., Sobro, I.F., and Hallenbeck, P.C. (2012a) Optimization of the hydrogen yield from single-stage photofermentation of glucose by *Rhodobacter capsulatus* JP91 using response surface methodology. *Biores Technol* **123:** 199–206.

Ghosh, D., Sobro, I.F., and Hallenbeck, P.C. (2012b) Stoichiometric conversion of biodiesel derived crude glycerol to hydrogen: response surface methodology study of the effects of light intensity and crude glycerol and glutamate concentration. *Biores Technol* **106:** 154–160.

Ghosh, D., Tourigny, A., and Hallenbeck, P.C. (2012c) Near stoichiometric reforming of biodiesel derived crude glycerol to hydrogen by photofermentation. *Int J Hydrogen Energy* **37:** 2273–2277.

Golomysova, A., Gomelsky, M., and Ivanov, P.S. (2010) Flux balance analysis of photoheterotrophic growth of purple nonsulfur bacteria relevant to biohydrogen production. *Int J Hydrogen Energy* **35:** 12751–12760.

Guo, X.M., Trably, E., Latrille, E., Carrère, H. and Steyer, J.P. (2010) Hydrogen production from agricultural waste by dark fermentation: a review. *Int J Hydrogen Energy* **35,** 10660–10673.

Hädicke, O., Grammel, H., and Klamt, S. (2011) Metabolic network modeling of redox balancing and biohydrogen production in purple nonsulfur bacteria. *BMC Syst Biol* **5:** 150.

Hallenbeck, P.C. (2011) Microbial paths to renewable hydrogen production. *Biofuels* **2:** 285–302.

Hallenbeck, P.C. (2012) Fundamentals of dark hydrogen production: multiple pathways and enzymes. In *State of the Art and Progress in Biohydrogen.* Azbar, N., and Levin, D. (eds). Sharjah, United Arab Emirates: Bentham Science Publishers, pp. 94–111.

Hallenbeck, P.C. (2013) Photofermentative biohydrogen production. In *Biohydrogen.* Pandey, A., Chang, J.-S., Hallenbeck, P.C., and Larroche, C. (eds). Burlington, MA, USA: Elsevier, pp. 145–159.

Hallenbeck, P.C. (ed) (2017) *Modern Topics in the Phototrophic Prokaryotes.* Cham, Switzerland: Springer.

Hallenbeck, P.C., and Ghosh, D. (2009) Advances in fermentative biohydrogen production: the way forward? *Trends Biotechnol* **27:** 287–297.

Hallenbeck, P.C., and Liu, Y. (2015) Recent advances in hydrogen production by photosynthetic bacteria. *Int J Hydrogen Energy* **41:** 4446–4454.

Han, H., Jia, Q., Liu, B., Yang, H., and Shen, J. (2013) Fermentative hydrogen production from acetate using *Rhodobacter sphaeroides* RV. *Int J Hydrogen Energy* **38:** 10773–10778.

Igeño, M.I., Del Moral, C.G., Castillo, F., and Caballero, J. (1995) Halotolerenace of the phototrophic bacterium *Rhodobacter capsulatus* E1F1 is dependent on the nitrogen source. *Appl Environ Microbiol* **61:** 2970–2975.

James, B.D., Baum, G.N., Perez, J. and Baum, K.N. (2009) *Technoeconomic boundary analysis of biological pathways to hydrogen production.* NREL Technical Monitor: Ali Jalalzadeh-Azar, A. Subcontract No. AFH-8-88601-01. Subcontract Report. NREL/SR-560-46674. Golden, CO, USA: NREL, pp 1–207.

Keskin, T., and Hallenbeck, P.C. (2012) Hydrogen production from sugar industry wastes using single-stage photofermentation. *Biores Technol* **112:** 131–136.

Kim, M.S., Kim, D.H., Son, H.N., Ten, L.N., and Lee, J.K. (2011) Enhancing photo-fermentative hydrogen production by *Rhodobacter sphaeroides* KD131 and its PHB synthase deleted-mutant from acetate and butyrate. *Int J Hydrogen Energy* **36:** 13964–13971.

Kubas, A., Orain, C., De Sancho, D., Saujet, L., Sensi, M., Gauquelin, C., *et al.* (2017) Mechanism of O_2 diffusion and reduction in FeFe hydrogenases. *Nat Chem* **9:** 88–95.

Lazaro, C.Z., Vich, D.V., Hirasawa, J.S., and Varesche, M.B.A. (2012) Hydrogen production and consumption of organic acids by a phototrophic microbial consortium. *Int J Hydrogen Energy* **37:** 11691–11700.

Ma, C., Wang, X., Guo, L., Wu, X., and Yang, H. (2012) Enhanced photo-fermentative hydrogen production by *Rhodobacter capsulatus* with pigment content manipulation. *Biores Technol* **118:** 490–495.

Mohan, S.V., and Pandey, A. (2013) Biohydrogen production: an introduction. In *Biohydrogen.* Pandey, A., Chang, J.-S., Hallenbeck, P.C., and Larroche, C. (eds). Burlington, MA, USA: Elsevier, pp. 1–24.

Nakada, E., Nishikata, S., Asada, Y., and Miyake, J. (1999) Photosynthetic bacterial hydrogen production combined with a fuel cell. *Int J Hydrogen Energy* **24:** 1053–1057.

Nath, K. (2009) Effect of light intensity and initial pH during hydrogen production by an integrated dark and photofermentation process. *Int J Hydrogen Energy* **34:** 7497–7501.

Nikolaidis, P., and Poullikkas, A. (2017) A comparative overview of hydrogen production processes. *Renew Sustain Energy Rev* **67:** 597–611.

Oh, Y.-K., Seol, E.-H., Kim, M.-S., and Park, S. (2004) Photoproduction of hydrogen from acetate by a chemoheterotrophic bacterium *Rhodopseudomonas palustris* P4. *Int J Hydrogen Energy* **29:** 1115–1121.

Oh, Y.K., Raj, S.M., Jung, G.Y., and Park, S. (2013) Metabolic engineering of microorganisms for biohydrogen production. In *Biohydrogen.* Pandey, A., Chang, J.-S., Hallenbeck, P.C., and Larroche, C. (eds). Burlington, MA, USA: Elsevier, pp. 45–65.

Rabis, A., Rodriguez, P., and Schmidt, J.J. (2012) Electrocatalysis for polymer electrolyte fuel cells. Recent achievements and future challenges. *ACS Catal* **2:** 864–890.

Redwood, M.D.. (2007) Bio-hydrogen production and biomass-supported palladium catalyst for energy production and waste minimization. PhD Thesis. Burlington, MA, UK: University of Birmingham.

Redwood, M.D., Orozco, R.L., Majewski, A.J., and Macaskie, L.E. (2012a) Electro-extractive fermentation for efficient biohydrogen production. *Biores Technol* **107:** 166–174.

Redwood, M.D., Orozco, R.L., Majewski, A.J., and Macaskie, L.E. (2012b) An integrated biohydrogen refinery: synergy of photofermentation, extractive fermentation and hydrothermal hydrolysis of food wastes. *Biores Technol* **119:** 384–392.

Ryznar-Luty, A., Krzywonos, M., Obis, E., and Miskiewicz, T. (2008) Aerobic biodegradation of vinasse by a mixed culture of bacteria of the genus *Bacillus.* Optimization of temperature, pH and oxygenation state. *Pol J Environ Stud* **17:** 101–112.

Saratale, G.D., Saratale, R.G., and Chang, J.-S. (2013) Biohydrogen from renewable resources. In *Biohydrogen.*

Pandey, A., Chang, J.-S., Hallenbeck, P.C., and Larroche, C. (eds). Burlington, MA, USA: Elsevier, pp. 185–221.

Show, K.Y., and Lee, D.J. (2013) Biioreactor and bioprocess design for biohydrogen production. In *Biohydrogen.* Pandey, A., Chang, J.-S., Hallenbeck, P.C., and Larroche, C. (eds.). Burlington, MA, USA: Elsevier, pp. 317–337.

Singh, J.S., Kumar, A., Rai, A.N., and Singh, D.P. (2016) Cyanobacteria: a precious bio-resource in agriculture, ecosystem and environmental sustainability. *Front Microbiol* **7:** 1–19. article 529

Strahan, D.. (ed.) (2017) *Clean cold and the global goals.* URL www.birmingham.ac.uk/Documents/college-eps/energy/Publications/Clean-Cold-and-the-Global-Goals.pdf.

Tiwari, A., and Pandey, A. (2012) Cyanobacterial hydrogen- a step towards clean environment. *Int J Hydrogen Energy* **37:** 139–150.

Uyar, B., Eroglu, I., Yücel, M., Gündüz, U., and Türker, L. (2007) Effect of light intensity, wavelength and illumination protocol on hydrogen production in photobioreactors. *Int J Hydrogen Energy* **32:** 4670–4677.

Xiaobing, W. (2012) Enhanced photo-fermentative hydrogen production from different organic substrate using hupL inactivated *Rhodopseudomonas palustris. Afr J Microbiol Res* **6:** 5362–5370.

Zhang, Y., Yang, H., and Guo, L. (2016) Enhancing photo-fermentative hydrogen production performance of *Rhodobacter capsulatus* by disrupting methylmalonate-semialdehyde dehydrogenase gene. *Int J Hydrogen Energy* **41:** 190–197.

Zürrer, H., and Bachofen, R. (1979) Hydrogen production by the photosynthetic bacterium *Rhodospirillum rubrum. Appl Environ Microbiol* **37:** 789–793.

11

Diversity and functions of the sheep faecal microbiota: a multi-omic characterization

Alessandro Tanca,[1] Cristina Fraumene,[1] Valeria Manghina,[1,2] Antonio Palomba,[1,2] Marcello Abbondio,[2] Massimo Deligios,[2] Daniela Pagnozzi,[1] Maria Filippa Addis[1] and Sergio Uzzau[1,2]*

[1] Porto Conte Ricerche, Science and Technology Park of Sardinia, Tramariglio, Alghero, Italy.
[2] Department of Biomedical Sciences, University of Sassari, Sassari, Italy.

Summary

Little is currently known on the microbial populations colonizing the sheep large intestine, despite their expected key role in host metabolism, physiology and immunity. This study reports the first characterization of the sheep faecal microbiota composition and functions, obtained through the application of a multi-omic strategy. An optimized protocol was first devised for DNA extraction and amplification from sheep stool samples. Then, 16S rDNA sequencing, shotgun metagenomics and shotgun metaproteomics were applied to unravel taxonomy, genetic potential and actively expressed functions and pathways respectively. Under a taxonomic perspective, the sheep faecal microbiota appeared globally comparable to that of other ruminants, with Firmicutes being the main phylum. In functional terms, we detected 2097 gene and 441 protein families, finding that the sheep faecal microbiota was primarily involved in catabolism. We investigated carbohydrate transport and degradation activities and identified phylum-specific pathways, such as methanogenesis for Euryarchaeota and acetogenesis for Firmicutes. Furthermore, our approach enabled the identification of proteins expressed by the eukaryotic component of the microbiota. Taken together, these findings unveil structure and role of the distal gut microbiota in sheep, and open the way to further studies aimed at elucidating its connections with management and dietary variables in sheep farming.

*For correspondence. E-mail uzzau@portocontericerche.it

Funding Information
This work was supported by Sardegna Ricerche – Science and Technology Park of Sardinia, Grant Program 'art. 26_2014' to Porto Conte Ricerche. VM, AP and MA were supported by Doctoral Fellowships from the International PhD Course in Life Sciences and Biotechnologies, University of Sassari.

Introduction

Sheep farming is widespread worldwide for the purpose of meat, milk, skin and/or wool production. According to the specific productive purpose, farmers select sheep breeds and, traditionally, efforts are continuously made to improve physical/genetic traits that, in turn, ameliorate the production performances.

Regardless of the genetic background and of the production purpose, a correct nutritional management plays a crucial role in warranting healthy and fertile sheep and productive dairy ewes. Sheep are grazers and eat a variety of plants, including grass, clover and weeds. In addition to pasture grazing, sheep are generally fed with hay and a controlled amount of grain. As for other dairy ruminants, energy stored in plant matter is eventually converted to protein food products (i.e., milk and meat) after a complex digestive process. The plant mass is first fermented by the ruminal microbial communities, and a large part of the organic matter is then degraded and partially adsorbed in the sheep four-chambered stomach. Bacteria involved in this process have a widest range of biochemical activities enabling digestion of cellulose, hemicellulose, starch and proteins. While storage polysaccharides (starch) can be promptly degraded in the rumen ecosystem, full degradation of structural polysaccharides (i.e. cellulose) and resistant starch is a longer and more complex process (Huntington, 1997; Krause *et al.*, 2003). In addition, a number of ruminal bacteria have been described to be involved in the isomerization and saturation of the dietary unsaturated fatty acids, leading to the sheep milk saturated fat composition (Huws *et al.*, 2011). Short-chain fatty acids (SCFAs) are also produced as a consequence of carbohydrate fermentation. Further, rumen microorganisms and food residues pass on to the abomasum and the small intestine where food degradation endures by microbial and host-secreted enzymes. Finally, undigested organic matter reaches the large intestine where it undergoes the last digestive processes by the colonic microbial population, before water and salt absorption by colonic mucosa completes the formation of the faecal pellets. Here, degradation of both resistant starch and fibres is expected to occur, as in monogastric vertebrates colon, as well as SCFA production and absorption by the colonic mucosa. A considerable amount of data have been collected on the composition and functions of sheep rumen microbiota (Shi *et al.*, 2014; Brilhante *et al.*, 2015; Morgavi *et al.*, 2015; Zeng *et al.*, 2015), but much less is currently known on the microbial populations that colonize the large intestine, despite their crucial role in the sheep intestinal metabolism.

As in other non-ruminant mammalian, including humans, the microbial population that colonizes the sheep large intestine is expected to be key in providing energy, antigens and metabolites that positively affect host metabolism, physiology and immunity. Basically, a well-balanced microbiota, with highly diverse taxonomic content and stability, appears of paramount importance throughout the whole digestive system, where a bidirectional driving force between microbial metabolic circuits and mucosal physiology allows to maintain a stable microbiota and a healthy gut and, consequently, an overall healthy and productive organism.

In keeping with these premises, we investigated composition and functions of the microbial populations associated with the final tract of the sheep large intestine, where the last stage of plant mass digestion occurs with a significant potential contribution to host energy harvesting and physiology homeostasis. To reach this aim, we employed an integrated, multi-omic strategy, comprising 16S rDNA and shotgun metagenomic sequencing, to unravel microbiota structure and genetic potential, as well as metaproteomics, to identify and characterize functions and pathways actively expressed by the sheep faecal microbial communities.

Results and discussion

Optimization of protocols for DNA extraction and sample clean-up for the analysis of the sheep faecal microbiota

Stool is a complex sample matrix with regard to DNA extraction. Sample pretreatment protocols, such as differential centrifugation (DC), have been reported to help bacterial DNA extraction from the faecal matrix (Apajalahti *et al.*, 1998), although direct lysis (DL) of the microbial matter by chemical or mechanical methods (or a combination of these) is the most widely applied strategy (Hart *et al.*, 2015; Wagner Mackenzie *et al.*, 2015). Several substances may be co-extracted having inhibitory effects on downstream analysis, with even massive influence on quality of the extracted DNA, and consequently on feasibility and outcome of DNA sequencing. This is especially key when dealing with stool from

sheep, whose distal colon reduces the faecal water content up to 65% (Hecker and Grovum, 1975). Here, we compared five different sample preparation/extraction methods from sheep faecal samples and used 16S rRNA gene amplification efficiency as a probe to determine the DNA suitability for downstream metagenomic analysis. Specifically, as described in the "Experimental procedures" section, samples were subjected to DC or DL; then, DNA was extracted from both preparations after enzymatic and mechanical lysis with QIAamp Fast DNA Stool or with the E.Z.N.A. Soil DNA Kit. Finally, samples were also subjected to DC and then extracted with the standard phenol/chloroform/isoamyl alcohol (25:24:1) method.

As a result, the combination of stool DC preparation followed by DNA extraction and purification with the E.Z.N.A. soil DNA kit, designed to remove with highest efficiency PCR inhibitors, was the only protocol capable of providing a satisfactory quantity (25 ± 0.28 ng g^{-1} stool sample) and the best quality of extracted DNA (100% of samples providing a 16S rRNA gene PCR amplification product; data not shown).

Sheep faecal microbiota composition

To assess the faecal microbiota composition in sheep, we analysed its metagenome, by sequencing the V4 region of the 16S rDNA (V4-MG) and the whole microbial DNA (shotgun metagenomics, S-MG), as well as its metaproteome (MP), by means of shotgun mass spectrometry. These were intended as complementary approaches, being based on the measurement of different molecules (DNA for V4-MG and S-MG, peptides for MP) with different strategies (targeted for V4-MG, shotgun for S-MG and MP). The whole DNA sequencing and peptide mass spectrometry metrics are presented in Table S1. Read counts (for MG approaches) and spectral counts (for MP) were used throughout the study to estimate the relative abundance of the taxonomic and functional features of the microbiota.

The taxonomic composition of the prokaryotic microbiota according to V4-MG, S-MG and MP results is shown in Fig. 1. As expected, Firmicutes and Bacteroidetes made over 80% of total bacteria in all cases. Moreover, Firmicutes was detected as the most represented phylum in all animals and with all approaches, followed by Bacteroidetes (Fig. 1A). However, the average Firmicutes-to-Bacteroidetes ratio (F/B) ranged from 6.0 for V4-MG down to 1.6 for MP, through 3.4 for S-MG. The archaeal Euryarchaeota was the fifth, seventh and third most abundant phylum according to V4-MG, S-MG and MP data respectively. While Firmicutes levels were well conserved among individuals (CV < 10% with all approaches), a higher variation could be observed for other important phyla, especially for Actinobacteria and Verrucomicrobia. When comparing our data with the existing studies reporting a metagenomic analysis of faecal samples from other ruminants, we could find a general predominance of Firmicutes over Bacteroidetes in cattle, with Ruminococcaceae and Lachnospiraceae being the most representative microbial families of the former phylum (Durso *et al.*, 2010, 2011; Shanks *et al.*, 2011; Kim *et al.*, 2014), in line with our results obtained in sheep. Conversely, Bacteroidetes was the most abundant phylum in the sheep rumen according to the literature (Castro-Carrera *et al.*, 2014; de la Fuente *et al.*, 2014; Kittelmann *et al.*, 2015; Lopes *et al.*, 2015; Morgavi *et al.*, 2015). Consistently, a marked increase in the F/B ratio from rumen to colon was recently described in cow (Mao *et al.*, 2015). Among minor phyla, we found a remarkable amount of functionally active Spirochaetes, mainly belonging to the genus *Treponema*, especially *T. saccharophilum*, early described as a large pectinolytic spirochaete present in the rumen (Paster and Canale-Parola, 1985). This phylum was recently observed as the fourth most abundant within the microbiota of the ruminant digestive tract (Peng *et al.*, 2015).

S-MG and MP approaches also allowed for the identification of fungal taxa, accounting for about 0.05% and 1.1% of the microbiota respectively. Ascomycota was detected in both cases as the most abundant fungal phylum.

Going down to the family level, 76, 385 and 171 different microbial families were detected with V4-MG, S-MG and MP respectively. Considering the 'core microbiota' (i.e. taxa found consistently in all animals analysed), V4-MG, S-MG and MP analyses led to find a total of 45, 168 and 50 microbial families, respectively, of which 19 in common between the three approaches. As shown in Fig. 1B, Ruminococcaceae, Lachnospiraceae and Clostridiaceae were consistently found with all approaches to be the first, second and third most abundant family within Firmicutes, respectively, consistent with previous studies in cows (Durso *et al.*, 2010, 2011; Shanks *et al.*, 2011; Kim *et al.*, 2014). On the other hand, bacterial families belonging to Bacteroidetes exhibited more variable distributions, with Prevotellaceae becoming the main family overall when considering protein expression data. Families from other phyla, such as Spirochaetaceae from Spirochaetes and Verrucomicrobiaceae from Verrucomicrobia, were also found among the top 20 abundant families with all techniques.

Based on our results, V4-MG, S-MG and MP provided generally comparable taxonomic distributions, with some slight but important differences, such as the F/B ratio and the relative abundance of Prevotellaceae. This is

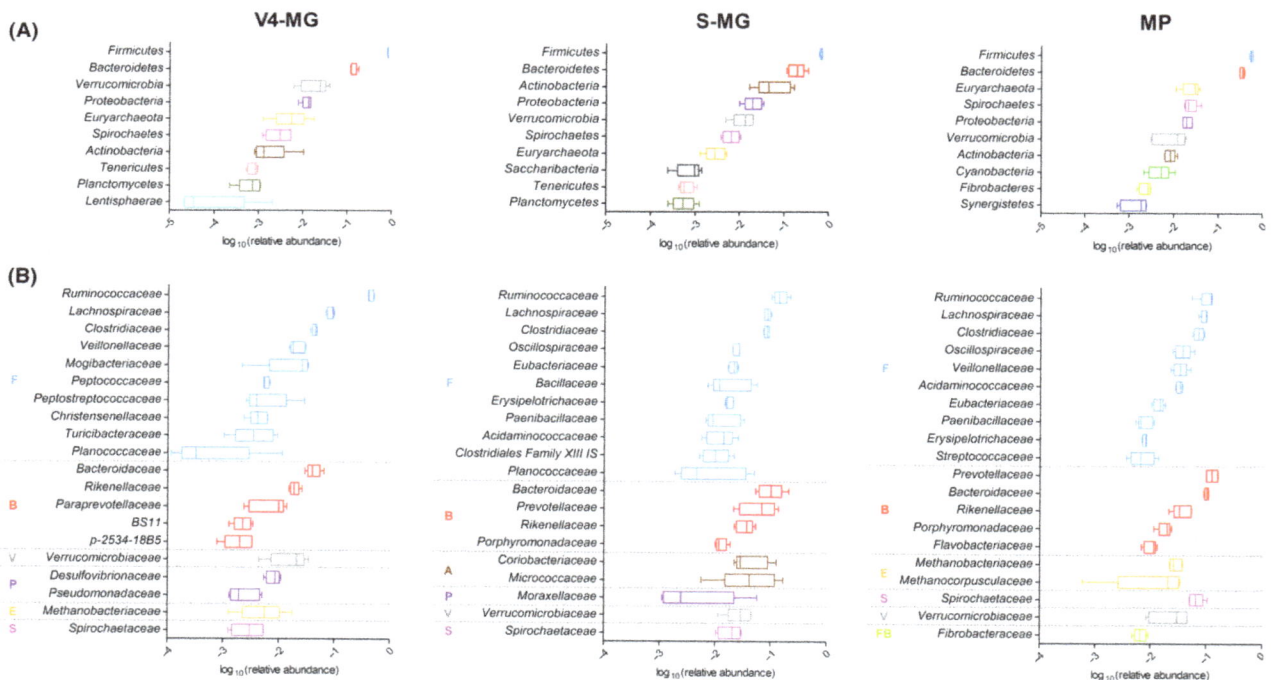

Fig. 1. Taxonomic composition of the sheep faecal prokaryotic microbiota, according to V4-16S rRNA (V4-MG, left), metagenomic (S-MG, centre) and metaproteomic (MP, right) results. (A) Tukey's box plot illustrating the microbiota composition at phylum level. The top 10 phyla are shown, ordered by decreasing mean relative abundance. (B) Tukey's box plot illustrating the microbiota composition at family level. The top 20 families are shown, grouped based on the relative phylum (A, Actinobacteria; B, Bacteroidetes; E, Euryarchaeota; F, Firmicutes; FB, Fibrobacteres; P, Proteobacteria; S, Spirochaetes; V, Verrucomicrobia) and further ordered by decreasing mean relative abundance.

largely expected when dealing with conceptually and technically different approaches, as previously outlined. In addition, microbial taxa bearing a higher number of 16S rRNA genes are expected to be overestimated when compared with those with a lower number according to V4-MG analysis (Vetrovsky and Baldrian, 2013), and species with smaller genome size are expected to be underestimated when compared with those with larger size according to S-MG analysis (Nayfach and Pollard, 2015). As a confirmation, in this work, when comparing V4-MG and S-MG results, phyla with larger genomes on average (i.e. Actinobacteria and Proteobacteria) reached a higher rank in abundance with S-MG, while those with smaller genomes (i.e. Verrucomicrobia and Tenericutes) were more represented with V4-MG. In addition, it is worth noting that some data analysis-related biases may considerably influence comparability of outputs, such as differences in taxonomic classification and update frequency among databases (Green-Genes versus NCBI).

The complete taxonomic distribution data (also comprising class, order and genus levels) for V4-MG, S-MG and MP are presented in Data S1–S3.

Assessment of functions potentially and actively expressed by the sheep faecal microbiota

Figure 2A illustrates the 20 most abundant gene and protein families found upon S-MG and MP analyses respectively. As explained above, the former data regard the functional potential of the microbiota, while the latter represent the key functions actually exerted by the faecal microorganisms. The metagenome was found to be rich in genes related to membrane transport of molecules (ABC transporter, ATPase, permease, SecA), DNA replication and repair (helicase, DNA polymerase, topoisomerase, MutS), transcription (RNA polymerase), translation (tRNA synthetases and translation factors) and protein folding (chaperones), plus a few encoding for metabolic enzymes. As revealed by MP data, among the most abundant protein families expressed by the faecal microbiota, we primarily found functions related to metabolism (eight enzymes), especially carbohydrate degradation, followed by protein synthesis and folding (translation factor, ribosomal proteins, chaperones). Protein families involved in transport and signalling, such as ABC transporters, TonB-dependent receptor, ATPases, flagellin, were also present. On the whole, 2097 gene families and 441 protein families were identified by S-MG and MP respectively. Considering the 'core functions' (i.e. associated with the microbiota of all the animals analysed), S-MG and MP analyses led to detect a total of 904 and 191 functional families, respectively, of which 152 were identified both as genetic trait and as expressed proteins ('core' protein families are listed in Table S2).

When considering function–taxonomy combinations at the phylum level (Fig. 2B), consistently with the sole taxonomic information, we found a higher representation of functions encoded/expressed by Firmicutes, compared with those encoded/expressed by Bacteroidetes. According to MP data, some functions from other phyla (namely from Euryarchaeota) were also considerably expressed. More interestingly, the most represented genes assigned to Firmicutes according to S-MG results were in most cases identical to the most represented genes assigned to Bacteroidetes, whereas, when considering the main expressed protein functions, several of them were not overlapping between these two phyla. To further investigate this 'phylum-specific' functional contribution, we sought for those protein families exclusively and unambiguously assigned to a single phylum, and detected in all samples (42 in total). As shown in Table 1, several relevant and considerably abundant protein functions were actually phylum specific, whereas the 60 phylum-specific 'core' genes were all detected at very low abundance (Table S3). Among protein functions, we detected the TonB-dependent receptor as specific for Bacteroidetes, the enzyme 5,10-methylenetetrahydromethanopterin reductase (Mer) for Euryarchaeota, involved in methanogenesis, as well as aldehyde oxidoreductase (belonging to the xanthine dehydrogenase family), formate-tetrahydrofolate ligase and carbon monoxide dehydrogenase (both involved in one-carbon metabolism) for Firmicutes.

Among other expressed protein functions of possible interest, we can cite flagellins, which mainly belonged to Firmicutes (from genera such as *Clostridium* and *Selenomonas*) and Spirochaetes (again, essentially *Treponema*). Notably, half of the functionally annotated Spirochaetes peptides identified in this study were from flagellar proteins. In fact, members of this phylum are known to have long flagella, enclosed in the periplasm and capable to confer them a unique motility (Wolgemuth, 2015). Moreover, we were able to identify SASPs (small, acid-soluble spore proteins), mainly of clostridial origin, in four of five samples, indicating the presence of endospores within the faecal microbiota of these animals.

As a further consideration, we note that the massive discrepancy found between S-MG and MP functional results, which should be mainly attributed to actual (and largely expected) differences between genetic potential and protein expression, may also be partially due to varying sample pretreatment strategies. As stated above, DNA extraction required DC steps to ensure efficient PCR amplification, while protein extraction was performed on the faecal material as is to avoid artifactual depletion of food-bound microbes (Tanca *et al.*, 2015).

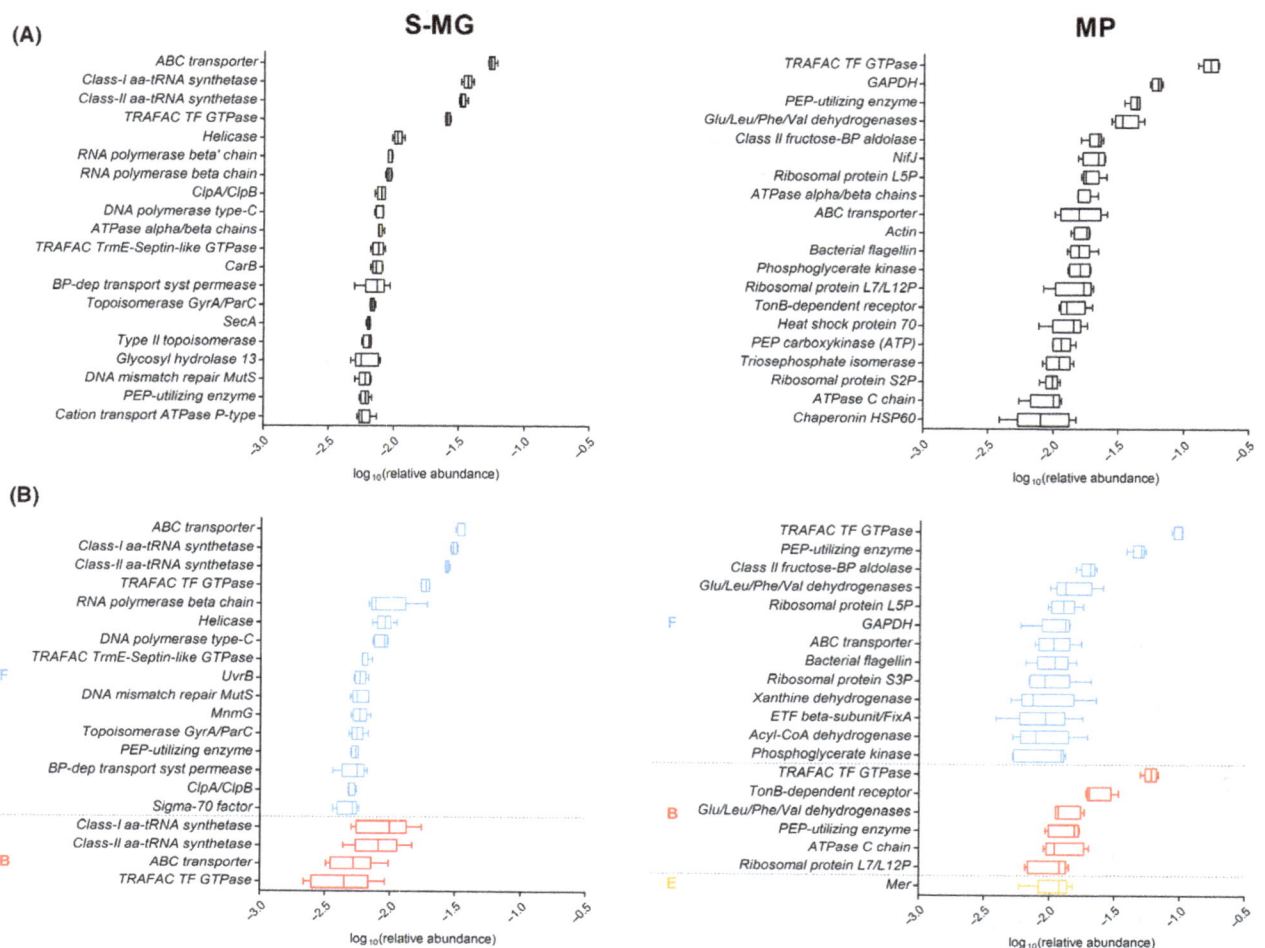

Fig. 2. Functional potential and activity of the sheep faecal microbiota, as measured by metagenomics (S-MG, left) and metaproteomics (MP, right) respectively. (A) Tukey's box plot illustrating the 20 most abundant gene (left) and protein families. (B) Tukey's box plot illustrating the 20 most abundant gene family–phylum (left) and protein family–phylum (right) combinations, grouped based on the relative phylum (B, Bacteroidetes; E, Euryarchaeota; F, Firmicutes) and further ordered by decreasing mean relative abundance.

Complete data concerning gene and protein families identified upon S-MG and MP analyses are presented in Data S2 and S3 respectively.

Microbial metabolic pathways in sheep gut

Gene and protein functional data were further grouped based on the UniProt 'pathway' annotation. Figure 3A illustrates the main metabolic pathways potentially and actively functioning in the sheep faecal microbiota, according to S-MG and MP data respectively. Several amino acid, nucleoside and carbohydrate biosynthetic routes were consistently highly represented both in the metagenome and in the metaproteome. However, while most of the main genes-according to S-MG data- were related to biosynthetic pathways, MP data revealed catabolic activities of the microbiota as clearly prominent in terms of active expression and relative abundance.

Furthermore, enzymes involved in methanogenesis were considerably more abundant than expected, considering their gene content as assessed by S-MG analysis.

Figure 3B illustrates pathway–taxonomy combinations at phylum level, revealing that, although the large majority of pathways covered in the metagenome were related to Firmicutes, several different phyla do actually participate in the metabolism at comparable extents, according to MP results. Phylum-specific 'core' pathways could also be identified, namely 1,2-propanediol degradation and butanoate metabolism for Firmicutes, starch degradation for Bacteroidetes and methanogenesis for Euryarchaeota. Among them, utilization of 1,2-propanediol is usually mediated by a bacterial microcompartment, in which a multiprotein shell encapsulates enzymes and cofactors for 1,2-propanediol catabolism, sequestering the reactive propionaldehyde to limit its cellular toxicity (Havemann et al., 2002). We

Table 1. Protein families assigned exclusively to a single phylum and detected in all samples, ordered by phylum and then by mean percentage abundance.

Protein family	Phylum	Sheep 1	Sheep 2	Sheep 3	Sheep 4	Sheep 5	Mean
TonB-dependent receptor	Bacteroidetes	2.021%	2.036%	1.912%	2.628%	3.400%	2.399%
Group II decarboxylase	Bacteroidetes	0.218%	0.375%	0.421%	0.113%	0.174%	0.260%
Ribosomal protein S1P	Bacteroidetes	0.273%	0.402%	0.153%	0.188%	0.262%	0.255%
NagA	Bacteroidetes	0.109%	0.161%	0.268%	0.188%	0.087%	0.163%
GHMP kinase	Bacteroidetes	0.055%	0.161%	0.115%	0.075%	0.174%	0.116%
ExbB/TolQ	Bacteroidetes	0.164%	0.054%	0.153%	0.038%	0.131%	0.108%
Class-I fumarase	Bacteroidetes	0.164%	0.080%	0.038%	0.075%	0.087%	0.089%
Gfo/Idh/MocA	Bacteroidetes	0.055%	0.054%	0.038%	0.038%	0.131%	0.063%
Eukaryotic mitochondrial porin	Basidiomycota	0.055%	0.027%	0.076%	0.038%	0.044%	0.048%
RuBisCO large chain	Cyanobacteria	0.218%	0.054%	0.191%	0.300%	0.305%	0.214%
Reaction centre PufL/M/PsbA/D	Cyanobacteria	0.055%	0.080%	0.115%	0.113%	0.044%	0.081%
Mer	Euryarchaeota	1.202%	0.589%	1.262%	1.201%	1.526%	1.156%
[NiFe]/[NiFeSe] hydrogenase large subunit	Euryarchaeota	0.546%	0.054%	0.574%	0.488%	1.133%	0.559%
MTD	Euryarchaeota	0.164%	0.080%	0.115%	0.225%	0.305%	0.178%
MtrA	Euryarchaeota	0.164%	0.080%	0.229%	0.150%	0.262%	0.177%
FrhB	Euryarchaeota	0.164%	0.027%	0.115%	0.150%	0.218%	0.135%
Archaeal histone HMF	Euryarchaeota	0.109%	0.107%	0.038%	0.075%	0.087%	0.083%
Ribosomal protein L12P	Euryarchaeota	0.055%	0.027%	0.076%	0.075%	0.131%	0.073%
N-Me-Phe pilin	Fibrobacteres	0.055%	0.027%	0.038%	0.113%	0.087%	0.064%
Xanthine dehydrogenase	Firmicutes	2.294%	0.509%	1.033%	0.751%	0.741%	1.065%
Formate–tetrahydrofolate ligase	Firmicutes	0.874%	0.589%	0.727%	0.526%	0.567%	0.656%
Ni-containing carbon monoxide dehydrogenase	Firmicutes	0.492%	0.509%	0.727%	0.563%	0.349%	0.528%
ETF alpha subunit/FixB	Firmicutes	0.492%	0.456%	0.421%	0.638%	0.262%	0.453%
Complex I 51 kDa subunit	Firmicutes	0.328%	0.643%	0.459%	0.263%	0.305%	0.400%
FldB/FldC dehydratase beta subunit	Firmicutes	0.819%	0.295%	0.306%	0.413%	0.044%	0.375%
Acetyl-CoA hydrolase/transferase	Firmicutes	0.328%	0.241%	0.306%	0.263%	0.218%	0.271%
Elongation factor P	Firmicutes	0.109%	0.241%	0.268%	0.188%	0.392%	0.240%
Diol/glycerol dehydratase small subunit	Firmicutes	0.273%	0.080%	0.153%	0.263%	0.305%	0.215%
Glycosyltransferase 1	Firmicutes	0.109%	0.161%	0.268%	0.150%	0.131%	0.164%
Glycosyl hydrolase 101	Firmicutes	0.055%	0.161%	0.229%	0.150%	0.044%	0.128%
Glyoxalase I	Firmicutes	0.164%	0.080%	0.153%	0.150%	0.087%	0.127%
Bacterial solute-binding protein 1	Firmicutes	0.164%	0.107%	0.153%	0.075%	0.131%	0.126%
Acyl-CoA mutase large subunit	Firmicutes	0.109%	0.107%	0.115%	0.113%	0.131%	0.115%
Diol/glycerol dehydratase medium subunit	Firmicutes	0.109%	0.080%	0.076%	0.150%	0.131%	0.109%
V-ATPase proteolipid subunit	Firmicutes	0.164%	0.080%	0.153%	0.038%	0.087%	0.104%
Glutamine synthetase	Firmicutes	0.055%	0.080%	0.076%	0.075%	0.131%	0.083%
Aldolase class II	Firmicutes	0.055%	0.134%	0.115%	0.038%	0.044%	0.077%
GSP E	Firmicutes	0.055%	0.027%	0.038%	0.038%	0.044%	0.040%
Hfq	Firmicutes	0.055%	0.027%	0.038%	0.038%	0.044%	0.040%
Peptidase S41A	Planctomycetes	0.055%	0.027%	0.038%	0.075%	0.044%	0.048%
Resistance–nodulation–cell division	Proteobacteria	0.055%	0.027%	0.038%	0.038%	0.044%	0.040%

found considerable amounts of propanediol utilization protein (PduA) mainly produced by members of Clostridia, in line with recent evidences correlating fucose and rhamnose metabolism with propanediol utilization microcompartments in *Clostridium phytofermentans* (Petit *et al.*, 2013). Furthermore, production of butyric acid by specific members of Firmicutes has been largely demonstrated as a key process in the host–microbiota cross-talk in mammalians, likely related to intestinal health (Pryde *et al.*, 2002; Hamer *et al.*, 2009; Vital *et al.*, 2014).

We then focused our attention specifically on carbon metabolism and mapped the identified proteins into the corresponding KEGG pathway in order to investigate which microbial players were mainly involved in each specific enzymatic step. While glycolytic reactions were revealed to be carried out in parallel by several different phyla, Fig. 4 clearly illustrates how other metabolic functions are exerted in a phylum-specific fashion within the sheep faecal microbiota. First, these data evidence, as expected, the unique contribution of Archaea to methanogenesis. Characterization and monitoring of methanogenesis in ruminants is receiving growing attention for zootechnical, environmental and ecological reasons (Kumar *et al.*, 2014; Shi *et al.*, 2014). Moreover, activity of methanogens in lambs' caecum, but not in the rumen, has been reported to be affected by specific diets, suggesting a possible compensation of rumen methane production with caecum methanogenesis (Popova *et al.*, 2013). In this work, while S-MG led to detect a minimum amount of archaeal genes, of which none directly involved in methanogenic reactions, MP allowed us to reconstruct almost entirely the methanogenic route (from 5,10-methenyltetrahydromethanopterin

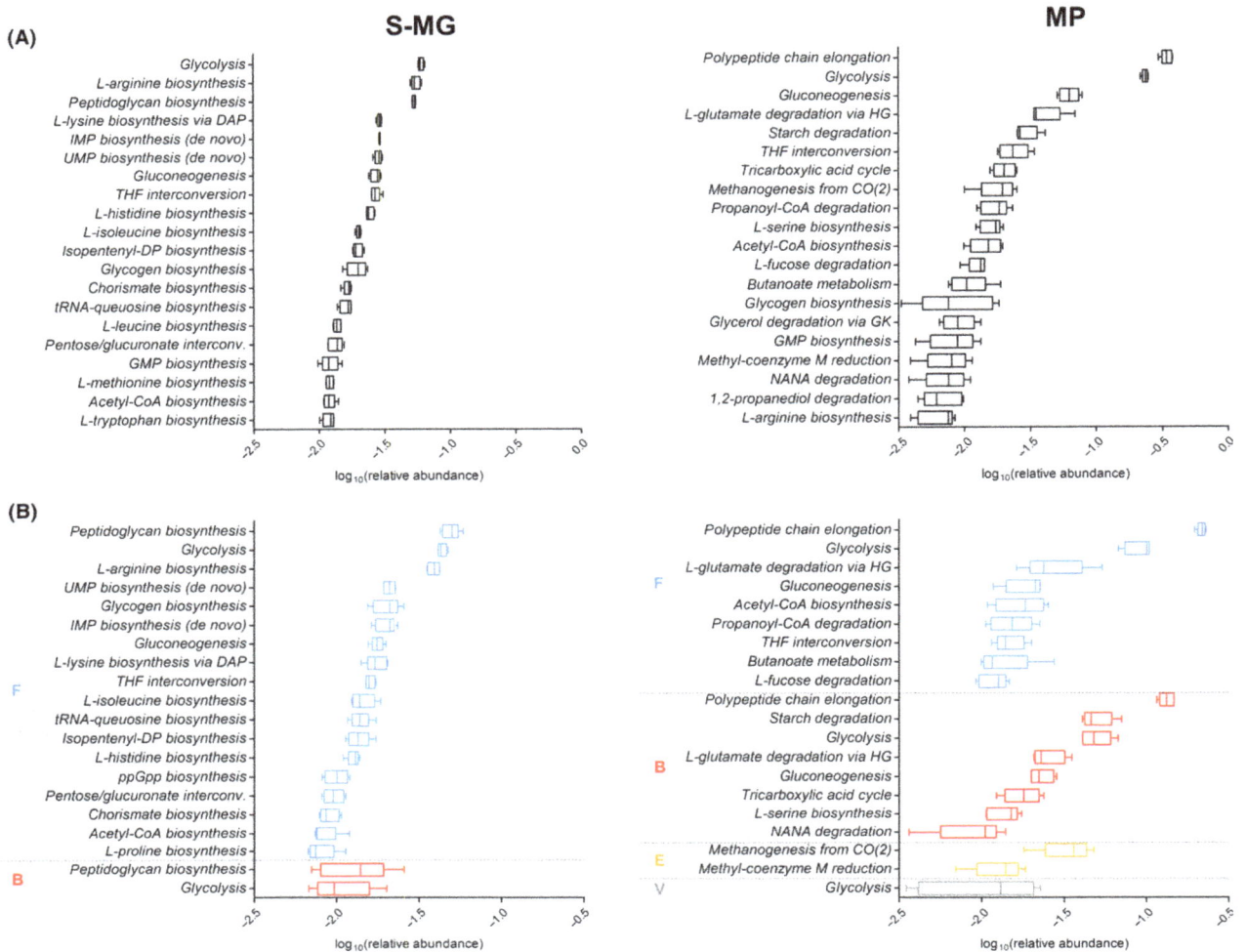

Fig. 3. Metabolic pathway potential and activity of the sheep faecal microbiota, as measured by metagenomics (S-MG, left) and metaproteomics (MP, right) respectively. (A) Tukey's box plot illustrating the 20 most relevant pathways, based on the related gene (left) and protein (right) abundance. (B) Tukey's box plot illustrating the 20 most relevant pathway–phylum combinations, based on the related gene (left) and protein (right) abundance. Pathways are grouped based on the relative phylum (B, Bacteroidetes; E, Euryarchaeota; F, Firmicutes; V, Verrucomicrobia) and further ordered by decreasing mean relative abundance.

to methane, including those enzymes responsible for methyl-coenzyme M reduction) in the microbiota of all animals analysed, and to assign the corresponding enzymatic functions to members of Methanobacteriaceae and Methanocorpusculaceae families, specifically *Methanobrevibacter ruminantium* and *Methanocorpusculum labreanum* respectively. This might be due to a massively higher amount of enzymes synthesized when compared with the number of archaeal genes and/or cells, as well as to greater difficulties in co-extraction of archaeal DNA compared with proteins. It is also important to note that the presence of publicly available archaeal sequences of methanogenic genes within the database used for MP analysis compensated the absence of experimental metagenomic sequences and allowed for an efficient peptide identification. Furthermore, these results confirm

the ability of metaproteomics to map enzyme expression for entire pathways, as already described in the case of methanogens (Kohrs *et al.*, 2014; Gunnigle *et al.*, 2015).

Moreover, we found that the acetogenic Wood–Ljungdahl pathway (from carbon dioxide to acetate, including the tetrahydrofolate interconversion steps) was entirely covered by Firmicutes members, mainly Clostridiales. This pathway is used by acetogens to convert hydrogen and carbon dioxide into acetic acid, and its key importance is related to the oxidation of the hydrogen generated during the fermentation of dietary macromolecules (Koropatkin *et al.*, 2012). In this study, we were able to identify all enzymatic players involved in this pathway, supporting its key relevance within microbial metabolism in sheep colon. More specifically, most enzymes

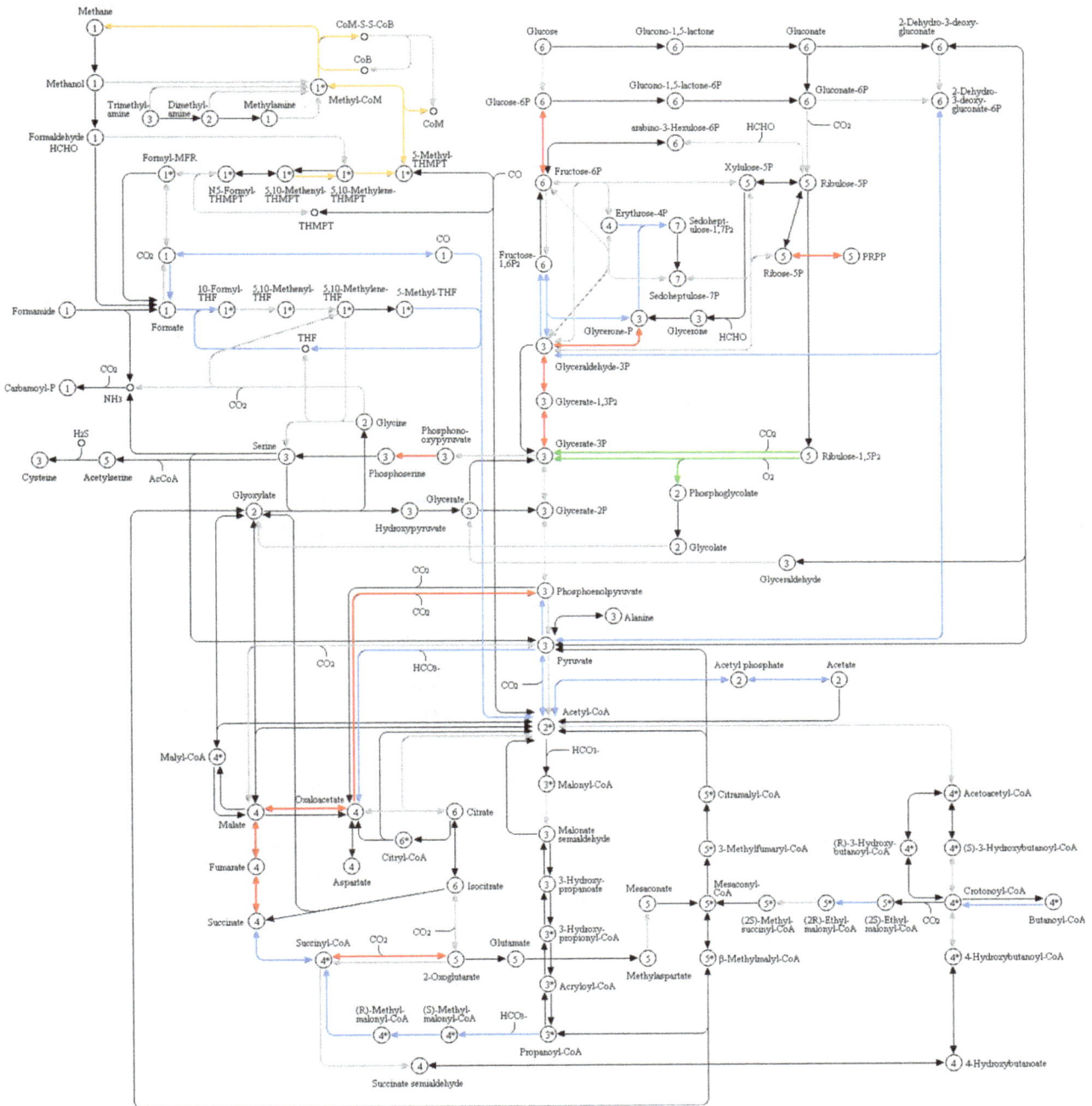

Fig. 4. Enzymatic functions identified by metaproteomics and mapped in the KEGG carbon metabolism pathway. Coloured arrows indicate enzymes detected in all animals, with the colour corresponding to the main phylum to which the function was assigned (red, Bacteroidetes; blue, Firmicutes; orange, Euryarchaeota; green, Cyanobacteria). Grey arrows indicate enzymes detected in at least one but not all animals, or not assigned unambiguously to at least one phylum.

involved in the tetrahydrofolate interconversion steps were taxonomically assigned not lower than the phylum level (always Firmicutes), indicating a high level of conservation within the Firmicutes members of the corresponding orthologous enzymes identified. Conversely,

the key players of the last two reactions (from acetyl-CoA to acetate, sequentially catalysed by phosphate acetyltransferase and acetate kinase) could be identified as being members of Clostridiales (including Lachnospiraceae and Oscillospiraceae), although a few peptides belonged to *Bacteroides* and *Prevotella* species from Bacteroidetes.

Furthermore, Bacteroidales were found to be responsible for most steps of the tricarboxylic acid cycle, as highlighted by a recent work on the effects of diet-induced obesity on the mouse gut microbiota (Denou *et al.*, 2016), whereas the enzyme RuBisCo, as expected, was detected only as expressed by photosynthetic Cyanobacteria. In addition, one key enzyme in the galactose metabolism, galactokinase, was found associated with Bacteroidetes members only. This enzyme was recently demonstrated as essential for a *Bacteroides* species to accomplish early colonization of the colonic microbiota (Yaung *et al.*, 2015).

Lastly, it is worth noting that the phylum-based distribution of the expressed carbon metabolism enzymes was different from the distribution of the same genes in the metagenome (Fig. S1), in line with the above-stated observations.

Focus on glycan import and degradation: ABC transporters, starch utilization system and glycosyl hydrolases

We then examined functions responsible for import and degradation of glycans, in view of the high content in glycans in the plant-based sheep diet, and to evaluate the (residual) relevance of such activities in the sheep colon after the massive digestion of plant material performed in the rumen.

First, we focused on genes and proteins classified as ABC transporters, especially those capable of transporting mono- and oligosaccharides, which are listed in Table 2. These comprise both generic multiple sugar transport systems, such as msmX (Ferreira and Sa-Nogueira, 2010), and more specific transporters, targeting mono- and disaccharides such as ribose and maltose (Oldham and Chen, 2011; Clifton *et al.*, 2015). In most cases, according to both S-MG and MP data, these functions were related to Firmicutes members (many different genera belonging to the order Clostridiales), followed by Actinobacteria (mainly Actinomycetales) for S-MG and Spirochaetes (especially *Treponema*) for MP.

As mentioned above, a large amount of peptides were assigned to the Bacteroidetes-specific TonB-dependent receptor family, namely to the starch utilization system (Sus) protein C, essential for complex carbohydrate degradation in Bacteroidetes (Reeves *et al.*, 1997; Martens *et al.*, 2009). The Sus proteins are located in the periplasm and the outer membrane and have the role of sequentially binding starch to the cell surface, degrading it into oligosaccharides and finally transporting the degradation products into the periplasmic space, where they are further digested into simpler sugars (e.g. glucose) and imported into the bacterial cell (Koropatkin *et al.*, 2012). Of note, the percentage of identified peptides classified as belonging to the Ton-B-dependent receptor family was 20-fold higher than the percentage of the corresponding genes sequenced, indicating a likely strong expression rate for this gene family.

Finally, we focused our attention on glycosyl hydrolases (GHs), because of their key role in degradation of plant biomass. We detected genes belonging to 56 different GH families in S-MG, of which 28 were found in all animals (listed in Table S4). The most represented family was GH 13, mainly composed by alpha amylases, with 70% of genes belonging to Firmicutes; then, we found GH 3 (mainly beta-glucosidases, 57% from Firmicutes and 41% from Bacteroidetes), GH 2 (mainly beta-galactosidases, assigned at 51% to Firmicutes and 44% to Bacteroidetes) and GH 51 (64% Firmicutes and 31%

Table 2. Carbohydrate ABC transporter genes and proteins identified in the faecal microbiota of all sheep by metagenomics and metaproteomics respectively.

Transported molecules	Identified gene(s)	Associated phyla	Identified protein(s)	Associated phyla
Aldouronate	*lplB, lplC*	Firmicutes	lplA	Firmicutes
Alpha glucoside	*aglK*		aglK	Spirochaetes
Arabinogalactan	*ganQ*	Firmicutes		
Arabinosaccharide	*araQ*	Firmicutes		
D-Allose	*alsA*			
D-Xylose	*xylG*			
L-Arabinose	*araG*			
Maltose/maltodextrin	*malK*		malK	Firmicutes
Methyl-galactoside	*mglA*	Firmicutes	mglB	Firmicutes
Multiple mono- and oligosaccharides	*msmX*	Firmicutes, Actinobacteria	msmX	Firmicutes
Multiple oligosaccharides	*gguA, gguS*	Firmicutes, Actinobacteria	gguS	Firmicutes
Myoinositol	*iatA*	Firmicutes		
Rhamnose	*rhaT*	Firmicutes		
Ribose/D-xylose	*rbsA, rbsC*	Firmicutes, Actinobacteria	rbsB	Firmicutes
sn-Glycerol-3-phosphate	*ugpC*	Firmicutes, Actinobacteria	ugpC	Firmicutes

Bacteroidetes). On the other hand, three GH families were found as expressed by the microbiota of all animals, according to MP data, namely GH 101 (clostridial endo-alpha-*N*-acetylgalactosaminidase), GH 13 (pullulanase from Lachnospiraceae) and GH 94 (cellobiose phosphorylase, assigned to various phyla). Interestingly, the relative abundance of GH peptides identified was about sixfold lower than that of the corresponding genes sequenced, indicating a likely poor expression of this functional gene family, probably because of the relatively low amount of complex (and still undigested) polysaccharides that reach the colon after the extensive degradation occurred within the upper tracts of the ruminant digestive system (Huntington *et al.*, 2006).

Non-microbial components detected in the faecal material through metaproteomics

The mass spectra generated in this work were also searched against a 'generic' database (without taxonomic filters towards microbial sequences) to achieve information about all organisms contained in the faecal samples. As a result, we found that besides spectra of microbial origin (about 30% of the total) approximately 45% of spectra were assigned to the host (phylum Chordata), while 13% could be attributed to plant material (with those assigned to Poaceae and Fabaceae families accounting for about 70% of them). Less than 1% of detected peptides were classified as belonging to further eukaryotic phyla as - in decreasing abundance order – Arthropoda, Mollusca, Nematoda, Annelida (including the *Hirudo* genus) and Platyhelminthes (including the *Fasciola* genus, comprising parasites of the small ruminant intestine). Even more interestingly, peptides from several protists known for intestinal tropism in sheep were also identified, including the following genera: *Entamoeba*, *Blastocystis*, *Andalucia*, *Giardia* and *Entodinium*, the last being a ciliate protozoan with a well-known capability to digest and ferment starch, producing SCFAs (Belzecki *et al.*, 2013).

In addition, based on S-MG data, we were able to detect in four animals a few sequences attributed to *Haemonchus* (about 0.005% of annotated reads; an identified peptide sequence was also assigned to this genus). Of note, *Haemonchus contortus* is described as the most important gastroenteric nematode of sheep in many regions of the world, causing haemonchosis (Getachew *et al.*, 2007).

Conclusions

To the best of our knowledge, the results presented in this study represent the first 'omic' characterization of the sheep faecal microbiota to date. The application of a multi-omic strategy, comprising V4-MG, S-MG and MP approaches, enabled us to take a comprehensive picture of both the taxonomic structure and the functional activity of the microbial communities inhabiting the distal tract of the ovine digestive system.

Under a taxonomic perspective, with Firmicutes being the main phylum, the sheep microbiota appeared globally comparable to that of other ruminants. In functional terms, we found that the sheep faecal microbiota was primarily involved in catabolism, and described several activities responsible for transport and degradation of carbohydrates. We also identified several phylum-specific pathways, such as methanogenesis for Euryarchaeota and acetogenesis for Firmicutes. Furthermore, our approach provided information regarding the eukaryotic part of the microbiota (e.g. fungi and protists).

These findings, and the whole data set provided here, can be useful for deepening our understanding on organization and role of the large intestine microbiota in small ruminants, and pave the way to further studies investigating the relationship between gut microbiota dynamics and management and dietary variables key for sheep farming.

Experimental procedures

Animal description and faecal sample collection

Faecal samples were collected from the rectal ampulla of five lactating Sarda sheep in November 2012. Contact with the surrounding environment was minimized by transferring the faecal material immediately from the rectum to the collection tube. Sheep came from the same flock, were free-grazing and fed a limited amount of commercial feed only during milking (max. 400 g per day) and were apparently healthy. All samples were immediately stored at −80°C until use. At the time of the analyses, samples were thawed at 4°C, and from each of them two stool fragments were collected for DNA and protein extraction respectively.

DNA sample preparation

Stool samples were subjected to DL or DC as described earlier (Apajalahti *et al.*, 1998; Tanca *et al.*, 2015). Briefly, faecal samples (weighing approximately 100 mg each) were resuspended in PBS and subjected to low-speed centrifugation to eliminate gross particulate material for a total of three rounds. The supernatants were then centrifuged at 20 000 *g* for 15 min, and the derivative pellets were subjected to DNA extraction. DNA extraction was performed in parallel using two commercial kits, namely the QIAamp Fast DNA Stool Kit (Qiagen, Hilden, Germany) and the E.Z.N.A. Soil DNA Kit

(Omega Bio-Tek, Norcross, GA, USA), according to the respective manufacturers' instructions. DC-pretreated samples were additionally subjected to DNA extraction according to the standard phenol/chloroform/isoamyl alcohol (25:24:1) method.

16S rDNA analysis

Primer design for universal amplification of the V4 region of 16S rDNA was based on a protocol published by Caporaso and co-workers (Caporaso *et al.*, 2011). PCR cycling conditions were as follows: 2 min at 94°C; 28 cycles of 30 s at 94°C, 30 s at 55°C, 2 min at 68°C; finally, 7 min at 72°C. PCR products were confirmed on 2% agarose gel (Sigma Aldrich, St Louis, MO, USA). Two separate 16S rRNA gene amplification reactions were performed, pooled together, cleaned up using AMPure XP (Beckman Coulter, Brea, CA, USA) magnetic beads and quantified with the Qubit HS assay using the Qubit fluorometer 2.0 (Life Technologies, Grand Island, NY, USA).

Libraries were constructed according to the Nextera XT kit (Illumina, San Diego, CA, USA). Sequence-ready libraries were normalized to ensure equal library representation in the pooled samples. DNA sequencing was performed with the Illumina HiScanSQ sequencer, using the paired-end method and 93 cycles of sequencing.

The Illumina demultiplexed paired reads were trimmed for the first 20 bp using FASTX, and the sequences with Nextera adapter contamination were identified using the UniVec database (ftp://ftp.ncbi.nlm.nih.gov/pub/UniVec/) and removed. Therefore, the paired reads with a minimum overlap of eight bases were merged using a specific QIIME script. OTU generation was performed using a QIIME pipeline based on USEARCH's OTU clustering recommendations (http://www.drive5.com/usearch/manual/otu_clustering.html). Reads were clustered at 97% identity using UCLUST to produce OTUs (Edgar, 2010). Taxonomy assignment of resulting OTUs was performed using the Greengenes 13_8 database (DeSantis *et al.*, 2006). With taxonomic lineages in hand, OTU tables were computed using the QIIME 1.9.0 software suite (Caporaso *et al.*, 2010; Kuczynski *et al.*, 2010). The relative proportion of read counts was used as a quantitative estimation of the abundance of each taxon.

Metagenome analysis

Libraries were constructed according to the Nextera XT kit and sequenced with the HiScanSQ sequencer (both from Illumina), using the paired-end method and 93 cycles of sequencing.

Read processing was carried out using tools from the USEARCH suite v.8.1.1861 (Edgar, 2010; Edgar and

Flyvbjerg, 2015), specifically merging of paired reads (fastq_mergepairs command, setting parameters as follows: fastq_truncqual 3, fastq_minovlen 8, fastq_maxdiffs 0) and quality filtering (fastq_filter command, with fastq_truncqual 15 and fastq_minlen 100).

Taxonomic annotation was performed using MEGAN v.5.10 (Huson and Mitra, 2012). Read sequences were preliminary subjected to DIAMOND (v.0.7.1) search against the NCBI-nr DB (2014/05 update), using the blastx command with default parameters (Buchfink *et al.*, 2015). Then, a lowest common ancestor classification was performed on DIAMOND results using MEGAN with default parameters.

Functional annotation was accomplished by DIAMOND blastx search (e-value threshold 10^{-5}) against bacterial sequences from the UniProt/Swiss-Prot database (release 2014_12) and subsequent retrieval of protein family, KEGG orthologous group and pathway information associated with each UniProt/Swiss-Prot accession number (UniProtConsortium, 2015).

The relative proportion of read counts was used as a quantitative estimation of the abundance of each taxon or function.

The metagenomic sequence data were deposited in the European Nucleotide Archive under the Project Accession Number PRJEB14312.

Protein sample preparation

Stool samples (average weight 356 ± 31 mg) were resuspended by vortexing in SDS-based extraction buffer and then heated and subjected to a combination of bead-beating and freeze–thawing steps as detailed elsewhere (Tanca *et al.*, 2014).

Protein extracts were subjected to on-filter reduction, alkylation and trypsin digestion according to the filter-aided sample preparation protocol (Wisniewski *et al.*, 2009), with slight modifications detailed elsewhere (Tanca *et al.*, 2013).

Metaproteome analysis

LC-MS/MS analysis was carried out using an LTQ-Orbitrap Velos mass spectrometer (Thermo Scientific, San Jose, CA, USA) interfaced with an UltiMate 3000 RSLCnano LC system (Thermo Scientific). The single-run 1D LC peptide separation was performed as previously described (Tanca *et al.*, 2014), loading 4 μg of peptide mixture per each sample and applying a 485 min separation gradient. The mass spectrometer was set up in a data-dependent MS/MS mode, with higher-energy collision dissociation as the fragmentation method, as detailed elsewhere (Tanca *et al.*, 2013). An average of 69 435 ± 870 spectra were acquired per sample.

Peptide identification was performed using the Proteome Discoverer informatic platform (version 1.4; Thermo Scientific), with Sequest-HT as search engine and Percolator for peptide validation (FDR < 1%). Search parameters were set as described previously (Tanca et al., 2014).

Parallel searches were performed using three different sequence databases. The first database was composed by the metagenomic sequences obtained in this study, both as raw reads and as assembled contigs (8 595 757 sequences). Paired reads were merged using the script join_paired_ends.py inside the QIIME package, v.1.9 (Caporaso et al., 2010) with a minimum overlap of 8 base pairs. The output sequences were filtered (with a fastq_truncqual option = 15) and clustered at 100% using USEARCH v. 5.2.236 (Edgar, 2010). Read assembly into contigs was carried out using Velvet v.1.2.10 (Zerbino and Birney, 2008), by setting 61 as k-mer length, 200 as insert length and 300 as minimum contig length. Open reading frames were found from both reads and contigs using FragGeneScan v.1.19, with the training for Illumina sequencing reads with about 0.5% error rate (Rho et al., 2010).

The second database was a selection of all bacterial, archaeal, fungal and gut microbiota sequences (79 203 800 sequences in total) from the 2015_02 release of the UniProtKB database.

The metaproteomic data (regarding the microbial component of the faecal material) were obtained by merging results of searches against the two above-mentioned databases.

A third database (specifically, the whole UniProtKB database, release 2014_12, 89 136 540 sequences) was finally employed to achieve information only concerning the non-microbial components of the sheep microbiota.

The relative proportion of spectral counts (peptide-spectrum matches) was used as a quantitative estimation of the abundance of each taxon or function.

The mass spectrometry proteomics data have been deposited to the ProteomeXchange Consortium via the PRIDE (Vizcaino et al., 2016) partner repository with the data set identifier PXD004524.

Taxonomic and functional assignments were performed as described above for metagenome sequences, except using the DIAMOND blastp command instead of blastx.

Conflict of Interest

None of the authors has any potential conflict of interest to declare.

References

Apajalahti, J.H., Sarkilahti, L.K., Maki, B.R., Heikkinen, J.P., Nurminen, P.H., and Holben, W.E. (1998) Effective recovery of bacterial DNA and percent-guanine-plus-cytosine-based analysis of community structure in the gastrointestinal tract of broiler chickens. *Appl Environ Microbiol* **64:** 4084–4088.

Belzecki, G., Miltko, R., Kwiatkowska, E., and Michalowski, T. (2013) The ability of rumen ciliates, Eudiplodinium maggii, Diploplastron affine, and Entodinium caudatum, to use the murein saccharides. *Folia Microbiol (Praha)* **58:** 463–468.

Brilhante, R.S., Silva, S.T., Castelo-Branco, D.S., Teixeira, C.E., Borges, L.C., Bittencourt, P.V., et al. (2015) Emergence of azole-resistant Candida albicans in small ruminants. *Mycopathologia* **180:** 277–280.

Buchfink, B., Xie, C., and Huson, D.H. (2015) Fast and sensitive protein alignment using DIAMOND. *Nat Methods* **12:** 59–60.

Caporaso, J.G., Kuczynski, J., Stombaugh, J., Bittinger, K., Bushman, F.D., Costello, E.K., et al. (2010) QIIME allows analysis of high-throughput community sequencing data. *Nat Methods* **7:** 335–336.

Caporaso, J.G., Lauber, C.L., Walters, W.A., Berg-Lyons, D., Lozupone, C.A., Turnbaugh, P.J., et al. (2011) Global patterns of 16S rRNA diversity at a depth of millions of sequences per sample. *Proc Natl Acad Sci U S A* **108 (Suppl** 1): 4516–4522.

Castro-Carrera, T., Toral, P.G., Frutos, P., McEwan, N.R., Hervas, G., Abecia, L., et al. (2014) Rumen bacterial community evaluated by 454 pyrosequencing and terminal restriction fragment length polymorphism analyses in dairy sheep fed marine algae. *J Dairy Sci* **97:** 1661–1669.

Clifton, M.C., Simon, M.J., Erramilli, S.K., Zhang, H., Zaitseva, J., Hermodson, M.A., and Stauffacher, C.V. (2015) In vitro reassembly of the ribose ATP-binding cassette transporter reveals a distinct set of transport complexes. *J Biol Chem* **290:** 5555–5565.

Denou, E., Marcinko, K., Surette, M.G., Steinberg, G.R., and Schertzer, J.D. (2016) High-intensity exercise training increases the diversity and metabolic capacity of the mouse distal gut microbiota during diet-induced obesity. *Am J Physiol Endocrinol Metab* **310:** E982–E993.

DeSantis, T.Z., Hugenholtz, P., Larsen, N., Rojas, M., Brodie, E.L., Keller, K., et al. (2006) Greengenes, a chimera-checked 16S rRNA gene database and workbench compatible with ARB. *Appl Environ Microbiol* **72:** 5069–5072.

Durso, L.M., Harhay, G.P., Smith, T.P., Bono, J.L., Desantis, T.Z., Harhay, D.M., et al. (2010) Animal-to-animal variation in fecal microbial diversity among beef cattle. *Appl Environ Microbiol* **76:** 4858–4862.

Durso, L.M., Harhay, G.P., Bono, J.L., and Smith, T.P. (2011) Virulence-associated and antibiotic resistance genes of microbial populations in cattle feces analyzed using a metagenomic approach. *J Microbiol Methods* **84:** 278–282.

Edgar, R.C. (2010) Search and clustering orders of magnitude faster than BLAST. *Bioinformatics* **26:** 2460–2461.

Edgar, R.C., and Flyvbjerg, H. (2015) Error filtering, pair assembly and error correction for next-generation sequencing reads. *Bioinformatics* **31:** 3476–3482.

Ferreira, M.J., and Sa-Nogueira, I. (2010) A multitask ATPase serving different ABC-type sugar importers in Bacillus subtilis. *J Bacteriol* **192:** 5312–5318.

de la Fuente, G., Belanche, A., Girwood, S.E., Pinloche, E., Wilkinson, T., and Newbold, C.J. (2014) Pros and cons of ion-torrent next generation sequencing versus terminal restriction fragment length polymorphism T-RFLP for studying the rumen bacterial community. *PLoS ONE* **9:** e101435.

Getachew, T., Dorchies, P., and Jacquiet, P. (2007) Trends and challenges in the effective and sustainable control of *Haemonchus contortus* infection in sheep. Review. *Parasite* **14:** 3–14.

Gunnigle, E., Nielsen, J.L., Fuszard, M., Botting, C.H., Sheahan, J., O'Flaherty, V., and Abram, F. (2015) Functional responses and adaptation of mesophilic microbial communities to psychrophilic anaerobic digestion. *FEMS Microbiol Ecol* **91:** PMID 26507125. DOI: 10.1093/femsec/fiv132.

Hamer, H.M., Jonkers, D.M., Bast, A., Vanhoutvin, S.A., Fischer, M.A., Kodde, A., *et al.* (2009) Butyrate modulates oxidative stress in the colonic mucosa of healthy humans. *Clin Nutr* **28:** 88–93.

Hart, M.L., Meyer, A., Johnson, P.J., and Ericsson, A.C. (2015) Comparative evaluation of DNA extraction methods from feces of multiple host species for downstream next-generation sequencing. *PLoS ONE* **10:** e0143334.

Havemann, G.D., Sampson, E.M., and Bobik, T.A. (2002) PduA is a shell protein of polyhedral organelles involved in coenzyme B(12)-dependent degradation of 1,2-propanediol in *Salmonella enterica* serovar typhimurium LT2. *J Bacteriol* **184:** 1253–1261.

Hecker, J.F., and Grovum, W.L. (1975) Rates of passage of digesta and water absorption along the larg intestines of sheep, cows and pigs. *Aust J Biol Sci* **28:** 161–167.

Huntington, G.B. (1997) Starch utilization by ruminants: from basics to the bunk. *J Anim Sci* **75:** 852–867.

Huntington, G.B., Harmon, D.L., and Richards, C.J. (2006) Sites, rates, and limits of starch digestion and glucose metabolism in growing cattle. *J Anim Sci* **84(Suppl):** E14–E24.

Huson, D.H., and Mitra, S. (2012) Introduction to the analysis of environmental sequences: metagenomics with MEGAN. *Methods Mol Biol* **856:** 415–429.

Huws, S.A., Kim, E.J., Lee, M.R., Scott, M.B., Tweed, J.K., Pinloche, E., *et al.* (2011) As yet uncultured bacteria phylogenetically classified as Prevotella, Lachnospiraceae incertae sedis and unclassified Bacteroidales, Clostridiales and Ruminococcaceae may play a predominant role in ruminal biohydrogenation. *Environ Microbiol* **13:** 1500–1512.

Kim, M., Kim, J., Kuehn, L.A., Bono, J.L., Berry, E.D., Kalchayanand, N., *et al.* (2014) Investigation of bacterial diversity in the feces of cattle fed different diets. *J Anim Sci* **92:** 683–694.

Kittelmann, S., Kirk, M.R., Jonker, A., McCulloch, A., and Janssen, P.H. (2015) Buccal swabbing as a noninvasive method to determine bacterial, archaeal, and eukaryotic microbial community structures in the rumen. *Appl Environ Microbiol* **81:** 7470–7483.

Kohrs, F., Heyer, R., Magnussen, A., Benndorf, D., Muth, T., Behne, A., *et al.* (2014) Sample prefractionation with liquid isoelectric focusing enables in depth microbial metaproteome analysis of mesophilic and thermophilic biogas plants. *Anaerobe* **29:** 59–67.

Koropatkin, N.M., Cameron, E.A., and Martens, E.C. (2012) How glycan metabolism shapes the human gut microbiota. *Nat Rev Microbiol* **10:** 323–335.

Krause, D.O., Denman, S.E., Mackie, R.I., Morrison, M., Rae, A.L., Attwood, G.T., and McSweeney, C.S. (2003) Opportunities to improve fiber degradation in the rumen: microbiology, ecology, and genomics. *FEMS Microbiol Rev* **27:** 663–693.

Kuczynski, J., Costello, E.K., Nemergut, D.R., Zaneveld, J., Lauber, C.L., Knights, D., *et al.* (2010) Direct sequencing of the human microbiome readily reveals community differences. *Genome Biol* **11:** 210.

Kumar, S., Choudhury, P.K., Carro, M.D., Griffith, G.W., Dagar, S.S., Puniya, M., *et al.* (2014) New aspects and strategies for methane mitigation from ruminants. *Appl Microbiol Biotechnol* **98:** 31–44.

Lopes, L.D., de Souza Lima, A.O., Taketani, R.G., Darias, P., da Silva, L.R., Romagnoli, E.M., *et al.* (2015) Exploring the sheep rumen microbiome for carbohydrate-active enzymes. *Antonie Van Leeuwenhoek* **108:** 15–30.

Mao, S., Zhang, M., Liu, J., and Zhu, W. (2015) Characterising the bacterial microbiota across the gastrointestinal tracts of dairy cattle: membership and potential function. *Sci Rep* **5:** 16116.

Martens, E.C., Koropatkin, N.M., Smith, T.J., and Gordon, J.I. (2009) Complex glycan catabolism by the human gut microbiota: the Bacteroidetes Sus-like paradigm. *J Biol Chem* **284:** 24673–24677.

Morgavi, D.P., Rathahao-Paris, E., Popova, M., Boccard, J., Nielsen, K.F., and Boudra, H. (2015) Rumen microbial communities influence metabolic phenotypes in lambs. *Front Microbiol* **6:** 1060.

Nayfach, S., and Pollard, K.S. (2015) Average genome size estimation improves comparative metagenomics and sheds light on the functional ecology of the human microbiome. *Genome Biol* **16:** 51.

Oldham, M.L., and Chen, J. (2011) Crystal structure of the maltose transporter in a pretranslocation intermediate state. *Science* **332:** 1202–1205.

Paster, B.J., and Canale-Parola, E. (1985) *Treponema saccharophilum* sp. nov., a large pectinolytic spirochete from the bovine rumen. *Appl Environ Microbiol* **50:** 212–219.

Peng, S., Yin, J., Liu, X., Jia, B., Chang, Z., Lu, H., *et al.* (2015) First insights into the microbial diversity in the omasum and reticulum of bovine using Illumina sequencing. *J Appl Genet* **56:** 393–401.

Petit, E., LaTouf, W.G., Coppi, M.V., Warnick, T.A., Currie, D., Romashko, I., *et al.* (2013) Involvement of a bacterial microcompartment in the metabolism of fucose and rhamnose by *Clostridium phytofermentans*. *PLoS ONE* **8:** e54337.

Popova, M., Morgavi, D.P., and Martin, C. (2013) Methanogens and methanogenesis in the rumens and ceca of lambs fed two different high-grain-content diets. *Appl Environ Microbiol* **79:** 1777–1786.

Pryde, S.E., Duncan, S.H., Hold, G.L., Stewart, C.S., and Flint, H.J. (2002) The microbiology of butyrate formation in the human colon. *FEMS Microbiol Lett* **217:** 133–139.

Reeves, A.R., Wang, G.R., and Salyers, A.A. (1997) Characterization of four outer membrane proteins that play a role in utilization of starch by *Bacteroides thetaiotaomicron*. *J Bacteriol* **179:** 643–649.

Rho, M., Tang, H., and Ye, Y. (2010) FragGeneScan: predicting genes in short and error-prone reads. *Nucleic Acids Res* **38:** e191.

Shanks, O.C., Kelty, C.A., Archibeque, S., Jenkins, M., Newton, R.J., McLellan, S.L., *et al.* (2011) Community structures of fecal bacteria in cattle from different animal feeding operations. *Appl Environ Microbiol* **77:** 2992–3001.

Shi, W., Moon, C.D., Leahy, S.C., Kang, D., Froula, J., Kittelmann, S., *et al.* (2014) Methane yield phenotypes linked to differential gene expression in the sheep rumen microbiome. *Genome Res* **24:** 1517–1525.

Tanca, A., Biosa, G., Pagnozzi, D., Addis, M.F., and Uzzau, S. (2013) Comparison of detergent-based sample preparation workflows for LTQ-Orbitrap analysis of the *Escherichia coli* proteome. *Proteomics* **13:** 2597–2607.

Tanca, A., Palomba, A., Pisanu, S., Deligios, M., Fraumene, C., Manghina, V., *et al.* (2014) A straightforward and efficient analytical pipeline for metaproteome characterization. *Microbiome* **2:** 49.

Tanca, A., Palomba, A., Pisanu, S., Addis, M.F., and Uzzau, S. (2015) Enrichment or depletion? The impact of stool pretreatment on metaproteomic characterization of the human gut microbiota. *Proteomics* **15:** 3474–3485.

UniProtConsortium (2015) UniProt: a hub for protein information. *Nucleic Acids Res* **43:** D204–D212.

Vetrovsky, T., and Baldrian, P. (2013) The variability of the 16S rRNA gene in bacterial genomes and its consequences for bacterial community analyses. *PLoS ONE* **8:** e57923.

Vital, M., Howe, A.C., and Tiedje, J.M. (2014) Revealing the bacterial butyrate synthesis pathways by analyzing (meta) genomic data. *mBio* **5:** e00889.

Vizcaino, J.A., Csordas, A., Del-Toro, N., Dianes, J.A., Griss, J., Lavidas, I., *et al.* (2016) 2016 update of the PRIDE database and its related tools. *Nucleic Acids Res* **44:** D447–D456.

Wagner Mackenzie, B., Waite, D.W., and Taylor, M.W. (2015) Evaluating variation in human gut microbiota profiles due to DNA extraction method and inter-subject differences. *Front Microbiol* **6:** 130.

Wisniewski, J.R., Zougman, A., Nagaraj, N., and Mann, M. (2009) Universal sample preparation method for proteome analysis. *Nat Methods* **6:** 359–362.

Wolgemuth, C.W. (2015) Flagellar motility of the pathogenic spirochetes. *Semin Cell Dev Biol* **46:** 104–112.

Yaung, S.J., Deng, L., Li, N., Braff, J.L., Church, G.M., Bry, L., *et al.* (2015) Improving microbial fitness in the mammalian gut by in vivo temporal functional metagenomics. *Mol Syst Biol* **11:** 788.

Zeng, Y., Zeng, D., Zhang, Y., Ni, X., Tang, Y., Zhu, H., *et al.* (2015) Characterization of the cellulolytic bacteria communities along the gastrointestinal tract of Chinese Mongolian sheep by using PCR-DGGE and real-time PCR analysis. *World J Microbiol Biotechnol* **31:** 1103–1113.

Zerbino, D.R., and Birney, E. (2008) Velvet: algorithms for de novo short read assembly using de Bruijn graphs. *Genome Res* **18:** 821–829.

12

Engineering *Mycobacterium smegmatis* for testosterone production

Lorena Fernández-Cabezón, Beatriz Galán and
José L. García*
*Department of Environmental Biology, Centro de
Investigaciones Biológicas, Consejo Superior de
Investigaciones Científicas, Ramiro de Maeztu 9, 28040
Madrid, Spain.*

Summary

A new biotechnological process for the production of testosterone (TS) has been developed to turn the model strain *Mycobacterium smegmatis* suitable for TS production to compete with the current chemical synthesis procedures. We have cloned and overexpressed two genes encoding microbial 17β-hydroxysteroid: NADP 17-oxidoreductase, from the bacterium *Comamonas testosteroni* and from the fungus *Cochliobolus lunatus*. The host strains were *M. smegmatis* wild type and a genetic engineered androst-4-ene-3,17-dione (AD) producing mutant. The performances of the four recombinant bacterial strains have been tested both in growing and resting-cell conditions using natural sterols and AD as substrates respectively. These strains were able to produce TS from sterols or AD with high yields. This work represents a proof of concept of the possibilities that offers this model bacterium for the production of pharmaceutical steroids using metabolic engineering approaches.

Introduction

Testosterone (TS) is one of the oldest drugs used in medicine and has a long efficacy and safety record for hormone replacement therapy in men with androgen deficiency. Currently, TS is chemically produced from androst-4-ene-3,17-dione (AD) (Ercoli and Ruggierii, 1953). In mammals, the synthesis of TS from AD is catalysed by the microsomal 17-ketosteroid reductase (17β-HSD; 17β-hydroxysteroid:NADP 17-oxidoreductase, EC 1.1.1.64) (Bogovich and Payne, 1980) (Fig. 1). Up to now, 14 different subtypes of 17β-HSD

*For correspondence. E-mail jlgarcia@cib.csic.es

have been identified in mammals and most of them belong to the short-chain dehydrogenase:reductase superfamily (SDR). They catalyse NAD(P)H/NAD(P)+-dependent reductions/oxidations at the C-17 position of different steroids (Peltoketo *et al.*, 1999; Moeller and Adamski, 2006, 2009; Marchais-Oberwinkler *et al.*, 2011). The majority of 17β-HSD enzymes are able to catalyse, at least to some extent, reverse reactions under *in vitro* conditions. In the presence of a substantial excess of a suitable cofactor and/or in the absence of the preferred cofactor, 17β-HSD can be compelled to catalyse both oxidative and reductive reactions. Based on this property, a process has been developed to produce *in vitro* TS from AD using the recombinant murine 17β-HSD type V (aldo-keto-reductase instead of SDR family) and glucose dehydrogenase as cofactor recycling enzyme (Fogal *et al.*, 2013). However, due to the high cost of the process, it is not currently used for industrial purposes.

On the other hand, enzymatic reduction of AD to TS by 17β-HSD has also been described in different microorganisms (Donova *et al.*, 2005), including bacteria (Schultz *et al.*, 1977; Payne and Talalay, 1985; Sarmah *et al.*, 1989; Liu *et al.*, 1994; Egorova *et al.*, 2002a,b, 2005), yeasts (Ward and Young, 1990; Singer *et al.*, 1991; Długoński and Wilmańska, 1998; Pajic *et al.*, 1999), filamentous fungi (Kristan and Rižner, 2012) and plants (Hamada and Kawabe, 1991). Moreover, a single-step microbial transformation process has been reported for the production of TS from sterols using several *Mycobacterium* sp. mutants (Wang *et al.*, 1982; Hung *et al.*, 1994; Liu *et al.*, 1994; Llanes *et al.*, 1995; Liu and Lo, 1997; Borrego *et al.*, 2000; Lo *et al.*, 2002; Mei *et al.*, 2005; Egorova *et al.*, 2009). During the course of this work, Dlugovitzky *et al.* (2015) have shown that *Mycobacterium smegmatis* PTCC 1307 was able to produce TS and other estrogens from tritiated precursors. However, TS has not been detected as a metabolic intermediate when *M. smegmatis* mc²155 is cultured in the presence of phytosterols or cholesterol, neither in the wild-type strain nor in the AD-producing strain (Galán *et al.*, unpublished), unlike in other mycobacterial species (Wang *et al.*, 1982; Smith *et al.*, 1993; Egorova *et al.*, 2002b). This observation suggests that *M. smegmatis* mc²155 does not contain a functional gene encoding a 17β-HSD or at least, it is not induced in the presence of these compounds.

Although several microbial 17β-HSD enzymes have been cloned and characterized (Abalain *et al.*, 1993; Rižner *et al.*, 1999; Chang *et al.*, 2010), none of them were used to develop genetically engineered bacteria to improve the biotechnological production of TS. These genes have been only expressed in *Escherichia coli*, a bacterium unable to efficiently transport sterols or AD, impairing the development of an industrial biotransformation processes.

The aim of this work was to develop recombinant bacteria overexpressing 17β-HSD genes that will be able to efficiently biotransform either natural sterols (e.g. phytosterols or cholesterol) or AD into TS in order to compete with the current chemical synthesis of TS by using a new biotechnological process. To fulfil this goal, we have cloned and overexpressed the genes encoding two

microbial 17β-HSDs, from the bacterium *Comamonas testosteroni* (Abalain *et al.*, 1993) and from the fungus *Cochliobolus lunatus* (Rižner *et al.*, 1999), using as hosts the wild-type *M. smegmatis* and an AD-producing mutant of this bacterium. The performances of the new created recombinant bacterial strains have been tested both in growing and resting-cell conditions using sterols and AD as substrates respectively (Fig. 2).

Results and discussion

Working hypothesis and selection of 17β-HSD encoding genes

Up to now, to our knowledge, there is not an example of any engineered bacterium able to produce TS from sterols or AD. In this sense, we decided to investigate if

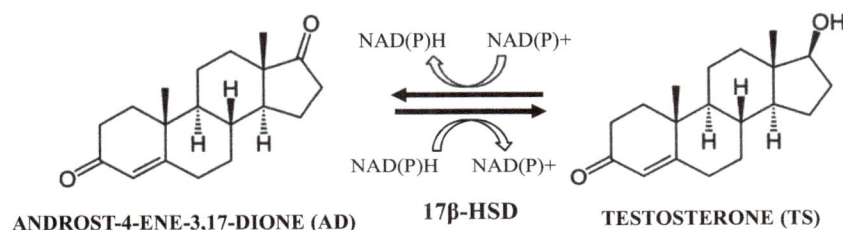

Fig. 1. Schematic representation of transformation process of AD into TS by 17β-HSD.

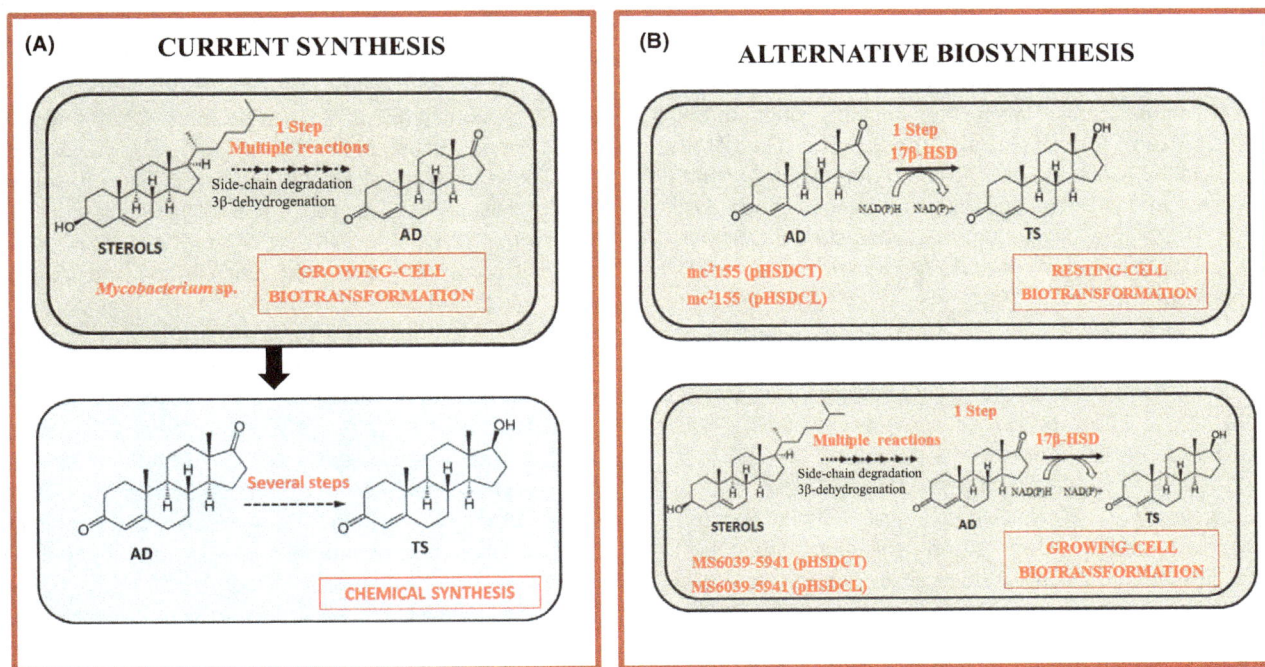

Fig. 2. Methods for TS synthesis. (A) Current synthesis of TS at the pharmaceutical industry. First, biotransformation process for the production of AD from sterols is carried out by *Mycobacterium* sp. Second, AD is transformed into TS by a chemical process. (B) Alternative production of TS proposed in this work by recombinant *M. smegmatis* strains overexpressing 17β-HSD-encoding genes. The biotransformation of AD into TS can be achieved by resting-cell in the strains *M. smegmatis* mc²155 (pHSDCT) and *M. smegmatis* mc²155 (pHSDCL). The production of TS from sterols can be realized by growing-cell biotransformations in the mutant strains *M. smegmatis* MS60369-5941 (pHSDCT) and *M. smegmatis* MS60369-5941 (pHSDCL).

M. smegmatis could be a suitable chassis for this purpose. The selection of *M. smegmatis* to achieve TS production is mainly based in two properties: first, it is not able to degrade AD and second, there are evidences that AD can be efficiently transported (L. Fernández-Cabezón *et al., unpublished*). Therefore, the circumvention of the bacterial mineralization of AD and TS during the biotransformation process is not a requirement. We have already evidenced that this fast-growing and non-pathogenic bacterium, which is able to transport and metabolize cholesterol and phytosterols, can be a suitable cell factory for the industrial production of steroid intermediates such as AD using sterols as feedstock (Galán *et al., unpublished*). Other steroid-metabolizing bacteria that are able to transport AD (e.g. *Comamonas, Rhodococcus* or *Gordonia*) cannot be in principle used as an alternative host because they degrade AD and TS efficiently (Tamaoka *et al.*, 1987; Cabrera *et al.*, 2000; Fernández de las Heras *et al.*, 2009; Li *et al.*, 2014). Taking into account that we were not able to identify in *M. smegmatis* mc²155, a functional gene encoding a 17β-HSD, the aim of this work was to overproduce a 17β-HSD obtained from a heterologous organism either in the wild-type or the AD-producing mutant strains. In this way, the *M. smegmatis* recombinant strains can be utilized to transform AD into TS by a resting-cell system or to produce TS from sterols by a fermentation process (Fig. 2).

As the genes encoding 17β-HSD enzymes from mycobacterial species have not been identified and these proteins have been only partially purified and characterized (Goren *et al.*, 1983; Egorova *et al.*, 2002a, 2005), we initially selected as enzyme candidates for metabolic engineering the well-described 17β-HSDs from the bacterium *C. testosteroni* (Schultz *et al.*, 1977; Lefebvre *et al.*,1979; Minard *et al.*, 1985; Genti-Raimondi *et al.*, 1991; Yin *et al.*, 1991; Abalain *et al.*, 1993; Benach *et al.*, 1996, 2002; Oppermann *et al.*, 1997; Cabrera *et al.*, 2000) and the fungus *C. lunatus* (Plemenitas *et al.*, 1988; Rižner *et al.*, 1996, 1999, 2000, 2001a,b; Rižner and Zakelj-Mavric, 2000; Zorko *et al.*, 2000; Kristan *et al.*, 2003, 2005, 2007a,b; Cassetta *et al.*, 2005; Ulrih and Lanisnik Rižner, 2006; Brunskole *et al.*, 2009; Svegelj *et al.*, 2012), because both enzymes present some relevant differences. Although they catalyse a reversible reaction and display similar reaction mechanisms, the reaction equilibrium of the fungal 17β-HSD is shifted towards reduction, whereas the bacterial enzyme is shifted towards oxidation, as this enzyme is mainly involved into the TS catabolism in *C. testosteroni* (Genti-Raimondi *et al.*, 1990; Cabrera *et al.*, 2000). Moreover, the fungal 17β-HSD prefers NADPH as coenzyme for reduction of AD, while preferences between $NAD^+/$

$NADP^+$ are not observed for oxidation of TS (Rižner and Zakelj-Mavric, 2000). The bacterial enzyme is in fact a 3β/17β-HSD, this is a bifunctional enzyme with a single catalytic site able to accommodate both the 3β- and 17β-activities, and uses NAD(H) as cofactor (Minard *et al.*, 1985; Benach *et al.*, 2002). These issues are relevant for developing a biotechnological process as the cellular content of NAD(H) and/or NADP(H) will determine the direction of the catalytic reaction and the yield of TS production. Remarkably, Xu *et al.* (2016) have described the Hsd4A protein of *M. neoaurum* ATCC 25795 as a dual-function enzyme, with both 17β-HSD and β-hydroxyacyl-CoA dehydrogenase activities *in vitro*. However, the 17β-HSD enzyme does not appear to be reversible *in vitro* as it is able to transform TS into AD but not AD into TS (Xu *et al.*, 2016).

As an alternative to the 17β-HSD enzymes of *C. testosteroni* and *C. lunatus*, we have tried to identify homologous enzymes in other microorganisms. For instance, we found mycobacterial proteins with a small identity (<40%) to the 17β-HSD enzyme from *C. testoteroni*. In particular, several putative homologue short-chain dehydrogenases were found in *M. smegmatis* mc²155, such as the 3-α-(or 20-β)-hydroxysteroid dehydrogenase (*MSMEG_3515*, 38% identity), and the cyclopentanol dehydrogenase (*MSMEG_6709*, 37% identity). Analysing the genomic context of such genes, non-relationship with steroid degradative enzymes was found. On the other hand, the 17β-HSD from *C. lunatus* presents a high sequence identity (60–95%) with proteins from the fungal *Leotiomyceta* group, which belongs to Ascomycota Phylum. These identities are not present in other representatives of this phylum, such as the *Saccharomyces* genus. However, 17β-HSD activity was detected in *Saccharomyces cerevisiae* and other fungi (Ward and Young, 1990; Długoński and Wilmańska, 1998). Outside *Leotiomyceta* group, identities of 42–48% are found in bacteria of different phyla (Cyanobacteria, different groups of *Proteobacteria, Firmicutes*, etc.), including members of actinobacteria (e.g. *Mycobacterium abscessus, Rhodococcus wratislaviensis* NBRC 100605 or several *Streptomyces* species). In *M. smegmatis*, only proteins with an identity lower than 36% are found. Consequently, the correlation between protein identity and 17 β-HSD activity, as well as their true biological role in the native organisms, cannot be easily established by *in silico* analysis. In fact, this observation has been also described in vertebrates whose 17 β-HSD enzymes show generally low sequence similarity (15–20%) (Moeller and Adamski, 2009). Therefore, on the light of these analyses, finally the 17β-HSD enzymes from *C. testosteroni* and *C. lunatus* were selected as the best choice to carry out our work.

Engineering heterologous 17β-HSDs in recombinant M. smegmatis strain

We have cloned the genes encoding the 17 β-HSDs from *C. testosteroni* and *C. lunatus* in pMV261, an *E. coli/M. smegmatis* shuttle vector, under the control of the constitutive *Phsp60* promoter, rendering the plasmids pHSDCT and pHSDCL respectively (see Experimental procedures). Both plasmids were transformed in both, the *M. smegmatis* wild-type strain and the AD-producing *M. smegmatis* MS6039-5941 mutant, generating four recombinant strains, i.e. *M. smegmatis* mc²155 (pHSDCT), *M. smegmatis* mc²155 (pHSDCL), *M. smegmatis* MS6039-5941 (pHSDCT) and *M. smegmatis* MS6039-5941 (pHSDCL) (Table 1).

Production of TS from AD by resting-cell biotransformation

The ability of the *M. smegmatis* (pHSDCT) and *M. smegmatis* (pHSDCL) recombinant strains to produce TS from AD was tested using a resting-cell assay. In the standard conditions (no additional carbon sources present in the reaction medium), small amount of TS was detected (Fig. 3). However, when the reaction medium was supplemented with 1% glucose, the substrate AD was efficiently transformed into TS using both strains (Fig. 3). The biotransformation was slightly more efficient using the recombinant bacteria harbouring the fungal gene [i.e. *M. smegmatis* (pHSDCL)]. When the reaction medium was supplemented with 1% glycerol instead of glucose, the biotransformation was slightly less efficient (Fig. 3). This result suggest that the intracellular NAD(P)$^+$/NAD(P)H ratio could be different in the presence of glucose or glycerol being critical for the process. In this sense, the effect of a carbon source supplementation and also the pH have been already demonstrated to be determinant in the reaction equilibrium for the TS production (Liu *et al.*, 1994; Llanes *et al.*, 1995; Liu and Lo, 1997; Egorova *et al.*, 2009).

The use of the recombinant *M. smegmatis* MS6039-5941 (pHSDCT) and *M. smegmatis* MS6039-5941 (pHSDCL) strains instead the wild-type recombinants does not provide any significant advantage in the resting-cell process because, as mentioned above, both the wild-type and MS6039-5941 mutant strains are unable to metabolize AD or TS.

Table 1. List of bacterial strains, plasmids and primers used in this study.

Strains or plasmids	Genotype and/or description	Source or reference
Strains		
Mycobacterium "smegmatis		
mc²155	*ept-1*, mc²6 mutant efficient for electroporation	Snapper *et al.* (1990)
MS6039-5941	mc²155 mutant Δ*MSMEG_6039* Δ*MSMEG_5941*	Galán *et al.* (unpublished)
mc²155 (pHSDCT)	mc²155 harbouring plasmid pHSDCT	This study
mc²155 (pHSDCL)	mc²155 harbouring plasmid pHSDCL	This study
MS6039-5941(pHSDCT)	MS6039-5941 harbouring plasmid pHSDCT	This study
MS6039-5941(pHSDCL)	MS6039-5941 harbouring plasmid pHSDCL	This study
Escherichia coli		This study
DH10B	F$^-$, *mcrA*, Δ (*mrrhsdRMS-mcrBC*), Φ80d*lacZ*ΔM15, Δ*lacX74*, *deoR*, *recA1*, *araD139*, Δ(*ara-leu*)7697, *galU*, *galK*, λ$^-$, *rpsL*, *endA1*, *nupG*	Invitrogen
DH10B (pUC57-17HSD)	DH10B strain harbouring plasmid pUC57-17HSD	This study
DH10B (pGEMT-HSDCT)	DH10B strain harbouring plasmid pGEMT-HSDCT	This study
DH10B (pHSDCT).	DH10B strain harbouring plasmid pHSDCT	This study
Comamonas testosteroni		
ATCC® 11996™		ATCC
Plasmids		
pMV261	*Mycobacterium/E. coli* shuttle vector with the kanamycin resistance *aph* gene from transposon Tn*903* and the promoter from the *hsp60* gene from *M. tuberculosis*	Stover *et al.* (1991)
pGEM®-T Easy	*E. coli* cloning vector; Amp^R; T7 and SP6 RNA polymerase promoters flanking a multiple cloning region within the α-peptide coding region of β-galactosidase for the identification of recombinants by blue/white screening	Promega
pGEMT-HSDCT	pGEMT-Easy harbouring the gene encoding the 17β-HSD from *C. testosteroni*	This study
pUC57-17HSD	pUC57 harbouring the synthetic gene encoding the 17β-HSD from *C. lunatus*	This study
pHSDCT	pMV261 harbouring the gene encoding the 17β-HSD from *C. testosteroni*	This study
pHSDCL	pMV261 harbouring the synthetic gene encoding the 17β-HSD from *C. lunatus*	This study
Primers		
HDHF	AGAGGAGATATACCATGGGCAGCAGCCATCATCATCATCATCACACAAATCG TTTGCAGGGTAAGG	This study
HDHR	AAGCTTCTATAGCCC CATGCCCAGAATCG	This study

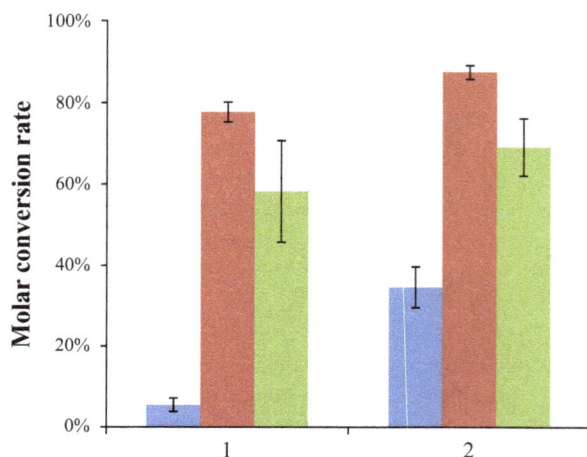

Fig. 3. Production of TS by resting-cell processes. The conversion of AD by the strains *Mycobacterium smegmatis* mc²155 (pHSDCT) (1) and *M. smegmatis* mc²155 (pHSDCL) (2) was tested in three culture conditions: no carbon source addition (blue), 1% glucose (red) and 1% glycerol (green). Average and standard deviation of two biological replicates at 24 h of culture are represented.

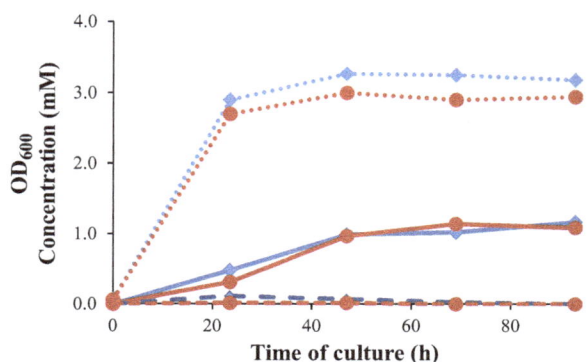

Fig. 4. Production of TS from cholesterol by growing-cell processes. The strains *Mycobacterium smegmatis* MS6039-5941 (pHSDCT) (red) and *M. smegmatis* MS6039-5941 (pHSDCL) (blue) were growth at minimal medium containing 18 mM glycerol (carbon and energy source) and 1.8 mM cholesterol (substrate). The AD concentration (continuous lines) and TS (dashed lines) and DO_{600} (dotted line) are represented. A representative experiment is shown.

Production of TS from sterols by growing-cells biotransformation

The ability to produce TS from natural sterols in a single biotransformation step was tested by growing *M. smegmatis* MS6039-5941 (pHSDCT) and *M. smegmatis* MS6039-5941 (pHSDCL) strains in minimal medium containing 18 mM glycerol as a main carbon and energy source and 1.8 mM cholesterol as a substrate. In this culture condition, the two recombinant strains transformed the cholesterol into AD but TS was only detected in small traces during the exponential growth phase (Fig. 4). Based on the resting-cell results, we tested the production of TS from sterols using 9 mM glucose instead of 18 mM glycerol as a carbon source obtaining similar results those using glycerol (data not shown). It is worth to mention that, in this culture condition, cell growth is only observed during the first 24 h due to the complete consumption of glycerol or glucose. After this time, the cholesterol is still transformed into AD but the cholesterol side-chain degradation might not supply enough carbon and/or energy to support cell division (Fig. 4). Moreover, the reduction of AD into TS is not efficient in nutrient-limited conditions and an alternative carbon source could be necessary for reactivating cell metabolism.

As the biotransformation of AD into TS appears to be dependent of the NAD(P)/NAD(P)H cofactor balance, we simulated a pseudo-resting-cell system carrying out two consecutive biotransformation steps at the same shake flask. The first step was carried out in the presence of 18 mM glycerol and 1.8 mM cholesterol. In these

conditions, we have observed that the cholesterol is mostly depleted and transformed into AD before 69 h of culture (data not shown). So at this moment, in a second step, we added glucose or glycerol and measured the production of TS after 24 h as it was done for the resting-cell assays. Using this approach, the two recombinant strains were able to transform sterols into TS more efficiently, although significant differences were observed probably due to kinetic differences between both 17β-HSD enzymes (Fig. 5). In the presence of 1% glucose, the average rate of TS to androstenes (AD and TS) was 77.6% and 28.6% using the strains MS6039-5941 (pHSDCL) and MS6039-5941 (pHSDCT) respectively. Similar results were obtained when glucose was used as an initial carbon source and glucose was added at 69 h of culture (data no shown). However, this average rate was slightly lower when an addition of glycerol instead of glucose was supplied to both strains (Fig. 5).

To test the influence of bacterial metabolic state on the TS production, other culture conditions were assayed. First, we tested the addition of glucose at the late-exponential growth phase (e.g. 23.5 h) (Fig. 6). Second, we added higher concentrations of carbon source at the beginning of the culture (e.g. 55.5 mM glucose (1% glucose) instead of 18 mM glycerol) (Fig. 6). In both culture conditions, the recombinant strains were able to produce TS from cholesterol but their behaviour was different. The strain MS6039-5941 (pHSDCL) produced higher TS yields and the reversion of TS to AD was not significant. However, the strain MS6039-5941 (pHSDCT) produced TS but in this case a notable reconversion of TS into AD was observed after 47 h of culture. The molar conversion rates and the TS to androstenes ratios are shown in Fig. 7. According to these results, the production of TS from sterols is achieved more efficiently

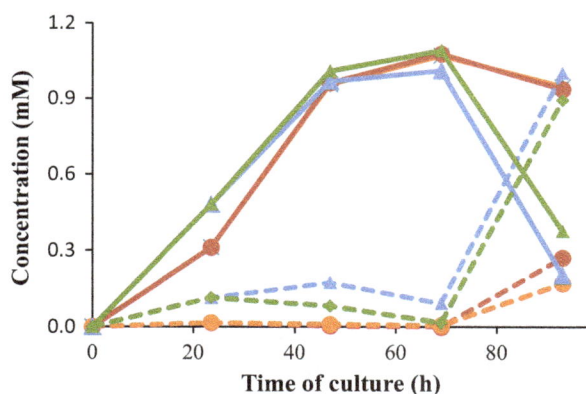

Fig. 5. Production of TS from cholesterol by a pseudo-resting-cell process. The strains *Mycobacterium smegmatis* MS6039-5941 (pHSDCT) and *M. smegmatis* MS6039-5941 (pHSDCL) were growth at minimal medium containing 18 mM glycerol (carbon and energy source) and 1.8 mM cholesterol (substrate) and, 1% glucose or glycerol was added at 69 h of culture. The following cultures were tested: MS6039-5941 (pHSDCT) with glucose (red); MS6039-5941 (pHSDCT) with glycerol (orange); MS6039-5941 (pHSDCL) with glucose (blue) and MS6039-5941 (pHSDCL) with glycerol (green). The AD concentration (continuous lines) and TS (dashed lines) are represented. A representative experiment is shown.

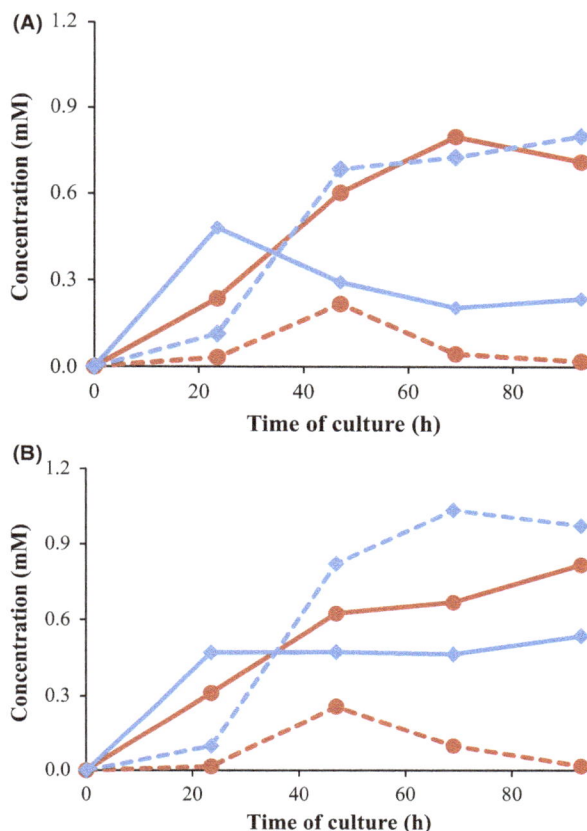

Fig. 6. Production of TS from cholesterol by growing-cell processes. The *Mycobacterium smegmatis* MS6039-5941 (pHSDCT) (red) and *M. smegmatis* MS6039-5941 (pHSDCL) (blue) strains were grown in minimal medium containing 1.8 mM cholesterol (substrate) and an alternative carbon source: (A) 18 mM glycerol with an addition of 1% glucose at 24 h; (B) 1% glucose without any addition The AD concentration (continuous lines) and TS (dashed lines) are represented. A representative experiment is shown

with the strain MS6039-5941 (pHSDCL) than with the strain MS6039-5941 (pHSDCT). Although the presence of an additional carbon source (i.e. glycerol or glucose) is required for the production of TS for both recombinant strains, the reconversion of TS into AD is not observed in the case of the fungal 17β-HSD-expressing strain once the carbon source has been consumed. This reversibility was described previously in other mycobacterial mutants producing TS from cholesterol (Liu and Lo, 1997). These results reinforce the use of the fungal enzyme as the best candidate to genetically engineer the recombinant strains to produce TS from sterols.

Concluding remarks

We have demonstrated that *M. smegmatis* is an excellent chassis to develop biotechnological processes for the biotransformation of sterols and their derivatives into valuable pharmaceutical compounds. The current genetic tools available to transform this organism have allowed for expressing stably two genes coding 17β-HSDs enzymes. Our recombinant strains were able to produce TS from AD and/or from sterols with high yields that are comparable with previously published data using mycobacterial strains obtained by conventional mutation procedures (Liu *et al.*, 1994; Liu and Lo, 1997). Moreover, our rational approach makes possible the introduction of heterologous 17β-HSD enzymes from diverse origins with different catalytic properties (e.g., substrate and cofactor specificity) offering more versatility than

conventional methods. Some of the 17β-HSD activities described in the conventional mutants appears to have different catalytic properties. In fact, in some mycobacterial strains, the double reduction of ADD [both of 17-keto group and 1(2)-double bound] was more effective for TS formation than a single reduction of 17-keto group of AD (Hung *et al.*, 1994; Egorova *et al.*, 2009).

This work represents a proof of concept of the possibilities of using this model bacteria and metabolic engineering approaches for the production of pharmaceutical steroids. However, additional efforts are needed to optimize the production of TS from sterols in a single biotransformation step. Two factors appear to be determinant for the improvement of this process: the reversibility of 17β-HSD enzymes and the cell metabolic state. To overcome the first of these factors, works to identify non-reversible 17β-HSDs and/or to modify the characterized enzymes by protein engineering can be explored. Several *in vitro* attempts have been already

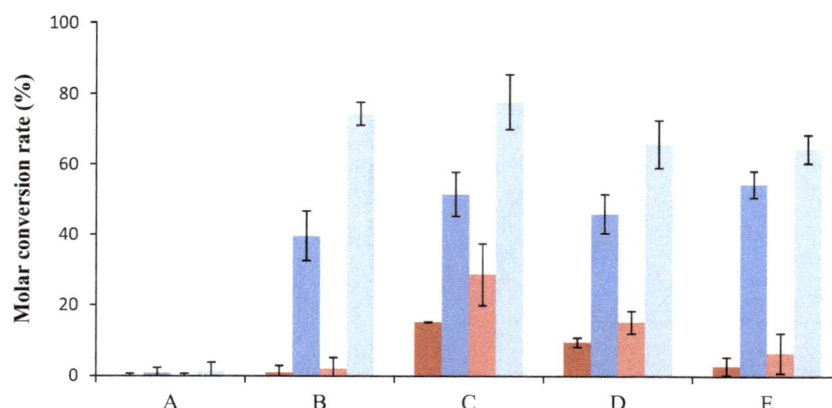

Fig. 7. Conversion rates of TS in growing-cell processes. The strains *Mycobacterium smegmatis* MS6039-5941 (pHSDCT) (red bars) and *M. smegmatis* MS6039-5941 (pHSDCL) (blue bars) were grown at minimal medium containing 1.8 mM cholesterol (substrate) and 18 mM glycerol (carbon and energy source). The following culture conditions were tested: (A) Without addition of any carbon source; (B) with addition of 1% glucose at 24 h; (C) with addition of 1% glucose at 69 h; (D) with addition of 1% glycerol at 69 h; (E) without addition and 1% glucose instead of glycerol as initial substrate. The molar conversion rate of TS (first and second bars) and the ratio of TS to androstenes (third and fourth bars) at the end of the culture (93 h) are shown. The molar conversion rate was calculated on the basis of 1.8 mM cholesterol added. Androstenes are calculated by adding AD and TS. Average and standard deviation of two biological replicates are represented.

done to design rationally 17β-HSD mutants from *C. testosteroni* and *C. lunatus* which present alterations in substrate specificity and/or coenzyme requirements, as well as improvements overall catalytic activity (Oppermann *et al.*, 1997; Kristan *et al.*, 2003, 2005, 2007a,b; Brunskole *et al.*, 2009; Svegelj *et al.*, 2012). To overcome the second factor, the control of bacterial metabolism by metabolic engineering and systems biology approaches would be useful. Factors such as the supplementation of carbon source, pH or mode of substrate addition, have been proved to be determinant for the reduction of AD to TS. Systems biology approaches could give more detailed information about microbial metabolism and how to increase TS yields.

Experimental procedures

Chemicals

4-Androstene-3,17-dione (AD) was purchased from TCI America. Chloroform, glucose and glycerol were purchased from Merck (Darmstardt, Germany). Methanol and acetonitrile of HPLC quality were purchased from Scharlau (Sentmenat, Spain). Cholesterol, TS, Tween 80, tyloxapol, ampicilin and kanamycin were from Sigma (Steinheim, Germany).

Bacterial strains, plasmids and culture conditions

The bacterial strains, plasmids and primers used in this work are listed in Table 1. *Mycobacterium smegmatis* strains were grown at 37°C in an orbital shaker at 200 rpm. Middlebrook 7H9 broth medium (Difco) supplemented with 0.4% glycerol and 0.05% Tween 80 was

used as rich medium. 7H9 broth without any supplement was used as minimal medium. 7H10 agar (Difco) plates supplemented with 10% albumin-dextrose-catalase (Becton Dickinson) were used for solid media. Kanamycin (20 μg ml^{-1}) was used for strain selection when appropriate.

Escherichia coli DH10B strain was used as host for cloning purposes. It was grown in LB medium at 37°C in an orbital shaker at 200 rpm. LB agar plates were also used for solid media. Ampicilin (100 μg ml^{-1}) or kanamycin (50 μg ml^{-1}) were used for plasmid selection and maintenance.

Comamonas testosteroni ATCC 11996 was grown in LB medium at 30°C in an orbital shaker at 200 rpm to extract genomic DNA.

DNA manipulations and sequencing

DNA manipulations and other molecular biology techniques were essentially as described by Sambrook and Russell (2001). Isolation of *C. testosteroni* genomic DNA was performed with the Bacteria Genomic Prep Mini Spin Kit (GE Healthcare). Oligonucleotides were purchased from Sigma-Aldrich. DNA amplification was performed on a Mastercycler Gradient (Eppendorf) using DNA polymerase I and Pfu polymerase from Biotools B. M. Labs. Reaction mixtures contained 1.5 mM MgCl$_2$ and 0.25 mM dNTPs. DNA fragments were purified with High Pure PCR System Product Purification Kit (Roche). Restriction enzymes were obtained from various suppliers and were used according to their specifications. Plasmid DNA was prepared with a High Pure Plasmid Isolation Kit (Roche Applied Science). *Escherichia coli* was transformed by the rubidium chloride method (Wirth et al., 1989). *Mycobacterium smegmatis* cells were transformed by electroporation (Gene Pulser; Bio-Rad) (Parish and Stoker, 1998). All cloned inserts and DNA fragments were confirmed by DNA sequencing through an ABI Prism 377 automated DNA sequencer (Applied Biosystems Inc.) at Secugen S.L. (Madrid, Spain).

Construction of plasmids pHSDCT and pHSDCL

To isolate the gene encoding the 3β/17β-HSD from *C. testosteroni* ATCC 11996, the genomic DNA was isolated and amplified the gene by PCR using the oligonucleotides HDHF and HDHR (Table 1). A sequence coding six histidines was inserted in the oligonucleotide HDHF, generating a novel gene version encoding a modified 17β-HSD from *C. testosteroni* that contains a polyhistidine-tag inserted in the N-terminal end. The amplified 812 bp fragment was cloned into pGEM®-T Easy (Promega) to generate the plasmid pGEMT-HSDCT using *E. coli* DH10B as host. The cloned fragment was further subcloned into the plasmid pMV261 able to replicate in *E. coli* and *Mycobacterium*. For this goal, the plasmid pGEMT-HSDCT was digested with EcoRI and HindIII and the fragment was ligated with the vector pMV261 cut with the same restriction enzymes generating the plasmid pHSDCT. This plasmid was transformed into *E. coli* DH10B to generate the recombinant strain *E. coli* DH10B (pHSDCT). The plasmid pHSDCT isolated from *E. coli* was transformed by electroporation into *M. smegmatis* mc²155 and the AD-producing mutant *M. smegmatis* MS6039-5941 competent cells, generating the recombinant strains *M. smegmatis* mc²-155 (pHSDCT) and *M. smegmatis* MS6039-5941 (pHSDCT) (Table 1).

The gene encoding the 17β-HSD from *C. lunatus* was chemically synthesized with an optimized codon usage for its expression in *Mycobacterium* (ATG: biosynthetics GmbH, Germany) (Fig. 8). The codon optimization was carried out using the program OPTI-MIZER (Puigbò et al., 2007) and the codon usage table found in KDRI (Kazusa DNA Research Institute, Japan). The synthetic gene was supplied cloned into the EcoRV site of the commercial vector pUC57 generating the plasmid pUC57-17HSD. This plasmid was transformed into *E. coli* DH10B generating the recombinant strain *E. coli* DH10B (pUC57-17HSD). The fragment containing the synthetic gene encoding the 17β-HSD was subcloned into the plasmid pMV261. For this purpose, the pUC57-17HSD plasmid was digested with EcoRI and HindIII and the fragment was ligated to the pMV261 vector digested with the same restriction enzymes. Thus, we created the plasmid pHSDCL which was transformed into *E. coli* DH10B to generate the recombinant *E. coli* DH10B (pHSDCL). The plasmid pHSDCL extracted from *E. coli* was transformed by electroporation into *M. smegmatis* mc²155 and *M. smegmatis* MS6039-5941 competent cells to generate the recombinant strains *M. smegmatis* mc²155 (pHSDCL) and *M. smegmatis* MS6039-5941 (pHSDCL) respectively.

```
   BamHI        EcoRI              RBS              NdeI
CGGGATCCCGGAATTCTGACCTAAGGAGGTGAATCATATGCCGCACGTGG
AGAACGCCTCGGAGACCTACATCCCGGGCCGCCTGGACGGCAAGGTGGCC
CTGGTGACCGGCTCGGGCCGCGGCATCGGCGCCGCCGTGGCCGTGCACCT
GGGCCGCCTGGGCGCCAAGGTGGTGGTGAACTACGCCAACTCGACCAAGG
ACGCCGAGAAGGTGGTGTCGGAGATCAAGGCCCTGGGCTCGGACGCCATC
GCCATCAAGGCCGACATCCGCCAGGTGCCGGAGATCGTGAAGCTGTTCGA
CCAGGCCGTGGCCCACTTCGGCCACCTGGACATCGCCGTGTCGAACTCGG
GCGTGGTGTCGTTCGGCCACCTGAAGGACGTGACCGAGGAGGAGTTCGAC
CGCGTGTTCTCGCTGAACACCCGCGGCCAGTTCTTCGTGGCCCGCGAGGC
CTACCGCCACCTGACCGAGGGCGGCCGCATCGTGCTGACCTCGTCGAACA
CCTCGAAGGACTTCTCGGTGCCGAAGCACTCGCTGTACTCGGGCTCGAAG
GGCGCCGTGGACTCGTTCGTGCGCATCTTCTCGAAGGACTGCGGCGACAA
GAAGATCACCGTGAACGCCGTGGCCCCCGGGCGGCACCGTGACCGACATGT
TCCACGAGGTGTCGCACCACTACATCCCGAACGGCACCTCGTACACCGCC
GAGCAGCGCCAGCAGATGGCCGCCCACGCCTCGCCGCTGCACCGCAACGG
CTGGCCGCAGGACGTGGCCAACGTGGTGGGCTTCCTGGTGTCGAAGGAGG
GCGAGTGGGTGAACGGCAAGGTGCTGACCCTGGACGGCGGCGCCGCCTAA
AAGCTTCCCGTCGACACTAGTC
HindIII      SalI    SpeI
```

Fig. 8. Gene sequence encoding the 17β-HSD from *C. lunatus* chemically synthesized with an optimized codon usage for its expression in *Mycobacterium*. The ribosome-binding site (RBS), start codon and stop codon are underlined. Several recognition sites by restriction enzymes were incorporated to facilitate subcloning tasks.

Resting-cell biotransformations

The recombinant strains were grown in rich medium at 37°C during 24 h. The cells were harvested by centrifugation at $5000 \times g$ for 20 min at 4°C and washed once with 0.85% NaCl. The biotransformation was carried out with an optical density (OD_{600}) of 18 in a 100 ml shake flask containing 40 ml of reaction mixture: 0.1 M phosphate buffer (pH 8.0), 2 mM AD (substrate) and 0.05% Tween 80. An additional carbon source was added in some cases. AD was incorporated into the medium as a solution with randomly methylated β-cyclodextrin (1:10.3, molar ratio) (Klein et al., 1995).

Growing-cell biotransformations

The precultures of the recombinant strains were grown in rich medium at 37°C during 24 h and used to inoculate 30 ml of 7H9 minimal medium containing 1.8 mM cholesterol (substrate), 18 mM glycerol (carbon and energy source), 3.6% tyloxapol and 20 μg ml^{-1} kanamycin. Different concentrations of glucose instead of glycerol were also tested. The cholesterol was dissolved in 10% tyloxapol prior to its addition to the minimal medium. Due to the low solubility of this steroid, a stock solution was warmed at 80°C in agitation, sonicated in a bath for 1 h and then autoclaved. The cultures were grown in 100 ml shake flasks at 37°C in an orbital shaker at 200 rpm. An addition of 1% glucose or 1% glycerol was added at different times of culture.

Analytical methods

The culture broth was extracted with chloroform (0.5:1, v/v) twice. The organic phase was evaporated and the residue was dissolved in acetonitrile. AD and TS were determined by reversed-phase HPLC using a Teknokroma mediterranea™ Sea$_{18}$ column (15 cm × 0.46 cm; 5 μM) and UV detection at 240 nm. Mobile phase was composed of methanol and water (75/25, v/v), flow rate 0.85 ml min^{-1}. AD and TS were used as standards. The conversion rate of TS was calculated on the basis of cholesterol added to the medium (growing-cell biotransformations) or AD measured into the sample (resting-cell biotransformations).

Bioinformatic analysis

Sequence alignments were carried out using CLUSTAL W (Thompson et al., 1994) and different BLAST algorithms from the National Centre of Biotechnology Information Server (NCBI) were also used.

Acknowledgements

The technical work of A. Valencia is greatly appreciated. This work was supported by grants from the Ministry of Science and Innovation (BFU2006-15214-C03-01, BFU2009-11545-C03-03) and Ministry of Economy and Competitiveness (BIO2012-39695-C02-01). L.F.C. was supported by a fellowship from the Spanish Ministry of Education, Culture and Sports.

Conflict of interest

None declared.

References

Abalain, J.H., Di Stefano, S., Amet, Y., Quemener, E., Abalain-Colloc, M.L., and Floch, H.H. (1993) Cloning, DNA sequencing and expression of (3-17)β hydroxysteroid dehydrogenase from Pseudomonas testosteroni. J Steroid Biochem Mol Biol 44: 133–139.

Benach, J., Knapp, S., Oppermann, U.C., Hägglund, O., Jörnvall, H., and Ladenstein, R. (1996) Crystallization and crystal packing of recombinant 3 (or 17) β-hydroxysteroid dehydrogenase from Comamonas testosteroni ATTC 11996. Eur J Biochem 236: 144–148.

Benach, J., Filling, C., Oppermann, U.C., Roversi, P., Bricogne, G., Berndt, K.D., et al. (2002) Structure of bacterial 3β/17β-hydroxysteroid dehydrogenase at 1.2 A resolution: a model for multiple steroid recognition. Biochemistry 41: 14659–14668.

Bogovich, K., and Payne, A.H. (1980) Purification of rat testicular microsomal 17-ketosteroid reductase. J Biol Chem 255: 5552–5559.

Borrego, S., Espinosa, E.E., Martí, E., and Fonseca, M. (2000) Conversion of cholesterol to testosterone by Mycobacterium sp. MB-3638. Revista CENIC Ciencias Biológicas 31: 17–20.

Brunskole, M., Kristan, K., Stojan, J., and Rižner, T.L. (2009) Mutations that affect coenzyme binding and dimer formation of fungal 17β-hydroxysteroid dehydrogenase. Mol Cell Endocrinol 301: 47–50.

Cabrera, J.E., Pruneda Paz, J.L., and Genti-Raimondi, S. (2000) Steroid-inducible transcription of the 3β/17β-hydroxysteroid dehydrogenase gene (3β/17β-hsd) in Comamonas testosteroni. J Steroid Biochem Mol Biol 73: 147–152.

Cassetta, A., Büdefeld, T., Rižner, T.L., Kristan, K., Stojan, J., and Lamba, D. (2005) Crystallization, X-ray diffraction analysis and phasing of 17β-hydroxysteroid dehydrogenase from the fungus Cochliobolus lunatus. Acta Crystallogr Sect F Struct Biol Cryst Commun 61: 1032–1034.

Chang, Y.H., Wang, Y.L., Lin, J.Y., Chuang, L.Y., and Hwang, C.C. (2010) Expression, purification, characterization of a human recombinant 17beta-hydroxysteroid dehydrogenase type 1 in Escherichia coli. Mol Biotechnol 44: 133–139.

Długoński, J., and Wilmańska, D. (1998) Deleterious effects of androstenedione on growth and cell morphology of

Schizosaccharomyces pombe. Antonie Van Leeuwenhoek
73: 189–194.

Dlugovitzky, D.G., Fontela, M.S., Martinel Lamas, D.J., Valdez, R.A., and Romano, M.C. (2015) *Mycobacterium smegmatis* synthesizes in vitro androgens and estrogens from different steroid precursors. *Can J Microbiol* **61:** 451–455.

Donova, M.V., Egorova, O.V., and Nikolayeva, V.M. (2005) Steroid 17β-reduction by microorganism – a review. *Process Biochem* **40:** 2253–2262.

Egorova, O.V., Nikolayeva, V.M., and Donova, M.V. (2002a) 17-Hydroxysteroid dehydrogenases of *Mycobacterium* sp. VKM Ac-1815D mutant strain. *J Steroid Biochem Mol Biol* **81:** 273–279.

Egorova, O.V., Gulevskaya, S.A., Puntus, I.F., Filonov, A.E., and Donova, M.V. (2002b) Production of androstenedione using mutants of *Mycobacterium* sp. *J Chem Tech Biotech* **77:** 141–147.

Egorova, O.V., Nikolayeva, V.M., Suzina, N.E., and Donova, M.V. (2005) Localization of 17β-hydroxysteroid dehydrogenase in *Mycobacterium* sp. VKM Ac-1815D mutant strain. *J Steroid Biochem Mol Biol* **94:** 519–525.

Egorova, O.V., Nikolayeva, V.M., Sukhodolskaya, G.V., and Donova, M.V. (2009) Transformation of C_{19}-steroids and testosterone production by sterol-transforming strains of *Mycobacterium* sp. *J Mol Catal B Enzym* **57:** 198–203.

Ercoli, A., and Ruggierii, P.D. (1953) An improved method of preparing testosterone, dihydrotestosterone and some of their esters. *J Am Chem Soc* **75:** 650–653.

Fernández de las Heras, L., García-Fernández, E., Navarro Llorens, M.J., Perera, J. and Drzyzga, O. (2009) Morphological, physiological, and molecular characterization of a newly isolated steroid-degrading actinomycete, identified as *Rhodococcus ruber* strain Chol-4. *Curr Microbiol* **59:** 548–553.

Fogal, S., Bergantino, E., Motterle, R., Castellin, A. and Arvotti, A. (2013) Process for the preparation of testosterone. Patent US 2013/8592178B2.

Genti-Raimondi, S., Tolmasky, M.E., Patrito, L.C., Flury, A., and Actis, L.A. (1991) Molecular cloning and expression of the β-hydroxysteroid dehydrogenase gene from *Pseudomonas testosteroni*. *Gene* **105:** 43–49.

Goren, T., Harnik, M., Rimon, S., and Aharonowitz, Y. (1983) 1-Ene-steroid reductase of *Mycobacterium* sp. NRRL B-3805. *J Steroid Biochem* **19:** 1789–1797.

Hamada, H., and Kawabe, S. (1991) Biotransformation of 4-androstene-3,17-dione by green cell suspension of *Marchantia polymorpha*: stereoselective reduction at carbon 17. *Life Sci* **48:** 613–615.

Hung, B., Falero, A., Llanes, N., Pérez, C., and Ramírez, M.A. (1994) Testoterone as biotransformation product in steroid conversion by *Mycobacterium* sp. *Biotechnol Lett* **16:** 497–500.

Klein, U., Gimpl, G., and Fahrenholz, F. (1995) Alteration of the myometrial plasma membrane cholesterol content with b-cyclodextrin modulates the binding affinity of the oxytocin receptor. *Biochemistry* **34:** 13784–13793.

Kristan, K., and Rižner, T.L. (2012) Steroid-transforming enzymes in fungi. *J Steroid Biochem Mol Biol* **129:** 79–91.

Kristan, K., Rižner, T.L., Stojan, J., Gerber, J.K., Kremmer, E., and Adamski, J. (2003) Significance of individual amino acid residues for coenzyme and substrate specificity of 17β-hydroxysteroid dehydrogenase from the fungus *Cochliobolus lunatus*. *Chem Biol Interact* **143–144:** 493–501.

Kristan, K., Stojan, J., Möller, G., Adamski, J., and Rižner, T.L. (2005) Coenzyme specificity in fungal 17beta-hydroxysteroid dehydrogenase. *Mol Cell Endocrinol* **241:** 80–87.

Kristan, K., Adamski, J., Rižner, T.L., and Stojan, J. (2007a) His164 regulates accessibility to the active site in fungal 17β-hydroxysteroid dehydrogenase. *Biochimie* **89:** 63–71.

Kristan, K., Stojan, J., Adamski, J., and Lanisnik Rižner, T. (2007b) Rational design of novel mutants of fungal 17β-hydroxysteroid dehydrogenase. *J Biotechnol* **129:** 123–130.

Lefebvre, Y.A., Schultz, R., Groman, E.V., and Watanabe, M. (1979) Localization of 3 β and 17 β-hydroxysteroid dehydrogenase in *Pseudomonas testosteroni*. *J Steroid Biochem* **10:** 523–528.

Li, W., Ge, F., Zhang, Q., Ren, Y., Yuan, J., He, J., *et al.* (2014) Identification of gene expression profiles in the actinomycete *Gordonia neofelifaecis* grown with different steroids. *Genome* **57:** 345–353.

Liu, W.H., and Lo, C.K. (1997) Production of testosterone from cholesterol using a single-step microbial transformation of *Mycobacterium* sp. *J Ind Microbiol Biotechnol* **19:** 269–272.

Liu, W.H., Kuo, C.W., Wu, K.L., Lee, C.Y., and Hsu, W.Y. (1994) Transformation of cholesterol to testosterone by *Mycobacterium* sp. *J Ind Microbiol* **13:** 167–171.

Llanes, N., Hung, B., Falero, A., Pérez, C., and Aguila, B. (1995) Glucose and lactose effect on AD and ADD bioconversion by *Mycobacterium* sp. *Biotechnol Lett* **17:** 1237–1240.

Lo, C.K., Pan, C.P. and Liu, W.H. (2002) Production of testosterone from phytosterol using a single-step microbial transformation by a mutant of *Mycobacterium* sp. *J Ind Microbiol Biotechnol* **28:** 280–283.

Marchais-Oberwinkler, S., Henn, C., Möller, G., Klein, T., Negri, M., Oster, A., *et al.* (2011) 17β-Hydroxysteroid dehydrogenases (17β-HSDs) as therapeutic targets: protein structures, functions, and recent progress in inhibitor development. *J Steroid Biochem Mol Biol* **125:** 66–82.

Mei, G., Lei, F., Liang-Fei, L., Dai-Jie, C. and Xìng, X. (2005) *Mycobacterium fortuitum* and its use in production of testosterone by conversion of microbe. Patent CN1670185 A.

Minard, P., Legoy, M.D. and Thomas, D. (1985) 3 β, 17 β-hydroxysteroid dehydrogenase of *Pseudomonas testosteroni*. Kinetic evidence for the bifunctional activity at a common catalytic site. *FEBS Lett* **188:** 85–90.

Moeller, G., and Adamski, J. (2006) Multifunctionality of human 17β-hydroxisteroid dehydrogenases. *J Steroid Biochem Mol Biol* **125:** 66–82.

Moeller, G., and Adamski, J. (2009) Integrated view on 17-beta-hydroxisteroid dehydrogenases. *Mol Cell Endocrinol* **25:** 7–19.

Oppermann, U.C., Filling, C., Berndt, K.D., Persson, B., Benach, J., Ladenstein, R., and Jörnvall, H. (1997) Active site directed mutagenesis of 3 β/17 β-hydroxysteroid dehydrogenase establishes differential effects on short-chain dehydrogenase/reductase reactions. *Biochemistry* **36:** 34–40.

Pajic, T., Vitas, M., Zigon, D., Pavko, A., Kelly, S.L., and Komel, R. (1999) Biotransformation of steroids by the fission yeast *Schizosaccharomyces pombe*. *Yeast* **15:** 639–645.

Parish, T., and Stoker, N.G. (1998) Electroporation of mycobacteria. *Methods Mol Biol* **101:** 129–144.

Payne, D.W., and Talalay, P. (1985) Isolation of novel microbial 3 alpha-, 3 beta-, and 17 beta-hydroxysteroid dehydrogenases. Purification, characterization, and analytical applications of a 17 beta-hydroxysteroid dehydrogenase from an *Alcaligenes* sp. *J Biol Chem* **260:** 13648–13655.

Peltoketo, H., Luu-The, V., Simard, J. and Adamski, J. (1999) 17β-Hydroxysteroid dehydrogenase (HSD)/17-ketosteroid reductase (KSR) family; nomenclature and main characteristics of the 17HSD/KSR enzymes. *J Mol Endocrinol* **23:** 1–11.

Plemenitas, A., Zakelj-Mayric, M., and Komel, R. (1988) Hydroxysteroid dehydrogenase of *Cochliobolus lunatus*. *J Steroid Biochem* **29:** 371–372.

Puigbò, P., Guzmán, E., Romeu, A., and Garcia-Vallvé, S. (2007) OPTIMIZER: a web server for optimizing the codon usage of DNA sequences. *Nucl Acids Res* **35:** W126–W131.

Rižner, T.L., and Zakelj-Mavric, M. (2000) Characterization of fungal 17β-hydroxysteroid dehydrogenases. *Comp Biochem Physiol B Biochem Mol Biol* **127:** 53–63.

Rižner, T.L., Zakelj-Mavric, M., Plemenitas, A., and Zorko, M. (1996) Purification and characterization of 17β-hydroxysteroid dehydrogenase from the filamentous fungus *Cochliobolus lunatus*. *J Steroid Biochem Mol Biol* **59:** 205–214.

Rižner, T.L., Moeller, G., Thole, H.H., Zakelj-Mavric, M., and Adamski, J. (1999) A novel 17beta-hydroxysteroid dehydrogenase in the fungus *Cochliobolus lunatus*: new insights into the evolution of steroid-hormone signalling. *Biochem* **337:** 425–431.

Rižner, T.L., Adamski, J., and Stojan, J. (2000) 17β-hydroxysteroid dehydrogenase from *Cochliobolus lunatus*: model structure and substrate specificity. *Arch Biochem Biophys* **15:** 255–262.

Rižner, T.L., Stojan, J., and Adamski, J. (2001a) 17β-hydroxysteroid dehydrogenase from the fungus *Cochliobolus lunatus*: structural and functional aspects. *Chem Biol Interact* **130–132:** 793–803.

Rižner, T.L., Stojan, J., and Adamski, J. (2001b) Searching for the physiological function of 17β-hydroxysteroid dehydrogenase from the fungus *Cochliobolus lunatus*: studies of substrate specificity and expression analysis. *Mol Cell Endocrinol* **171:** 193–198.

Sambrook, J., and Russell, D.W. (2001) *Molecular Cloning: A Laboratory Manual*, 3rd edn. New York, NY, USA: Cold Spring Harbor Laboratory Press.

Sarmah, U., Roy, M.K., and Singh, H.D. (1989) Steroid transformation by a strain of *Arthrobacter oxydans* incapable of steroid ring degradation. *J Basic Microbiol* **29:** 85–92.

Schultz, R.M., Groman, E.V., and Engel, L.L. (1977) 3(17)β-Hydroxysteroid dehydrogenase of *Pseudomonas testosteroni*. A convenient purification and demonstration of multiple molecular forms. *J Biol Chem* **252:** 3775–3783.

Singer, Y., Shity, H., and Bar, R. (1991) Microbial transformations in a cyclodextrin medium. Part 2. Reduction of androstenedione to testosterone by *Saccharomyces cerevisiae*. *Appl Microbiol Biotechnol* **35:** 731–737.

Smith, M., Zahnley, J., Pfeifer, D., and Goff, D. (1993) Growth and cholesterol oxidation by *Mycobacterium* species in Tween 80 medium. *Appl Environ Microbiol* **59:** 1425–1429.

Snapper, S.B., Melton, R.E., Mustafa, S., Kieser, T., and Jacobs, W.R. Jr (1990) Isolation and characterization of efficient plasmid transformation mutants of *Mycobacterium smegmatis*. *Mol Microbiol* **4:** 1911–1919.

Stover, C.K., de la Cruz, V.F., Fuerst, T.R., Burlein, J.E., Benson, L.A., Bennett, L.T., *et al.* (1991) New use of BCG for recombinant vaccines. *Nature* **351:** 456–460.

Svegelj, M.B., Stojan, J., and Rižner, T.L. (2012) The role of Ala231 and Trp227 in the substrate specificities of fungal 17β-hydroxysteroid dehydrogenase and trihydroxynaphthalene reductase: steroids versus smaller substrates. *J Steroid Biochem Mol Biol* **129:** 92–98.

Tamaoka, J., Ha, D.M., and Komagata, K. (1987) Reclassification of *Pseudomonas acidovorans* den Dooren de Jong 1926 and *Pseudomonas testosteroni* Marcus and Talahay 1956 as *Comamonas acidovorans* comb. nov. and *Comamonas testosteroni* comb. nov. with an emended description of the genus *Comamonas*. *Int J Syst Bacteriol* **37:** 52–59.

Thompson, J.D., Higgins, D.G., and Gibson, T.J. (1994) CLUSTAL W: improving the sensitivity of progressive multiple sequence alignment through sequence weighting, position-specific gap penalties and weight matrix choice. *Nucl Acids Res* **22:** 4673–4680.

Ulrih, N.P., and Lanisnik Rižner, T. (2006) Conformational stability of 17 β-hydroxysteroid dehydrogenase from the fungus *Cochliobolus lunatus*. *FEBS J* **273:** 3927–3937.

Wang, K.C., Gau, C.S., and Chen, R.L. (1982) Microbial oxidation of sterol. I. Conversion of cholesterol and sitosterol to 17-hydroxy steroids. *J Taiwan Pharm Assoc* **34:** 129–137.

Ward, O.P., and Young, C.S. (1990) Reductive biotransformations of organic compounds by cells or enzymes of yeast. *Enzyme Microb Technol* **12:** 482–493.

Wirth, R., Friesenegger, A., and Fieldler, S. (1989) Transformation of various species of gram-negative bacteria belonging to 11 different genera by electroporation. *Mol Gen Genet* **216:** 175–177.

Xu, L.Q., Liu, Y.J., Yao, K., Liu, H.H., Tao, X.Y., Wang, F.Q., and Wei, D.Z. (2016) Unraveling and engineering the production of 23,24-bisnorcholenic steroids in sterol metabolism. *Sci Rep* **6:** 21928.

Yin, S.J., Vagelopoulos, N., Lundquist, G., and Jörnvall, H. (1991) *Pseudomonas* 3 β-hydroxysteroid dehydrogenase. Primary structure and relationships to other steroid dehydrogenases. *Eur J Biochem* **197:** 359–365.

Zorko, M., Gottlieb, H.E., and Zakelj-Mavric, M. (2000) Pluripotency of 17β-hydroxysteroid dehydrogenase from the filamentous fungus *Cochliobolus lunatus*. *Steroids* **65:** 46–53.

13

Biosynthesis of micro- and nanocrystals of Pb (II), Hg (II) and Cd (II) sulfides in four *Candida* species: a comparative study of *in vivo* and *in vitro* approaches

Mayra Cuéllar-Cruz,[1,*] Daniela Lucio-Hernández,[1] Isabel Martínez-Ángeles,[2] Nicola Demitri,[3] Maurizio Polentarutti,[3] María J. Rosales-Hoz[4] and Abel Moreno[2,**]

[1]*Departamento de Biología, División de Ciencias Naturales y Exactas, Campus Guanajuato, Universidad de Guanajuato, Noria Alta S/N, Col. Noria Alta, C.P. 36050, Guanajuato, México.*
[2]*Departamento de Química de Biomacromoléculas, Instituto de Química, Universidad Nacional Autónoma de México, Av. Universidad 3000, Ciudad Universitaria, Ciudad de México, 04510, México.*
[3]*Elettra – Sincrotone Trieste, S.S. 14 km 163.5 in Area Science Park, 34149, Basovizza – Trieste, Italy.*
[4]*Departamento de Química, Centro de Investigación y de Estudios Avanzados del I.P.N., Apdo. Postal 14-740, 07000, México, D.F., México.*

Summary

Nature produces biominerals (biogenic minerals) that are synthesized as complex structures, in terms of their physicochemical properties. These biominerals are composed of minerals and biological macromolecules. They are produced by living organisms and are usually formed through a combination of chemical, biochemical and biophysical processes. Microorganisms like *Candida* in the presence of heavy metals can biomineralize those metals to form microcrystals (MCs) and nanocrystals (NCs). In this work, MCs and NCs of PbS, HgS or $HgCl_2$ as well as CdS are synthesized both *in vitro* (gels) and *in vivo* by four *Candida* species. Our *in vivo* results show that, in the presence of Pb^{2+}, *Candida* cells are able to replicate and form extracellular PbS MCs, whereas in the presence of Hg^{2+} and Cd^{2+}, they did synthesize intercellular MCs from HgS or $HgCl_2$ and CdS

NCs respectively. The MCs and NCs biologically obtained in *Candida* were compared with those PbS, HgS and CdS crystals synthetically obtained *in vitro* through the gel method (grown either in agarose or in sodium metasilicate hydrogels). This is, to our knowledge, the first time that the biosynthesis of the various MCs and NCs (presented in several species of *Candida*) has been reported. This biosynthesis is differentially regulated in each of these pathogens, which allows them to adapt and survive in different physiological and environmental habitats.

For Correspondence. *E-mail mcuellar@ugto.mx (M.C.C.). **E-mail carcamo@unam.mx

Funding Information
Universidad de Guanajuato (Grant/Award Number: 'Proyecto-Institucional-831/2016').

Introduction

Natural gemstones (sapphires, diamonds, emeralds, rubies) and minerals (calcite, quartz, pyrite, galena), grown under Earth's surface for millions of years, are remarkable for their shining beauty, fascinating colours, extraordinary geometry and shape. These extraordinary characteristics have enticed men into wearing them as luxurious ornaments, amulets or even into using them in ancient medicine or in high-technology applications as we do nowadays. However, the most extraordinary minerals, in terms of technological applications and biomedical properties, are those forming complex structures that appear in biological systems. These are called biominerals, as they are composed of minerals and biological macromolecules and are produced by living organisms in the formation of bones, teeth, stromatolites, eggshells, nacre shells, among others. Biomineralization studies the properties, structure and formation of inorganic solids deposited in biological systems, and it is also relevant to the Earth's environmental and evolution processes on practically all scales. The impact of biomineralization is recorded on a global scale and stretches far back in the history of life (Mann and Weiner, 1999). There is one approach that helps to emulate nature in growing these extraordinary minerals. This approach is called the crystal growth in gel technique. This gel technique favours the growing of crystals as instantaneous mineralogy. This mean that, where nature takes millions of years to grow those beautiful minerals, we will be able to do the same in just a few months, weeks or even days (Henisch and Garciaruiz, 1986a,b; Garciaruiz, 1991). On the other hand, the most exciting crystal growth method of technologically and biomedical important materials in the form of microcrystals (MCs), nanocrystals (NCs) or as nanoparticles (NPs) are those biogenic minerals produced by different microorganisms from bacteria to fungi. However, these minerals of biological origin are sometimes difficult to grow *in vitro* that is probably why they have never being synthesized in a laboratory (Mann and Ozin, 1996). It is through biomineralization that these organisms synthesize MCs, NCs or nanoparticles (NPs). These MCs or NCs were reported to have properties of semiconductor quantum crystallites (Williams *et al.*, 1996). Microcrystals and nanocrystals (MNCs) are usually formed through a combination of chemical, biochemical and biophysical processes. Biosynthesis of MCs or NCs by microorganisms is therefore a great alternative to conventional systems of crystal growth, as MCs and NCs produced through biosynthesis are relatively easy to reproduce and more stable compared with those synthesized *in vitro* (Inouye *et al.*, 1982; Borrelli *et al.*, 1987; Herron *et al.*, 1989; Williams *et al.*, 1996). In addition, stable MCs and NCs have several useful applications in different areas. In nanotechnology, for instance, they can be used as sensors (Kowshik *et al.*, 2002; Krumov *et al.*, 2007), they can also be used as quantum semiconductors (Williams *et al.*, 1996), as markers, in biological systems (Krumov *et al.*, 2007) and also in health care, cosmetics, chemical industries and space (Korbekandi *et al.*, 2014), as well as in environmental remediation due to their ability to sequester and detoxify intracellular cadmium ions (Dameron *et al.*, 1989). Other studies have reported that yeasts such as *Saccharomyces cerevisiae* (Prasad and Jha, 2010), *Schizosaccharomyces pombe* and *C. glabrata* are good producers of NCs in the presence of heavy metals and peptides with the general structure (γ-Glu-Cys)$_n$-Gly; Dameron *et al.*, 1989). Even though *C. glabrata* has been used in the biosynthesis of CdS NPs, not other *Candida* species so far have been reported capable of biosynthesizing NCs like this yeast, nor has it been reported whether these microorganisms are capable of producing MCs and/or NCs in the presence of other chemical elements such as mercury (II) or lead (II). MCs formed in the presence of these elements are of special interest, because they can be used to monitor the concentration of these heavy metals in contaminated environments (Pham *et al.*, 2015). Furthermore, it has also been reported that HgS NCs are technologically important materials (Jeong *et al.*, 2014). However, their instability has hampered their usefulness (Yang *et al.*, 2015). NCs chemically synthesized from Cd:HgS/CdS have proved to be stable and highly fluorescent compared with the HgS NCs also obtained by chemical synthesis (Yang *et al.*, 2015). Other research groups have used other synthesis strategies for AgS NCs, but all with a high degree of

complexity (Han *et al.*, 2014; Jeong *et al.*, 2014). The PbS NCs were chemically synthesized, and like HgS NCs, the chemical synthesis was also complex and needed to be well controlled *in vitro* (Lim *et al.*, 2014; Lee *et al.*, 2016). Obtaining stable MCs or NCs from PbS is of special interest due to their widespread use in photo detectors and solar cells (Lim *et al.*, 2014; Ko *et al.*, 2016; Lee *et al.*, 2016). Although NCs have been synthesized *in vitro* in the presence of mercury (II) and lead (II), our working group is interested in synthesizing MCs or NCs of PbS, HgS and CdS not just *in vitro* but also *in vivo*, using gels and microorganisms of *Candida* respectively. This method will allow us to obtain MNCs with high reproducibility and stability, thus enhancing their number of applications in future.

The goal of this study was to evaluate the formation of MCs or NCs of lead (II), mercury (II) and cadmium (II) sulfide *in vitro* (gels) and *in vivo* in four species of *Candida* (*C. albicans*, *C. glabrata*, *C. krusei* and *C. parapsilosis*).

Results and discussion

Growth of Candida *is inhibited in the presence of Hg^{2+} and Cd^{2+}, but not in the presence of Pb^{2+}*

In order to determine how *C. albicans*, *C. glabrata*, *C. krusei* and *C. parapsilosis* respond to Pb^{2+}, Hg^{2+} or Cd^{2+}, as the first step in the evaluation of their capacity to synthesize MCs or NCs, the cells were exposed to each of these metals. We observed that these four species of *Candida* were able to replicate in the presence of Pb^{2+}. However, no growth/replication was observed in the presence of Hg^{2+} or Cd^{2+}. In order to corroborate that the growth is indeed affected by the presence of Hg^{2+} or Cd^{2+}, but not for Pb^{2+}, we had to test the susceptibility of these yeasts through serial dilutions and growth curves. For the purpose of the susceptibility testing, cells were placed in the presence of metals at different concentrations, for 90 min, as described in the methods section. A representative figure of the results obtained with two species *Candida* is shown (see Fig. 1A–C). The cells of *C. albicans*, *C. glabrata*, *C. krusei* and *C. parapsilosis* in the presence of Pb^{2+} are able to withstand up to 2.0 mM of the metals (Fig. 1A, Table S1). *C. albicans* cells can grow (in the presence of Hg^{2+}) up to 2.0 mM, while *C. glabrata*, *C. krusei* and *C. parapsilosis* can only grow up to 1.0 mM in the presence of this metal (Fig. 1B, Table S1). In the presence of Cd^{2+}, the cell growth in the four species of *Candida* is observed up to 2.0 mM, but this growth is considerably affected compared with control cells (Fig. 1C, Table S1). Altogether, these tests show that the four species of *Candida* are able to withstand 1.0 mM of Pb^{2+}, Cd^{2+} or Hg^{2+} for 90 min. Although all *Candida* species are able

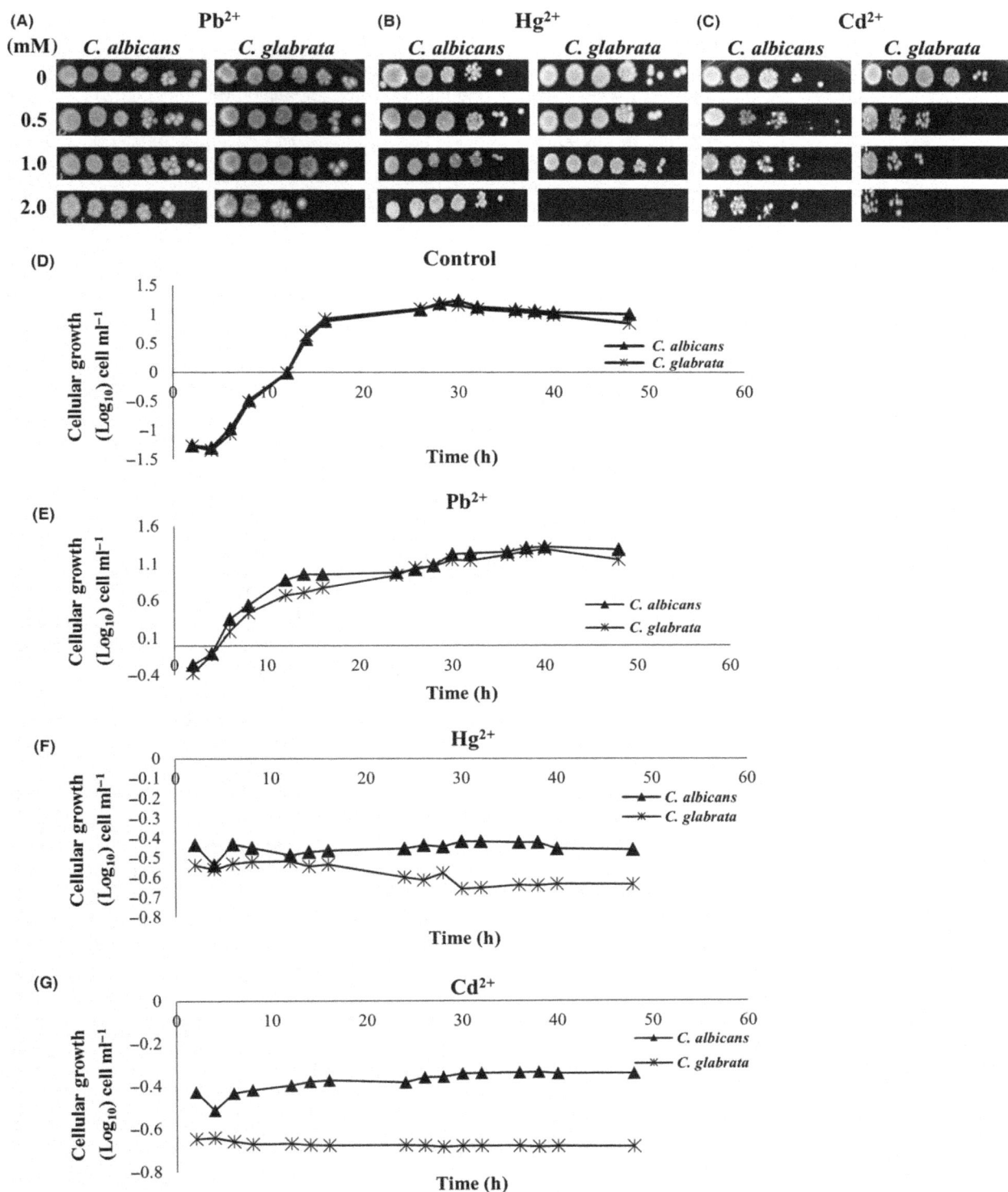

Fig. 1. Response of *Candida* species to different metals.
(i) Susceptibility tests: cultures of the two *Candida* species were diluted to an $OD_{600\ nm}$ 0.5 with sterile deionized water. Aliquots of dilutions were incubated at 28°C in the presence of the indicated concentrations of (A) Pb^{2+}, (B) Hg^{2+} or (C) Cd^{2+}. After 90 min, water was removed by low-speed centrifugation and the corresponding cell pellets were resuspended in sterile deionized water to an $OD_{600\ nm}$ 0.5.
(ii). Growth curves of *C. albicans* and *C. glabrata*. (D) *Candida* cells were not exposed to heavy metals (controls). Cells were exposed at a concentration of 1.0 mM (E) Pb^{2+}, (F) Hg^{2+} or (G) Cd^{2+}. From each culture in the presence of metal, an aliquot was taken every 2 h, over 48 h, and the $OD_{600\ nm}$ was measured in a spectrophotometer. The log_{10} of the $OD_{600\ nm}$ was plotted against time in hours.

to grow at a concentration of 1.0 mM of any of the three metals evaluated, the ability of any *Candida* species to withstand a higher concentration of these metals will depend on the adaptability of each of these yeasts to the different physiological niches in order to survive in the host (Serrano-Fujarte *et al.*, 2015). In this regard, each species of *Candida* was reported to respond differentially whether they were in the presence of different azole antifungals or in the presence of different reactive oxygen species (ROS; Cuellar-Cruz *et al.*, 2008; Ramirez-Quijas *et al.*, 2015; Serrano-Fujarte *et al.*, 2015). This indicates that the differential resistance to toxic metals may be related to the mechanism of adaptation that each *Candida* species develops for surviving in the environment, even though these yeasts can withstand a concentration of 1.0 mM of any of the analysed metals for a short period (Fig. 1A–C). The most interesting thing here is to find out whether the cells will be able to duplicate themselves. For this purpose, we established growth curves for each of the yeast, from 0 to 48 h, in the presence of each of these four cations, and aliquots were taken manually every two hours. As seen in the presence of Pb^{2+}, both *C. albicans* and *C. glabrata* duplicated from the exponential phase up to the stationary phase (Fig. 1E). The same behaviour was observed for *C. krusei* and *C. parapsilosis* (data not shown). These results indicate that apparently, Pb^{2+} does not affect cell growth of *Candida* (Fig. 1D, E). However, cells of the four species of *Candida* did not replicate in the presence of Hg^{2+} or Cd^{2+} compared with that replication of control cells (Fig. 1D, F, G).

These data show that even when the three metals are divalent cations and could act in the same way at the cellular level, *Candida* species have developed a selective detoxification mechanism for each cation, which allowed them to survive in the various habitats to which they have had to adapt. Most identified species of *Candida* are prevalent in rich soil and aquatic habitats that have been contaminated with heavy metals (Hagler and Mendonca-Hagler, 1981; Suihko and Hoekstra, 1999; Lopez-Archilla *et al.*, 2004). *C. albicans*, *C. glabrata*, *C. krusei* and *C. parapsilosis* are opportunistic human pathogens, which can be isolated from these habitats or from the bloodstream or from any of the organs in human host. Therefore, the fact that *Candida* species are able to withstand up to 2.0 mM of Pb^{2+} (Fig. 1A) and are normally able to duplicate in the presence of this metal (Fig. 1E) compared with Hg^{2+} or Cd^{2+} (Fig. 1F, G) indicates that these yeasts, in both natural habitats and in the human body, are mainly exposed to Pb^{2+}. It is this exposure to Pb^{2+} that makes them to develop selective mechanisms to resist this metal.

Other studies have reported that in yeast, the selectivity and capacity of metal uptake depend on cell age,

composition of the growth medium, contact time, pH, temperature (Bishnoi and Garima, 2005) and the composition of its cell wall (CW). In *Candida,* the CW retains the metal through interaction with amino, hydroxyl, phosphate, sulfhydryl and carboxyl groups, forming a coordinated covalent bond. Pb^{2+} has high affinity for sulfhydryl groups, by competing with calcium, and tends to replace Zn^{2+} in enzymes such as HIV nucleocapsid protein (HIV-CCHC) that can cause the complete inhibition of these sulfhydryl groups (Payne *et al.*, 1999). It was also observed that lead has the ability to coordinate with nucleic acids (Da Costa and Sigel, 2000). Hg^{2+} was reported to be preferentially methylated by microorganisms forming methylmercury $(CH_3)_2Hg$ or CH_3Hg^+. Any of these alkylmercury species has a high affinity to form covalent bonds with sulfur (Chwastowska *et al.*, 1999). This chemical behaviour explains most of the biological properties of mercury. The affinity for sulfhydryl groups promotes replacement of some metal ions such as Zn^{2+} and Cu^{2+} in various enzymes. Moreover, mercury is also coordinated with relative ease to phosphate groups and heterocyclic bases in nucleic acids (Onyido *et al.*, 2004). Cd^{2+} like Pb^{2+} and Hg^{2+} also binds sulfhydryl groups, which inhibits the activity of sensitive enzymes. Interestingly, Cd^{2+} and Hg^{2+} cause considerable oxidative stress (OS), which is, in many cases, the basis of cell genotoxicity (Luo *et al.*, 1996; Kwak *et al.*, 2003). These metals can react with molecular oxygen to generate bis-glutathione (GS-SG), the cation of metal and hydrogen peroxide (Kachur *et al.,* 1998a). Because reduction of GS-SG requires the participation of NADPH and because the metal cations are linked immediately to other molecules of GSH, cations of heavy metals cause considerable OS (Xiang and Oliver, 1998; Jacob *et al.*, 2001). As reservoir of cysteine, glutathione is an important antioxidant system in the cell (Elskens *et al.*, 1991). It is also involved in metabolic processes such as cell communication, metal transport and in the regulation of the redox state of proteins for degradation processes and folding (Meister, 1995; Anderson, 1997; Pastore *et al.*, 2003). Due to the non-availability of glutathione, the enzymatic antioxidant mechanisms cannot detoxify the ROS generated during OS, and thus, ROS damages cellular biomolecules such as nucleic acids, lipids and proteins (Klaunig *et al.*, 1998). Metallothioneins belong to a group of proteins that play an important role in the toxicity of mercury and cadmium. They are responsible for protecting cells from these metals (Nordberg and Nordberg, 2000). However, we have found that when *Candida* cells are in OS, both the synthesis of metallothioneins and the synthesis of the antioxidant enzyme systems (catalase, superoxide dismutases, glutathione peroxidases, thioredoxins, glutaredoxins) stop. This damages the DNA, and therefore, just the cellular repairing systems alone will not suffice to repair it (Hoeijmakers, 2001; Jin *et al.*, 2003; McMurray and Tainer, 2003). For this reason, our results suggest that possibly *Candida* species are not able to replicate in the presence of Hg^{2+} or Cd^{2+} (Fig. 1F, G). This is not the case for Pb^{2+} cations, probably because they do not generate large concentrations of ROS and cannot be detoxified by the yeast's antioxidant mechanisms (Fig. 1E).

Candida *species form extracellular crystals with* Pb^{2+} *and intracellular crystals with* Hg^{2+} *or* Cd^{2+}

In order to assess whether the cellular structure of *Candida* is modified in the presence of Pb^{2+}, Hg^{2+} or Cd^{2+} (so that the cells cannot be replicated, but can still survive), these four species of *Candida* were grown in the presence of 1.0 mM of each of three metals, which were visualized using scanning electron microscopy (SEM). The most representative images were selected from a large collection and are shown in Fig. 2. Photomicrographs revealed the formation of extracellular and intracellular micro- or nanocrystals by *Candida* species in the presence of Pb^{2+}, Hg^{2+} or Cd^{2+}. The extracellular crystals were observed attached to the cell wall (CW), considering that the CW is the outermost structure of *Candida* and the first to interact with heavy metals. Meanwhile, intracellular crystals were observed like lights inside the yeast. All crystals below 1 μm in size are considered nanocrystals (NCs) and larger to this value are considered microcrystals (MCs).

As in the presence of Pb^{2+}, *C. albicans*, *C. glabrata*, *C. krusei* and *C. parapsilosis* produced mainly extracellular MCs, which were attached to the CW (Fig. 2B, F, J, N). The synthesis of extracellular MCs of lead may be due to the mechanisms developed by *Candida*, which permitted to form these MCs in the CW (Fig. 2B, F, J, N) and thus protect the biomolecules of the cell membrane, so that its cell viability is not affected (Fig. 1E). One of these mechanisms involves biosorption of Pb^{2+} by the CW, as this is the site where metal retention is performed by a physicochemical interaction with amino and hydroxyl groups of the chitin in the CW, such as phosphate, sulfhydryl and carboxyl groups. This interaction results in the formation of a coordinate covalent bond, in which the metal ion acts as a central atom with empty orbitals capable of accepting electron pairs (Aslangul *et al.*, 1972). Another mechanism that explains why lead MCs are formed on the outside of the CW is the extracellular precipitation mechanism, in which the microorganism (in this case *Candida* in the presence of Pb^{2+}) activates the synthesis of extracellular components of low molecular weight peptides that act as chelating agents (Baldrian, 2003). Another

chelating system or immobilization of the metals may be compounds such as oxalates, sulfides and organic polycarboxylic acids (Gadd and White, 1993; Sierra-Alvarez, 2007). Thus, the chelating agents form a complex or immobilize the soluble metal ions to insoluble compounds, decreasing the bioavailability of these metals, but increasing their tolerance. A third mechanism by which *C. albicans* could synthesize lead MCs in the CW has been reported elsewhere in fungi such as those of *Pleurotus ostreatus*, *Phanerochaete chrysosporium* and *Trametes versicolor*. This mechanism produced an extracellular hyphal sheath mainly composed of polysaccharides (β-1,3 and β-1,6 glucans) that trap metal ions, thereby providing a barrier against the metal. Another way in which these yeasts adapt to different conditions is through a mutation in the genes that encode any of the lead transport systems. By doing so, they do not bond with this cation and the cells will be able to tolerate this metal (Fig. 1E) and to keep the lead outside of the cell together with other compounds in the CW (Fig. 2B, F, J, N). Other microorganisms with mutated gene transport system called fast and unspecific were reported to have mutants tolerant of metals. For example, Cor A mutants were tolerant of Co (II; Bui *et al.*, 1999; Pfeiffer *et al.*, 2002) and Pit mutants were tolerant of arsenate (Rosen, 1999). Data show that *Candida* species may have a mutation in the lead transportation system, which confers resistance to this metal. Another mechanism by which *Candida* species can maintain extracellularly lead may result from carriers that are in the membrane contributing to their resistance to this metal. So far, in *Candida* species, there have not been reported any facilitators of this cation that allows efflux. However, lead transporters can be the ATP-dependent transporters (ABC) and facilitators (MDR), known as CDR and MDR, respectively, which have been reported as protein transporters for multiple drugs (Ramage *et al.*, 2002; Mukherjee *et al.*, 2003; Anderson, 2005; Cuellar-Cruz *et al.*, 2012). These transporters may be responsible for lead efflux in *Candida*, as reported in bacteria. The most studied ATPase type P is the protein CadA encoded in plasmid pI258 from *Staphylococcus aureus* (Yoon and Silver, 1991). In the genome of *Cupriavidus metallidurans* strain CH34, the presence of genes encoding ATPases type P is notable, which are collectively involved in the homeostasis or in the resistance to cadmium, copper, lead and zinc (Nies, 2003). The formation of extracellular MCs by the four species of *Candida* is the first evidence that these pathogenic fungi have been exposed to environments with high concentrations of lead, such as the bloodstream or organs of the host. The second evidence is their adaptability to this hostile environment in order to survive. These data open the possibility of studying and

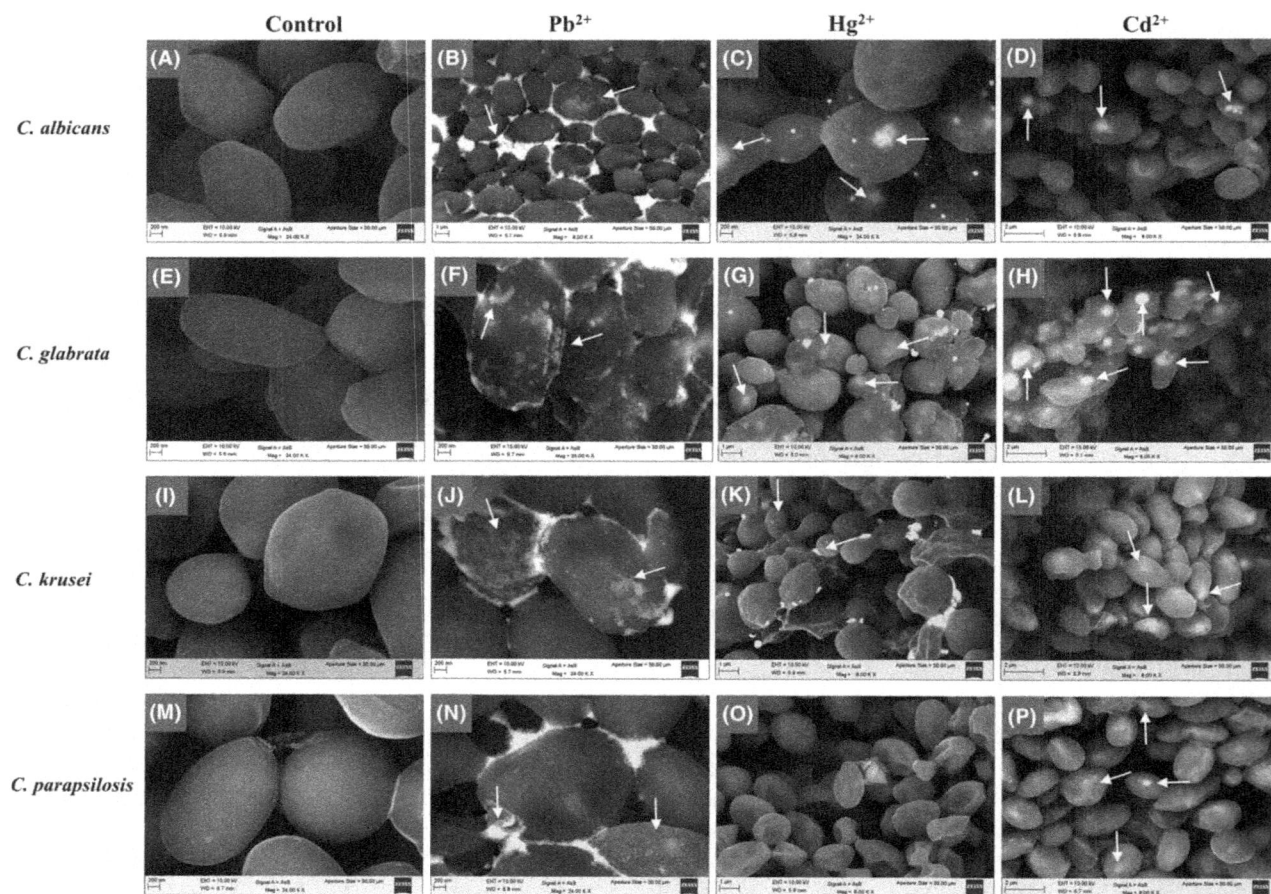

Fig. 2. Formation of extracellular and intracellular micro- or nanocrystals by *Candida* species in the presence of Pb^{2+}, Hg^{2+} or Cd^{2+}. *Candida* cells were exposed to the different metals and analysed by SEM as described in the previous methods section. Control cells (A, E, I, M), treated with Pb^{2+} (B, F, J, N), with Hg^{2+} (C, G, K, O) or with Cd^{2+} (D, H, L, P). Scale bar is included in each photomicrograph. Arrows indicate the extracellular and intracellular micro- and nanocrystals formed in the treated cells with respect to the control cells. All crystals below 1 μm in size are considered nanocrystals and larger to this value are considered microcrystals.

determining not just the mechanisms of assimilation but also the resistance of *Candida* to heavy metals, therefore enlightening our understanding of the pathogenesis and virulence of these yeasts in human host. In the case of Hg^{2+}, *C. albicans*, *C. glabrata* and *C. krusei* form mainly intracellular MCs (Fig. 2C, G, K), in contrast to *C. parapsilosis* that has not been observed to form MCs, although the cells show apparent structural damage (Fig. 2O). However, in the presence of Cd^{2+}, four species of *Candida* were found to form intracellular NCs (Fig. 2D, H, L, P). These intracellular NCs were observed like lights inside the yeasts (Fig. 2). Unlike lead, the formation of intracellular MCs or NCs from mercury or cadmium, respectively, was probably due to the fact that *Candida* species are not normally exposed to these metals. This is why their mechanisms have not resistance to enable them to maintain these extracellular cations. Even though *Candida* is unable to replicate in the presence of Hg^{2+} or Cd^{2+} (Fig. 1F, G), it can,

nonetheless, form MCs and NCs (Fig. 2C, D, G, H, K, L, P) because its metabolism appears to be mainly directed to synthesizing MCs or NCs. As a first step in the formation of intracellular MCs or NCs, *Candida* needs to import metals to the cytoplasm through any of the systems that capture heavy metal cations (Nies and Silver, 1995). Once inside the cell, the excess of metals can form coordinate bonds with anions that block functional groups of enzymes to inhibit the transport system by moving essential metals from native binding sites and by disrupting the integrity of the cell membrane (Nies, 2003). Thus, the facility of Cd^{2+} to move to the cytoplasm enables the interaction with the membrane transporters involved in the capture of Zn^{2+}, Ca^{2+} or Fe^{2+}, displacing these metals (Bridges and Zalups, 2005). Therefore, cadmium will be able to replace these cations and form complex coordinate covalent bonds with biomolecules and sulfhydryl groups. This may result either in the inhibition of the activity of sensitive enzymes or in the reaction with molecular oxygen (Kachur et al., 1998). Calcium channels provide one of the main entrances of Cd^{2+} to the cell, considering that Cd^{2+} and Ca^{2+} have similar ionic radius (Goyer, 2003; Mendez-Armenta and Rios, 2007; Flora et al., 2008).

The alteration in the homeostasis of intracellular calcium leads the cell to release mitochondrial and endoplasmic reticulum calcium, thereby producing alterations in the metabolism. Hg^{2+} can enter the cytoplasm and be methylated by the microorganism, forming methyl mercury (Pan-Hou and Imura, 1982). This methyl mercury complex is structurally similar to methionine, so that its transport into the cell is possible through the protein transporter of neutral amino acids. It has been reported that when the cell gradually accumulates metal cations, there tends to be a homeostasis of heavy metals within the cell. Thus, the cell adapts by counteracting the effects of high concentrations of metal ions (Trajanovska et al., 1997). This process (a type of homeostasis and adaptation of Candida species to heavy metals) is directly related to the formation of intracellular MCs and NCs. In the biosorption of Cd^{2+} or Hg^{2+}, the Candida CW plays an important role as it facilitates the entry of these cations. Studies have also reported that, at a pH between four and six, the negative charge of the fungal wall favours ionic approach with metal ions so these will enter the cell (Loukidou et al., 2004). Data show that in Candida species, the pH of the CW changes depending on the metal that is present. The fact that C. parapsilosis cannot form MCs in the presence of Hg^{2+} may be due, in part, to the composition of its CW, which differs from that of C. albicans, C. glabrata and C. krusei (Silva et al., 2012). The apparent damage to the CW may be due to mercury generating high concentrations of OS and ROS. Previous studies show that OS generates damage to the CW of Candida species (Ramirez-Quijas et al., 2015). Moreover, the exporter proteins of Candida species are specific for certain cations; subsequently, lead (Pb^{2+}) will not be able to bind to these systems.

We have additionally performed a qualitative analysis of the elements present using energy-dispersive spectroscopy (EDS) in order to confirm that the MCs and NCs observed by SEM were indeed MCs or NCs of Pb^{2+}, Hg^{2+} or Cd^{2+} sulfides. The representative EDS signal in C. albicans and C. krusei (in the analysed MCs and NCs) was found either for lead, or for mercury or for cadmium as appropriate in both the extracellular and intracellular MCs and NCs (Fig. 3). These results show that Candida species have specific input mechanisms and resistance to Pb^{2+}, Cd^{2+} or Hg^{2+}. This suggests a typical biomineralization process (biogenic crystallization) that achieves homeostasis at a high concentration of these heavy metals.

Tailored synthesis of PbS, HgS, HgCl₂ and CdS crystals by Candida species

The ability to form MCs or NCs is apparently specific to some species of Candida. SEM showed that some species of Candida such as C. albicans and C. glabrata have higher affinity for metals than for others. It is this higher affinity for metal that enables them to form MCs or NCs (Figs 2 and 3). In order to elucidate the chemical composition of the MCs and NCs formed by each of the Candida species for each individual cation, MCs and NCs were analysed through synchrotron radiation XRD-XRF. Synchrotron radiation is by now the only tool to determine the chemical composition of low amount samples, low concentrations and also to analyse crystals of the size obtained in this study. Usually, the home powder X-ray diffractometers do not have the appropriate sample holders for a limited proportion of mass of these samples. The fluorescence spectra from synchrotron radiation confirmed the presence of expected bioaccumulated heavy metals in lyophilized cells, with superimposable patterns among different cell lines (Fig. 4, Table S2). Furthermore, endogenous zinc and other common metals (K, Fe) have been found, even in blank samples (Fig. 4A).

X-ray powder patterns collected from Candida cells, not exposed to heavy metals (blanks), showed broad peaks that are in agreement with previous data published for glucans extracted from CW (Lowman et al., 2014) and 'poorly ordered' lipid phases (Fidan et al., 2014; giving the broad peak at ~4.3 Å). Blank patterns of the four Candida species analysed are plotted and compared properly (Fig. 5A).

The presence of heavy atoms (which belong to crystalline metallic micro- or nanoparticles) in the cells added more sharp signals to the blank background. For lead-loaded cells (Fig. 5B), the sharp signals nicely matched the peak positions expected for crystalline cubic Fm3m galena (PbS) phase (Noda et al., 1987) samples from different cell lines that were equivalent. These results show that MCs which appeared in the presence of Pb^{2+} in the four species of Candida were formed by PbS (Figs 4B and 5B). Galena (PbS), lead sulfide with halite-type structure, is the most important lead mineral in the Earth's crust. PbS is further interesting because it is a naturally occurring semiconductor. The atomic arrangement of galena is the same as that of NaCl, that is cubic closed-packed, with Pb atoms in the octahedral interstices. If a Born ionic model is assumed for galena, periodic bond chain (PBC) analysis of crystal morphology (phase) gives similar results as those for the NaCl case.

The mechanism by which Candida species promote the synthesis of PbS MCs extracellularly involves the CW in the formation of covalent coordinate bonds

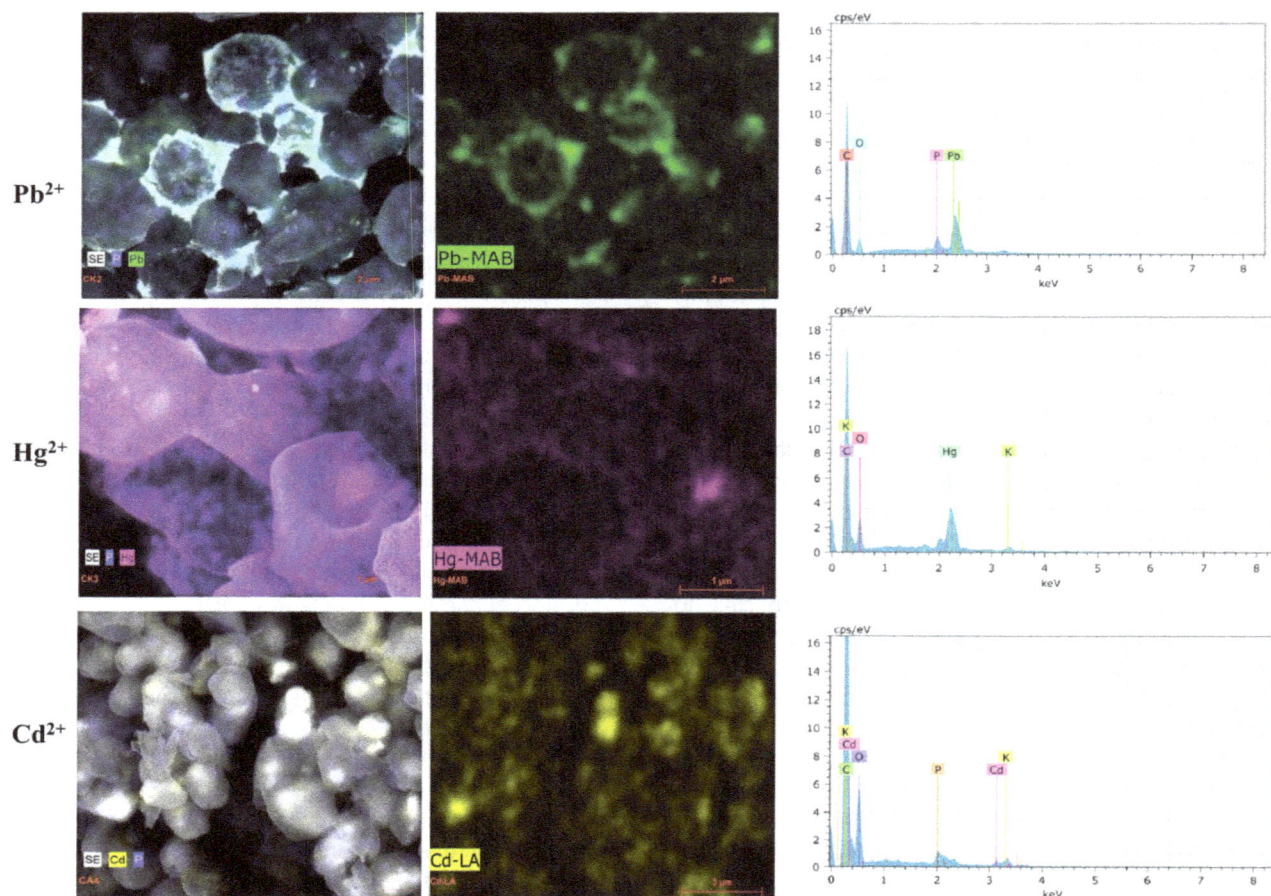

Fig. 3. Qualitative analysis of elements present in the micro- or nanocrystals by energy-dispersive spectroscopy (EDS). The samples of *Candida* species were observed by SEM and were analysed qualitatively in order to determine their main components. As shown, micro- or nanocrystals are formed depending on the metal that they were exposed.

between the metal ion with other components of the CW, favouring an acid pH. This process depends on the degree of protonation of the CW (Gupta *et al.*, 2000). The photomicrograph of the SEM shows that *C. albicans* cells are found only as yeast form and not as hyphae in acidic pH (Fig. 2B). However, under extracellular alkaline pH, they will have a hyphal growth as shown *in vitro* (Davis *et al.*, 2003). Furthermore, it has been reported elsewhere that the ability of *C. albicans* to respond to changes in extracellular pH is controlled, in part, by changes in the gene expression. The *PHR2* gene encodes for a CW protein involved in binding of β-1,3 and β-1,6 glucans expressed in acidic conditions (Muhls-chlegel and Fonzi, 1997). In addition, transcriptional analyses in *C. albicans* have shown that in response to the changes in extracellular pH, nearly five hundred genes are regulated, either by alkalization or by acidification of the medium. Of these, 267 genes are activated in response to pH and, among these, a significant number are related to iron metabolism (Bensen *et al.*, 2004).

Some of the genes involved in iron metabolism may also be involved in the metabolism of lead, which could partly promote the synthesis of PbS MCs in the CW of *Candida* species. Additionally, *Candida* species ensure an extracellular acidic environment in the presence of Pb^{2+} and ATPase of the membrane Pma1. A similar mechanism in *S. cerevisiae* has also been reported to have proton export activity (Vanderrest *et al.*, 1995).

On the other hand, synchrotron radiation analysis of the MCs formed in the presence of Hg^{2+} showed different chemical composition for each of the *Candida* species. The diffraction patterns also showed significant differences for mercury. Among the different samples (Fig. 5D, E), XRD peaks for *C. glabrata*, *C. krusei* and *C. parapsilosis* can be interpreted as cinnabar (HgS) trigonal $P3_221$ crystals (Auvray and Genet, 1973; Clever *et al.*, 1985). It should also be noticed that *C. parapsilosis* signals are much weaker compared with other cell lines (Fig. 5D). This suggests that mercury is less efficient for *in vivo* crystallization. This result is in

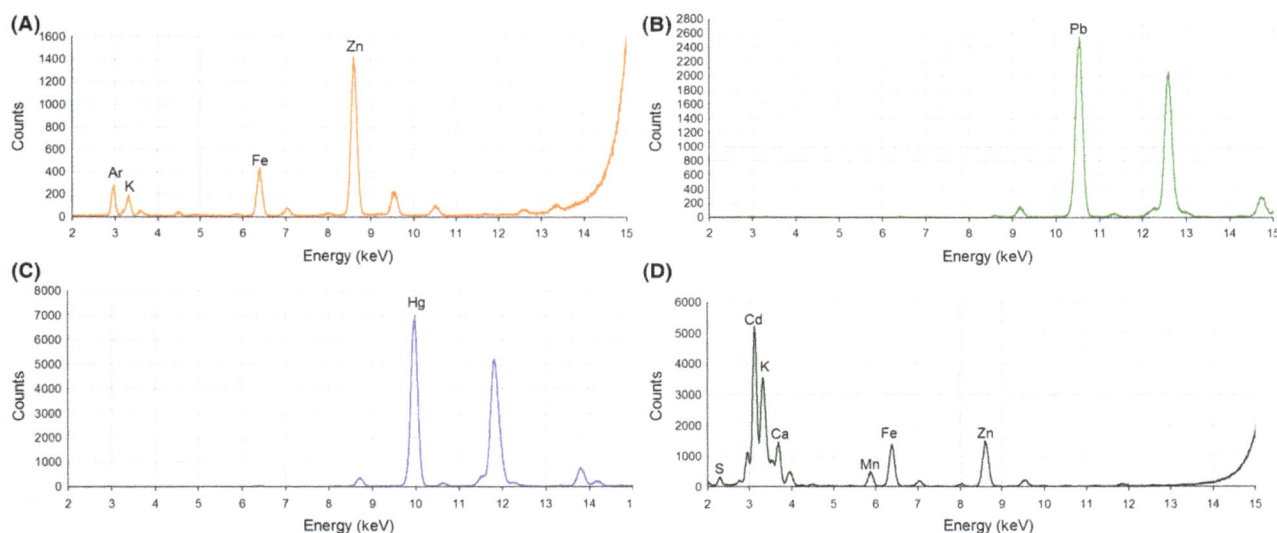

Fig. 4. Major fluorescence peaks interpretation for (A) *Candida* blank samples packed in capillary tubes – elements labelled on corresponding K_α lines.
B. Peaks for a *Candida* lead-loaded samples belong to Pb; each element is labelled on its L_α lines.
C. Peaks for a *Candida* mercury-loaded sample belong to Hg; each element is labelled on its L_α lines.
D. Peaks interpretation for a *Candida* cadmium – elements are labelled on corresponding K_α or L_α lines.

agreement with that found in the SEM, where no MCs of mercury were observed for this particular *Candida* species (Fig. 2O). Mercury-loaded *C. albicans* shows a completely different powder profile, suggesting the presence of calomel particles ($HgCl_2$, with a tetragonal *I4/mmm* space group; Barnes and Bosch, 2006; see Fig. 5E).

The synthesis of the HgS MCs formation of *C. glabrata*, *C. krusei* and *C. parapsilosis* is performed as described in other microorganisms (Wood *et al.*, 1968; Landner, 1972; Bisogni and Lawrence, 1973). The biomineralization of HgS MCs in *C. glabrata*, *C. krusei* and *C. parapsilosis* requires an acidic pH, which can be maintained in an intracellular manner through any of the mechanisms described in other fungi. One such mechanism is through the Nha1 antiporter, which in *S. cerevisiae* has the principal function of continuous recycling of potassium cations through the plasma membrane, in homeostasis of intracellular K^+ and at a specific pH value. Its contribution to the detoxification of Na^+ and Li^+ ions is important, but not crucial. Another antiporter involved in intracellular pH is Kha1, which is an antiporter of Na^+/H^+ and is located in the membrane of the Golgi apparatus. As in the case of other intracellular transporters, cations of alkali/H^+ metal yeast, it is involved in regulation of potassium within the organelle and pH homeostasis. Nha1 and Kha1 (together with the vacuolar H^+-ATPase) are involved in the intraorganellar pH and in the balanced alkali metal cations (Arino *et al.*, 2010; Cyert and Philpott, 2013).

In the particular case of $HgCl_2$ MCs formed by *C. albicans*, mercuric ions are biomineralized in a different mechanism. This difference in the way of mineralizing certain heavy metals is due to *C. albicans* that instead of taking S^{2-} ions uses the anion Cl^- to detoxify Hg^{2+}. The existence of various mechanisms that these yeasts have developed to deal with toxic metals indicates that they have been subjected to different niches.

Synchrotron radiation diffraction, on cadmium-loaded cells, has been measured on small quantities of lyophilized cells. The high toxicity of this metal for *Candida*, which inhibited cell proliferation (Fig. 1G), suggests the absence of efficient detoxification mechanisms for this metal in *Candida*. Furthermore, the bioaccumulated metal gives very small diffraction peaks, which could be explained by the presence of tiny ('quantum') crystalline particles, as previously reported in the literature (Williams *et al.*, 1996). X-ray diffraction peaks are more pronounced in the *C. glabrata* pattern and their positions correspond to the expected positions of peaks for a cubic $F\bar{4}3m$ Hawleyite (CdS) phase (Malik *et al.*, 1993; Barnes and Bosch, 2006; Shakouri-Arani and Salavati-Niasari, 2014; Fig. 5C). The cubic crystals of Hawleyite (CdS) usually show halite-type structure as that found for galena previously mentioned. Hawleyite is a rare sulfide and usually appears as a bright yellow coating on sphalerite; it is usually precipitated and confused with the mineral greenockite, which crystallizes in a hexagonal group and appears as an orange-yellow colour. Slight shifts in the angular position could suggest

(A)

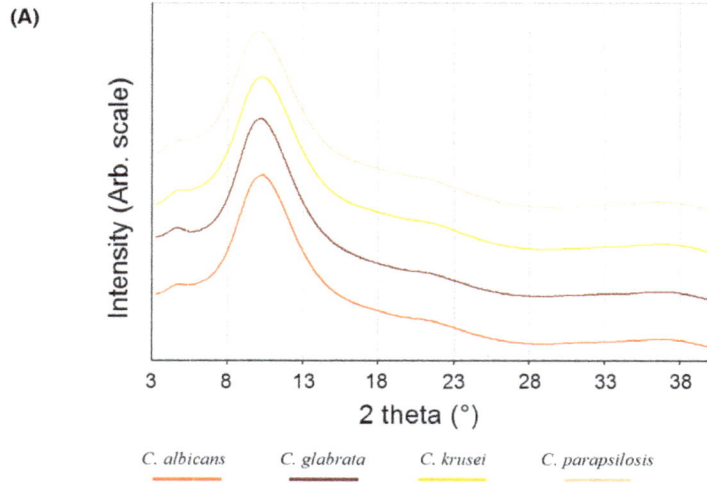

C. albicans   *C. glabrata*   *C. krusei*   *C. parapsilosis*

(B)

PbS

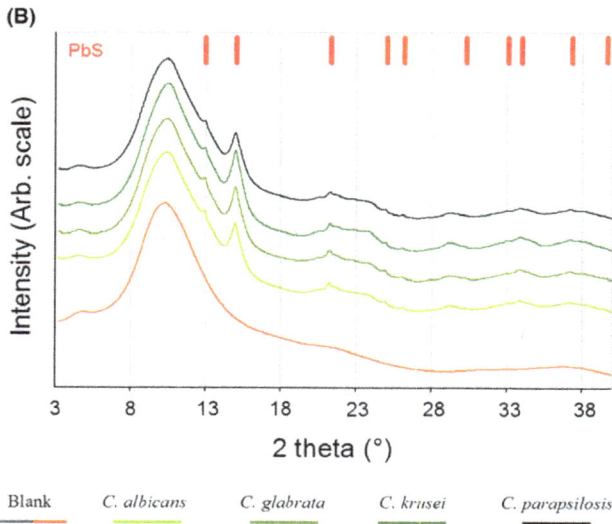

(C)

CdS

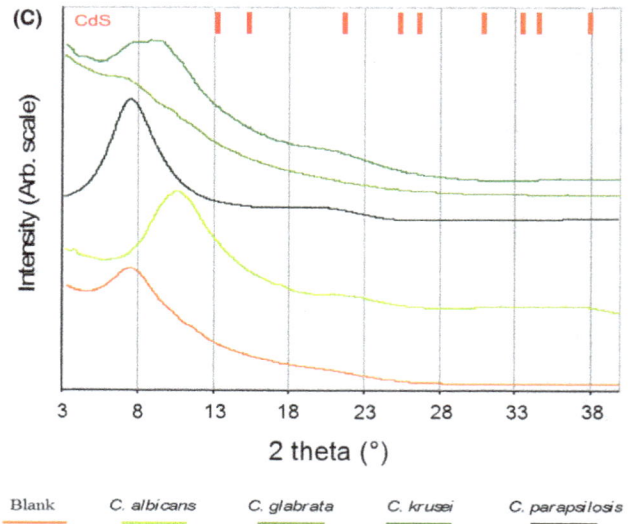

Blank   *C. albicans*   *C. glabrata*   *C. krusei*   *C. parapsilosis*   Blank   *C. albicans*   *C. glabrata*   *C. krusei*   *C. parapsilosis*

(D)

HgS

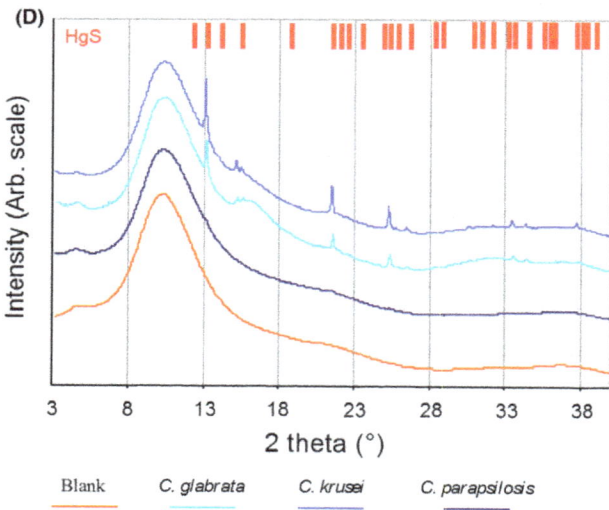

(E)

HgCl₂

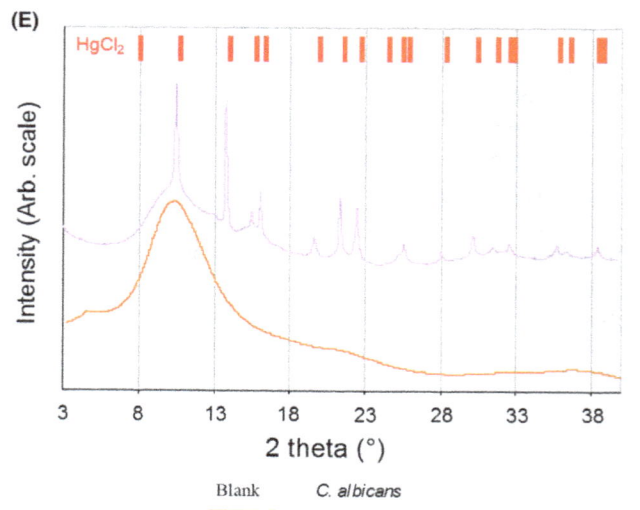

Blank   *C. glabrata*   *C. krusei*   *C. parapsilosis*   Blank   *C. albicans*

Fig. 5. X-ray powder patterns collected on *Candida* cells.
A. Cells not exposed to heavy metals (controls).
B. *Candida* cells exposed to Pb^{2+}. Red bars represent expected positions of cubic $Fm\bar{3}m$ PbS.
C. *Candida* cells exposed to Cd^{2+}. Red bars represent expected positions of cubic $F\bar{4}3m$ CdS.
D. *C. glabrata*, *C. krusei* or *C. parapsilosis* cells exposed to Hg^{2+}. Red bars represent expected positions of trigonal $P3_221$ HgS (Auvray and Genet, 1973).
E. *C. albicans* cells exposed to Hg^{2+}. Red bars represent expected positions of tetragonal $I4/mmm$ $HgCl_2$ (Adams *et al.*, 1970; Boggon and Shapiro, 2000). Patterns vertically shifted for clarity.

differences in lattice parameters of CdS, which may be due to crystallization of Cd-Zn sulfide solid mixtures (Noor *et al.*, 2010). The formation of CdS NCs has already been reported for *C. glabrata* and *S. pombe* (Dameron *et al.*, 1989). The intracellular formation mechanism is performed when Cd^{2+} ions in the cytoplasm are complexed with γ-Glu peptides (Grill *et al.*, 1986; Hayashi *et al.*, 1986). These complex metal–γ-peptides incorporate sulfide ions arising from a Cd-mediated enhancement of cellular sulfide generation (Murasugi *et al.*, 1984; Mehra *et al.*, 1988; Reese and Winge, 1988). In the formation of HgS and $HgCl_2$ MCs or CdS NCs, the process requires an acidic intracellular pH, which *Candida* maintains in the previously described mechanisms.

These results show that each species of *Candida* has developed specific and differential mechanisms to detoxify each of the metals to which they have been subjected in the various environmental and physiological habitats. The mineralization *in vivo* is one of the principal mechanisms by which cell homeostasis, with the cation, occurs through the formation of MCs and NCs (Fig. 6). Thus, in conclusion, biomineralization through crystallization is a fundamental chemical mechanism produced from microorganisms to higher organisms in order to perform basic functions, and it can be used as a defence mechanism against toxic elements, such as that produced by *Candida* species that form micro- or nanocrystals to reach homeostasis in the presence of toxic metals.

In order to complement the *in vivo* experiments, we decided to synthesize sulfides of different metals (Pb^{2+}, Hg^{2+} and Cd^{2+}) *in vitro*. We based our synthesis strategy in the cell growth information where four types of *Candida* species were exposed to these heavy metals. The *in vitro* experiments showed that the synthesis of the PbS, HgS and CdS is feasible when the pH of the reaction is acidic through diffusing-reacting systems, either agarose gel or those gels obtained from the neutralization of sodium metasilicate.

Synthesis of XS (X: Pb^{2+}, Hg^{2+} and Cd^{2+}) and crystal growth experiments

Years of experimenting with different crystals has confirmed that by minimizing the convective transport of mass, we will generally obtain higher-quality crystals,

with improved mechanical and optical properties, reduced density of defects and of larger size. There are two approaches that suppress or at least reduce the convection in crystal growth that demonstrated this assumption. One is the crystallization in space (Littke and John, 1984; Long *et al.*, 1994; Kundrot *et al.*, 2001), and the other is the crystal growth in gels, which produces crystals of high quality for high-resolution X-ray crystallography compared with those crystals obtained in solution (Lorber *et al.*, 2009). The way of reducing the natural convection of solutions under Earth gravity is by incorporating jellified media into the solutions (Garciaruiz, 1991). Although the gel growth method is known since the end of the nineteenth century from the experiments performed by Liesegang (1896) about periodic precipitation (cited elsewhere; Henisch and Garciaruiz, 1986a,b), it has not been sufficiently explored in the growth of inorganic/organic crystals (Robert and Lefaucheux, 1988).

For the *in vitro* synthesis of sulfides of heavy metals, a Granada Crystallization Box (Garcia-Ruiz *et al.*, 2002) from Triana Sci and Tech (GCB, Spain) was used by means of the gel growth method in three-layer configuration (Fig. 7). The first layer of the gel contained one of the components, which was Na_2S (reactant A) of the synthesis after which a sandwich layer was prepared (without any reactants). The second component (reactant B) of the reaction was poured onto the top of the sandwich gel (third gel layer). The components of the reaction counter-diffused each other to react in the central part of the sandwich gel, which was the second gel layer without any of the reactants of the reaction. MCs of PbS, HgS and CdS were synthesized according to the reactions described in the paragraphs below (these sulfides are virtually insoluble in water). The *in vitro* experiments showed that the synthesis of the PbS, HgS and CdS is feasible when the pH of the reaction is acidic through diffusing-reacting systems, either agarose gel or those gels obtained from the neutralization of sodium metasilicate.

The reactant A (0.1 M Na_2S initially prepared in an acidic buffered solution using phosphoric acid at pH 2) was introduced to the first gel layer mixing 1:1 with 1% (w/v) agarose. The reactant B (each of the aqueous solutions of XNO_3 salts, where X: Pb^{2+}, Hg^{2+} and Cd^{2+}) was incorporated to the third gel layer following the same procedure. For crystals of galena (PbS) grown in gels, harvested crystals were evaluated by powder X-ray

In vivo

Non-treated cells

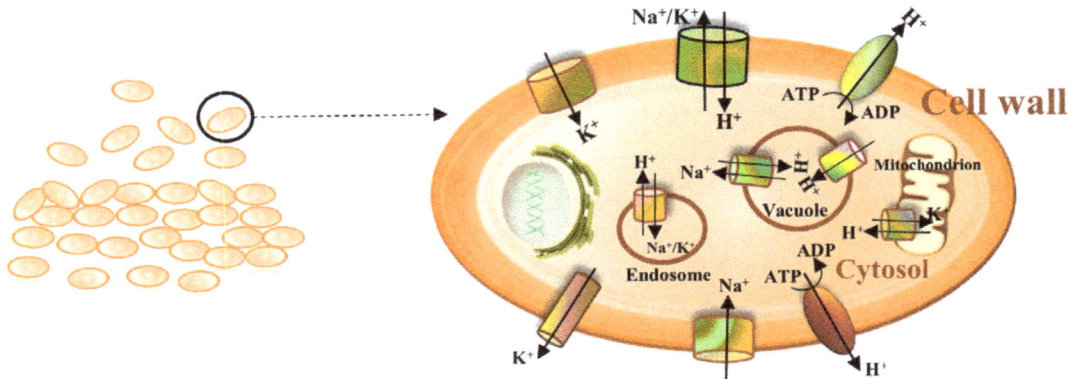

Metal-treated cells

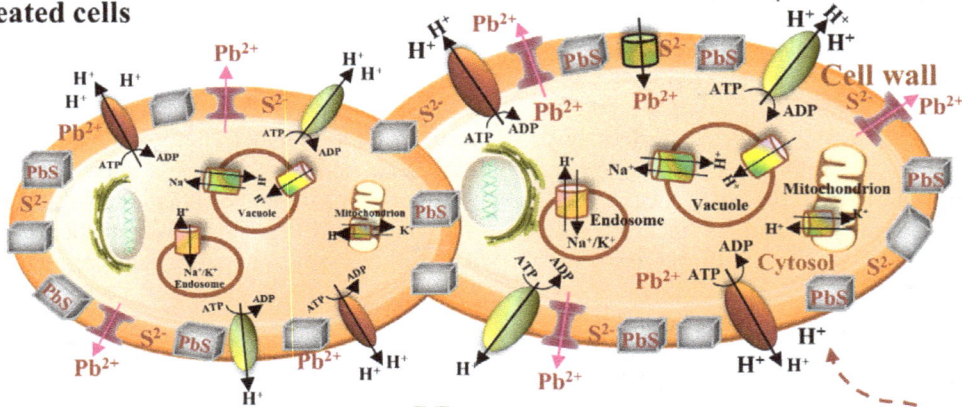

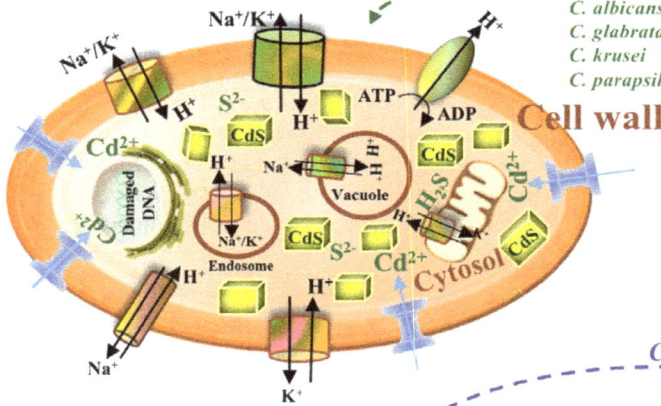

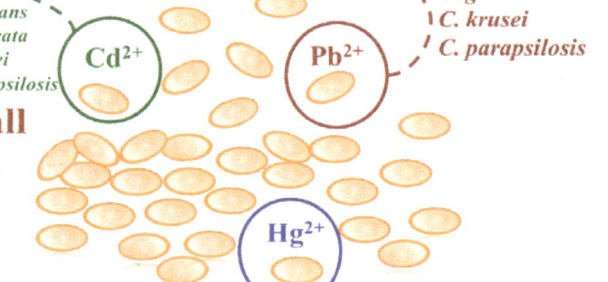

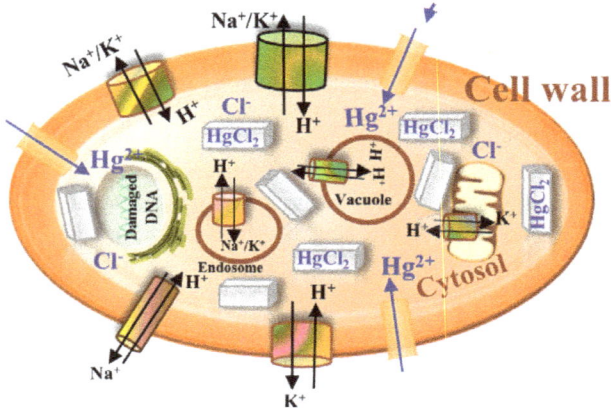

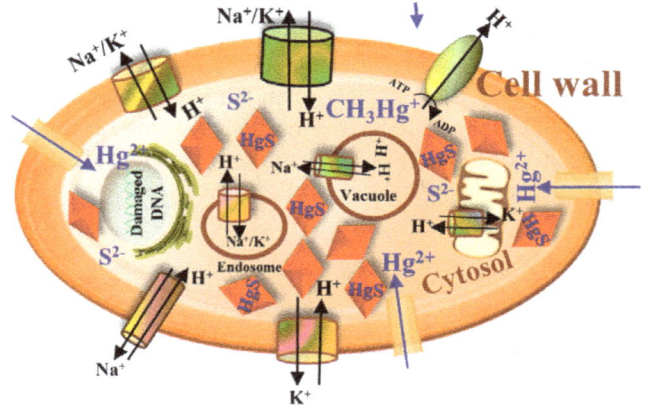

Fig. 6. Proposed mechanism by which *Candida* species synthesize micro- or nanocrystals from PbS, HgS, HgCl₂ or CdS. Control cells are both metabolized and replicated. Pb²⁺-treated cells are able to replicate and to detoxify metal-forming extracellular PbS MCs. Although cells treated with either Hg²⁺ or Cd²⁺ failed to replicate, they achieved homeostasis by intracellular formation of MCs or NCs from HgS, HgCl₂ or CdS respectively. In the presence of mercury salts, *C. albicans* used to form MCs of HgCl₂; though, *C. glabrata* and *C. krusei* synthesized MCs of HgS. *C. parapsilosis* was not able to synthesize MCs of mercury. In the formation of both extracellular and intracellular MCs and NCs, cells of *Candida* need to maintain an acidic pH value either intracellular or extracellular.

diffraction and EDS-SEM. These microcrystals matched the peak positions expected for crystalline cubic *Fm3m* of galena phase in X-ray diffraction and characteristic peaks for the metal and sulfur in EDS-SEM. These crystals showed a classical cubic-shaped habitats and a grey metallic colour (Fig. 7A) as reported elsewhere (Garciaruiz, 1986; Noda *et al.*, 1987). These galena crystals grown *in vitro* showed a characteristic crystal growth pathway: first, they started to grow a dendritic hollow shape, followed by a quasi-cubic shape (when the dendrite was completely filled) up to a perfect cubic shape (Garciaruiz, 1986). They finally reached a classical cubic shape at the end of the sandwich gel layer, and compared with the other sulfides, these galena crystals were well-formed cubes.

Though, we followed the same method of crystal growth for synthetic crystals of cinnabar (α-HgS, dark-red colour mixed with black powders (β-HgS), appear at the beginning of the crystallization reaction. These cinnabar (α-HgS) red crystals were characterized by powder X-ray diffraction showing a trigonal *P3₂21* space group (Auvray and Genet, 1973), whereas EDS-SEM showed the characteristic peaks for Hg and sulfur. In the lower part of the sandwich gel layer, red tiny spheres at the beginning up to dendrites of this mercury (II) sulfide were observed (Fig. 7B).

Finally, crystals of cadmium (II) sulfide were obtained in a fine yellow powder at the beginning of the reaction in the sandwich gel layer. When both concentrations were counter-diffused and properly equilibrated, tiny octahedral-shaped crystals were obtained (Fig. 7C). The expected positions of peaks from powder X-ray diffraction corresponded to characteristic diffractions for a cubic *F43m* Hawleyite phase (Shakouri-Arani and Salavati-Niasari, 2014) and EDS-SEM showed the characteristic peaks for cadmium and sulfur.

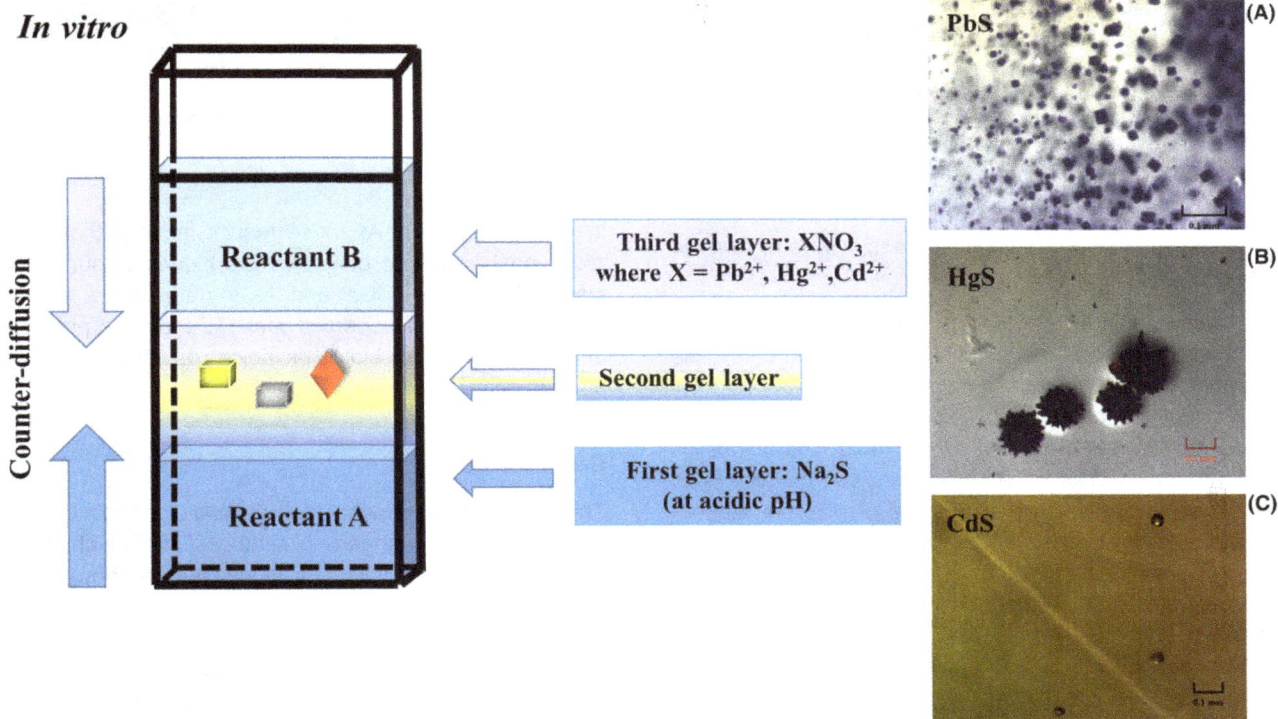

Fig. 7. Crystallization of XS (X:Pb (II), Hg (II) and Cd (II)) in hydrogels. The figure on the left shows the experimental set-up used for growing these *in vitro*-grown crystals. The sandwich gel part is the place where all crystals are obtained by the counter-diffusion of all the ions. The right image corresponds to the synthetically obtained crystals of these three sulfides: (A) the upper part shows the grey metallic cubic-shaped PbS crystals, (B) HgS a red-dark spherulites and (C) tetragonal-shaped crystals of CdS at the end of the sandwich layer of the gel.

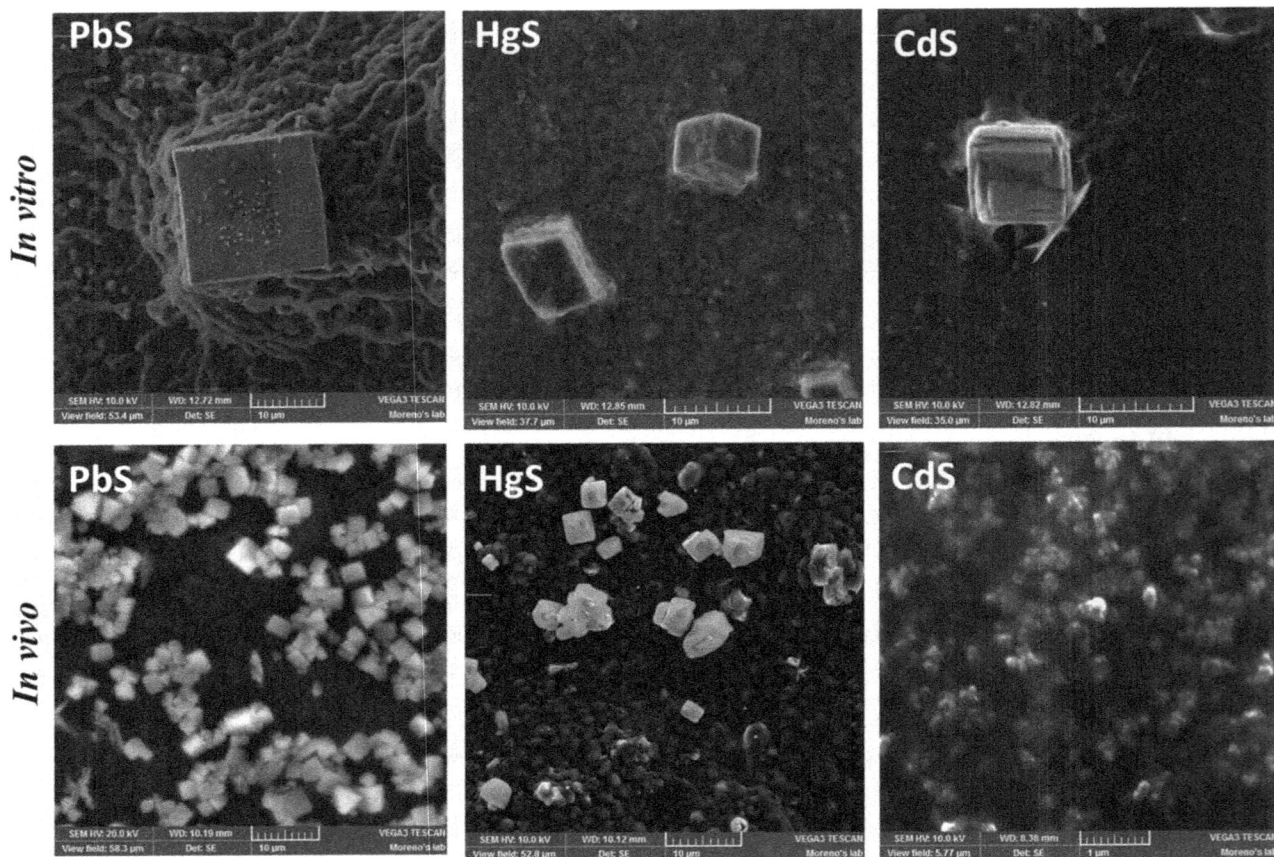

Fig. 8. Comparison of heavy metals sulfides.
The *in vitro* crystals in the upper part show the SEM micrographs of these sulfides of Pb (II), Hg (II) and Cd (II) obtained in hydrogels respectively. The lower part of the figure shows the most representative microcrystals of PbS and HgS and nanocrystals of CdS obtained in these four species of *Candida*.

MCs and NCs synthesized in vivo *and* in vitro *exhibit the same crystalline morphology*

As illustrated in Figs 7 and 8, MCs of PbS and HgS synthesized by *Candida* cells show the same crystalline morphology and shapes than those obtained *in vitro*. However, those formed *in vivo* produced higher amounts, but different in size. These characteristics make *Candida* cells useful producers of PbS and HgS, which can be utilized in a number of different technological areas as described previously (Williams *et al.*, 1996; Kowshik *et al.*, 2002; Krumov *et al.*, 2007; Korbekandi *et al.*, 2014). Interestingly, *Candida* also synthesizes CdS crystals that look like clusters in the form of bunches (Fig. 8). They appear in the form of lights inside the cell when observed by SEM (Fig. 2). CdS crystals obtained in *Candida* species are produced in considerable amounts although of smaller size. However, they exhibit the same crystalline morphology as those obtained *in vitro*. As we described along this contribution, PbS, HgS and CdS can be technologically

important materials. All these results indicate that *Candida* yeasts can not only produce higher amounts but also more stable MCs and NCs than those formed *in vitro*. All these properties make *Candida* a potential alternative in future biotechnological developments and applications.

Conclusions

This contribution showed how *Candida* species have a singular way of synthesizing either sulfides or chlorides of different heavy metals. Our *in vivo* results showed that in the presence of Pb^{2+}, *Candida* cells are able to replicate and form extracellular PbS MCs, whereas in the presence of Hg^{2+} and Cd^{2+}, the cells fail to replicate. However, they did synthesize intercellular MCs or NCs from HgS or $HgCl_2$ and CdS respectively. This shows that these *Candida* species adapt differentially to the environment, counteracting the effect of the heavy metals. The synthesis of MCs and NCs is a form of homeostasis and adaptation of these fungi. These crystals grown *in vitro* were bigger in size (in some cases) than those observed in *Candida*. However, the synthesis in gels is anything but simple. This is due to the number of chemicals used to produce the hydrogel as well as for the reactants used to complete this synthetic reaction. On the other hand, in *Candida*, there are plenty of biological machineries that are activated by genes in the synthesis of these micro- or nanocrystals of heavy metal sulfides. We can finally infer from these results that in the near future, these *Candida* species could be used either as biologically based decontaminants or as specific heavy metals removal.

Experimental procedures

Strains and culture conditions

The strains of *C. albicans*, *C. glabrata*, *C. krusei* and *C. parapsilosis* used in this study are clinical isolates from blood cultures of the collection of Department of Microbiology, ENCB-IPN, México. Yeast strains were cultured on yeast peptone (YP; yeast extract, 1%; peptone, 2% glucose), and 2% agar was added to solidify the media (Ausubel *et al.*, 2001). The formation of MCs and NCs was induced by the addition of lead nitrate, mercury nitrate or cadmium nitrate (all obtained from Sigma-Aldrich Toluca, Mexico) to the culture medium, prior to yeast inoculation.

Susceptibility testing of Pb(II), Hg(II) or Cd(II)

Cells in the exponential phase of each *Candida* species at OD_{600nm} 1.0 were divided and exposed to different concentrations (0, 0.5, 1.0, 1.5 or 2.0 mM) of each of the heavy metals Pb^{2+}, Hg^{2+} or Cd^{2+}. The next step was to incubate them with shaking at 28°C for 48 h to allow the cells to reach the stationary phase. Subsequently, aliquots of the control and treated cultures with different metals were taken and adjusted to an $OD_{600\ nm}$ 0.5. With these aliquots, exponential dilutions were made in 96 well boxes and each dilution was spotted on to YPD plates and incubated at 28°C for 48 h (Cuellar-Cruz *et al.*, 2008). The plates were photographed in the plate reader Gene genius bioimaging system, Syngene. The experiments were performed in duplicate.

Growth curve

A preinoculum of each of the four species of *Candida* was added to 25 ml of YPD liquid medium and incubated at 28°C with constant stirring. From this culture every 2 h for 48 h, the $OD_{600\ nm}$ was measured using a Genesys 20 (Thermo Scientific) spectrophotometer. The cells exposed to heavy metals (Pb^{2+}, Cd^{2+} or Hg^{2+}) had a final concentration of 1.0 mM of the corresponding compound. Finally, the $OD_{600\ nm}$ was plotted against time in hours (h). Two independent experiments were performed.

Isolation of MCs and NCs after lysis of Candida protoplasts

To isolate MCs and NCs, yeast protoplasts were obtained as follows: cells of the four *Candida* species treated with Pb^{2+}, Hg^{2+} or Cd^{2+} were pelleted by centrifugation at 3500 × g for 15 min at 4°C, and the pellets were washed four times with sterile deionized water, resuspended in water and counted. Aliquots of the cell suspension were resuspended at a final $OD_{600\ nm}$ of 1.0 in 1.0 ml of lysis buffer containing 50 mM Tris–HCl, pH 7.2, 0.8 M sorbitol, 0.8 M KCl, 10 mM $MgSO_4$, 15 mM β-mercaptoethanol and 0.25 mg ml^{-1} lyticase (all reagents from Sigma-Aldrich) and incubated at 37°C. After 3 h, cells were observed with an Zeiss Axiostar microscope (Carl Zeiss, Jena, Germany) to assess protoplast formation. This was about 90%. Protoplasts were collected and gently lysed by resuspending those in 500 μl of sterile deionized water, and MCs or NCs formed *in vivo* were pelleted and separated from cellular debris by centrifugation at 120 × *g* for 3 min.

Scanning electron microscopy (SEM)

The cells of *C. albicans*, *C. glabrata*, *C. krusei* and *C. parapsilosis*, after exposure to Pb^{2+}, Hg^{2+} or Cd^{2+}, were centrifuged at 3500 × *g* for 10 min at 4°C, and cell pellet was thoroughly washed four times with sterile deionized water. The cells were then lyophilized in a Tousimis Auto-Samdri critical point dryer 815 for 4 h. The dried samples were covered with a layer of colloidal gold. Then, the samples were observed in the scanning electron microscope, model EVO HD15, high-definition ZEISS®. Finally, the samples were photographed using the secondary electron detector (SE1) at 15 kV in high vacuum conditions at a working distance of 4 mm. We used, additionally, a SEM from TESCAN (Brno, Czech Republic) model VEGA3 SB, for obtaining the images of the crystals synthesized *in vitro* through a secondary electron detector (SE) from 10 to 20 kV in high vacuum conditions (work distance of 10 mm). All crystalline samples were sputtered with a gold film in order to increase the resolution imaging and contrast.

X-ray powder diffraction (XRPD)

X-ray powder diffraction (XRPD) analysis was performed at the X-ray diffraction beamline (XRD1) of the Elettra Synchrotron, Trieste (Italy; Lausi *et al.*, 2015). Powder diffraction patterns were collected, in transmission mode, at room temperature (25°C) with a monochromatic wave-length of 0.77491 Å (16 keV) and 200 × 200 μm^2 spot size, using a Pilatus 2M hybrid-pixel area detector. Samples of lyophilized *Candida* cells were packed in borosilicate capillaries (700 μm diameter and 10 μm wall thickness). Blank samples were analysed in the same way, collecting data on lyophilized cells not exposed to heavy metals, but treated with the same protocol (growth parameters, washing and lyophilization steps). The small amount of cadmium-loaded samples available (limited by high metal toxicity) prevented the usage of capillaries; these powders were therefore 'glued' on a 300 μm circular Kapton loop using N-paratone. Blank pattern for these samples refers to diffraction of the empty loop containing similar amount of oil. Bidimensional powder patterns have been integrated using Fit2D program (Hammersley *et al.*, 1996; Hammersley, 2016), after preliminary calibration of hardware set-up, using a capillary filled with boron lanthanide (LaB_6) standard reference powder (NIST 660a). Fluorescence spectra have been recorded for all the samples, during diffraction data acquisition on a Silicon drift Amptek X-123SDD detector, perpendicular to X-ray beam.

Synthesis of crystals in vitro *and crystal growth in different kinds of gels*

Crystals of PbS, HgS and CdS were synthesized by the gel method using two types of hydrogels: the first one was a low melting point agarose, and the second one was a gel obtained by the neutralization of sodium metasilicate solution with acetic acid. In the following paragraphs, we describe each of the gel preparations and recipes used in the chemical synthesis of these crystals:

Agarose preparation. Agarose gel 1.0% (w/v) stock solution of low melting point agarose ($T_{gel} = 20$°C, Hampton Research Cod. HR8-092) can be prepared following the conventional procedure: dissolve 0.1 g agarose in 10 ml of water heated at 90°C until obtaining a transparent solution when stirring. Then, this solution is passed through a 0.22 μm porosity membrane filter for removing all dust particles or insoluble fibres of agarose. The gel solution is kept at 10°C in the fridge in order to avoid any contamination. Prior to the crystallization in agarose, the gel can be heated at 90°C by using a heating plate in order to melt the agarose. The mixture 1:1 of 1% (w/v) agarose and 0.1 M Na_2S (prepared in buffer phosphate pH 2.0) is obtained. This helped us to obtain the first gel layer with one of the reactants included in the agarose (reactant A). Then, a second layer of 0.5% (w/v) agarose in water (ready for crystal growth experiments) is poured onto the top the first gel

layer. It is important to emphasize that this sandwich gel layer must not contain any of the two reactants of the precipitating reaction. The second reactant (called B and located in the third gel layer, which is a jellified solution of the heavy metal nitrate (XNO_3: X can be Pb^{2+}, Hg^{2+} and Cd^{2+})) is poured onto the top of the second gel layer. The sulfur ions counter-diffuse through the second layer of the gel. The second reactant (XNO_3) of the reaction is also counter-diffused through the second layer to produce the crystals in the middle of the sandwich gel layer. Although, generally speaking, agarose has been the most popular gel for the crystallization of different substances, there are other types of gels that have also been used for the crystallization of organic/organometallic substances when organic solvents are needed (Choquesillo-Lazarte and Garcia-Ruiz, 2011).

Gel made of sodium metasilicate. Commercial sodium metasilicate stock solution (Na_2SiO_3, $\rho = 1.39$ g ml^{-1}, Cat. No. 33,844-3, Aldrich) was diluted with water to prepare a working solution with a density of 1.06 g ml^{-1} using the relationship $V_{SS} = (0.06\ V_a)/(1.39-\rho^T)$, where V_{SS} is the volume of stock solution required to prepare a final volume V_a and ρ^T is the density of water at temperature T (Moreno *et al.*, 1999). The sodium hydroxide contained in this sodium metasilicate solution ($\rho = 1.06$ g ml^{-1}) was used to neutralize a 1 M acetic acid solution in order to obtain a monosilicic acid (H_4SiO_4) solution that polymerizes as a polysiloxane hydrogel. The same chemical gel can be prepared by the hydrolysis of either tetramethyl orthosilicate (TMOS) or tetraethyl orthosilicate (TEOS) as described elsewhere (Robert and Lefaucheux, 1988). Protection of skin and eyes is recommended during the handling of siloxanes as they are corrosive liquids. Glassware must be thoroughly rinsed with ethanol prior to cleaning them with water. As previously mentioned for agarose, the reactants can be incorporated to the silica gel by first diffusing them through the gel. The second layer, free of reactants, is then put in a sandwich configuration. Finally, the synthesis of these sulfides (located in the third gel layer of these heavy metals: Pb^{2+}, Hg^{2+} and Cd^{2+}) will occur by counter-diffusion processes in the sandwich gel layer.

Acknowledgements

We are grateful to Dr. Ricardo Navarro and MSc Paulina Lozano-Sotomayor from the Laboratorio Nacional, Universidad de Guanajuato, México, for the facilities and technical assistance with the SEM photographs. The authors (M.C.C. and A. M.) acknowledge the XRD1-Hard X-ray Diffraction Beamline and Structural

Biology Laboratory of the Elettra Synchrotron, in Italy, for the support and beamtime awarded to collect data from the different micro- or nanocrystals. We give our appreciation to Prof. Everardo López-Romero of the Universidad de Guanajuato for his support of cell lysis. This work was carried out with the financial support granted to Dr. M. Cuéllar-Cruz by Proyecto-Institucional-831/2016 from Universidad de Guanajuato, México. The authors acknowledge Prof. John Dye and Ms. Antonia Sánchez-Marín for the English revision of this contribution.

Conflict of interest

The authors declare that they have no competing interest.

References

Adams, M.J., Hodgkin, D.C. and Raeburn, U.A. (1970) Crystal structure of a complex of mercury(II) chloride and histidine hydrochloride. *J Chem Soc A* **263**, 2–2635.

Anderson, M.E. (1997) Glutathione and glutathione delivery compounds. *Adv Pharmacol* **38**: 65–78.

Anderson, J.B. (2005) Evolution of antifungal-drug resistance: mechanisms and pathogen fitness. *Nat Rev Microbiol* **3**: 547–556.

Arino, J., Ramos, J., and Sychrova, H. (2010) Alkali metal cation transport and homeostasis in yeasts. *Microbiol Mol Biol Rev* **74**: 95–120.

Aslangul, C., Gallais, F., Daudel, R., and Veillard, A. (1972) Study on localized distribution of electrons in molecule containing at time one covalent bond and one coordinate bond, established between boron intersphere and nitrogen intersphere. *C R Hebd Seances Acad Sci C* **273**: 33.

Ausubel, F., Brent, R., Kingston, R.E., Moore, D.D., Seidman, J.G., Smith, J.A., *et al.* (2001) *Current Protocols in Molecular Biology.* New York, NY: John Wiley & Sons.

Auvray, P., and Genet, F. (1973) Refining crystal-structure of cinnabar (Alpha Hgs). *Bull Soc Fr Mineral Crist* **96**: 218–219.

Baldrian, P. (2003) Interactions of heavy metals with white-rot fungi. *Enzyme Microb Technol* **32**: 78–91.

Barnes, C.L., and Bosch, E. (2006) Synthesis and X-ray crystal structure of a complex formed by reaction of 1,2-bis(2 '-pyridylethynyl)benzene and mercury(II) chloride. *J Chem Crystallogr* **36**: 563–566.

Bensen, E.S., Martin, S.J., Li, M., Berman, J., and Davis, D.A. (2004) Transcriptional profiling in *Candida albicans* reveals new adaptive responses to extracellular pH and functions for Rim101p. *Mol Microbiol* **54**: 1335–1351.

Bishnoi, N.R., and Garima, A. (2005) Fungus – an alternative for bioremediation of heavy metal containing wastewater: a review. *J Sci Ind Res* **64**: 93–100.

Bisogni, J.J. and Lawrence, A.W. (1973) Kinetics of microbially mediated methylation of mercury in aerobic and anaerobic aquatic conditions. U.S. Nat. Tech. Inform. Serv., PBRep. No. 222025/9, 195.

Boggon, T.J., and Shapiro, L. (2000) Screening for phasing atoms in protein crystallography. *Structure* **8**: R143–R149.

Borrelli, N.F., Hall, D.W., Holland, H.J., and Smith, D.W. (1987) Quantum confinement effects of semiconducting microcrystallites in glass. *J Appl Phys* **61**: 5399–5409.

Bridges, C.C., and Zalups, R.K. (2005) Molecular and ionic mimicry and the transport of toxic metals. *Toxicol Appl Pharmacol* **204**: 274–308.

Bui, D.M., Gregan, J., Jarosch, E., Ragnini, A., and Schweyen, R.J. (1999) The bacterial magnesium transporter CorA can functionally substitute for its putative homologue Mrs2p in the yeast inner mitochondrial membrane. *J Biol Chem* **274**: 20438–20443.

Choquesillo-Lazarte, D., and Garcia-Ruiz, J.M. (2011) Poly (ethylene) oxide for small-molecule crystal growth in gelled organic solvents. *J Appl Crystallogr* **44**: 172–176.

Chwastowska, J., Rogowska, A., Sterlinska, E., and Dudek, J. (1999) Chelating 2-mercaptobenzothiazole loaded resin. Application to the separation of inorganic and alkylmercury species for their atomic absorption spectrometry determination in natural waters. *Talanta* **49**: 837–842.

Clever, H.L., Johnson, S.A., and Derrick, M.E. (1985) The solubility of mercury and some sparingly soluble mercury salts in water and aqueous-electrolyte solutions. *J Phys Chem Ref Data* **14**: 631–681.

Cuellar-Cruz, M., Briones-Martin-Del-Campo, M., Canas-Villamar, I., Montalvo-Arredondo, J., Riego-Ruiz, L., Castano, I., and Penas, A. (2008) High resistance to oxidative stress in the fungal pathogen *Candida glabrata* is mediated by a single catalase, Cta1p, and is controlled by the transcription factors Yap1p, Skn7p, Msn2p, and Msn4p. *Eukaryot Cell* **7**: 814–825.

Cuellar-Cruz, M., Vega-Gonzalez, A., Mendoza-Novelo, B., Lopez-Romero, E., Ruiz-Baca, E., Quintanar-Escorza, M.A., and Villagomez-Castro, J.C. (2012) The effect of biomaterials and antifungals on biofilm formation by *Candida* species: a review. *Eur J Clin Microbiol Infect Dis* **31**: 2513–2527.

Cyert, M.S., and Philpott, C.C. (2013) Regulation of cation balance in *Saccharomyces cerevisiae*. *Genetics* **193**: 677–713.

Da Costa, C.P., and Sigel, H. (2000) Lead(II)-binding properties of the 5 '-monophosphates of adenosine (AMP(2-)), inosine (IMP2-), and guanosine (GMP(2-)) in aqueous solution. Evidence for nucleobase-lead(II) interactions. *Inorg Chem* **39**: 5985–5993.

Dameron, C.T., Reese, R.N., Mehra, R.K., Kortan, A.R., Carroll, P.J., Steigerwald, M.L., *et al.* (1989) Biosynthesis of cadmium-sulfide quantum semiconductor crystallites. *Nature* **338**: 596–597.

Davis, S.A., Dujardin, E., and Mann, S. (2003) Biomolecular inorganic materials chemistry. *Curr Opin Solid State Mater Sci* **7**: 273–281.

Elskens, M.T., Jaspers, C.J., and Penninckx, M.J. (1991) Glutathione as an endogenous sulfur source in the yeast *Saccharomyces cerevisiae*. *J Gen Microbiol* **137**: 637–644.

Fidan, I., Kalkanci, A., Yesilyurt, E., and Erdal, B. (2014) In vitro effects of *Candida albicans* and *Aspergillus fumigatus* on dendritic cells and the role of beta glucan in this effect. *Adv Clin Exp Med* **23:** 17–24.

Flora, S.J.S., Mittal, M., and Mehta, A. (2008) Heavy metal induced oxidative stress & its possible reversal by chelation therapy. *Indian J Med Res* **128:** 501–523.

Gadd, G.M., and White, C. (1993) Microbial treatment of metal pollution – a working biotechnology. *Trends Biotechnol* **11:** 353–359.

Garciaruiz, J.M. (1986) Growth history of Pbs single crystals at room temperature. *J Cryst Growth* **75:** 441–453.

Garciaruiz, J.M. (1991) The uses of crystal-growth in gels and other diffusing-reacting systems. In *Crystalline Materials: Growth and Characterization*. Rodriguez-Clemente, R. (ed.). pp. 87–106.

Garcia-Ruiz, J.M., Gonzalez-Ramirez, L.A., Gavira, J.A., and Otalora, F. (2002) Granada Crystallisation Box: a new device for protein crystallisation by counter-diffusion techniques. *Acta Crystallogr Sect D Biol Crystallogr* **58:** 1638–1642.

Goyer, R.A. (2003) Response to comments by professor duffus regarding chapter 23, toxic effects of metals, in cassarett and doull's toxicology, 6th edition, C. D. Klaassen, Editor. New York: McGraw Hill, 2001. *Arch Environ Health* **58,** 265–266.

Grill, E., Winnacker, E.L., and Zenk, M.H. (1986) Synthesis of 7 different homologous phytochelatins in metal-exposed *Schizosaccharomyces pombe* cells. *FEBS Lett* **197:** 115–120.

Gupta, R., Ahuja, P., Khan, S., Saxena, R.K., and Mohapatra, H. (2000) Microbial biosorbents: meeting challenges of heavy metal pollution in aqueous solutions. *Curr Sci* **78:** 967–973.

Hagler, A.N., and Mendonca-Hagler, L.C. (1981) Yeasts from marine and estuarine waters with different levels of pollution in the state of rio de janeiro, Brazil. *Appl Environ Microbiol* **41:** 173–178.

Hammersley, A.P. (2016) FIT2D: a multi-purpose data reduction, analysis and visualization program. *J Appl Crystallogr* **49:** 646–652.

Hammersley, A.P., Svensson, S.O., Hanfland, M., Fitch, A.N., and Hausermann, D. (1996) Two-dimensional detector software: from real detector to idealised image or two-theta scan. *High Pressure Res* **14:** 235–248.

Han, L., Hou, P.F., Feng, Y., Liu, H., Li, J.L., Peng, Z.J., and Yang, J. (2014) Phase transfer-based synthesis of HgS nanocrystals. *Dalton Trans* **43:** 11981–11987.

Hayashi, Y., Nakagawa, C.W., and Murasugi, A. (1986) Unique properties of Cd-binding peptides induced in fission yeast, *Schizosaccharomyces pombe*. *Environ Health Persp* **65:** 13–19.

Henisch, H.K., and Garciaruiz, J.M. (1986a) Crystal-growth in gels and liesegang ring formation: 1. Diffusion relationships. *J Cryst Growth* **75:** 195–202.

Henisch, H.K., and Garciaruiz, J.M. (1986b) Crystal-Growth in gels and liesegang ring formation: 2. Crystallization criteria and successive precipitation. *J Cryst Growth* **75:** 203–211.

Herron, N., Wang, Y., Eddy, M.M., Stucky, G.D., Cox, D.E., Moller, K., and Bein, T. (1989) Structure and optical-properties of CdS superclusters in zeolite hosts. *J Am Chem Soc* **111:** 530–540.

Hoeijmakers, J.H. (2001) Genome maintenance mechanisms for preventing cancer. *Nature* **411:** 366–374.

Inouye, K., Endo, R., Otsuka, Y., Miyashiro, K., Kaneko, K., and Ishikawa, T. (1982) Oxygenation of ferrous-ions in reversed micelle and reversed micro-emulsion. *J Phys Chem* **86:** 1465–1469.

Jacob, C., Courbot, M., Brun, A., Steinman, H.M., Jacquot, J.P., Botton, B., and Chalot, M. (2001) Molecular cloning, characterization and regulation by cadmium of a superoxide dismutase from the ectomycorrhizal fungus *Paxillus involutus*. *Eur J Biochem* **268:** 3223–3232.

Jeong, K.S., Deng, Z.Y., Keuleyan, S., Liu, H., and Guyot-Sionnest, P. (2014) Air-stable n-doped colloidal HgS quantum dots. *J Phys Chem Lett* **5:** 1139–1143.

Jin, Y.H., Clark, A.B., Slebos, R.J.C., Al-Refai, H., Taylor, J.A., Kunkel, T.A., *et al.* (2003) Cadmium is a mutagen that acts by inhibiting mismatch repair. *Nat Genet* **34:** 326–329.

Kachur, A.V., Koch, C.J., and Biaglow, J.E. (1998a) Mechanism of copper-catalyzed oxidation of glutathione. *Free Radical Res* **28:** 259–269.

Kachur, A.V., Tuttle, S.W., and Biaglow, J.E. (1998b) Autoxidation of ferrous ion complexes: a method for the generation of hydroxyl radicals. *Radiat Res* **150:** 475–482.

Klaunig, J.E., Xu, Y., Isenberg, J.S., Bachowski, S., Kolaja, K.L., Jiang, J.Z., *et al.* (1998) The role of oxidative stress in chemical carcinogenesis. *Environ Health Perspect* **106:** 289–295.

Ko, D.K., Maurano, A., Suh, S.K., Kim, D., Hwang, G.W., Grossmann, J.C., *et al.* (2016) Photovoltaic performance of PbS quantum dots treated with metal salts. *ACS Nano* **10:** 3382–3388.

Korbekandi, H., Jouneghani, R.M., Mohseni, S., Pourhossein, M., and Iravani, S. (2014) Synthesis of silver nanoparticles using biotransformations by *Saccharomyces boulardii*. *Green Process Synth* **3:** 271–277.

Kowshik, M., Deshmukh, N., Vogel, W., Urban, J., Kulkarni, S.K., and Paknikar, K.M. (2002) Microbial synthesis of semiconductor CdS nanoparticles, their characterization, and their use in the fabrication of an ideal diode. *Biotechnol Bioeng* **78:** 583–588.

Krumov, N., Oder, S., Perner-Nochta, I., Angelov, A., and Posten, C. (2007) Accumulation of CdS nanoparticles by yeasts in a fed-batch bioprocess. *J Biotechnol* **132:** 481–486.

Kundrot, C.E., Judge, R.A., Pusey, M.L., and Snell, E.H. (2001) Microgravity and macromolecular crystallography. *Cryst Growth Des* **1:** 87–99.

Kwak, Y.H., Lee, D.S., and Kim, H.B. (2003) *Vibrio harveyi* nitroreductase is also a chromate reductase. *Appl Environ Microbiol* **69:** 4390–4395.

Landner, L. (1972) The biological alkylation of mercury. *Biochem J* **130:** 67P–69P.

Lausi, A., Polentarutti, M., Onesti, S., Plaisier, J.R., Busetto, E., Bais, G., *et al.* (2015) Status of the crystallography beamlines at Elettra. *Eur Phys J Plus* **130:** 43.

Lee, J.W., Kim, D.Y., Baek, S., Yu, H., and So, F. (2016) Photodetectors: inorganic UV-Visible-SWIR broadband photodetector based on monodisperse PbS nanocrystals (Small 10/2016). *Small* **12:** 1246.

Lim, S., Kim, Y., Lee, J., Han, C.J., Kang, J., and Kim, J. (2014) Investigation of colloidal PbS quantum dot-based solar cells with near infrared emission. *J Nanosci Nanotechnol* **14:** 9346–9350.

Littke, W., and John, C. (1984) Materials: protein single crystal growth under microgravity. *Science* **225:** 203–204.

Long, M.M., Delucas, L.J., Smith, C., Carson, M., Moore, K., Harrington, M.D., *et al.* (1994) Protein Crystal-Growth in microgravity – temperature induced large scale crystallization of insulin. *Microgravity Sci Technol* **7:** 196–202.

Lopez-Archilla, A.I., Gonzalez, A.E., Terron, M.C., and Amils, R. (2004) Ecological study of the fungal populations of the acidic Tinto river in southwestern Spain. *Can J Microbiol* **50:** 923–934.

Lorber, B., Sauter, C., Theobald-Dietrich, A., Moreno, A., Schellenberger, P., Robert, M.C., *et al.* (2009) Crystal growth of proteins, nucleic acids, and viruses in gels. *Prog Biophys Mol Biol* **101:** 13–25.

Loukidou, M.X., Zouboulis, A.I., Karapantsios, T.D., and Matis, K.A. (2004) Equilibrium and kinetic modeling of chromium(VI) biosorption by *Aeromonas caviae*. *Colloids Surf A Physicochem Eng Asp* **242:** 93–104.

Lowman, D.W., Greene, R.R., Bearden, D.W., Kruppa, M.D., *et al.* (2014) Novel structural features in *Candida albicans* hyphal glucan provide a basis for differential innate immune recognition of hyphae versus yeast. *J Biol Chem* **289:** 3432–3443.

Luo, H., Lu, Y.D., Shi, X.L., Mao, Y., and Dalal, N.S. (1996) Chromium(IV)-mediated Fenton-like reaction causes DNA damage: implication to genotoxicity of chromate. *Ann Clin Lab Sci* **26:** 185–191.

Malik, M.A., Motevalli, M., Saeed, T., and Obrien, P. (1993) Methylzinc or methylcadmium-N, N, N-Trimethylpropylenediaminedithiocarbamates – precursors for zinc or cadmium-sulfide the X-Ray crystal structure of methylcadmiumtrimethylpropylenediaminedithiocarbamate benzene solvate. *Adv Mater* **5:** 653–654.

Mann, S., and Ozin, G.A. (1996) Synthesis of inorganic materials with complex form. *Nature* **382:** 313–318.

Mann, S., and Weiner, S. (1999) Biomineralization: structural questions at all length scales. *J Struct Biol* **126:** 179–181.

McMurray, C.T., and Tainer, J.A. (2003) Cancer, cadmium and genome integrity. *Nat Genet* **34:** 239–241.

Mehra, R.K., Tarbet, E.B., Gray, W.R., and Winge, D.R. (1988) Metal-specific synthesis of two metallothioneins and gamma-glutamyl peptides in *Candida glabrata*. *Proc Natl Acad Sci USA* **85:** 8815–8819.

Meister, A. (1995) Glutathione biosynthesis and its inhibition. *Biothiols, Pt B* **252:** 26–30.

Mendez-Armenta, M., and Rios, C. (2007) Cadmium neurotoxicity. *Environ Toxicol Pharmacol* **23:** 350–358.

Moreno, A., Juarez-Martinez, G., Hernandez-Perez, T., Batina, N., Mundo, M., and McPherson, A. (1999) Physical and chemical properties of gels - Application to protein nucleation control in the gel acupuncture technique. *J Cryst Growth* **205:** 375–381.

Muhlschlegel, F.A., and Fonzi, W.A. (1997) *PHR2* of *Candida albicans* encodes a functional homolog of the pH-regulated gene PHR1 with an inverted pattern of pH-dependent expression. *Mol Cell Biol* **17:** 5960–5967.

Mukherjee, P.K., Chandra, J., Kuhn, D.A., and Ghannoum, M.A. (2003) Mechanism of fluconazole resistance in *Candida albicans* biofilms: phase-specific role of efflux pumps and membrane sterols. *Infect Immun* **71:** 4333–4340.

Murasugi, A., Nakagawa, C.W., and Hayashi, Y. (1984) Formation of cadmium binding peptide allomorphs in fission yeast. *J Biochem* **96:** 1375–1379.

Nies, D.H. (2003) Efflux-mediated heavy metal resistance in prokaryotes. *FEMS Microbiol Rev* **27:** 313–339.

Nies, D.H., and Silver, S. (1995) Ion efflux systems involved in bacterial metal resistances. *J Ind Microbiol* **14:** 186–199.

Noda, Y., Masumoto, K., Ohba, S., Saito, Y., Toriumi, K., Iwata, Y., and Shibuya, I. (1987) Temperature-dependence of atomic thermal parameters of lead chalcogenides, Pbs, Pbse and Pbte. *Acta Crystallogr C* **43:** 1443–1445.

Noor, N.A., Ikram, N., Ali, S., Nazir, S., Alay-e-Abbas, S.M., and Shaukat, A. (2010) First-principles calculations of structural, electronic and optical properties of CdxZn1-xS alloys. *J Alloys Compd* **507:** 356–363.

Nordberg, M., and Nordberg, G.F. (2000) Toxicological aspects of metallothionein. *Cell Mol Biol* **46:** 451–463.

Onyido, I., Norris, A.R., and Buncel, E. (2004) Biomolecule-mercury interactions: modalities of DNA base-mercury binding mechanisms. Remediation strategies. *Chem Rev* **104:** 5911–5929.

Pan-Hou, H.S., and Imura, N. (1982) Involvement of mercury methylation in microbial mercury detoxication. *Arch Microbiol* **131:** 176–177.

Pastore, A., Federici, G., Bertini, E., and Piemonte, F. (2003) Analysis of glutathione: implication in redox and detoxification. *Clin Chim Acta* **333:** 19–39.

Payne, J.C., ter Horst, M.A., and Godwin, H.A. (1999) Lead fingers: Pb^{2+} binding to structural zinc-binding domains determined directly by monitoring lead-thiolate charge-transfer bands. *J Am Chem Soc* **121:** 6850–6855.

Pfeiffer, J., Guhl, J., Waidner, B., Kist, M., and Bereswill, S. (2002) Magnesium uptake by CorA is essential for viability of the gastric pathogen *Helicobacter pylori*. *Infect Immun* **70:** 3930–3934.

Pham, A.L.T., Johnson, C., Manley, D., and Hsu-Kim, H. (2015) Influence of sulfide nanoparticles on dissolved mercury and zinc quantification by diffusive gradient in thin-film passive samplers. *Environ Sci Technol* **49:** 12897–12903.

Prasad, K., and Jha, A.K. (2010) Biosynthesis of CdS nanoparticles: an improved green and rapid procedure. *J Colloid Interface Sci* **342:** 68–72.

Ramage, G., Bachmann, S., Patterson, T.F., Wickes, B.L., and Lopez-Ribot, J.L. (2002) Investigation of multidrug efflux pumps in relation to fluconazole resistance in *Candida albicans* biofilms. *J Antimicrob Chemother* **49:** 973–980.

Ramirez-Quijas, M.D., Zazueta-Sandoval, R., Obregon-Herrera, A., Lopez-Romero, E., and Cuellar-Cruz, M. (2015) Effect of oxidative stress on cell wall morphology in four pathogenic *Candida* species. *Mycol Prog* **14:** 8.

Reese, R.N., and Winge, D.R. (1988) Sulfide stabilization of the cadmium-gamma-glutamyl peptide complex of *Schizosaccharomyces pombe*. *J Biol Chem* **263:** 12832–12835.

Robert, M.C., and Lefaucheux, F. (1988) Crystal-growth in gels – principle and applications. *J Cryst Growth* **90**: 358–367.

Rosen, B.P. (1999) The role of efflux in bacterial resistance to soft metals and metalloids. *Essays Biochem* **34**: 1–15.

Serrano-Fujarte, I., Lopez-Romero, E., Reyna-Lopez, G.E., Martinez-Gamez, M.A., Vega-Gonzalez, A., and Cuellar-Cruz, M. (2015) Influence of culture media on biofilm formation by *Candida* species and response of sessile cells to antifungals and oxidative stress. *Biomed Res Int* **2015**: Article ID 783639, 15 pages.

Shakouri-Arani, M., and Salavati-Niasari, M. (2014) Synthesis and characterization of cadmium sulfide nanocrystals in the presence of a new sulfur source via a simple solvothermal method. *New J Chem* **38**: 1179–1185.

Sierra-Alvarez, R. (2007) Fungal bioleaching of metals in preservative-treated wood. *Process Biochem* **42**: 798–804.

Silva, S., Negri, M., Henriques, M., Oliveira, R., Williams, D.W., and Azeredo, J. (2012) *Candida glabrata*, *Candida parapsilosis* and *Candida tropicalis*: biology, epidemiology, pathogenicity and antifungal resistance. *FEMS Microbiol Rev* **36**: 288–305.

Suihko, M.L., and Hoekstra, E.S. (1999) Fungi present in some recycled fibre pulps and paperboards. *Nor Pulp Pap Res J* **14**: 199–203.

Trajanovska, S., Britz, M.L., and Bhave, M. (1997) Detection of heavy metal ion resistance genes in gram-positive and gram-negative bacteria isolated from a lead-contaminated site. *Biodegradation* **8**: 113–124.

Vanderrest, M.E., Kamminga, A.H., Nakano, A., Anraku, Y., Poolman, B., and Konings, W.N. (1995) The plasma-membrane of *Saccharomyces cerevisiae* - structure, function, and biogenesis. *Microbiol Rev* **59**: 304–322.

Williams, P., KeshavarzMoore, E., and Dunnill, P. (1996) Efficient production of microbially synthesized cadmium sulfide quantum semiconductor crystallites. *Enzyme Microb Technol* **19**: 208–213.

Wood, J.M., Scott-Kennedy, F., and Rosen, C.G. (1968) Synthesis of methylmercury compounds by extracts of a methanogenic bacterium. *Nature* **220**: 173–174.

Xiang, C., and Oliver, D.J. (1998) Glutathione metabolic genes coordinately respond to heavy metals and jasmonic acid in Arabidopsis. *Plant Cell* **10**: 1539–1550.

Yang, J., Hu, Y.P., Luo, J., Zhu, Y.H., and Yu, J.S. (2015) Highly Fluorescent, Near-infrared-emitting Cd^{2+}-tuned HgS nanocrystals with optical applications. *Langmuir* **31**: 3500–3509.

Yoon, K.P., and Silver, S. (1991) A second gene in the *Staphylococcus aureus* cadA cadmium resistance determinant of plasmid pI258. *J Bacteriol* **173**: 7636–7642.

Effect of temperature and colonization of *Legionella pneumophila* and *Vermamoeba vermiformis* on bacterial community composition of copper drinking water biofilms

Helen Y. Buse,[1,*] Pan Ji,[2] Vicente Gomez-Alvarez,[1] Amy Pruden,[2] Marc A. Edwards[2] and Nicholas J. Ashbolt[3]

[1]*Pegasus Technical Services, Inc c/o US EPA, 26 W Martin Luther King Drive NG-16, Cincinnati, OH 45268, USA.*

[2]*Department of Civil and Environmental Engineering, Virginia Tech, Blacksburg, VA, USA.*

[3]*School of Public Health, University of Alberta, Edmonton, AB T6G 2G7, Canada.*

Summary

It is unclear how the water-based pathogen, *Legionella pneumophila* (Lp), and associated free-living amoeba (FLA) hosts change or are changed by the microbial composition of drinking water (DW) biofilm communities. Thus, this study characterized the bacterial community structure over a 7-month period within mature (> 600-day-old) copper DW biofilms in reactors simulating premise plumbing and assessed the impact of temperature and introduction of Lp and its FLA host, *Vermamoeba vermiformis* (Vv), co-cultures (LpVv). Sequence and quantitative PCR (qPCR) analyses indicated a correlation between LpVv introduction and increases in *Legionella* spp. levels at room temperature (RT), while at 37°C, Lp became the dominant *Legionella* spp. qPCR analysis suggested Vv presence may not be directly associated with Lp biofilm growth at RT and 37°C, but may contribute to or be associated with non-Lp legionellae persistence at RT. Two-way PERMANOVA and PCoA revealed that temperature was a major driver

of microbiome diversity. Biofilm community composition also changed over the seven-month period and could be associated with significant shifts in dissolved oxygen, alkalinity and various metals in the influent DW. Hence, temperature, biofilm age, DW quality and transient intrusions/amplification of pathogens and FLA hosts may significantly impact biofilm microbiomes and modulate pathogen levels over extended periods.

Introduction

Biofilms are an important source of microorganisms within drinking water (DW) systems, where sessile growth confers a growth advantage compared to their planktonic counterparts [reviewed in Berry *et al.*, 2006]. Drinking water biofilms are ubiquitous within distribution systems and in premise (i.e. building) plumbing, providing an ideal microenvironment for a diverse array of microorganisms, with their development influenced by water chemistry (Ji *et al.*, 2015); disinfectant residual (Shen *et al.*, 2016); pipe material/corrosion products (Yu *et al.*, 2010; Buse *et al.*, 2013b; Lin *et al.*, 2013; Buse *et al.*, 2014); biofilm age (Martiny *et al.*, 2003; Lee *et al.*, 2005); and composition of upstream microbiomes (Pinto *et al.*, 2012; Lu *et al.*, 2014). Drinking water biofilms are viewed as a source of pathogens as they provide a reservoir for their accumulation/growth and subsequent release. Opportunistic, water-based pathogens, such as representative *Legionella* spp., *Mycobacterium* spp. and *Pseudomonas* spp., have all been reported to grow within DW biofilms (Falkinham *et al.*, 2001; Wingender and Flemming, 2004; Feazel *et al.*, 2009; Buse *et al.*, 2014; Lu *et al.*, 2014). In particular, *L. pneumophila* may now account for some 66% of DW-related outbreaks in the United States, noting a 286% increase in legionellosis cases between 2000 and 2014 (0.42–1.62 cases per 100 000 persons; Garrison *et al.*, 2016). Similarly in Europe, between 2012 and 2014, an increase in Legionnaires' disease cases (5852–6941 respectively) was also observed with the majority of those due to *L. pneumophila* and hot- and cold-water systems and cooling towers being implicated as likely sources (ECDC, 2016). Coupled with the significant healthcare costs due to

*For correspondence. E-mail buse.helen@epa.gov

Funding Information
Alfred P. Sloan Foundation Microbiology of the Built Environment programme, National Science Foundation CBET award (1336650), Virginia Tech Institute for Critical Technology and Applied Science, and Alberta Innovates-Health Solutions.

Legionnaires' disease and non-tuberculous mycobacteria (with over 400 and 6000 hospital cases and associated healthcare costs of $9.4 and 46 million between 2004 and 2007 in the United States respectively; Collier et al., 2012), these water-based pathogens are a major concern from economic and public health standpoints.

Drinking water biofilm development has been shown to differ based on plumbing surface material such as chlorinated, plasticized and unplasticized polyvinylchloride (cPVC, pPVC and uPVC, respectively), polybutylene, cast iron, cross-linked polyethylene (PEX), copper (Cu), stainless steel (SS) and glass (Rogers and Keevil, 1992; Rogers et al., 1994; Armon et al., 1997; Kuiper et al., 2004; Lehtola et al., 2007; Moritz et al., 2010; Valster et al., 2010; Messi et al., 2011; Lin et al., 2013). Unfortunately, most biofilm-related studies to date are on the order of days to weeks with water chemistries not representative of DW, and longer-term comparisons of L. pneumophila-associated biofilm growth between surface materials are both rare and often conflicting. After laboratory inoculation of L. pneumophila into biofilm reactors operated at 20°C, culturable L. pneumophila can persist for up to 4 weeks in PEX- and Cu-grown DW biofilms and up to 4 months in uPVC-grown DW biofilms (Moritz et al., 2010; Buse et al., 2013b). However, at the same temperature, L. pneumophila culturability was not supported within Cu DW biofilms compared to growth within polybutylene and cPVC DW biofilms; yet at 40°C, L. pneumophila was culturable for up to 3 weeks within Cu DW biofilms in the same study (Rogers et al., 1994). Survival of L. pneumophila was enhanced on PVC-grown relative to glass-grown DW biofilms (Armon et al., 1997), while colonization of Cu-grown DW biofilms occurred more readily and persistently than within uPVC-grown DW biofilms (Buse et al., 2013). Collectively, these studies indicate that legionellae-related biofilm development is a common, yet complex phenomenon that is heavily impacted by temperature, surface material, water quality and age, among other factors. Thus, it is difficult to interpret key factors impacting L. pneumophila biofilm colonization; however, characterizing the broader bacterial composition of biofilms could draw further insight. This could allow comparisons to be made across studies, providing principles of L. pneumophila ecological interactions and potentially explaining otherwise aberrant observations regarding the factors influencing its colonization of biofilms.

Free-living amoebae (FLA), such as Acanthamoeba spp. and Vermamoeba (formerly Hartmannella) vermiformis, are also a public health concern as they are known hosts for L. pneumophila (Harb et al., 2000; Lau and Ashbolt, 2009) and have been co-isolated from L. pneumophila-contaminated systems (Yamamoto et al., 1992; Thomas et al., 2006, 2008). Acanthamoeba spp.

are less abundant in freshwater and DW systems compared to V. vermiformis (Valster et al., 2010; Buse et al., 2013a), with the latter demonstrated to be the prominent host for L. pneumophila in DW derived from hospital and recreational building hot-water tanks (Wadowsky et al., 1988; Fields et al., 1989), within polyethylene- and pPVC-grown DW biofilms (Kuiper et al., 2004; Valster et al., 2010) and within SS biofilms composed of P. aeruginosa, Klebsiella pneumoniae and Flavobacterium spp. (Murga et al., 2001). A previous study reported high infectivity and intracellular replication of L. pneumophila strain Chicago-2 within the strain CDC-19 of permissive amoeba host, V. vermiformis (Buse and Ashbolt, 2011); thus, both were chosen for use in this study.

The aim of this study was to profile bacterial diversity and relative abundance within complex, mature DW biofilms (initially grown on Cu surfaces at pH > 8.2 for over 600 days) and track taxonomic changes as a function of operating temperature and introduction of L. pneumophila and V. vermiformis leading to colonization over an extended period (7 months). The overall goal of this work was to advance understanding of complex microbial niches conducive to water-based pathogen colonization and survival within DW biofilms.

Results

Quantification of Legionella spp., L. pneumophila and V. vermiformis within Cu DW biofilms

All control (Con) and LpVv reactor biofilm samples contained levels of Legionella spp. detectable by quantitative PCR (qPCR; Fig. 1A and B respectively). Legionella spp. were consistently detected in all Cu biofilms from Con_RT reactors, ranging 1.9–2.9 $\log_{10}$ cell equivalents (CE) cm^{-2} from day 0 (before LpVv inoculation) to month 7 (except month 5, where levels dropped to 1.1 $\log_{10}$ CE cm^{-2}; Fig. 1A, blue circles). In contrast, Legionella spp. were detected only at day 0 to week 1, month 2, month 6 and month 7, with levels between 1.2 and 1.7 $\log_{10}$ CE cm^{-2}, with a spike at month 6 of 2.7 $\log_{10}$ CE cm^{-2} in Con_37°C biofilms (Fig. 1A, red circles). Notably, after the introduction of LpVv co-cultures into the reactors (E2; Fig. S1), Legionella spp. were detected in all biofilm samples at each temperature, with LpVv_RT biofilm samples displaying a more consistent Legionella spp. level of between 2.8 and 4.3 $\log_{10}$ CE cm^{-2} for all time points, while levels within the LpVv_37°C-derived biofilms fluctuated between 1.2 and 4.2 $\log_{10}$ CE cm^{-2} throughout the experiment (Fig. 1B). These observations based on qPCR were corroborated by the relative abundance of Legionella spp.-associated sequences within the biofilm samples from each reactor (Fig. 2).

Legionella pneumophila was not detected in control reactor biofilm samples as measured by qPCR (except

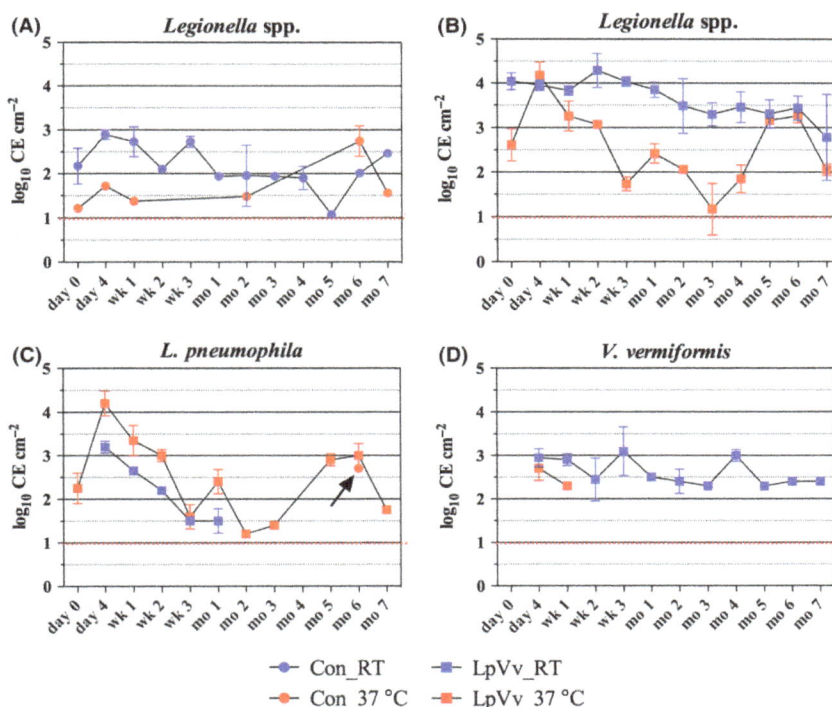

Fig. 1. Quantitative PCR (qPCR) analysis of *Legionella* spp., *Legionella pneumophila* and *Vermamoeba vermiformis*. Biofilm samples were subjected to qPCR assays targeting the 16S rRNA gene of *Legionella* spp. (A, B), the *sidF* gene of *L. pneumophila* (C) and the 18S rRNA gene of *V. vermiformis* (D). Data represent the mean for duplicate samples with standard error mean bars.
Legionella spp. quantification data for control (Con) and inoculated (LpVv) biofilm samples are shown separately (A and B respectively).
Legionella pneumophila was not detected in the control reactors except at the 6-month time point (indicated by the black arrow, C). The limit of quantification of 1 $\log_{10}$ CE cm^{-2} is indicated by the red dotted line.

at month 6 where duplicate biofilm samples contained Lp levels between 2.4 and 3.0 $\log_{10}$ CE cm^{-2}; Fig. 1C, red circles, black arrow). *Legionella pneumophila* levels in LpVv_RT biofilm samples were transient, detectable only between day 4 and month 1, with levels decreasing steadily post-inoculation from 3.2 to 1.5 $\log_{10}$ CE cm^{-2} (Fig. 1C, blue squares). Furthermore, Lp levels in LpVv_RT biofilms were most likely due to the LpVv co-culture inoculation as no Lp was detected at day 0 (before LpVv inoculation), which is in contrast to Lp levels in the LpVv_37°C biofilms, where starting Lp levels at day 0 (before LpVv inoculation) were 2.2 $\log_{10}$ CE cm^{-2} (Fig. 1C, red squares). Lp levels in LpVv_37°C biofilms proceeded to fluctuate between 1.2 and 4.2 $\log_{10}$ CE cm^{-2} in a similar pattern as observed for *Legionella* spp. levels in the same samples (Fig. 1B and C, red squares). Again, these qPCR-based observations were corroborated by the relative abundance of *Legionella* spp.-associated sequences within the biofilm samples from each reactor (Fig. 2). It has been previously demonstrated that partial 16S rRNA gene sequencing can identify *Legionella* isolates to the genus level and also allow for the differentiation of *L. pneumophila* from non-pneumophila *Legionella* species (Wilson *et al.*, 2007). Thus, in this study, the ability of the partial 16S

rRNA gene sequences to differentiate *L. pneumophila* from non-pneumophila *Legionella* species was confirmed by comparing sequences against the ribosomal databases SILVA, Greengenes and RDP, using 100% cut-off and similarity values for the analysis which yielded consistent results for each database.

Notably, Vv was not detected in any of the control biofilm samples, and in inoculated reactors, Vv was detected only transiently in the LpVv_37°C biofilms at day 4 and week 1 (Fig. 1D, red squares, 2.7 and 2.3 $\log_{10}$ CE cm^{-2} respectively). In contrast, Vv levels were consistently and persistently high in LpVv_RT biofilms (Fig. 1D, blue squares; 2.4–3.1 $\log_{10}$ CE cm^{-2}).

Bacterial diversity and relative abundance in Cu DW biofilms

Bacterial phylum-level taxonomic assignments revealed different compositions within each of the four reactors with members of the phylum *Proteobacteria* representing the dominant population within biofilm samples (Table 1). Specifically, the relative abundance of α-*Proteobacteria* in each reactor ranged from 41% to 57%; however, β-*Proteobacteria* were about twice as abundant in biofilms incubated at RT (18–19%) than at 37°C (9–11%)

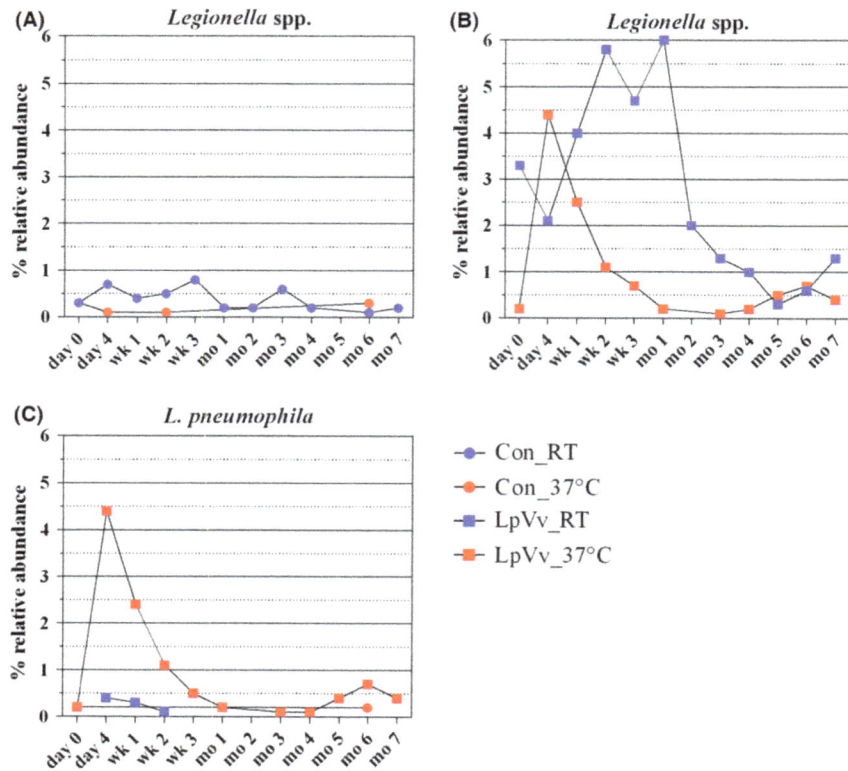

Fig. 2. Relative abundance of *Legionella* spp.- and *L. pneumophila*-associated sequences in biofilm samples.
The relative abundance of *Legionella* spp. (A and B) and *L. pneumophila* (C) OTU taxonomic assignments is shown based on 16S rRNA gene sequences derived from Con_RT (blue circle), Con_37°C (red circle), LpVv_RT (blue square) and LpVv_37°C (red square) Cu DW biofilms at each time point.

Table 1. Distribution of bacterial phyla within samples from each reactor.

Phylum	Relative abundance (%)			
	Con_RT	Con_37°C	LpVv_RT	LpVv_37°C
Proteobacteria	79.0	74.9	81.3	62.1
α-*Proteobacteria*	50.7	48.5	56.6	40.9
β-*Proteobacteria*	18.0	9.1	19.4	10.7
γ-*Proteobacteria*	9.6	15.1	4.9	10.0
δ-*Proteobacteria*	0.7	2.2	0.4	0.4
Actinobacteria	9.0	8.6	3.5	6.8
Acidobacteria	0.1	0.1	0.1	21.2
Bacteroidetes	4.9	2.0	10.8	2.0
Gemmatimonadetes	0.5	4.1	0.6	1.9
Planctomycetes	0.4	3.1	0.4	1.4
Others < 1%	6.2	7.3	3.4	4.7
Total	100.0	100.0	100.0	100.0

Relative abundance is shown for each reactor for all time points. Classes of *Proteobacteria* and their relative abundance are shown in grey text.

while, conversely, γ-*Proteobacteria* were about twice as abundant in biofilm incubated at 37°C (10–15%) than at RT (5–10%; Table 1). Notably, when compared to the other three reactors, *Acidobacteria*- and *Bacteroidetes*-associated sequences were more dominant in LpVv_37°C (21 versus 0.1%) and LpVv_RT (11 versus 2–5%) respectively, indicating that, in the presence of Lp and Vv, higher temperatures supported a greater abundance of the *Acidobacteria* sequences while an abundance of *Bacteroidetes* bacteria-associated sequences was more supported at lower temperatures. In Con_37°C biofilms, the absence of LpVv and higher temperatures seemed to support a higher diversity of bacteria including *Gemmatimonadetes*, *Planctomycetes* and δ-*Proteobacteria* sequences being more abundant compared to samples from the other three reactors (Table 1). Moreover, introduction of LpVv into Cu DW biofilms seemed to result in a slight decrease in the group of bacterial operational taxonomic units (OTUs) that were detected at < 1% each (Table 1; LpVv samples 3–5% versus Con samples 6–7%), an observation that is corroborated by the decrease in richness and diversity indices of the LpVv biofilm samples (Fig. 3).

Effect of temperature and LpVv inoculation on the microbiome of Cu DW biofilms

Compared to Con biofilm samples at either temperature, there was an association between LpVv inoculation and

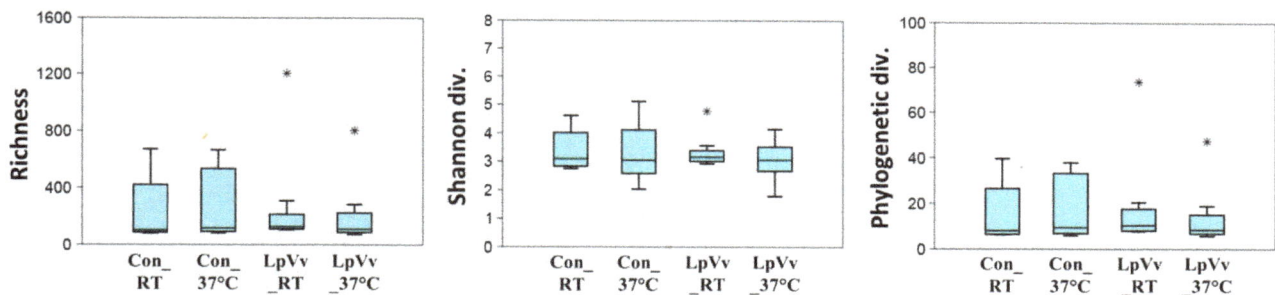

Fig. 3. Alpha diversity indices for sequences derived from biofilm samples. Richness, Shannon diversity and phylogenetic diversity indices are shown for all biofilm-derived sequences from each of the four reactors. The black dots represent libraries that were considered outliers during diversity analyses (LpVv_RT: month 5 and LpVv_37°C: week 1 samples).

Table 2. Results of two-way PERMANOVA with permutation N: 9999.

Source	Sum of squares	df	Mean square	F	P
Temperature RT versus 37°C	2.333	1.0	2.333	20.639	0.0001
Variable Con versus LpVv	0.690	1.0	0.690	6.103	0.0001
Interaction Temperature versus Variable	0.574	1.0	0.574	5.080	0.0002
Residual	4.974	44.0	0.113		
Total	8.572	47.0			

a lower number of species (richness index: LpVv samples mean *c.* 180 versus Con samples mean *c.* 325) and reduced diversity of the bacterial community (Shannon and phylogenetic diversity indices: LpVv samples mean *c.* 3.3 and 11 versus Con samples mean *c.* 3.6 and 20 respectively; Fig. 3). Although LpVv introduction into the biofilm samples influenced overall community composition, principal coordinate analysis (PCoA) of the biofilm sequences indicated that temperature was the major driver of diversity, contributing to 31% of the total variation in biofilm samples at all time points compared to LpVv inoculation contributing 11% of the total variation (Fig. S2). Two-way PERMANOVA of the samples also indicated that temperature played a major role in biofilm microbiome composition (F = 20.6, P = 0.0001) compared to pathogen and host inoculation (F = 6.1, P = 0.0001), while the combined effect of temperature and LpVv inoculation was also statistically relevant in determining overall diversity (F = 5.1, P = 0.0002; Table 2). A non-metric multidimensional scaling (NMDS) ordination plot also indicated temperature had a more significant impact on the microbiomes (observed as the tight clustering of RT compared to 37°C biofilm samples) than LpVv inoculation (observed as individual reactor samples forming secondary clusters; Fig. 4A). Despite the relatively high OTU number identified from all biofilm samples in this study (> 2,700), only 20 OTUs were

responsible for *c.* 66% of dissimilarity (based on SIM-PER analysis) between biofilm samples (Fig. 4B and C).

Nine of the 20 major OTUs identified were more abundant in Con_RT and LpVv_RT biofilm samples than in 37°C-derived biofilms: class β-*Proteobacteria*, order *Sphingomonadales*, family *Chitinophagaceae* and *Comamonadaceae* (two OTUs), and genus *Bradyrhizobium*, *Rhodobacter* (two OTUs) and *Reyranella* (Fig. 4B, upper left oval cluster, and C). Five of the 20 major OTUs identified were more abundant in Con_37°C and LpVv_37°C biofilm samples than in RT-derived biofilms: family *Acetobacteraceae* and genus *Erythromicrobium*, *Gemmatimonas*, *Phenylobacterium* and *Sphingopyxis* (Fig. 4B, upper right oval cluster, and C). Unique OTUs, which were defined as being abundant in biofilm samples from a single reactor, included sequences associated with the genus *Sphingomonas* in Con_RT samples, *Pseudomonas* in Con_37°C and family Ellin6075 in LpVv_37°C samples (Fig. 4B, yellow boxes, and C). Ellin6075, affiliated with the bacterial phylum *Acidobacteria*, had a relative abundance of 21% in LpVv_37°C samples, but was not identified in the other three reactors (Fig. 4C), and could also be contributing exclusively to the relative abundance of 21% of *Acidobacteria* bacterial phylum sequences observed for LpVv_37°C samples (Table 1). Notably, a relatively high abundance (3–8%) of *Mycobacterium*-associated sequences was found in all reactors and they did not cluster with any particular reactor samples in the NMDS plots indicating the organism's ability to stably colonize DW Cu biofilms regardless of temperature and pathogen/host colonization (Fig. 4B, black box, and C).

To visualize community changes under the respective inoculation and temperature conditions over time, taxonomic assignments expressed as a percentage of total OTUs for each biofilm samples were plotted (Fig. 5). As stated above, temperature was the primary driver, followed by LpVv inoculation as the secondary driver, of microbiome diversity, which was visualized in the OTU profiles between the samples within each reactor

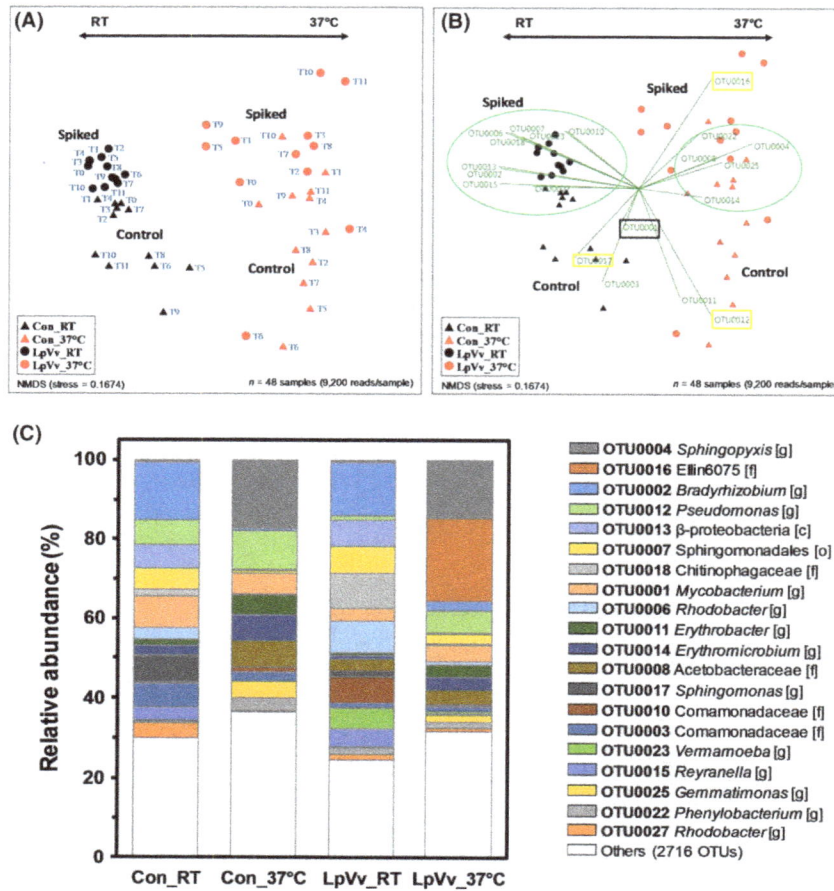

Fig. 4. Non-metric multidimensional scaling (NMDS) ordination plots and major OTU contributors for biofilm samples.
A. NMDS plot representing the relationship for each reactor (symbols for each shown in the lower right-hand corner square) between incubation temperature (red versus black symbols) and time points biofilms were sampled (T0–T11) with duplicates combined ($n = 48$).
B. Green arrows indicate the orientation and contribution (i.e. greatest change) of OTUs between all the biofilm samples from each reactor. Yellow boxes indicate OTUs specific to a single reactor, and the black box indicates the OTU that is found in all reactor samples. OTUs that are found more abundantly in RT and 37 C samples are highlighted in the upper left and right ovals respectively. The distribution and taxonomy of the major OTU contributors are shown in (C).

(Fig. 5). The temporal OTU profiles revealed that not only were *Acidobacteriaceae* (Ellin6075)-associated sequences highly abundant in LpVv_37°C samples, but abundance increased with time after inoculation (Fig. 5D, dark green bars). Moreover, starting at month 1, OTU profiles substantially shifted in each reactor as compared to their respective day 0–week 3 profiles, most notably for sequences associated with the majority of the top 20 contributing OTUs, in addition to sequences associated with the genera *Gemmata* in Con_37°C and LpVv_37°C samples and *Rhodoplanes* in LpVv_RT samples (Fig. 5). The change in OTU profiles after 1 month may also be explained by several significant changes in water quality characteristics over time such as an increase in per cent dissolved oxygen (DO) and decreases in alkalinity, barium (Ba) and sulfur (S), and fluctuations in aluminium (Al), calcium (Ca), sodium (Na), sulfur (S) and zinc (Zn) concentrations (Table 3).

For example, percent DO at 6–7 months were significantly higher than the levels at day 4 to month 4 (ANOVA, $P < 0.05$), while Ca, Na, S and Zn levels fluctuated significantly throughout the 510-day biofilm accumulation stage up until the 7-month post-inoculation time point (Table 3).

Discussion

Impact of L. pneumophila *colonization within copper DW biofilms*

Copper (Cu) is a commonly used pipe material within premise plumbing systems (Council, 2006). The antimicrobial properties of Cu on *L. pneumophila* inactivation have been demonstrated previously (Lin *et al.*, 1996; Biurrun *et al.*, 1999), with the combined effects of copper–silver ionization considered an effective control strategy for *L. pneumophila* in US hospital water systems

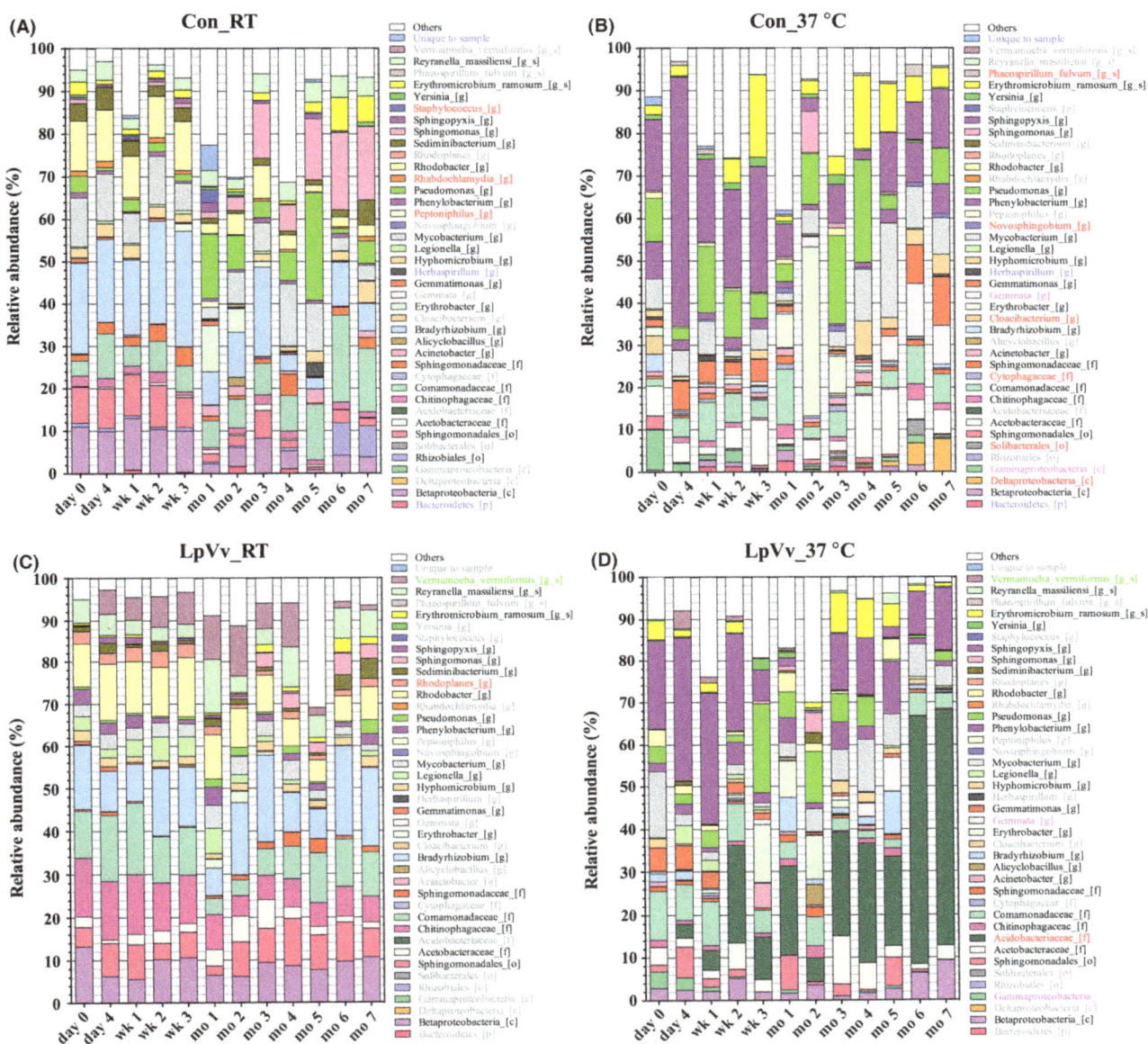

Fig. 5. OTU taxonomic assignments expressed as percentage of total assigned sequences for each sample. 16S rRNA gene sequences derived from Con_RT (A), Con_37°C (B), LpVv_RT (C) and LpVv_37°C (D) Cu DW biofilms for each time point. In each legend, OTUs in coloured text indicate the following: absence or low abundance of sequences (grey); unique or abundant sequences for that reactor (red); sequences found in abundance in Con reactors only (blue); sequences found in abundance in LpVv reactors only (green); and sequences found in abundance in reactors incubated at 37°C only (purple).

(Lin *et al*., 2011). However, it has been reported that the antimicrobial properties of Cu can be negatively impacted by biofilm age, accumulation and presence of corrosion products. As measured by total bacterial counts, early biofilm growth (≤ 30 days) on Cu surfaces was lower than on PVC or PEX material (Mathys *et al*., 2008; Morvay *et al*., 2011), while in older, mature biofilms (18 months old), Cu biofilm growth was two- to eightfold higher than PVC, PE and SS biofilms (Wingender and Flemming, 2004). Moreover, *Legionella* densities were initially higher in < 250-day-old SS than in Cu

biofilms, but after 24 months, similar levels of *Legionella* were observed within SS and Cu biofilms suggesting the inhibitory effect of Cu on *Legionella* may be transient and reduced by the accumulation of corrosion products on the Cu surfaces over time (van der Kooij *et al*., 2005). Although there is currently no standard definition of a mature DW biofilm (with studies labelling 1- to 2-year-old (Boe-Hansen *et al*., 2002; Wingender and Flemming, 2004) and 20-year-old (Henne *et al*., 2012) DW biofilms as mature), the consensus seems to be that mature biofilms consist of a complex and functional

Table 3. Water quality measurements of CDC biofilm reactor feed water.

Parameters	Units	Start to E1 (510 days)	E1 to E2 (101 days) T0	Post-inoculation (time point[s])						
				1 month (T1–T5)	2 months (T6)	3 months (T7)	4 months (T8)	5 months (T9)	6 months (T10)	7 months (T11)
pH		8.43 ± 0.16	8.25 ± 0.37	8.45 ± 0.18	8.58 ± 0.05	8.57 ± 0.06	8.65 ± 0.06	8.55 ± 0.05	8.38 ± 0.05	8.38 ± 0.03
Temperature	°C	19.84 ± 0.71	20.60 ± 0.52	20.61 ± 0.38	20.37 ± 0.21	19.89 ± 0.60	18.40 ± 0.80	18.61 ± 0.78	18.89 ± 0.81	18.67 ± 0.34
Conductivity	mS cm^{-1}	0.37 ± 0.08	0.365 ± 0.02	0.345 ± 0.02	0.395 ± 0.02	0.426 ± 0.01	0.455 ± 0.05	0.473 ± 0.05	0.322 ± 0.01	0.364 ± 0.01
DO	%	78.1 ± 8.2	69.7 ± 8.5	65.4 ± 2.4	68.4 ± 3.2	67.7 ± 2.3	72.5 ± 2.4[a]	75.7 ± 2.6[a,b,c]	81.4 ± 1.8[a,b,c,d]	82.8 ± 2.5[a,b,c,d]
Free Cl$_2$	mg l^{-1}	0.40 ± 0.23	0.44 ± 0.07	0.49 ± 0.08	0.61 ± 0.05	0.60 ± 0.05	0.68 ± 0.14	0.59 ± 0.12	0.34 ± 0.12	0.41 ± 0.06
Total Cl$_2$		0.44 ± 0.22	0.47 ± 0.07	0.609 ± 0.05	0.637 ± 0.04	0.628 ± 0.04	0.714 ± 0.14	0.633 ± 0.11	0.391 ± 0.12	0.458 ± 0.06
Alkalinity		68.6 ± 9.8	68.5 ± 3.9	78.7 ± 6.2	79.6 ± 5.2	76.8 ± 0.8	83.0 ± 6.0[c]	87.8 ± 5.1	58.6 ± 2.3[a,b,c,d,e]	63.3 ± 2.4[a,b,c,d,e]
TOC		1.02 ± 0.41	0.96 ± 0.42	1.02 ± 0.30	0.68 ± 0.26	0.54 ± 0.03	0.86 ± 0.15	1.01 ± 0.15	1.07 ± 0.10	0.92 ± 0.07
NO$_3$		0.82 ± 0.17	0.84 ± 0.18	0.88 ± 0.14	0.72 ± 0.04	0.73 ± 0.03	0.89 ± 0.11	0.92 ± 0.09	0.87 ± 0.13	0.97 ± 0.05
PO$_4$		0.22 ± 0.06	0.19 ± 0.04	0.21 ± 0.04	0.22 ± 0.02	0.15 ± 0.05	0.15 ± 0.04	0.15 ± 0.02	0.10 ± 0.02	0.11 ± 0.03
Ca		31.1 ± 3.0	28.9 ± 1.0	28.89 ± 1.8	32.01 ± 1.9	34.09 ± 1.8	35.63 ± 5.1[a]	37.96 ± 2.8	27.34 ± 1.1[c,d,e]	30.4 ± 1.3
K		2.3 ± 0.5	2.0 ± 0.9	2.5 ± 0.2	2.6 ± 0.1	2.6 ± 0.1	3.0 ± 0.4	3.2 ± 0.3	2.2 ± 0.1	1.9 ± 0.1
Mg		10.0 ± 2.0	9.3 ± 0.6	8.4 ± 0.7	9.9 ± 0.5	10.3 ± 0.5	11.4 ± 1.3	12.2 ± 1.0	8.1 ± 0.4	9.1 ± 0.5
Na		24.6 ± 8.4	23.1 ± 2.1	19.1 ± 1.1	24.7 ± 1.9	28.8 ± 1.4[a]	30.9 ± 4.9[a,b]	32.2 ± 5.1[a]	19.2 ± 1.4[c,d,e]	22.5 ± 1.5[d]
P		0.16 ± 0.03	0.17 ± 0.08	0.17 ± 0.01	0.17 ± 0.01	0.16 ± 0.01	0.16 ± 0.01	0.17 ± 0.01	0.13 ± 0.01	0.13 ± 0.01
S		23.3 ± 5.7	21.2 ± 2.1	17.7 ± 0.6	22.2 ± 1.6	26.7 ± 2.2[a]	27.0 ± 4.2[a]	27.9 ± 3.0[a]	17.9 ± 1.3[c,d,e]	19.5 ± 0.8[c,d]
Si		2.2 ± 0.7	2.1 ± 0.3	3.0 ± 0.1	2.6 ± 0.1	2.3 ± 0.1	2.3 ± 0.0	2.2 ± 0.1	2.8 ± 0.2	2.9 ± 0.1
Sr	μg l^{-1}	0.22 ± 0.06	0.22 ± 0.02	0.19 ± 0.01	0.22 ± 0.01	0.23 ± 0.02	0.28 ± 0.02	0.28 ± 0.04	0.17 ± 0.01	0.19 ± 0.01
Al		45.9 ± 20.0	54.1 ± 13.9	67.2 ± 7.1	74.2 ± 11.6[a]	61.6 ± 11.2[b]	42.4 ± 4.9[a,b,c]	45.4 ± 3.3[a,b,c]	41.5 ± 2.2[b,c]	31.0 ± 5.1[a,b,c,d,e,f]
Ba		38.6 ± 7.4	36.7 ± 4.2	35.9 ± 1.5	41.1 ± 2.1	43.6 ± 1.5[a]	42.3 ± 3.8[a]	42.4 ± 4.6	33.2 ± 1.5[b,c,d]	32.3 ± 1.3[b,c,d]
Cu		26.0 ± 4.9	20.7 ± 2.2	17.8 ± 0.8	17.7 ± 1.2	16.0 ± 0.9	16.6 ± 1.8	19.9 ± 0.9	19.0 ± 1.1	21.2 ± 2.3
Fe		21.2 ± 24	14.7 ± 4.0	11.9 ± 2.8	16.6 ± 5.9	14.9 ± 1.4	16.9 ± 6.9	14.6 ± 1.5	15.2 ± 1.5	12.6 ± 10.9
Pb		3.3 ± 2.2	2.4 ± 1.1	3.6 ± 0.8	2.3 ± 1.6	1.3 ± 1.5	2.3 ± 0.3	2.2 ± 0.2	1.8 ± 1	3.3 ± 2.2
Zn		89.8 ± 25.3	90.3 ± 12.0	66.1 ± 4.3	73.7 ± 4.9[a]	80.7 ± 5.4[a,b]	84.6 ± 12.9[a,b]	113.8 ± 21.6[a,b,c,d]	132.5 ± 6.5[b,c,d,e]	141.6 ± 7.6[a,b,c,d,e,f]

Total organic carbon, TOC; nitrate, NO$_3$; calcium, Ca; potassium, K: magnesium, Mg; sodium, Na; sulfur, S; silicon, Si; phosphate, PO$_4$; aluminium, Al; borium, Ba; copper, Cu; iron, Fe; phosphorous, P; lead, Pb; strontium, Sr; zinc, Zn; alkalinity as measured by CaCO$_3$ concentration.

Values expressed as mean ± standard deviation.

Two-way ANOVA for each parameter:

[a] $P < 0.05$ for T1-T5 versus T6, T7, T8, T9, T10 and T11.

[b] $P < 0.05$ for T6 versus T7, T8, T9, T10 and T11.

[c] $P < 0.05$ for T7 versus T8, T9, T10 and T11.

[d] $P < 0.01$ for T8 versus T9, T10 and T11.

[e] $P < 0.05$ for T9 versus T10 and T11.

[f] $P < 0.05$ for T10 versus T11.

microbial community that exhibits physiological coopera-tion and metabolic efficiency all of which is gradually developed (Costerton et al., 1995; Dunne, 2002). Thus, mature DW biofilms may be a more accurate representa-tion of biofilms within building systems, and hence more relevant in examining *Legionella* colonization in premise plumbing systems.

Using approximately 20-month-old biofilms, the results from this study indicated that Cu DW biofilms inoculated with LpVv co-cultures exhibited temperature-dependent colonization of Lp at high and persistent levels for up to 7 months (Fig. 1C). Specifically, colonization of Lp within Cu biofilms incubated at RT resulted in an increase in the overall levels of *Legionella* spp. members; yet colo-nization of Lp within biofilms incubated at 37°C resulted in the dominance of Lp over other *Legionella* spp. Non-*L. pneumophila* members of the *Legionella* genus seem to be more adept at lower temperature growth (Figs 1A and 2), while higher temperatures appear to select for Lp growth, an observation which has been previously reported (Wullings and van der Kooij, 2006). Thus, although the temperature-dependent growth of Lp in DW has been described in other studies (Rogers et al., 1994; Rhoads et al., 2015), an interesting observation from this study was the influence of Lp biofilm coloniza-tion on other *Legionella* genus members, which had not been directly demonstrated previously. Additionally, this study further corroborates the general observation that Cu pipe is not universally antimicrobial against *Legio-nella* spp. nor *L. pneumophila*, especially at the higher pH values (> pH 8.2; Table 3; Lin 2002) tested in the current study.

Effect of V. vermiformis *inoculation into biofilms*

Forty-eight-hour co-culture of Lp and its FLA host, Vv, was introduced into biofilms based on previous studies indicating upregulation of Lp virulence after intra-amoeba passage and during their co-inoculations into mice (Brie-land et al., 1996, 1997). Thus, it was hypothesized that increased colonization of DW biofilms could be a func-tion of phenotypic changes induced during Lp passage and co-culture with Vv hosts, which was exploited in this study. While Vv levels were relatively high and persistent within LpVv_RT biofilms, Lp levels in the same samples steadily decreased from day 4 to month 1, indicating par-asitization and interactions with Vv did not sustain Lp levels at RT (Fig. 1C and D). This is supported by previ-ous data stating that at < 20°C, *L. pneumophila* was actively digested and/or eliminated from the FLA, *Acan-thamoeba polyphaga*, while at temperature above 25°C, the amoebae was parasitized by Lp (Ohno et al., 2008). Moreover, Vv cells were not detected past 1 week post-inoculation at 37°C (Fig. 1D), while persistent Lp levels

were observed in the LpVv_37°C biofilms (Fig. 1C), sug-gesting Lp biofilm colonization at this temperature was most likely not due to the presence of Vv cells. The qPCR detection of Lp only at month 6 (Fig. 1C, black arrow) and identification of Lp-associated sequences at day 0, before LpVv inoculation and month 6 (Fig. 2B, red circle; 0.2% relative abundance) in Con_37°C biofilm samples may also be further indication that this tempera-ture was amenable to Lp biofilm growth independent of Vv interactions. These results suggest that stable colo-nization of Vv within Cu DW biofilms may be strongly dri-ven by temperature; however, the complete, initial parasitization of Vv cells at 37°C by Lp bacteria (and potentially other endemic biofilm amoebae) and their effect on Lp throughout the rest of the experiment cannot be excluded. Additionally, contrary to previous reports indicating the abundance of Vv in both chlorinated and chloraminated DW systems (Thomas et al., 2006; Cor-saro et al., 2010; Valster et al., 2011; Wang et al., 2012), as well as within the building system used in this study (Buse et al., 2013), Vv cells were not detected in any of the control biofilm samples, which could indicate a preference for the bulk water phase and a need for high cellular densities for stable and persistent biofilm colonization.

Temperature and colonization of L. pneumophila *and* V. vermiformis *are major drivers for overall biofilm community composition*

Principal coordinate analysis, two-way PERMANOVA, and NMDS analyses all indicated that temperature was the major driver of biofilm microbiome composition fol-lowed by LpVv colonization (Fig. S2, Table 2, and Fig. 3A, respectively). The communities inoculated with LpVv at RT and 37°C maintained a more stable compo-sition, compared to control biofilms, based on their lower coefficient of variation as indicated in Figs 3 and 4B. FLA grazing on stream water-derived biofilms and *Klebsiella pneumoniae*, *Pseudomonas fluorescens* and *Staphylococcus epidermidis* biofilms were shown to alter overall morphology and microbial composition of their respective biofilms (Huws et al., 2005; Böhme et al., 2009). However, in this study, the predatory impact of Vv biofilm colonization on the bacterial community was not clear from the bacterial OTU distribution data espe-cially given the fact that the actively feeding, trophozoite or metabolically dormant cyst forms cannot be distin-guished using qPCR or sequencing analyses, nor were other amoebae investigated in this study. It must be noted that OTU 0023 was identified as Vv-associated sequences (Figs 4C and 5C and D, green highlighted text), yet only 16S rRNA gene amplification was per-formed. Thus, the Vv-associated sequences could have been derived either from intracellular bacterial organisms previously associated with Vv cells or from mitochondrial Vv sequences that were not taxonomically labelled as such and thus were not removed by MOTHUR. This OTU was left in the analysis as it was one of the top 20 major OTU indicators of microbial composition (high relative abundance in LpVv_RT samples; Fig. 4C) and was previously associated with bacterial sequences derived from two different sources of groundwater (Pedersen et al., 1997; Benzine et al., 2013).

Notably, Mycobacterium avium, M. intracellulare and M. abscessus, other opportunistic water-based pathogens, have been shown to readily adhere and stably colonize DW biofilms grown on glass, stainless steel, PVC and zinc-galvanized steel coupons (Mullis and Falkinham, 2013). Their stable colonization and relatively high abundance within DW biofilms were supported in this study, for Cu surfaces, as Mycobacterium was found to be one of the top 20 indicator OTUs impacting overall diversity with relative abundance across all samples unaffected by temperature or LpVv colonization (Fig. 4C).

Although the water quality parameters collected weekly in this study corresponded to the influent water, and not water collected from within each reactor, several parameters that were shown to change significantly over the study period have been previously correlated with Legionella occurrence and abundance within DW systems [reviewed in Buse et al., 2012]. Specifically, DW containing > 50 μg Cu l^{-1}, < 100 μg l^{-1} Zn, < 20 μg l^{-1} Fe and < 6 μg l^{-1} Mn was found to be protective against legionellae colonization, while in hot DW samples, the measured mean levels of DO (66%), TOC (6 ppm), Ca (25 mg l^{-1}), sulfate (33 mg l^{-1}) and nitrate (1 mg l^{-1}) were positively correlated with Legionella densities (Buse et al., 2012). Moreover, variations in bulk DW community profiles have been associated with water quality parameters such as concentrations of disinfectant, phosphorus, sulfate and Mg (Ji et al., 2015). Thus, significant changes in the various influent water quality parameters measured in this study (Table 3), in concert with LpVv (and other possible micro-eukaryotes) colonization, could have influenced biofilm microbiome composition throughout the study period. However, no statistical correlations could be drawn, as influent water was analysed rather than water collected from each CDC reactor; thus, future research will aim to address the putative importance of water quality on the biofilm microbiome.

Illumina sequencing of both bulk water and biofilm bacterial communities derived from chlorinated and chloraminated treated water was previously shown to be dominated by the phyla Acidobacteria, Actinobacteria, Bacteroidetes, Gemmatimonadetes, Planctomyces and Proteobacteria (Pinto et al., 2012; Buse et al., 2014;

Gomez-Alvarez et al., 2014), which was similar to the results obtained in this study (Table 1). Compared to earlier DNA sequencing methods, such as Sanger-based sequencing, there are an increasing number of studies utilizing next-generation sequencing technology to improve sequencing depth to better characterize the complex microbial composition of DW biofilms. Nonetheless, both methods have been able to identify sequences associated with the genera Legionella, Mycobacteria and/or Pseudomonas within chlorinated and chloraminated DW-derived biofilms grown on concrete/steel, Cu, cast iron, polycarbonate, PVC and SS surfaces (McBain et al., 2003; Zhang et al., 2012; Lin et al., 2013; Gomez-Alvarez et al., 2014; Chao et al., 2015) indicating that DW biofilms may be a potential and continual source of environmental (water-based) pathogens (Ashbolt, 2015).

Conclusion

Collectively, this study demonstrated that L. pneumophila can colonize mature DW biofilms grown on copper surfaces at relatively high pH levels (> pH 8.2) in which the antimicrobial properties of copper can be negatively impacted. Once incorporated, L. pneumophila may persist up to 1 month at cold-water temperatures for premise plumbing (20°C) and up to 7 months at higher temperatures (37°C), which may pose a significant public health risk due to the extended colonization potential of these pathogens within premise plumbing. Temperature and inoculation of L. pneumophila and V. vermiformis were strong drivers of biofilm microbiome diversity, with major changes in bacterial composition between 1 and 7 months, compared to those between day 0 and 1 month. Notably, the aged copper used in this study did not universally exclude Legionella spp. nor L. pneumophila colonization, and V. vermiformis was not a prerequisite for proliferation or persistence of L. pneumophila. Moreover, L. pneumophila seemed to influence the presence and abundance of other Legionella spp., as well as other genera within the biofilm microbiome, further underscoring the need to elucidate L. pneumophila behaviour and their ecological role within mature DW biofilms in premise plumbing.

Experimental procedures

Bacterial and amoebal co-culture preparation

Legionella pneumophila strain Chicago-2 (Lp) is a clinical isolate derived from a human lung (American Type Culture Collection [ATCC] 33215), while V. vermiformis strain CDC-19 (Vv) is an environmental isolate derived from a hospital cooling tower drain (ATCC 50237). Lp and Vv were grown and enumerated as previously described (Buse and Ashbolt, 2011). Briefly, Lp was grown overnight with continuous shaking at 37°C in buffered yeast extract (BYE) broth and Vv cells were grown as monolayers at 30°C in ATCC 1034 medium. Lp and Vv cultures were washed and diluted to desired concentrations in Page's amoeba saline (PAS). Lp densities, as measured by colony-forming units (CFU), were determined by taking a small aliquot of the bacterial suspension, performing serial dilutions, and plating on buffered charcoal yeast extract (BCYE) agar plates (BD Diagnostics, Franklin Lakes, NJ, USA) which were incubated for 48 h at 37°C. Vv densities were enumerated using light microscopy and an improved Neubauer hemocytometer (Hausser Scientific, Horsham, PA, USA).

Previous studies indicated that during Lp infection of Vv cells, Lp densities peaked after 48 h of co-culture (Buse and Ashbolt, 2011). Thus, for Lp and Vv inoculation into Centers for Disease Control and Prevention (CDC) biofilm reactors (Biosurface Technologies Corp., Bozeman, MT, USA), co-cultures were prepared by washing overnight cultures of Lp and 7-day-old Vv cultures in PAS, co-inoculating each into 75-cm^2 tissue culture flasks to generate a 15:1 bacteria-to-amoeba ratio (6.93 $\log_{10}$ CFU ml^{-1} of Lp and 5.74 $\log_{10}$ cells ml^{-1} of Vv), and incubating the flasks for 48 h at 32°C. After incubation, flasks were harvested and Lp and Vv enumerated as described above (7.27 $\log_{10}$ CFU ml^{-1} of Lp and 5.67 $\log_{10}$ cells ml^{-1} of Vv).

CDC biofilm reactor set-up and sample collection

Four CDC biofilm reactors containing copper (Cu) coupons (surface area of 1.27 cm^2) were fed with ambient (19.8 ± 0.8°C) DW from a light-protected storage tank (23 l) at 150 ml h^{-1} (2.7-h hydraulic residence time) using a peristaltic pump and Norprene™ food-grade tubing (Cole Parmer, Vernon Hills, IL, USA). The Cu coupons used in this study were made from C10100 alloy (99.99% Cu), which is similar in composition to the Cu alloy C12200 used in Cu plumbing (99.9% Cu and silver combined and 0.015–0.04% phosphorus as per requirements of the American Society for Testing and Materials B88 Standard Specification for Seamless Copper Water Tube). In-premise flow and stagnation periods were simulated by placing reactors on a magnetic stir plate set at approximately 100–125 rpm, which was activated for 30 min every 2 h [start to event 1 (E1)] and then every 2 h during a 10-h window on weekdays [E1 to the inoculation event 2 (E2); Fig. S1]. Mature DW biofilms were allowed to establish on the Cu coupon surfaces for 510 days under ambient/room temperature (RT) conditions. At E1, temperatures within two of the four reactors, which served as control/mock-inoculated (Con) and Lp and Vv (LpVv)-inoculated reactors, were elevated to 37°C for the duration of the experiment (labelled Con_37°C and LpVv_37°C, respectively) with the two

remaining reactors, designated Con_RT and LpVv_RT, maintained at RT. DW biofilms were then allowed to adjust to the new temperature conditions for 101 days before LpVv inoculation (E2). Prior to E2, duplicate coupons from each reactor were collected to record established biofilm characteristics [time 0 (T0)] as a baseline comparison to post-inoculation samples (T1–T11).

For LpVv co-culture inoculations (E2), reactors were disconnected from the peristaltic feed pump, with effluent lines clamped shut, and placed in a biological safety cabinet where the total DW volume was removed (approximately 400 ml) and replaced with 368 ml of 0.22 μm filtered, autoclaved tap water (fatH$_2$O). The control reactors, Con_RT and Con_37°C, were inoculated with 32 ml fatH$_2$O, while the LpVv_RT and LpVv_37°C reactors were inoculated with 32 ml of the LpVv co-culture suspension prepared as described above. Reactors were replaced onto the magnetic stir plates and incubated at either RT or 37°C. Microbial adhesion was allowed to occur for 24 h with 30 min of mixing every 2 h. After the 24-h incubation, each reactor was connected back to the peristaltic feed pump with effluent clamps released to allow flow. Two coupons (replicate 1, R1; and replicate 2, R2) were removed from each reactor at day 4 (T1), week 1 (T2), week 2 (T3), week 3 (T4) and month 1 (T5), 2 (T6), 3 (T7), 4 (T8), 5 (T9), 6 (T10) and 7 (T11) post-inoculation (Fig. S1). Biofilms were collected from coupon surfaces as previously described (Buse *et al.*, 2013b; Buse *et al.*, 2014). Briefly, a sterile wooden stick was used to scrape the surface and was then rinsed in 300 μl of fatH$_2$O. The coupon surface was washed twice with 100 μl fatH$_2$O resulting in a final volume of 500 μl suspended biofilm material.

Water quality measurements

An aliquot of 250 ml of the tank water feeding the CDC reactors was sampled every week prior to and 7 months after Lp and Vv inoculation ($n = 109$ and 28 respectively). Water samples were analysed for free and total chlorine (Cl$_2$), pH, temperature, conductivity, and per cent dissolved oxygen (DO; Table 3). The pH was measured using a glass electrode UltraBASIC-10 meter (Denver Instrument, USA) and temperature, conductivity, and per cent dissolved oxygen using a YSI 566 Multi Probe System (YSI Environmental, Yellow Springs, OH, USA). Free and total Cl$_2$ were measured using a TNT866 Chlorine Kit (Hach Company, Loveland, CO, USA) and the Hach® DR2800 spectrophotometer (Hach Company). An additional 550 ml of tank water was also submitted for weekly water quality analysis to assay total organic carbon (TOC; EPA Method 415.3 rev1.1), trace metals Cu, Fe, Mg, P, and Zn (EPA Method 200.7), phosphate (PO4; EPA Method 365.1), and nitrate (NO3;

EPA Method 353.2) by the National Risk Management Research Laboratory at the US Environmental Protection Agency in Cincinnati, OH (Table 3).

DNA isolation and quantitative PCR

DNA was extracted from biofilm suspensions using the T&C buffer and MasterPure Complete DNA Purification Kit™ (Epicentre Biotechnologies, Madison, WI, USA) as described previously (Buse *et al.*, 2014). Biofilm DNA samples were analysed using the Applied Biosystems 7900 HT Fast Real-Time PCR System (Applied Biosystems, Foster City, CA, USA). Each sample was analysed in duplicate in a reaction mixture (20 µl final volume) containing the following: 1 × Power SYBR® Green PCR Master Mix (Applied Biosystems); 500 nM of each forward and reverse Lp-specific primers targeting the *sidF* gene (Lu *et al.*, 2013) or Vv-specific primers targeting the 18S rRNA gene Hv1227/1728R (Kuiper *et al.*, 2006); and 5 g bovine serum albumin. For Lp, the qPCR conditions consisted of an initial denaturation step of 10 min at 95°C, followed by 40 cycles of 15 s at 95°C, 30 s at 64°C, and 30 s at 72°C with a final hold at 72°C for 2 min. For Vv, cycling conditions included pre-denaturation at 95°C for 3 min, 40 cycles of 20 s at 95°C, 30 s at 56°C, and 40 s at 72°C with a final extension step at 72°C for 10 min. Fluorescent detection was performed at the annealing phase and during subsequent dissociation curve analysis to confirm that a single product had been amplified. TaqMan qPCR assay for *Legionella* spp. detection was performed using the 16S rRNA gene primers 16S-LegF1c/16S-LegR1c and probe 16S-LegP1 (Lu *et al.*, 2015). The cycling conditions consisted of a pre-incubation step at 50°C for 2 min and a pre-denaturation step at 95°C for 10 min, followed by 40 cycles of denaturation at 95°C for 10 s, annealing at 50°C for 30 s, and extension at 70°C for 30 s.

All qPCR assays were performed in a 96-well plate containing DNA standards and no-template controls. The threshold cycles (Ct) were calculated using the SEQUENCE DETECTION SYSTEMS software v 2.3 (Life Technologies, Carlsbad, CA, USA). The reaction efficiency was calculated using the following equation: efficiency $\% = 100 \times (10(-1/slope) - 1)$ (Applied Biosystems). Experiments were performed with undiluted and 10-fold-diluted template DNA in duplicate to verify the absence of qPCR inhibition. Data were expressed as cell equivalents (CE), based on culture densities (as measured by colony-forming units), which were used to generate standard curves against the surface area of each reactor coupon (CE cm^{-2}). All standards were generated in triplicate and the best-fit standard curve used in data analysis. The limit of quantification (LOQ) for each qPCR assay was 1 $\log_{10}$ CE cm^{-2}.

Illumina sequencing

Biofilm samples were subjected to PCR amplification using the universal bacterial/archaeal primer set F515 (5′ GTG CCA GCM GCC GCG GTA A 3′)/R806 (5′ GGA CTA CHV GGG TWT CTA AT 3′, barcoded, #0–95) targeting the V4 region of 16S rRNA gene, which was determined to yield optimal community clustering with this particular read length (Caporaso *et al.*, 2011). Samples were prepared following the Earth Microbiome Project 16S rRNA Amplification Protocol version 4_13 (ftp://ftp.metagenomics.anl.gov/data/misc/EMP/Supplementary File1_barcoded_primers_515F_806R.txt) with the exceptions of using molecular biology grade water (Quality Biological, Gaithersburg, MD, USA) and the QIAquick PCR Purification Kit (Qiagen, Valencia, CA, USA). Of the 96 biofilm samples, 90 were pooled on an equal molar basis (200 ng), and the remaining six samples with low-yield PCR products were pooled based on 'maximum volume' criteria (maximum volume of the 90 samples, which is 50 µl). Paired-end 250-cycle Illumina sequencing was performed by the Genomics Research Laboratory at the Virginia Bioinformatics Institute (Blacksburg, VA, USA).

Sequence data analyses

Reads processing was performed using the software MOTHUR v1.36.1 (https://www.mothur.org; Schloss *et al.*, 2009) following the protocol of Kozich *et al.* (2013). Briefly, fastq files with forward and reverse reads were used to form contigs. Reads were screened and removed if they (i) had a length < 252 bp, (ii) contained ambiguous bases (Ns), (iii) contained homopolymers greater than seven bases, (iv) were identified as chimera, or (v) were classified as unknown, chloroplasts or mitochondria. Reads were aligned and grouped with 97% sequence identity as the cut-off point for each operational taxonomic unit (OTU). Taxonomic classification was obtained using the Greengenes reference taxonomy database release gg_13_8_99. A total of 1 721 964 16S rRNA gene reads were analysed in this study. Prior to community analysis, samples were rarefied to the smallest data set of 9200 reads (Gihring *et al.*, 2012). The Illumina reads have been deposited in the National Center for Biotechnology Information (NCBI) database BioProject number PRJNA214912: SRA accession #SAMN04528999-SAMN04529093.

Statistical analysis

Means and standard error means were calculated using duplicate samples collected for all biofilm and effluent water sample types at each time point. Statistical significance for water quality parameters was determined using ANOVA followed by the Tukey post-test (PRISM 6; GraphPad Software, La Jolla, CA, USA). Alpha diversity indices (supplemental information) and PERMANOVA of Illumina sequences were calculated using PAST version 2.03 (http://folk.uio.no/ohammer/past/; Hammer *et al.*, 2001). A non-metric multidimensional scaling (NMDS) ordination plot based on the Jensen–Shannon distance (JSD) was also generated using the PAST software with robustness of the ordination plot evaluated for goodness of fit using the Shepard diagram (Wickelmaier, 2003).

Acknowledgements

We would like to thank Ian T. Struewing and Sharon Yelton for their technical assistance with the daily operations of the CDC biofilm reactors. A portion of this effort was financially supported by the Alfred P. Sloan Foundation Microbiology of the Built Environment programme, National Science Foundation CBET award #1336650, Virginia Tech Institute for Critical Technology and Applied Science, and Alberta Innovates-Health Solutions.

Conflict of Interest

None declared.

References

Armon, R., Starosvetzky, J., Arbel, T., and Green, M. (1997) Survival of *Legionella pneumophila* and *Salmonella typhimurium* in biofilm systems. *Water Sci Technol* **35**: 293–300.

Ashbolt, N.J. (2015) Environmental (saprozoic) pathogens of engineered water systems: understanding their ecology for risk assessment and management. *Pathogens* **4**, 390–405.

Benzine, J., Shelobolina, E., Xiong, M.Y., Kennedy, D.W., McKinley, J.P., Lin, X., and Roden, E.E. (2013) Fe-phyllosilicate redox cycling organisms from a redox transition zone in Hanford 300 Area sediments. *Front Microbiol* **4**: 388.

Berry, D., Xi, C., and Raskin, L. (2006) Microbial ecology of drinking water distribution systems. *Curr Opin Biotechnol* **17**: 297–302.

Biurrun, A., Caballero, L., Pelaz, C., Leon, E., and Gago, A. (1999) Treatment of a *Legionella pneumophila*-colonized water distribution system using copper-silver ionization and continuous chlorination. *Infect Control Hosp Epidemiol* **20**: 426–428.

Boe-Hansen, R., Albrechtsen, H.-J., Arvin, E., and Jørgensen, C. (2002) Bulk water phase and biofilm growth in drinking water at low nutrient conditions. *Water Res* **36**: 4477–4486.

Böhme, A., Risse-Buhl, U., and Kusel, K. (2009) Protists with different feeding modes change biofilm morphology. *FEMS Microbiol Ecol* **69**: 158–169.

Brieland, J., McClain, M., Heath, L., Chrisp, C., Huffnagle, G., LeGendre, M., *et al.* (1996) Coinoculation with *Hartmannella vermiformis* enhances replicative *Legionella pneumophila* lung infection in a murine model of Legionnaires' disease. *Infect Immun* **64**: 2449–2456.

Brieland, J.K., Fantone, J.C., Remick, D.G., LeGendre, M., McClain, M., and Engleberg, N.C. (1997) The role of *Legionella pneumophila*-infected *Hartmannella vermiformis* as an infectious particle in a murine model of Legionnaire's disease. *Infect Immun* **65**: 5330–5333.

Buse, H.Y., and Ashbolt, N.J. (2011) Differential growth of *Legionella pneumophila* strains within a range of amoebae at various temperatures associated with in-premise plumbing. *Lett Appl Microbiol* **53**: 217–224.

Buse, H.Y., Schoen, M.E., and Ashbolt, N.J. (2012) Legionellae in engineered systems and use of quantitative microbial risk assessment to predict exposure. *Water Res* **46**: 921–933.

Buse, H.Y., Lu, J., Struewing, I.T., and Ashbolt, N.J. (2013a) Eukaryotic diversity in premise drinking water using 18S rDNA sequencing: implications for health risks. *Environ Sci Pollut Res Int* **20**: 6351–6366.

Buse, H.Y., Lu, J., Struewing, I.T., and Ashbolt, N.J. (2013b) Preferential colonization and release of *Legionella pneumophila* from mature drinking water biofilms grown on copper versus unplasticized polyvinylchloride coupons. *Int J Hyg Environ Health* **217**: 219–225.

Buse, H.Y., Lu, J., Lu, X., Mou, X., and Ashbolt, N.J. (2014) Microbial diversities (16S and 18S rRNA gene pyrosequencing) and environmental pathogens within drinking water biofilms grown on the common premise plumbing materials unplasticized polyvinylchloride and copper. *FEMS Microbiol Ecol* **88**: 280–295.

Caporaso, J.G., Lauber, C.L., Walters, W.A., Berg-Lyons, D., Lozupone, C.A., Turnbaugh, P.J., *et al.* (2011) Global patterns of 16S rRNA diversity at a depth of millions of sequences per sample. *Proc Natl Acad Sci USA* **108** (Suppl. 1): 4516–4522.

Chao, Y., Mao, Y., Wang, Z., and Zhang, T. (2015) Diversity and functions of bacterial community in drinking water biofilms revealed by high-throughput sequencing. *Sci Rep* **5**: 10044.

Collier, S.A., Stockman, L.J., Hicks, L.A., Garrison, L.E., Zhou, F.J., and Beach, M.J. (2012) Direct healthcare costs of selected diseases primarily or partially transmitted by water. *Epidemiol Infect* **140**: 2003–2013.

Corsaro, D., Pages, G.S., Catalan, V., Loret, J.F., and Greub, G. (2010) Biodiversity of amoebae and amoeba-associated bacteria in water treatment plants. *Int J Hyg Environ Health* **213**: 158–166.

Costerton, J.W., Lewandowski, Z., Caldwell, D.E., Korber, D.R., and Lappin-Scott, H.M. (1995) Microbial biofilms. *Annu Rev Microbiol* **49**: 1–45.

Council, N.R. (2006) *Drinking Water Distribution Systems: Assessing and Reducing Risks*. Washington, DC: The National Academy Press.

Dunne, W.M. Jr (2002) Bacterial adhesion: seen any good biofilms lately? *Clin Microbiol Rev* **15**: 155–166.

ECDC (2016) European Centre for Disease Prevention and Control. Annual Epidemiological Report 2016 – Legionnaires' Disease. [Internet – Cited 2016 09].

Falkinham, J.O. 3rd, Norton, C.D., and LeChevallier, M.W. (2001) Factors influencing numbers of *Mycobacterium* avium, *Mycobacterium intracellulare*, and other Mycobacteria in drinking water distribution systems. *Appl Environ Microbiol* **67**: 1225–1231.

Feazel, L.M., Baumgartner, L.K., Peterson, K.L., Frank, D.N., Harris, J.K., and Pace, N.R. (2009) Opportunistic pathogens enriched in showerhead biofilms. *Proc Natl Acad Sci USA* **106**: 16393–16399.

Fields, B.S., Sanden, G.N., Barbaree, J.M., Morrill, W.E., Wadowsky, R.M., White, E.H., and Feeley, J.C. (1989) Intracellular multiplication of *Legionella pneumophila* in amoebae isolated from hospital hot water tanks. *Curr Microbiol* **18**: 131–137.

Garrison, L.E., Kunz, J.M., Cooley, L.A., Moore, M.R., Lucas, C., Schrag, S., et al. (2016) Vital Signs: deficiencies in Environmental Control Identified in Outbreaks of Legionnaires' Disease — North America, 2000–2014. *MMWR Morb Mortal Wkly Rep* **65**: 576–584.

Gihring, T.M., Green, S.J., and Schadt, C.W. (2012) Massively parallel rRNA gene sequencing exacerbates the potential for biased community diversity comparisons due to variable library sizes. *Environ Microbiol* **14**: 285–290.

Gomez-Alvarez, V., Schrantz, K.A., Pressman, J.G., and Wahman, D.G. (2014) Biofilm community dynamics in bench-scale annular reactors simulating arrestment of chloraminated drinking water nitrification. *Environ Sci Technol* **48**: 5448–5457.

Hammer, Ø., Harper, D.A.T., and Ryan, P.D. (2001) PAST: Paleontological statistics software package for education and data analysis. *Paleontological Electronica* **4**: 1–9.

Harb, O.S., Gao, L.Y., and Abu Kwaik, Y. (2000) From protozoa to mammalian cells: a new paradigm in the life cycle of intracellular bacterial pathogens. *Environ Microbiol* **2**: 251–265.

Henne, K., Kahlisch, L., Brettar, I., and Höfle, M.G. (2012) Analysis of structure and composition of bacterial core communities in mature drinking water biofilms and bulk water of a citywide network in Germany. *Appl Environ Microbiol* **78**: 3530–3538.

Huws, S.A., McBain, A.J., and Gilbert, P. (2005) Protozoan grazing and its impact upon population dynamics in biofilm communities. *J Appl Microbiol* **98**: 238–244.

Ji, P., Parks, J., Edwards, M.A., and Pruden, A. (2015) Impact of water chemistry, pipe material and stagnation on the building plumbing microbiome. *PLoS ONE* **10**: e0141087.

van der Kooij, D., Veenendaal, H.R., and Scheffer, W.J. (2005) Biofilm formation and multiplication of *Legionella* in a model warm water system with pipes of copper, stainless steel and cross-linked polyethylene. *Water Res* **39**: 2789–2798.

Kozich, J.J., Westcott, S.L., Baxter, N.T., Highlander, S.K., and Schloss, P.D. (2013) Development of a dual-index sequencing strategy and curation pipeline for analyzing amplicon sequence data on the MiSeq Illumina sequencing platform. *Appl Environ Microbiol* **79**: 5112–5120.

Kuiper, M.W., Wullings, B.A., Akkermans, A.D., Beumer, R.R., and van der Kooij, D. (2004) Intracellular proliferation of *Legionella pneumophila* in *Hartmannella vermiformis* in aquatic biofilms grown on plasticized polyvinyl chloride. *Appl Environ Microbiol* **70**: 6826–6833.

Kuiper, M.W., Valster, R.M., Wullings, B.A., Boonstra, H., Smidt, H., and van der Kooij, D. (2006) Quantitative detection of the free-living amoeba *Hartmannella vermiformis* in surface water by using real-time PCR. *Appl Environ Microbiol* **72**: 5750–5756.

Lau, H.Y., and Ashbolt, N.J. (2009) The role of biofilms and protozoa in *Legionella* pathogenesis: implications for drinking water. *J Appl Microbiol* **107**: 368–378.

Lee, D.-G., Lee, J.-H., and Kim, S.-J. (2005) Diversity and dynamics of bacterial species in a biofilm at the end of the Seoul water distribution system. *World J Microbiol Biotechnol* **21**: 155–162.

Lehtola, M.J., Torvinen, E., Kusnetsov, J., Pitkanen, T., Maunula, L., von Bonsdorff, C.H., et al. (2007) Survival of *Mycobacterium avium*, *Legionella pneumophila*, *Escherichia coli*, and caliciviruses in drinking water-associated biofilms grown under high-shear turbulent flow. *Appl Environ Microbiol* **73**: 2854–2859.

Lin, Y.E., Vidic, R.D., Stout, J.E., and Yu, V.L. (1996) Individual and combined effects of copper and silver ions on inactivation of *Legionella pneumophila*. *Water Res* **30**: 1905–1913.

Lin, Y.-s.E., Vidic, R.D., Stout, J.E. and Yu, V.L. (2002) Negative effect of high pH on biocidal efficacy of copper and silver ions in controlling *Legionella pneumophila*, *Appl Environ Microbiol* **68**: 2711–2715.

Lin, Y.E., Stout, J.E., and Yu, V.L. (2011) Controlling *legionella* in hospital drinking water: an evidence-based review of disinfection methods. *Infect Control Hosp Epidemiol* **32**: 166–173.

Lin, W., Yu, Z., Chen, X., Liu, R., and Zhang, H. (2013) Molecular characterization of natural biofilms from household taps with different materials: PVC, stainless steel, and cast iron in drinking water distribution system. *Appl Microbiol Biotechnol* **97**: 8393–8401.

Lu, J., Struewing, I., Buse, H.Y., Kou, J., Shuman, H.A., Faucher, S.P., and Ashbolt, N.J. (2013) *Legionella pneumophila* transcriptional response following exposure to CuO nanoparticles. *Appl Environ Microbiol* **79**: 2713–2720.

Lu, J., Buse, H., Gomez-Alvarez, V., Struewing, I., Santo Domingo, J., and Ashbolt, N.J. (2014) Impact of drinking water conditions and copper materials on downstream biofilm microbial communities and *Legionella pneumophila* colonization. *J Appl Microbiol* **117**: 905–918.

Lu, J., Struewing, I., Yelton, S., and Ashbolt, N. (2015) Molecular survey of occurrence and quantity of *Legionella* spp., *Mycobacterium* spp., *Pseudomonas aeruginosa* and amoeba hosts in municipal drinking water storage tank sediments. *J Appl Microbiol* **119**: 278–288.

Martiny, A.C., Jorgensen, T.M., Albrechtsen, H.J., Arvin, E., and Molin, S. (2003) Long-term succession of structure and diversity of a biofilm formed in a model drinking water distribution system. *Appl Environ Microbiol* **69**: 6899–6907.

Mathys, W., Stanke, J., Harmuth, M., and Junge-Mathys, E. (2008) Occurrence of *Legionella* in hot water systems of single-family residences in suburbs of two German cities with special reference to solar and district heating. *Int J Hyg Environ Health* **211**: 179–185.

McBain, A.J., Bartolo, R.G., Catrenich, C.E., Charbonneau, D., Ledder, R.G., Rickard, A.H., et al. (2003) Microbial characterization of biofilms in domestic drains and the establishment of stable biofilm microcosms. *Appl Environ Microbiol* **69:** 177–185.

Messi, P., Anacarso, I., Bargellini, A., Bondi, M., Marchesi, I., de Niederhausern, S., and Borella, P. (2011) Ecological behaviour of three serogroups of *Legionella pneumophila* within a model plumbing system. *Biofouling* **27:** 165–172.

Moritz, M.M., Flemming, H.C., and Wingender, J. (2010) Integration of *Pseudomonas aeruginosa* and *Legionella pneumophila* in drinking water biofilms grown on domestic plumbing materials. *Int J Hyg Environ Health* **213:** 190–197.

Morvay, A.A., Decun, M., Scurtu, M., Sala, C., Morar, A., and Sarandan, M. (2011) Biofilm formation on materials commonly used in household drinking water systems. *Water Sci Technol* **11:** 252–257.

Mullis, S.N., and Falkinham, J.O. 3rd (2013) Adherence and biofilm formation of *Mycobacterium avium*, *Mycobacterium intracellulare* and *Mycobacterium abscessus* to household plumbing materials. *J Appl Microbiol* **115:** 908–914.

Murga, R., Forster, T.S., Brown, E., Pruckler, J.M., Fields, B.S., and Donlan, R.M. (2001) Role of biofilms in the survival of *Legionella pneumophila* in a model potable-water system. *Microbiology* **147:** 3121–3126.

Ohno, A., Kato, N., Sakamoto, R., Kimura, S., and Yamaguchi, K. (2008) Temperature-Dependent Parasitic Relationship between *Legionella pneumophila* and a Free-living Amoeba (*Acanthamoeba castellanii*). *Appl Environ Microbiol* **74:** 4585–4588.

Pedersen, K., Hallbeck, L., Arlinger, J., Erlandson, A.-C., and Jahromi, N. (1997) Investigation of the potential for microbial contamination of deep granitic aquifers during drilling using 16S rRNA gene sequencing and culturing methods. *J Microbiol Methods* **30:** 179–192.

Pinto, A.J., Xi, C., and Raskin, L. (2012) Bacterial community structure in the drinking water microbiome is governed by filtration processes. *Environ Sci Technol* **46:** 8851–8859.

Rhoads, W.J., Ji, P., Pruden, A., and Edwards, M.A. (2015) Water heater temperature set point and water use patterns influence *Legionella pneumophila* and associated microorganisms at the tap. *Microbiome* **3:** 67.

Rogers, J., and Keevil, C.W. (1992) Immunogold and fluorescein immunolabelling of *Legionella pneumophila* within an aquatic biofilm visualized by using episcopic differential interference contrast microscopy. *Appl Environ Microbiol* **58:** 2326–2330.

Rogers, J., Dowsett, A.B., Dennis, P.J., Lee, J.V., and Keevil, C.W. (1994) Influence of temperature and plumbing material selection on biofilm formation and growth of *Legionella pneumophila* in a model potable water system containing complex microbial flora. *Appl Environ Microbiol* **60:** 1585–1592.

Schloss, P.D., Westcott, S.L., Ryabin, T., Hall, J.R., Hartmann, M., Hollister, E.B., *et al.* (2009) Introducing mothur: open-source, platform-independent, community-supported software for describing and comparing microbial communities. *Appl Environ Microbiol* **75:** 7537–7541.

Shen, Y., Huang, C., Monroy, G.L., Janjaroen, D., Derlon, N., Lin, J., *et al.* (2016) Response of simulated drinking water biofilm mechanical and structural properties to long-term disinfectant exposure. *Environ Sci Technol* **50:** 1779–1787.

Thomas, V., Herrera-Rimann, K., Blanc, D.S., and Greub, G. (2006) Biodiversity of amoebae and amoeba-resisting bacteria in a hospital water network. *Appl Environ Microbiol* **72:** 2428–2438.

Thomas, V., Loret, J.F., Jousset, M., and Greub, G. (2008) Biodiversity of amoebae and amoebae-resisting bacteria in a drinking water treatment plant. *Environ Microbiol* **10:** 2728–2745.

Valster, R.M., Wullings, B.A., and van der Kooij, D. (2010) Detection of protozoan hosts for *Legionella pneumophila* in engineered water systems by using a biofilm batch test. *Appl Environ Microbiol* **76:** 7144–7153.

Valster, R.M., Wullings, B.A., van den Berg, R., and van der Kooij, D. (2011) Relationships between free-living protozoa, cultivable *Legionella* spp., and water quality characteristics in three drinking water supplies in the Caribbean. *Appl Environ Microbiol* **77:** 7321–7328.

Wadowsky, R.M., Butler, L.J., Cook, M.K., Verma, S.M., Paul, M.A., Fields, B.S., *et al.* (1988) Growth-supporting activity for *Legionella pneumophila* in tap water cultures and implication of hartmannellid amoebae as growth factors. *Appl Environ Microbiol* **54:** 2677–2682.

Wang, H., Edwards, M., Falkinham, J.O. 3rd, and Pruden, A. (2012) Molecular Survey of the Occurrence of *Legionella* spp., *Mycobacterium* spp., *Pseudomonas aeruginosa*, and Amoeba Hosts in Two Chloraminated Drinking Water Distribution Systems. *Appl Environ Microbiol* **78:** 6285–6294.

Wickelmaier, F. (2003) An introduction to MDS. In: *Reports from the Sound Quality Research Unit (SQRU), No. 7.*

Wilson, D.A., Reischl, U., Hall, G.S., and Procop, G.W. (2007) Use of partial 16S rRNA gene sequencing for identification of *Legionella pneumophila* and non-pneumophila *Legionella* spp. *J Clin Microbiol* **45:** 257–258.

Wingender, J., and Flemming, H.C. (2004) Contamination potential of drinking water distribution network biofilms. *Water Sci Technol* **49:** 277–286.

Wullings, B.A., and van der Kooij, D. (2006) Occurrence and genetic diversity of uncultured *Legionella* spp. in drinking water treated at temperatures below 15 degrees C. *Appl Environ Microbiol* **72:** 157–166.

Yamamoto, H., Sugiura, M., Kusunoki, S., Ezaki, T., Ikedo, M., and Yabuuchi, E. (1992) Factors stimulating propagation of legionellae in cooling tower water. *Appl Environ Microbiol* **58:** 1394–1397.

Yu, J., Kim, D., and Lee, T. (2010) Microbial diversity in biofilms on water distribution pipes of different materials. *Water Sci Technol* **61:** 163–171.

Zhang, M., Liu, W., Nie, X., Li, C., Gu, J., and Zhang, C. (2012) Molecular analysis of bacterial communities in biofilms of a drinking water clearwell. *Microbes Environ* **27:** 443–448.

Production of a novel medium chain length poly(3-hydroxyalkanoate) using unprocessed biodiesel waste and its evaluation as a tissue engineering scaffold

Pooja Basnett,[1] Barbara Lukasiewicz,[1] Elena Marcello,[1] Harpreet K. Gura,[2] Jonathan C. Knowles[3,4] and Ipsita Roy[1,*]

[1]Faculty of Science and Technology, University of Westminster, London, UK.

[2]Aarhus University, Denmark.

[3]Eastman Dental Institute, University College London, London, UK.

[4]Department of Nanobiomedical Science & BK21 Plus NBM Global Research Center for Regenerative Medicine, Dankook University, Cheonan 330-714, Republic of Korea.

Summary

This study demonstrated the utilization of unprocessed biodiesel waste as a carbon feedstock for *Pseudomonas mendocina* CH50, for the production of PHAs. A PHA yield of 39.5% CDM was obtained using 5% (v/v) biodiesel waste substrate. Chemical analysis confirmed that the polymer produced was poly(3-hydroxyhexanoate-*co*-3-hydroxyoctanoate-*co*-3-hydroxydecanoate-*co*-3-hydroxydodecanoate) or P(3HHx-3HO-3HD-3HDD). P(3HHx-3HO-3HD-3HDD) was further characterized and evaluated for its use as a tissue engineering scaffold (TES). This study demonstrated that P(3HHx-3HO-3HD-3HDD) was biocompatible with the C2C12 (myoblast) cell line. In fact, the % cell proliferation of C2C12 on the P(3HHx-3HO-3HD-3HDD) scaffold was 72% higher than the standard tissue culture plastic confirming

*For correspondence. E-mail royi@westminster.ac.uk

Funding Information
The authors would like to thank ReBioStent project – European Union's Seventh Programme for research, technological development and demonstration under grant agreement number 604251. The authors would also like to extend their gratitude to the University of Westminster for the support. The authors would also like to thank HyMedPoly project- Horizon 2020 under the grant agreement number 643050.

that this novel PHA was indeed a promising new material for soft tissue engineering.

Introduction

Biomaterials have played a significant role in improving health care over the years. They form an interface with the biological systems to treat diseased tissues or organs while restoring their function (Langer and Tirrell, 2004; Ulery *et al.*, 2011). The last two decades have witnessed a major paradigm shift from metals to biocompatible and biodegradable polymers for various medical applications such as the development of therapeutic devices and implants. The use of biodegradable polymer implants circumvents biocompatibility issues and complexities of revision surgeries associated with permanent metal implants (Nair and Laurencin, 2007). Most recently, a scaffold-based tissue engineering approach has demonstrated remarkable promise to create functional biological substitutes to restore tissue function. The underlying concept of this approach is to culture the patient's cells on a biomimetic scaffold made of an ideal biomaterial and deliver the tissue-engineered construct into the patient, inducing tissue formation. The scaffold should mimic the native cellular milieu and provide a structural framework to encourage the cells to organize into a functional tissue. Natural biopolymers are highly biocompatible, biodegradable and can be produced using renewable resources making them an ideal biomaterial for scaffold fabrication (Yang *et al.*, 2001). One such family of natural biodegradable polymers that has gained fresh impetus in recent years is the polyhydroxyalkanotes.

Polyhydroxyalkanotes (PHAs) are biodegradable polymers or bioplastics with a great structural diversity, produced by bacterial fermentation, normally under nutrient limiting conditions. They accumulate within the bacteria as intracellular storage compounds and can be degraded into carbon dioxide and water by a wide range of microorganisms that produce extracellular PHA depolymerases (Khanna and Srivastava, 2005). PHAs are composed of (R)-hydroxyalkanoates with diverse number of carbon atoms in the monomers (Fig. 1). The monomeric composition of PHAs is regulated by the metabolic pool present in the bacteria and the substrate specificity of

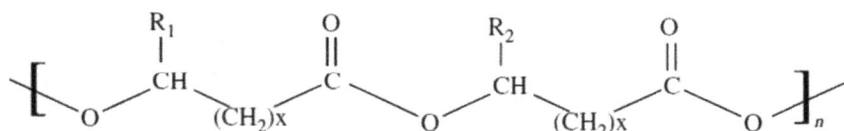

Fig. 1. General structure of PHA (Akaraonye *et al.*, 2010).

the concerned PHA synthase. Depending on the number of carbon atoms present in their monomeric units, PHAs can be classified into short chain length PHAs (scl-PHAs) containing three to five carbon atoms or medium chain length PHAs (mcl-PHAs) containing six to 13 carbon atoms in their monomeric units (Basnett and Roy, 2010).

Properties of PHAs vary based on their type, most scl-PHAs except for poly(4-hydroxybutyrate) are brittle and have a high melting temperature, whereas mcl-PHAs are highly elastomeric in nature; have low melting temperature and low glass transition temperature. PHAs can be either semi-crystalline (scl-PHAs and most of mcl-PHAs) or completely amorphous in nature (some of the mcl-PHAs). Different PHAs exhibit specific material properties, which can be used for particular applications. Amorphous PHAs are easily identified by differential scanning calorimetry (DSC) as they lack a distinct melting point but have a prominent glass transition temperature. Of all the various types of PHAs that exist, amorphous PHAs have not yet been very well explored. However, recently, Metabolix (Cambridge, Massachusetts, United States) have launched a new line of amorphous PHAs as an additive to PVC and PLA for improved material performance and processability, for application in food packaging (Additives for Polymers, 2013; Additives for Polymers, 2014). Polyhydroxyalkanoates have also been widely studied for a range of biomedical applications especially in the field of tissue engineering and medical devices. Scl-PHAs, such as P(3HB), P(4HB) and their copolymers have been used for hard tissue regeneration, such as bone tissue engineering or drug delivery systems, whereas mcl-PHAs, such as P(3HO), have been applied for soft tissue regeneration, such as wound healing or cardiac tissue engineering (Lizarraga-Valderrama *et al.*, 2016). The degradation products of PHAs are non-acidic in nature and therefore do not evoke an inflammatory response. They degrade via surface erosion, therefore, their degradation is controlled, thereby maintaining the integrity of the scaffold under *in vivo* conditions. Hence, PHAs have emerged as promising potential medical materials (Valappil *et al.*, 2006; Rai *et al.*, 2011a,b; Ali and Jamil, 2016). Furthermore, PHAs have also been identified as one of the safest bioplastics options for the environment. Life cycle assessment (LCA) is a tool that was developed to determine the ecological effect of a product during its life cycle. LCA

highlighted that PHAs were one of the safest options amongst the currently available bioplastics (Álvarez-Chávez *et al.*, 2012). Narodoslawsky *et al.* (2015) have demonstrated the ecological potential of PHAs in their critical review, highlighting the use of industrial and ecological waste products as well as clean energy, such as electricity, for the production of polymers.

Due to their attractive properties such as biodegradability and biocompatibility as well as a surge in the demand for green polymers, there has been considerable interest in the commercialization of PHAs in the recent times. One of the major challenges that have deterred their commercial exploitation is their high production cost. Studies have shown that bioplastics constitute only 0.1% of the total polymer market in Europe (Elvers *et al.*, 2016). Researchers have used various strategies to lower the cost of production, which include the use of metabolically engineered strains, various fermentation strategies, efficient downstream processing and transgenic plants. Facts state that the cost of the carbon source utilized in the production of PHAs accounts for 31% of the total production cost (Kim and Rhee, 2000; Marangoni *et al.*, 2001, 2002). Therefore, one of the most efficient approaches in making PHA production a cost effective process would be to utilize cheap carbon sources. The choice of biodiesel waste as a cheap feedstock for PHA production has gained momentum in the last few years. Biodiesel waste is known to be rich in glycerol content. With the dramatic increase in the biodiesel production globally, crude glycerol is being generated in large quantities. It is estimated that around 1 tonne of crude glycerol is generated for every 10 tonnes of biodiesel. Purification of this crude glycerol, a by-product of biodiesel production, is, however, an expensive process (Bormann and Roth, 1999; Ashby *et al.*, 2004; Ashby *et al.*, 2005). Therefore, direct utilization of the untreated biodiesel waste as a feedstock for the production of PHAs could improve the economical viability of both PHAs as well as the biodiesel industry. In this work, a study was carried out where biodiesel waste (crude glycerol phase) was used as the sole carbon source for the production of novel PHAs. This study followed the pattern of informed waste management where waste biodiesel was converted into a value-added product. In addition, we have also evaluated the use of the novel PHA produced for biomedical application.

Results and discussion

Production of PHAs by Pseudomonas mendocina *CH50 using 50 g l^{-1} of biodiesel waste as the sole carbon source*

Temporal profiling of the PHA production process in two 5 l bioreactors was carried out as described in the experimental procedures. Various parameters such as the pH, glycerol and nitrogen concentration in the media were monitored (Figure 2).

Optical density (O.D.) increased gradually until 10 h (lag phase) after which it increased rapidly until the end of the fermentation at 48 h, reaching a value of 7.15. The biomass concentration increased during the course of the fermentation and reached its highest value of 4.1 g l^{-1} around 45 h. The starting biomass concentration at 0 h was very high and did not correspond to the optical density value. Samples withdrawn at 0 h were washed with water followed by 10% ethanol. However, media components (impurities) could not be separated from the biomass giving an erroneous value. Characterization of the biodiesel waste to detect its exact composition could provide an insight into the choice of solvent to enable its dissolution. This could help in accurate estimation of the biomass concentration in the initial hours of the fermentation. Apart from 0 h, biomass values were generally low throughout the course of the fermentation. They did not correspond to the optical density values and the glycerol consumption. Previous studies have shown that *P. mendocina* is known to simultaneously produce extracellular alginate during PHA production when cultured using glucose as the carbon source (Guo *et al.*, 2011, 2012). More specifically, in a study carried out by Chanasit *et al.*, 2016; alginic acid was produced extracellularly when *P. mendocina* PSU was cultured using biodiesel liquid waste as the sole carbon source (Chanasit *et al.*, 2016). This could explain the utilization of the substrate; higher optical density and low biomass yield observed in this study. In future, the supernatant could be analysed to detect the production of extracellular polysaccharides.

Nitrogen was the limiting factor in this experiment. It is known that nitrogen limitation coupled with excess carbon triggers PHA production (Khanna and Srivastava, 2005). The concentration of nitrogen reduced from its initial value of 0.5 g l^{-1} to 0.02 g l^{-1} within 12 h of fermentation indicating that a nitrogen-limiting environment was maintained during the course of the fermentation. However, nitrogen in the form of ammonium ions was measured using phenol hypochlorite method, which only allows quantification of the inorganic nitrogen present in the media. Further increase in the optical density and biomass beyond 12 h indicates that there was an additional source of nitrogen available, probably from the biodiesel waste, which was being utilized by the bacterial cells for their growth and

metabolism. Shrivastav *et al.* (2010) reported the production of P(3HB) by novel bacterial strains using jatropha biodiesel waste as a carbon source. Teeka *et al.* (2012) isolated *Novosphingobium* sp. and utilized it for the production of P(3HB) using crude glycerol. They achieved a yield of around 45% cell dry mass (CDM). Chanasit *et al.* (2014) isolated a *Bacillus* sp. ST1C and used it for the production of P(3HB) using biodiesel waste. They achieved a yield of around 72.31% CDM. Several studies indicated that native PHA producers, under nutrient limiting conditions and in the presence of excess glycerol, converted the latter to P(3HB), which is one of the most widely studied PHAs. The pH of the culture medium was adjusted to 7.0 before the inoculation. pH values reduced gradually reaching a value of around 6.6 at the end of the fermentation. pH was not controlled during the fermentation process. This decrease could be due to the production of acid extracellularly into the fermentation media. As described above, several studies have shown that *Pseudomonas* species produce alginic acid simultaneously during PHA production (Guo *et al.*, 2011, 2012; Chanasit *et al.*, 2016).

Glycerol concentration was monitored to investigate the utilization of glycerol by the bacteria. It decreased from its initial concentration of 43 g l^{-1} to 14 g l^{-1} by the end of the fermentation. According to literature, most microorganisms are able to metabolize glycerol for their growth and development. In this particular study, glycerol consumption rate correlated with the increase in the optical density as shown in Fig. 2. The maximum cell dry mass (CDM) obtained was 4 g l^{-1} at 48 h of cultivation. In addition to glycerol, waste by-products generated from biodiesel production contain several impurities such as methanol, salts, saturated fatty acids such as lauric acid, palmitic acid, stearic acid, relatively high amounts of unsaturated fatty acids such as oleic acid, linoleic acid and alkyl esters (Ribeiro *et al.*, 2015). Teeka *et al.* (2012) observed that the novel bacterial strain AIK7 exhibited enhanced growth and higher polymer yield (41% cell dry mass) using biodiesel waste as the sole carbon source compared to pure glycerol. In another study conducted by Ashby *et al.*, they found that *Pseudomonas* sp. were able to utilize the free fatty acids and alkyl esters present in biodiesel waste as additional carbon sources to enhance their growth. Ashby *et al.* (2004) observed a yield of 42% CDM when *Pseudomonas corrugata* was grown on biodiesel waste. Poblete-Castro *et al.*, studied the production of mcl-PHAs via batch mode by different *Pseudomonas* strains using crude glycerol as the carbon source. *Pseudomonas putida* KT2440 accumulated the highest amount of mcl-PHAs amongst other *Pseudomonas* strains with a yield of around 34% CDM. This strain was then engineered by deleting the *phaZ* gene responsible

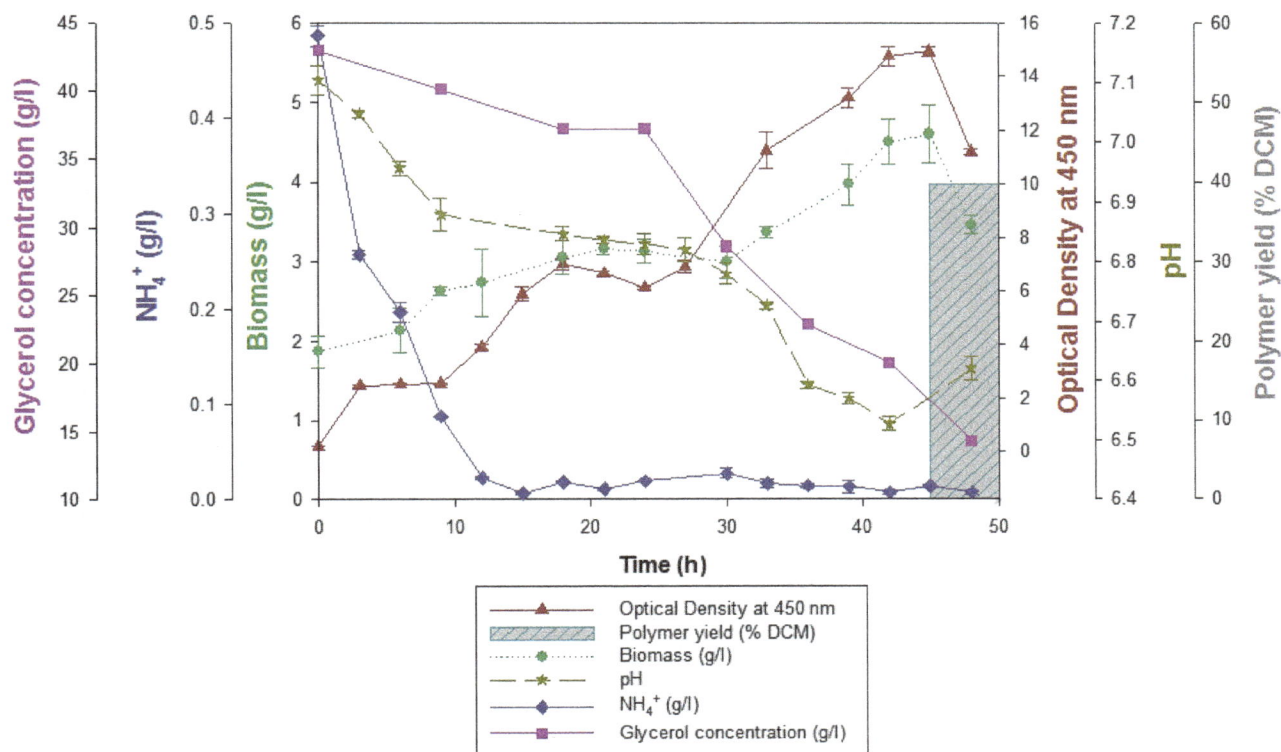

Fig. 2. Temporal profile of PHA production by *Pseudomonas mendocina* when cultured using 50 g l^{-1} biodiesel waste as the sole carbon source.

for the depolymerization of PHAs, resulting in an improved PHA yield of around 47% CDM (Poblete-Castro *et al.*, 2012). In this study, PHA yield of 39.5% CDM was obtained. This could be further enhanced by employing various optimization strategies such as fed-batch approaches with different feeding regimes and high cell density cultivation methods (Poblete-Castro *et al.*, 2012). Jiang *et al.* (2013) achieved a yield of about 75.5% CDM using the fed-batch approach, where three different carbon sources were used as a feed to trigger the synthesis of the polymer with a particular monomer unit composition (Figs 3–5).

Characterization of the PHAs produced

Gas chromatography mass spectrometry. Monomeric composition of the polymer produced was identified using gas chromatography mass spectrometry (GC-MS).

The mass spectrum (Figure 3) of the peak at a retention time (R_t) of 7.76 min was identical to that of the methyl ester of 3-hydroxyoctanoic acid, and the peak at a retention time (R_t) of 9.34 min was identical to that of the methyl ester of 3-hydroxydecanoic acid in the MS (NIST) library. The peak at a retention time (R_t) of 5.98 min was identical to that of the methyl ester of 3-hydroxyhexanoic acid, and the peak at a retention time (R_t) of 10.7 min

was identical to that of the methyl ester of 3-hydroxydodecanoic acid in the MS (NIST) library, confirming that the polymer produced was a copolymer of 3-hydroxyhexanoate, 3-hydroxyoctanoate, 3-hydroxydecanoate and 3-hydroxydodecanoate or P(3HHx-*co*-3HO-*co*-3HD-*co*-3HDD) – 2.3 mol% HHx, 27.8 mol% HO, 55.7 mol% HD and 14.2 mol% HDD. Methyl benzoate was used as an internal standard and was represented by the peak at a retention time of 6.4 min. 3-Hydroxydecanoate (C_{10}) was found to be the dominant monomer. Recent studies have shown the monomer content of the PHAs are determined by various enzymes that are capable of metabolizing fatty acids to 3-hydroxyacyl-CoA precursors, which are utilized by the PHA synthase enzymes. Several researchers have made similar observations where 3-hydroxydecanoate was found to be the dominant monomer in the mcl-PHA copolymer when *Pseudomonas* sp was grown on glycerol, glucose or gluconate (Poblete-Castro *et al.*, 2012, 2013). Studies carried out by Poblete-Castro *et al.* have shown that *P. putida* LS46 produced mcl-PHAs with a high amount of 3-hydroxydecanoate (C_{10}) units when cultured in waste glycerol. It has been reported that the *phaG* gene has affinity towards (R)-3-hydroxydecanoyl–ACP in both *P. putida* LS46 and *P. putida* KT2440. A high level of *phaG* expression was observed during

(A)

(B)

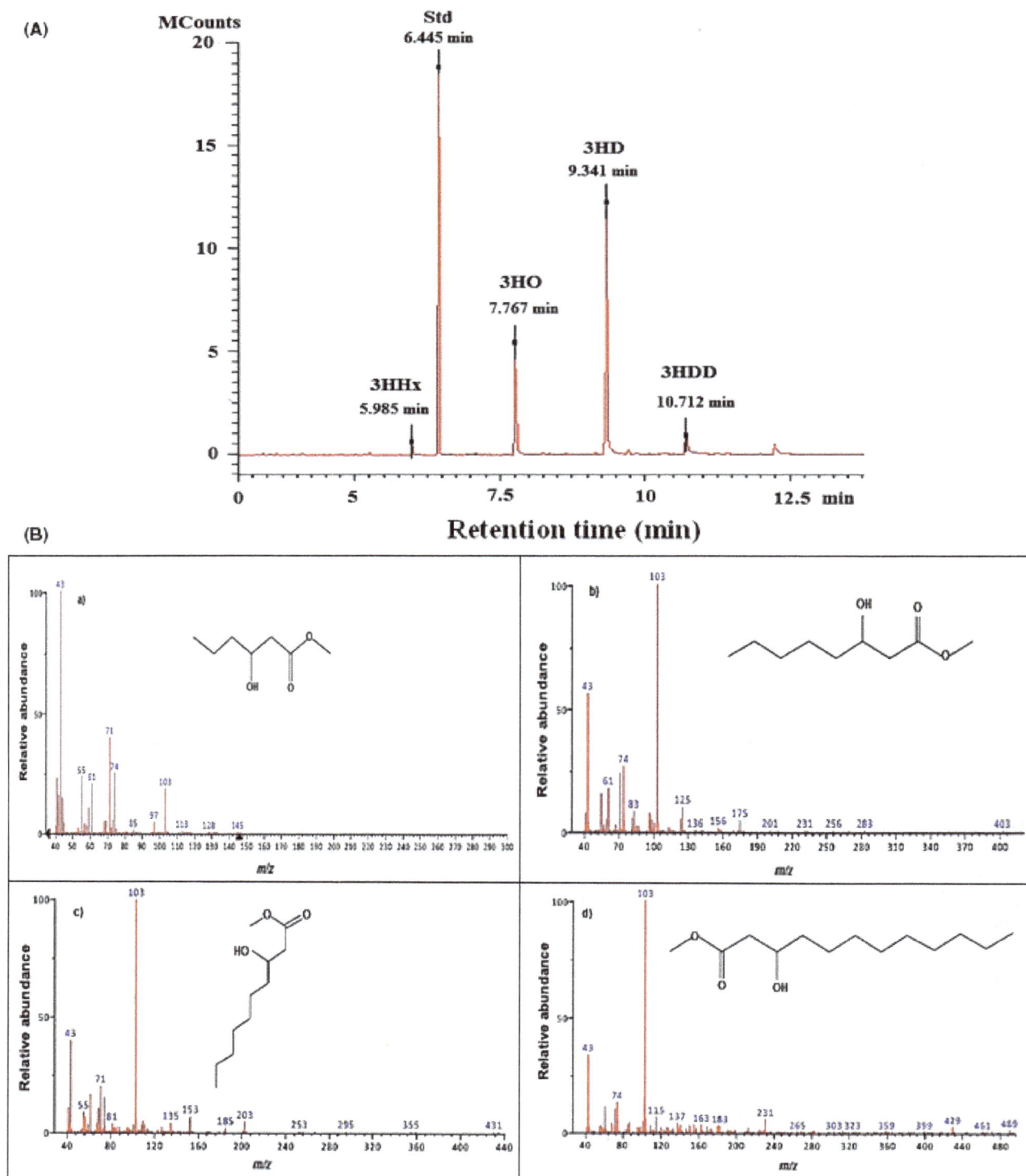

Fig. 3. (A) Gas chromatogram and (B) MS spectra of the polymer produced using biodiesel waste as the sole carbon source.

mcl-PHA production by *P. putida* LS46 using waste glycerol. It is known that (R)-3-hydroxyacyl–ACP is restricted to 10 carbon substrates, and therefore, high amount of 3-hydroxydecanoate (C_{10}) is produced during the

biosynthesis of mcl-PHAs by *Pseudomonas* species (Poblete-Castro *et al.*, 2012, 2013).

Production of P(3HHx-3HO-3HD-3HDD) has been reported earlier. However, the mol% of each monomer obtained in this study is novel. In a study conducted by Cerrone *et al.*, mannitol rich ensiled grass press juice (EGPJ) was used a sole carbon source for the production of PHAs under high cell density cultivation. Different strains of *Pseudomonas* sp were investigated for this study. They observed that *P. putida* W619 produced P(3HHx-3HO-3HD-3HDD) – 6 mol% HHx, 19 mol% HO, 68 mol% HD and 5 mol% HDD using EGPJ. In addition, *P. putida* W619 and *Pseudomonas chlororaphis* IMD555 both produced P(3HHx-3HO-3HD-3HDD) using simulated EGPJ. Monomer composition of P(3HHx-3HO-3HD-3HDD) produced by *P. putida* W619 was found to be 1 mol% HHx, 14 mol% HO, 63 mol% HD and 21 mol% HDD, whereas the composition of P(3HHx-3HO-3HD-3HDD) produced by *P. chlororaphis* IMD555 was measured to be 5 mol% HHx, 10 mol% HO, 54 mol% HD and 32 mol% HDD (Cerrone *et al.*, 2015). In another study conducted by Mozejko *et al.*, pulsed feeding strategy was employed to enhance the production of mcl-PHAs from waste rapeseed oil by *Pseudomonas* sp. G101. They identified the monomeric composition of the mcl-PHA produced as P(3HHx-3HO-3HD-3HDD). The monomeric structure remained unaffected by the cultivation time and feeding strategy, however, the mol% of the monomers varied (Możejko and Ciesielski, 2014). In this particular study, 3- hydroxyoctanoate was found to be the dominant monomer. Hence, the mole% of the 3HHx, 3HO, 3HD and 3HDD monomer in P(3HHx-3HO-3HD-3HDD) differed in each organism (summarized in Table 1). None of the previously reported copolymers containing 3HHx, 3HO, 3HD and 3HDD monomers have the same mole % as the mcl-PHA copolymer produced in this work; hence, a new/novel polymer was produced with a unique monomer content, which would in turn lead to unique material properties.

Table 1. Summary of the mol% of P(3HHx-3HO-3HD-3HDD) obtained in various studies.

| Organism | Carbon source | Mol % | | | |
		3HHx	3HO	3HD	3HDD
Pseudomonas mendocina CH50[a]	Biodiesel waste	2.3	27.8	55.7	14.2
P. putida W619	EGPJ	6	19	68	5.0
P. putida W619	Simulated EGPJ	1	14	63	21.0
P. chlororaphis IMD555	Simulated EGPJ	5	10	54	32.0
Pseudomonas sp. G101	Waste rapeseed oil- (Pulsed feeding at 48 h)	8.24	52.90	35.85	3.01
Pseudomonas sp. G101	Waste rapeseed oil(continuous feeding at 48 h)	11.40	48.54	29.82	10.24

a. Mol% of P(3HHx-3HO-3HD-3HDD) obtained in this study.

Nuclear magnetic resonance. Structural identification of the polymer was confirmed using nuclear magnetic resonance (NMR).

The ^{1}H NMR spectrum of P(3HHx-*co*-3HO-*co*-3HD-*co*-3HDD) as shown in Fig. 4A showed signals of the methine protons (CH) at 5.2 ppm demonstrating -O-(CH-)CH$_2$- bonding at carbon number 3. The peak present at 0.89 ppm corresponded to the terminal methyl group (CH$_3$). The multiplet resonance existing at 2.51 ppm corresponded to the methylene protons (CH$_2$), –CH-(CH$_2$)- CO- at carbon atom 2. The signal at 1.52 ppm corresponded to the methylene protons -CH$_2$-(CH$_2$)- CH- at carbon atom 4. The signal at 1.26 ppm corresponded to all other methylene protons of the saturated side-chain. CDCl$_3$ was used as the solvent, which resulted in a peak at 7.3 ppm. NMR analysis of the mcl-PHA copolymer produced using biodiesel waste as the sole carbon source also confirmed the GC-MS data. Hence, the mcl-PHA was structurally identified to be a P(3HHx-3HO-3HD-3HDD) copolymer.

Other biodiesel products were investigated for the production of mcl-PHAs. The PHAs produced showed different monomeric composition other than the one obtained in this study. Muhr *et al.* (2013a) used saturated biodiesel fractions (SFAE) originating from animal waste lipids for PHA production by *Pseudomonas citronellolis*. P(3HHx-3HHp-3HO-3HN-3HD) was obtained, and 3-hydroxyoctanoate and 3-hydroxydecanoate were the dominant monomers. Another study by the same group investigated the production by *P. chlororaphis* using the same biodiesel product. In this case, P(3HHx-3HHp-3HO-3HN-3HD-3HHD) was produced (Muhr *et al.*, 2013b).

Differential scanning calorimetry

Thermal properties of P(3HHx-3HO-3HD-3HDD) including the melting temperature (T_m) and glass transition temperature (T_g) were determined using DSC.

The T_g (Figure 5) of the copolymer was −21°C, which is typical for the MCL-PHAs. No melting event was observed during the first and the second heating cycles indicating that the copolymer was amorphous in nature. To confirm this result, further analysis was carried out. The polymer was stored for a period of four weeks before carrying out the DSC studies. There was no melting event confirming that P(3HHx-3HO-3HD-3HDD) was an amorphous polymer. This indicated that the monomer units were randomly oriented at the molecular level. Muangwong *et al.* (2016) carried out an extensive study by employing novel bacteria isolated from the soil such as ASC1, *Acinetobacter* sp. (94.9% similarity); ASC2, *Pseudomonas* sp. (99.2% similarity); ASC3, *Enterobacter* sp. (99.2% similarity); and ASC4, *Bacillus* sp. (98.4% similarity) for the production of mcl-PHAs using crude

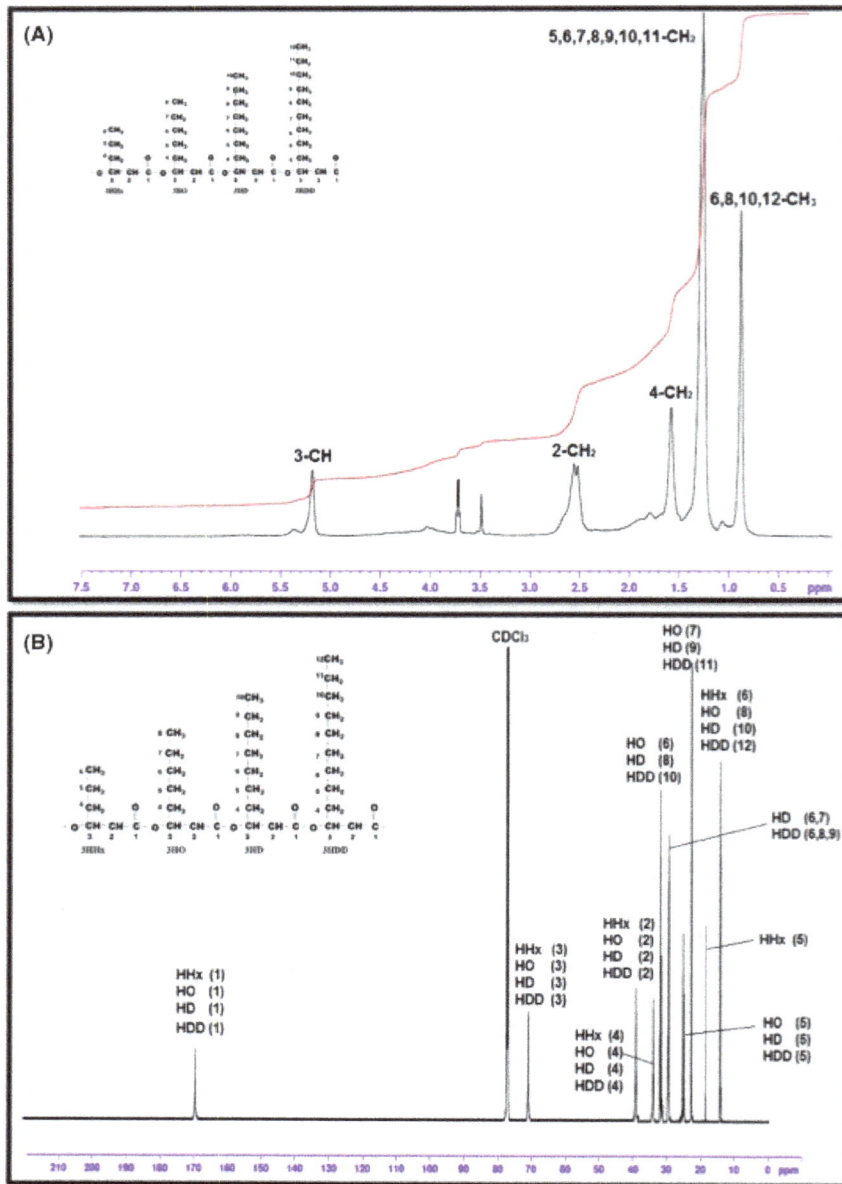

Fig. 4. (A) ^{1}H NMR spectra and (B): ^{13}C spectra of P(3HHx-co-3HO-co-3HD-co-3HDD).

glycerol as the carbon source. All the mcl-PHAs produced were amorphous in nature. Similar findings were reported by Song *et al.*, when *Pseudomonas* sp. strain DR2 was cultivated using waste vegetable oil for mcl-PHA production. They associated the amorphous nature of the polymer produced with the absence of 3-hydroxybutyrate monomer units. It is a known fact that the length of the aliphatic chains of a single monomer unit affects the crystallinity of the polymer. Presence of longer aliphatic chains in the monomer units interrupts their arrangement at the molecular level (Song *et al.*, 2008). Preusting *et al.* (1990) also concluded that the presence

of unsaturated groups interfered with the systematic arrangement of the polymeric chains, resulting in the synthesis of an amorphous polymer.

Synthesis of P(3HHx-3HO-3HD-3HDD) films by solvent casting and their characterization

P(3HHx-3HO-3HD-3HDD) films were prepared using the solvent cast method. These solvent cast films were thoroughly characterized for their surface properties before being tested for their compatibility with mammalian cells.

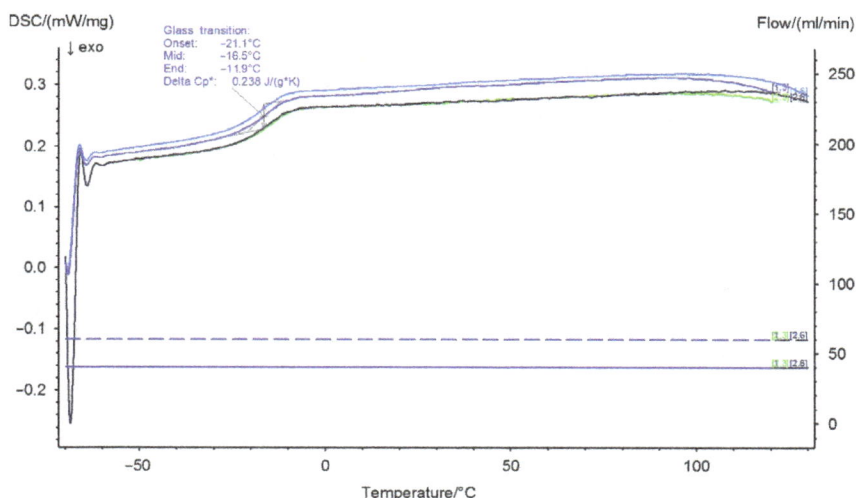

Fig. 5. The DSC spectra of P(3HHx-3HO-3HD-3HDD) produced using biodiesel waste as the carbon source.

Scanning electron microscopy

The surface topography of the P(3HHx-3HO-3HD-3HDD) film samples was studied using scanning electron microscopy (SEM). Surface topography of the P(3HHx-3HO-3HD-3HDD) film is shown in Fig. 6.

Scanning electron microscopy studies revealed that the surface of the P(3HHx-3HO-3HD-3HDD) film was smooth devoid of pores and protrusions; however, there were undulating areas on the air-exposed surface of the films, formed during the solvent casting process. This was due to the highly viscous nature of the polymer solution and was observed in all the P(3HHx-3HO-3HD-3HDD) solvent cast films. Previous studies have also shown that mcl-PHA matrix generally exhibits a smooth appearance (Rai, 2010; Renard *et al.*, 2011; Basnett *et al.*, 2012). Low crystallization rate of the polymer during the solvent evaporation process would have allowed the polymer to reorganize steadily avoiding the formation of pores and protrusions on the surface (Renard *et al.*, 2011).

Surface roughness. Surface roughness of the P(3HHx-3HO-3HD-3HDD) film was measured using the laser profilometer. Surface roughness values of the P(3HHx-3HO-3HD-3HDD) film were calculated to be $R_q = 1.77 \pm 0.29$ μm as shown in Fig. 7.

R_q value of 10 wt% P(3HHx-3HO-3HD-3HDD) film was significantly higher compared to the R_q values obtained for other mcl-PHA neat films, calculated to be 0.238 μm (Basnett *et al.*, 2012) or 0.272 μm (Rai *et al.*, 2011a,b). The high R_q value could be due to the uneven surface of the film formed due to the viscous nature of the polymer solution. This was consistent in all the three films that were cast. Surface topography is one of the crucial factors that is considered while evaluating any biomaterial for biomedical applications. Existing paradigm suggests that most cells prefer to attach on a rough surface compared to the smooth surface. Uneven or rough surfaces provide the cells with several adhesion points that allow them to migrate and proliferate (Hallab *et al.*, 2001). However, it has been suggested that there is an optimal surface roughness value beyond which it could have a detrimental effect on cell adhesion mechanism (Francis *et al.*, 2016). In addition to surface roughness, two other factors that play an important role in determining the biocompatibility of a material are hydrophilicity and protein adsorption (Hao and Lawrence, 2004).

Water contact angle measurements. Static contact angle study was carried out to measure the hydrophobicity of the P(3HHx-3HO-3HD-3HDD) film samples. The tests were carried out in triplicates on three different solvent cast films. Water contact angle (θ) of the P(3HHx-3HO-3HD-3HDD) film was measured to be 77.3 ± 5. It is a known fact that the surface of a material is considered to be hydrophobic if the water contact angle (θ) is higher than 70° (Basnett *et al.*, 2012). Therefore, the surface of the P(3HHx-3HO-3HD-3HDD) films was considered to be hydrophobic. Hydrophobicity is associated with the PHAs due to the presence of alkyl pendant groups in their side-chains. This is in agreement with the other researchers who have observed that the surface of a neat PHA film is inherently hydrophobic. In other studies, the static contact angle values for neat P(3HO) film was within the range of 77.3 ± 1.0 to 101 ± 0.8 (Rai *et al.*, 2011b; Bagdadi *et al.*, 2016). Contact angle values for P(3HHx-3HO-3HD-3HDD) obtained in this study (77.3 ± 5.0) were within the range described in the

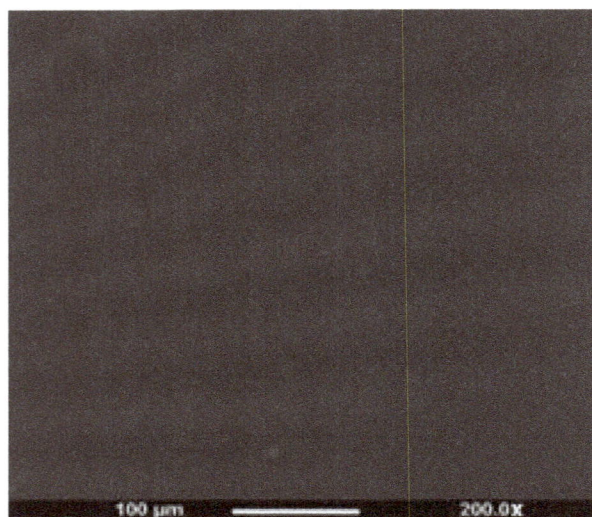

Fig. 6. Scanning electron microscopy images of the surface of the solvent cast P(3HHx-3HO-3HD-3HDD) film.

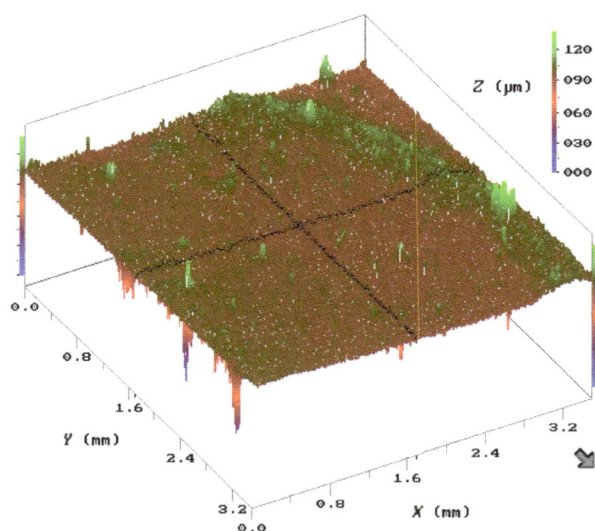

Fig. 7. Surface scan of the P(3HHx-3HO-3HD-3HDD) film using the Laser profilometer ($n = 3$).

literature. However, the θ value (77.3 ± 5) obtained for P(3HHx-3HO-3HD-3HDD) films was lower than that observed for P(3HO) films [θ values (104 ± 2)] as described by Renard *et al.* (2011). This decrease in the θ value could be due to the surface characteristics of the solvent cast P(3HHx-3HO-3HD-3HDD) films (Hao and Lawrence, 2006a,b). In an experiment conducted by Kubiak *et al.,* the influence of surface roughness on the wettability properties of the engineering surfaces was studied. They observed that θ values were strongly affected by the surface roughness. They found that θ values were reduced with intermediate roughness as the

water droplet spread along the grooves (Kubiak *et al.*, 2011). Hydrophilicity is considered to be a desirable property in a biomaterial or a tissue engineering scaffold. Several studies have demonstrated that most cells prefer to attach and proliferate on hydrophilic surfaces (Hallab *et al.*, 2001). Bearing in mind the application of P(3HHx-3HO-3HD-3HDD) films as a tissue engineering scaffold, this decrease in the hydrophobicity of the P(3HHx-3HO-3HD-3HDD) film as compared to the P(3HO) film is a positive outcome. P(3HO) is established as a highly biocompatible PHA (Rai *et al.*, 2011a,b). Hence, this result indicates that P(3Hx-3HO-3HD-3HDD) should promote cell adhesion and proliferation as well as P(3HO) and perhaps perform even better than P(3HO).

Protein adsorption. Another important factor that influences the biocompatibility of a scaffold is protein adsorption (Rechendorff *et al.*, 2006). Bicinchoninic acid assay was carried to quantify the total protein adsorption on the P(3HHx-3HO-3HD-3HDD) film samples. Overall protein adsorbed on the P(3HHx-3HO-3HD-3HDD) film samples was 1.6 ± 0.1 mg cm^{-2}. For this particular study, bovine serum albumin (BSA) was the protein of choice. In a detailed study carried out by Jeyachandran *et al.*, adsorption of BSA on both hydrophilic and hydrophobic surface was widely investigated. They inferred that the BSA surface coverage was saturated at 53% on the hydrophobic surfaces, whereas on the hydrophilic surfaces, BSA coverage was almost up to 95%. In addition, the interaction of the BSA molecules on the hydrophilic surfaces was much stronger compared to the hydrophobic surfaces (Jeyachandran *et al.*, 2009). The amount of protein adsorbed on the P(3HHx-3HO-3HD-3HDD) film samples in this study was significantly higher compared to values reported in the literature for the other PHA films of similar dimension and incubation time of 24 h. For example, BSA adsorption on neat P(3HO) film was previously observed to be 0.083 mg cm^{-2} (Rai, 2010) as compared to the value of 1.6 ± 0.1 mg cm^{-2} observed for the P(3HHx-3HO-3HD-3HDD) film. The difference was significant ($**P = 0.0018$). This outcome is coherent with the slightly lower θ value followed by the increased R_q value obtained for P(3HHx-3HO-3HD-3HDD) films compared to other neat mcl-PHA films. This finding thus supports the fact that protein adsorption is higher on hydrophilic and rough surfaces (Hao and Lawrence, 2006a,b). This higher protein adsorption observed on the P(3HHx-3HO-3HD-3HDD) films might confer higher biocompatibility to this novel material.

It is a known fact that mammalian cells are anchorage-dependent. They attach onto protein-rich surfaces and pro-liferate; therefore, protein adsorption is an important phe-nomenon while considering various biomaterials in TE

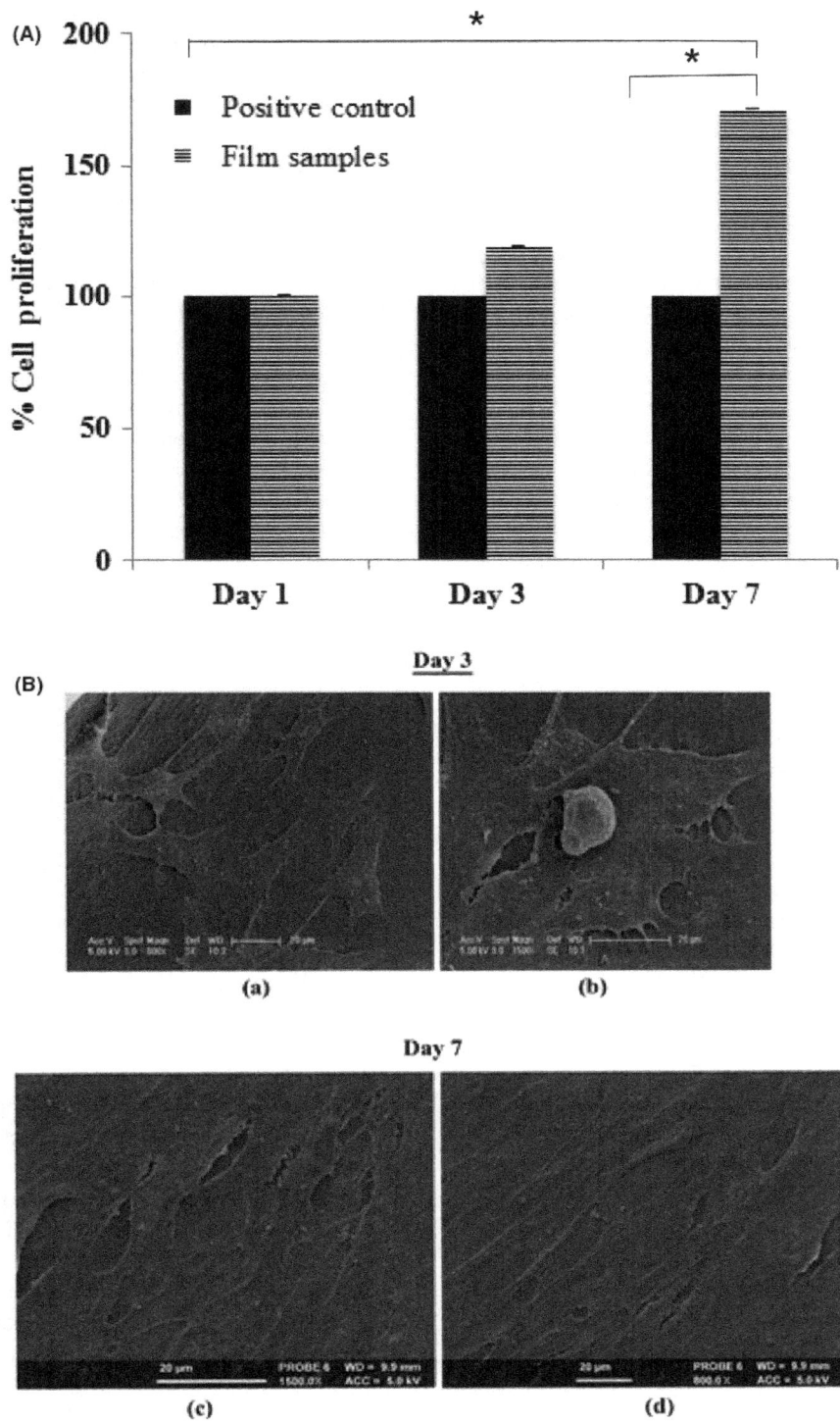

Fig. 8. (A) Cell proliferation study of the C2C12 cells on the P(3HHx-3HO-3HD-3HDD) film samples on day 1, 3 and 7 ($n = 3$). All the film samples were measured relative to the C2C12 cells on the tissue culture plate (TCP), which were normalized to 100% ($n = 3$). (B) SEM images of the C2C12 cells on the P(3HHx-3HO-3HD-3HDD) film samples on day 3 (a–b) and 7 (c–d). Size bar: 20 μm.

applications (Chen *et al.*, 2009). Sato and Murahara (2004) made an interesting observation that protein adsorption was higher on fabricated surfaces which were rough and less hydrophobic. Several techniques such as laser technology, microcontact printing and photopatterning have been implemented to fabricate the surface of the polymer films to enhance protein adsorption and cell adhesion (Hao and Lawrence, 2006a,b; Chen et al., 2009).

In vitro *biocompatibility studies.* Preliminary *in vitro* cell culture studies were carried out on P(3HHx-3HO-3HD-3HDD) films using the C2C12 cell line. C2C12, which is a myoblast cell line isolated from mouse muscle, was chosen to investigate the use of P(3HHx-3HO-3HD-3HDD) films as a potential scaffold for cardiovascular/skeletomuscular tissue engineering. From the literature, it is understood that C2C12 cells gradually fuse to form myotubes or plurinucleate syncytia. Myotubes then differentiate to attain functional features of a muscle cell (Burattini et al., 2004). Bearing in mind the potential application of P(3HHx-3HO-3HD-3HDD) films as cardiovascular/skeletomuscular tissue engineering scaffolds, differentiation of the C2C12 cells into cells of myogenic lineage would be of interest. The MTT assay was used to measure the biochemical activity of the cells seeded on the films. Attachment and proliferation of C2C12 cells on the film samples were studied over a period of 1, 3 and 7 days. Standard tissue culture plastic was used as the positive control for this experiment. The results of these biocompatibility studies have been summarized in Fig. 8.

The growth of C2C12 cells on both film samples and the TCP was comparable at the end of the 1st day of seeding. The % cell proliferation on the film samples was $101 \pm 0.3\%$ compared to the TCP at the end of day 1. It is known that the cells attach to the surface of the substrate or scaffold by focal contacts. At the end of day 3, there was a gradual increase in cell growth on the P(3HHx-3HO-3HD-3HDD) film samples. The % cell proliferation on the film samples was 119 ± 0.43. However, there was a significant increase in the growth of the C2C12 cells on the P(3HHx-3HO-3HD-3HDD) film samples compared to the standard TCP at the end of day 7. The % cell proliferation was found to be $172 \pm 0.05\%$ ($*P = 0.0129$). The % cell proliferation on day 7 was also significantly higher in comparison with % cell proliferation on day 1 ($*P = 0.0137$). The data confirmed that P(3HHx-3HO-3HD-3HDD), an mcl-PHA copolymer produced using biodiesel waste as the sole carbon source, supported the adhesion and proliferation of C2C12 cells. In a study conducted by Bagdadi et al. (2016), % cell proliferation of $21.11 \pm 5.29\%$ was observed when C2C12 cells were cultured on neat P(3HO) films for 24 h (Bagdadi et al., 2016). In another study, human keratinocyte cell line (HaCaT) was cultured on the neat P(3HO) film samples (Rai et al., 2011a,b). Highest % cell proliferation of 73.09% was obtained after 7 days of incubation (Rai et al., 2011a,b). Hence, it is known that PHAs are able to support cell adhesion and proliferation and are hence highly biocompatible (Rai et al., 2011a,b; Basnett et al., 2013). Data obtained in this study were consistent with the reports cited in literature by several other researchers (Misra et al., 2006; Basnett et al., 2013). Hence, from this particular study, it has been established that the PHA produced using unprocessed biodiesel waste as the carbon source was suitable to be used as a substrate for mammalian cell growth and hence as a scaffold material for soft tissue engineering.

Scanning electron microscopy. C2C12 cells were seeded onto the P(3HHx-3HO-3HD-3HDD) film samples. The cells were fixed and viewed under the SEM at day 3 and 7 as shown in Fig. 8. For this particular study, C2C12 cells were cultured in basal media (DMEM media with 10% fetal calf serum), and therefore, these cells were not expected to undergo differentiation. SEM images revealed the adhesion of C2C12 cells on the P(3HHx-3HO-3HD-3HDD) film samples on day 3. The film samples were covered with a confluent layer of cells by day 7, a confirmation of cell proliferation and growth. This also confirmed the results of the MTT assay. The morphology of the cells transitioned from cuboidal shape on day 3 to elongated cells on day 7. From the literature, it is known that the transition in cell shape is indicative of myogenic differentiation and is usually followed by the assembly of structures (actin and myosin filaments) involved in contraction (Burattini et al., 2004). This will be investigated in future by culturing C2C12 on P(3HHx-3HO-3HD-3HDD) films in low serum differentiation media. Presence of confluent, elongated C2C12 cells on P(3HHx-3HO-3HD-3HDD) films confirmed that it was biocompatible and supported cell growth and proliferation.

In conclusion, a novel mcl-PHA, P(3HHx-3HO-3HD-3HDD) was produced using unprocessed biodiesel waste as the sole carbon source. This finding could lead to the use of inexpensive substrates such as biodiesel waste as a replacement for expensive commercially available carbon sources for PHA production, paving the way for its economical production and hence successful commercialization. This study successfully demonstrated that P(3HHx-3HO-3HD-3HDD) was highly biocompatible in nature and supported cell adhesion and proliferation. Due to its amorphous nature, P(3HHx-3HO-3HD-3HDD) could also be used as an additive or a compatibilizer to obtain suitable blends and composites, further extending their application in the field of tissue engineering. Hence in this work, a completely novel mcl-PHA with great potential in soft tissue engineering was produced using unprocessed biodiesel waste, converting waste to a high-value medical product.

Experimental procedures

Bacterial strains, media and culture conditions

Pseudomonas mendocina CH50 was obtained from the National Collection of Industrial and Marine Bacteria

(NCIMB). They were cultured using sterile nutrient broth medium at 30°C at 200 rpm for 16 h in a shaking incubator. Stock cultures were stored in 40% glycerol at −80°C.

Cell lines, media and culture conditions

C2C12 (myoblast) cell line was obtained from the culture collection of the University of Westminster, London, UK. They were cultured using Dulbecco's modified Eagle's medium (DMEM) at 37°C at 5% CO_2 in a humidified incubator.

Chemicals

Chemicals used in this study were purchased from Sigma-Aldrich (Dorset, UK) and VWR (Poole, UK). Chemicals for the cell culture studies were obtained from Lonza (Slough, UK). Live versus Dead staining kit was purchased from Life Technologies, Paisley, UK. Biodiesel waste was obtained from a company called Purefuels, London. They convert waste frying oil into biodiesel via the alkaline esterification method.

Production of PHAs by Pseudomonas mendocina CH50 using 50 g l^{-1} of biodiesel waste as the sole carbon source

Production of PHAs by *P. mendocina* CH50 using 50 g l^{-1} of biodiesel waste (crude glycerol phase) was carried out in 5 l bioreactors (320 series Fermac, Electrolab, Tewkesbury, UK). The working volume used in this study was 3.5 l. Fermentation was carried out in batch mode in two stages. The seed culture was prepared using a single colony of *P. mendocina* CH50 to inoculate sterile nutrient broth. This was incubated for 16 h at 30°C, 200 rpm. Ten per cent (v/v) of the inoculum was used to inoculate the second stage seed culture (mineral salt medium-MSM) which was incubated at 30°C, 200 rpm for 24 h. This stage is also known as the acclimatization stage. Ten per cent (v/v) of the second stage seed culture was used to inoculate the final PHA production media (MSM media; Basnett *et al.*, 2012; Rai *et al.*, 2011a,b). All the media components including the biodiesel waste were sterilized at 121°C for 15 min prior to inoculation. The air flow rate was set at 1vvm, and the stirrer speed was set to 200 rpm. pH of all MSM media components, including biodiesel waste, was adjusted to 7.0 at the beginning of the fermentation. The culture was grown for 48 h, at 30°C, 200 rpm. Temporal profile of the production of PHAs by *P. mendocina* CH50 using 50 g l^{-1} of biodiesel waste phase was obtained by monitoring various parameters such as optical density, biomass, nitrogen and carbon concentration at regular intervals throughout the course of the fermentation.

Optical density measurements, biomass estimation and nitrogen estimation

One millilitre of culture was withdrawn after every 3 h in triplicates during the course of the fermentation. O.D. measurements were carried out at 450 nm using Spectrophotometer SB038 (Cadex, Richmond, Canada) and recorded to monitor the growth of the organism. During profiling, 1 ml of sample was withdrawn in a preweighed Eppendorf tubes and centrifuged at 12,000 rpm for 10 min. Pellet obtained was washed with water, then washed with 10% ethanol and centrifuged at 12,000 rpm (Heraeus Multifuge X3R Centrifuge, Thermo Scientific) for 10 min to remove impurities. The pellet obtained was freeze-dried. The weight of the Eppendorf tubes containing the dried biomass was measured, and the difference was recorded as the biomass weight. The cell pellet obtained after centrifugation was freeze-dried and weighed. The supernatant obtained after centrifugation was used to measure the pH and also analyse the amount of nitrogen, in the form of ammonia ions, using the phenol hypochlorite method (Weatherburn, 1967).

Total carbon estimation using the chemical oxygen demand

Closed reflux titrimetric analysis was carried out to determine the chemical oxygen demand (COD) of the samples as described in Environment Agency (UK) Standard method 5220D. Two millilitres samples were used for the analysis. Diluted samples were added to Ficodox (3.8 ml), which is a mixed COD reagent. This mixture was treated on a preheated heating block for 1.5 h at 150°C in closed digestion tubes. A 0.025 M ferrous ammonium sulfate (FAS) was used as the titrant along with 2-3 drops of Ferroin indicator solution (Fisher Scientific, Loughborough, UK). This solution was used to titrimetrically determine the residual potassium dichromate present in the Ficodox digestate following the digestion of the sample. The COD deduction was calculated using the following equation:

$$COD(mg\,l^{-1}) = (V_b - V_s) * DF * M * 4000$$

where V_b and V_s are ferrous ammonium sulphate (FAS) titrant volumes for the blank and the sample respectively, DF is the sample dilution factor and M is the molarity of FAS titrant (Fernando *et al.*, 2012).

Polymer extraction from biomass

The polymer was extracted from the dried biomass by using chloroform/sodium hypochlorite dispersion method (Hahn *et al.*, 1994). Lyophilized biomass was placed in conical flasks. Sodium hypochlorite solution and chloroform were added into biomass and incubated for 2 h at 30°C in the orbital shaker (140 rpm). After incubation, the slurry was centrifuged for 20 min at 3900 rpm (Heraeus Multifuge X3R Centrifuge, Thermo Scientific) to allow phase separation. The organic layer containing chloroform and dissolved polymer was collected, filtered and concentrated. The polymer was precipitated using ice-cold methanol. Polymer mass fraction was calculated as a percentage of the cell dry mass, using the formula:

$$\text{Polymer mass fraction}(\%DCM) = (\text{polymer weight/biomass}) * 100$$

Characterization of the PHAs produced

Gas chromatography mass spectrometry. Monomeric composition of the PHA produced using both biodiesel waste as the sole carbon source was identified using GC-MS as described in Furrer *et al.*, 2007. Prior to the GC-MS analysis, polymer samples were methanolysed. GCMS analysis was carried out using a Varian GS/MS system consisting of Chrompack CP-3800 gas chromatograph and Saturn 200 MS/MS block.

Nuclear magnetic resonance. Structural identification of the polymer was carried out using ^{13}C and ^{1}H NMR spectroscopy (Bruker Avance III 600 Cryo). Twenty milligrams of the purified polymer sample was dissolved in 1 ml of deuterated chloroform ($CDCl_3$) and analysed. All NMR spectra were measured at 298 K. Solution ^{1}H and ^{13}C NMR spectra were recorded using Bruker NMR spectrometer AVANCE III 600 (Coventry, UK) equipped with a ^{1}H,^{13}C-cryoprobe. Data acquisition and processing were performed using standard Bruker TOPSPIN (version 2.1) software. ^{1}H and ^{13}C chemical shifts were calibrated using residual solvent peak (^{1}H 7.26 ppm, ^{13}C 77.15 ppm for chloroform). This was carried out at University College London, UK (UCL, Department of Chemistry).

Differential scanning calorimetry. Thermal properties such as the melting temperature (T_m) and glass transition temperature (T_g) of the PHA produced using biodiesel waste were determined using DSC (DSC 214 Polyma, Netzsch, Germany equipped with Intracooler IC70 cooling system). Five milligram of the polymer sample was used for this analysis. Samples were heated, cooled and then heated again between −50°C and 100° at a heating rate of 20°C min^{-1}. Obtained thermograms were analysed using PROTEUS 7.0 software.

Preparation of solvent cast PHA films. One gram of the PHA produced using biodiesel waste as the carbon source was dissolved in 10 ml of chloroform, and the PHA solution was mixed thoroughly before being poured into glass petri dishes. They were allowed to dry at the room temperature (Basnett *et al.*, 2013). Films were stored for 3 weeks at room temperature prior analysis.

Characterization of the solvent cast PHA films

Scanning electron microscopy. Surface topography of the PHA films was studied using SEM. The film samples were mounted on aluminium stubs and gold-coated before being viewed under SEM (JEOL 5610LV-SEM). This study was carried out at the Eastman Dental Hospital, University College London.

Surface roughness. Surface roughness of the PHA film samples was measured using a laser-based optical non-contact 3D profilometer (Sony Proscan 1000 Laser Profilometer, Tokyo, Japan). This was carried out at the Eastman Dental Institute, University College London, UK.

Water contact angle measurements. Static contact angle study was carried out to measure the hydrophobicity of the PHA films. This was carried out using the KSV Cam 200 optical contact meter as described in Basnett *et al.*, 2013. Ten microlitres of water was added on the surface of the sample, and a total of 10 images were captured with a frame interval of one second. This experiment was carried out in triplicates. This analysis was carried out at the Eastman Dental Institute, University College London, UK.

Protein adsorption assay. Protein adsorption tests were carried out by incubating 1 cm^2 film samples in 400 µl of undiluted fetal bovine serum (FBS) at 37°C for 24 h in the 24-well tissue culture plate. After incubation, the samples were rinsed three times with phosphate buffer saline (PBS) and incubated in 1 ml of 2% sodium dodecyl sulphate (SDS) in PBS, for 24 h, at room temperature, under vigorous shaking. This was carried out to extract proteins attached to the surface of the polymer. The overall concentration of the proteins adsorbed on to the PHA film samples was quantified using the Bicinchoninic acid assay (Rai, 2010). This assay was carried out in triplicates.

In vitro *biocompatibility tests*

Cell seeding and proliferation studies. C2C12 cells (cell density – 25,000 cells ml^{-1}) were seeded on to the UV sterilized PHA films. They were cultured for a total period of 7 days. Cell viability tests were carried out at the end of day 1, day 3 and day 7 using the MTT assay. Standard tissue culture plastic was used as the positive control.

$$\%\text{Cell Proliferation} = \frac{\text{Mean absorbance of sample}}{\text{Mean absorbance of control}} * 100$$

Scanning electron microscopy. C2C12 seeded on the PHA films was fixed in 2% paraformaldehyde. These cell-seeded films were dehydrated; gold-coated and viewed using SEM.

Acknowledgements

The authors would like to thank ReBioStent project – European Union's Seventh Programme for research, technological development and demonstration under grant agreement number 604251. The authors would also like to thank HyMedPoly project - Horizon 2020 under the grant agreement no 643050. The authors would also like to extend their gratitude to the University of Westminster for the support. The authors would also like to thank Dr. Nicola Mordan and Dr. Graham Palmer from Eastman Dental Institute for their contribution in microstructural characterization of the polymer. The authors would also like to extend their gratitude to Dr. Rinat Nigmatullin for his help with the thermal characterization and to the technical staff at the University of Westminster. Finally, authors would like to thank Purefuels Ltd and Dr. Godfrey Kayzee for providing biodiesel waste for this study.

Conflict of Interest

None declared.

References

Akaraonye, E., Keshavarz, T., and Roy, I. (2010) Production of polyhydroxyalkanoates: the future green materials of choice. *J Chem Technol Biotechnol* **85**: 732–743.

Ali, I., and Jamil, N. (2016) Polyhydroxyalkanoates: current applications in the medical field. *Front Biol* **1**: 19–27.

Álvarez-Chávez, C.R., Edwards, S., Moure-Eraso, R., and Geiser, K. (2012) Sustainability of bio-based plastics: general comparative analysis and recommendations for improvement. *J Clean Prod* **23**: 47–56.

Ashby, R.D., Nuñez, A., Solaiman, D.K., and Foglia, T.A. (2005) Sophorolipid biosynthesis from a biodiesel co-product stream. *J Am Oil Chem Soc* **82**: 625–630.

Ashby, R.D., Solaiman, D.K., and Foglia, T.A. (2004) Bacterial poly (hydroxyalkanoate) polymer production from the biodiesel co-product stream. *J Polym Environ* **12**: 105–112.

Bagdadi, A.V., Safari, M., Dubey, P., Basnett, P., Sofokleous, P., Humphrey, E., and Knowles, J.C. (2016) Poly (3-hydroxyoctanoate), a promising new material for cardiac tissue engineering. *J Tissue Eng Regen Med* [Epub ahead of print]. doi: 10.1002/term.2318.

Basnett, P., and Roy, I. (2010) Microbial production of biodegradable polymers and their role in cardiac stent development. *Current Res Technol Educat Topics Applied Microbiol Microbial Biotechnol*: 405–1415.

Basnett, P., Knowles, J.C., Pishbin, F., Smith, C., Keshavarz, T., Boccaccini, A.R., and Roy, I. (2012) Novel biodegradable and biocompatible poly (3-hydroxyoctanoate)/bacterial cellulose composites. *Adv Eng Mater* **14**: B330–B343.

Basnett, P., Ching, K.Y., Stolz, M., Knowles, J.C., Boccaccini, A.R., Smith, C., *et al.* (2013) Novel Poly (3-hydroxyoctanoate)/Poly (3-hydroxybutyrate) blends for medical applications. *React Funct Polym* **73**: 1340–1348.

Bormann, E.J., and Roth, M. (1999) The production of polyhydroxybutyrate by *Methylobacterium rhodesianum* and Ralstonia eutropha in media containing glycerol and casein hydrolysates. *Biotechnol Lett* **21(12)**: 1059–1063.

Burattini, S., Ferri, P., Battistelli, M., Curci, R., Luchetti, F., and Falcieri, E. (2004) C2C12 murine myoblasts as a model of skeletal muscle development: morpho-functional characterization. *EJH* **48**: 223.

Cerrone, F., Davis, R., Kenny, S.T., Woods, T., O'Donovan, A., Gupta, V.K., and O'Connor, K. (2015) Use of a mannitol rich ensiled grass press juice (EGPJ) as a sole carbon source for polyhydroxyalkanoates (PHAs) production through high cell density cultivation. *Bioresour Technol* **191**: 45–52.

Chanasit, W., Sueree, L., Hodgson, B., and Umsakul, K. (2014) The production of poly (3-hydroxybutyrate) [P (3HB)] by a newly isolated Bacillus sp. ST1C using liquid waste from biodiesel production. *Ann Microbiol* **64**: 1157–1166.

Chanasit, W., Hodgson, B., Sudesh, K., and Umsakul, K. (2016) Efficient production of polyhydroxyalkanoates (PHAs) from *Pseudomonas mendocina* PSU using a biodiesel liquid waste (BLW) as the sole carbon source. *Biosci Biotechnol Biochem* **80**: 1440–1450.

Chen, H., Song, W., Zhou, F., Wu, Z., Huang, H., Zhang, J., *et al.* (2009) The effect of surface microtopography of poly (dimethylsiloxane) on protein adsorption, platelet and cell adhesion. *Colloids Surf B* **71**: 275–281.

Elvers, D., Song, C.H., Steinbüchel, A., and Leker, J. (2016) Technology trends in biodegradable polymers: evidence from patent analysis. *Polym Rev* **56**: 84–606.

Fernando, E., Keshavarz, T., and Kyazze, G. (2012) Enhanced bio-decolourisation of acid orange 7 by *Shewanella oneidensis* through co-metabolism in a microbial fuel cell. *Int Biodeterior Biodegradation* **72**: 1–9.

Francis, L., Meng, D., Locke, I.C., Knowles, J.C., Mordan, N., Salih, V., and Roy, I. (2016) Novel poly (3-hydroxybutyrate) composite films containing bioactive glass nanoparticles for wound healing applications. *Polym Int* **65**: 661–674.

Furrer, P., Hany, R., Rentsch, D., Grubelnik, A., Ruth, K., Panke, S., and Zinn, M. (2007) Quantitative analysis of bacterial medium-chain-length poly ([R]-3-hydroxyalkanoates) by gas chromatography. *J Chromatogr A* **1143**: 199–206.

Guo, W., Song, C., Kong, M., Geng, W., Wang, Y., and Wang, S. (2011) Simultaneous production and characterization of medium-chain-length polyhydroxyalkanoates and alginate oligosaccharides by *Pseudomonas mendocina* NK-01. *Appl Microbiol Biotechnol* **92**: 791–801.

Guo, W., Feng, J., Geng, W., Song, C., Wang, Y., Chen, N., and Wang, S. (2012) Augmented production of alginate oligosaccharides by the *Pseudomonas mendocina* NK-01 mutant. *Carbohydr Res* **352**: 109–116.

Hahn, S.K., Chang, Y.K., Kim, B.S., and Chang, H.N. (1994) Optimization of microbial poly (3-hydroxybutyrate) recover using dispersions of sodium hypochlorite solution and chloroform. *Biotechnol Bioeng* **44**: 256–261.

Hallab, N.J., Bundy, K.J., O'Connor, K., Moses, R.L., and Jacobs, J.J. (2001) Evaluation of metallic and polymeric biomaterial surface energy and surface roughness characteristics for directed cell adhesion. *Tissue Eng* **7**: 55–71.

Hao, L., and Lawrence, J. (2004) The adsorption of human serum albumin (HSA) on CO2 laser modified magnesia partially stabilised zirconia (MgO-PSZ). *Colloids Surf B* **34**: 87–94.

Hao, L., and Lawrence, J. (2006a) Effects of Nd: YAG laser treatment on the wettability characteristics of a zirconia-based bioceramic. *Opt Lasers Eng* **44**: 803–814.

Hao, L., and Lawrence, J. (2006b) Albumin and fibronectin protein adsorption on CO2-lasermodified biograde stainless steel. *Proc Inst Mech Eng H* **220**: 47–55.

Jeyachandran, Y.L., Mielczarski, E., Rai, B., and Mielczarski, J.A. (2009) Quantitative and qualitative evaluation of adsorption/desorption of bovine serum albumin on hydrophilic and hydrophobic surfaces. *Langmuir* **25**: 11614–11620.

Jiang, X.J., Sun, Z., Ramsay, J.A., and Ramsay, B.A. (2013) Fed-batch production of MCL-PHA with elevated 3-hydroxynonanoate content. *AMB Express* **3**: 50.

Khanna, S., and Srivastava, A.K. (2005) Recent advances in microbial polyhydroxyalkanoates. *Process Biochem* **40**: 607–619.

Kim, Y.B., and Rhee, Y.H. (2000) Evaluation of various carbon substrates for the biosynthesis of polyhydroxyalkanoates bearing functional groups by *Pseudomonas putida*. *Int J Biol Macromol* **28**: 23–29.

Kubiak, K.J., Wilson, M.C.T., Mathia, T.G., and Carval, P. (2011) Wettability versus roughness of engineering surfaces. *Wear* **271**: 523–528.

Langer, R., and Tirrell, D.A. (2004) Designing materials for biology and medicine. *Nature* **428**: 487–492.

Lizarraga-Valderrama, L.R., Panchal, B., Thomas, C., Boccaccini, A.R., and Roy, I.. (2016) Biomedical applications of polyhydroxyalkanoates. In *Biomaterials From Nature for Advanced Devices and Therapies*. Neves, N.M., Reis, R.I. (eds). Hoboken: Society for Biomaterials Willey, pp. 339–383.

Marangoni, C., Furigo, A. Jr, and Aragão, G.M. (2001) The influence of substrate source on the growth of Ralstonia eutropha, aiming at the production of polyhydroxyalkanoate. *Braz J Chem Eng* **18**: 175–180.

Marangoni, C., Furigo, A., and de Aragão, G.M. (2002) Production of poly (3-hydroxybutyrate-co-3-hydroxyvalerate) by Ralstonia eutropha in whey and inverted sugar with propionic acid feeding. *Process Biochem* **38**: 137–141.

Metabolix launches bio-based additive to enhance recycled PVC performance (2013). *Additives for Polymers*, 12: 4–5. https://doi.org/10.1016/s0306-3747(13)70186-6.

Metabolix develops PHA performance modifiers for PLA (2014) *Additives for Polymers*, 2014 (5): 2–3. https://doi.org/10.1016/s0306-3747(14)70068-5.

Misra, S.K., Valappil, S.P., Roy, I., and Boccaccini, A.R. (2006) Polyhydroxyalkanoate (PHA)/inorganic phase composites for tissue engineering applications. *Biomacromol* **7**: 2249–2258.

Możejko, J., and Ciesielski, S. (2014) Pulsed feeding strategy is more favorable to medium-chain-length polyhydroxyalkanoates production from waste rapeseed oil. *Biotechnol Progr* **30**: 1243–1246.

Muangwong, A., Boontip, T., Pachimsawat, J., and Napathorn, S.C. (2016) Medium chain length polyhydroxyalkanoates consisting primarily of unsaturated 3-hydroxy-5-cis-dodecanoate synthesized by newly isolated bacteria using crude glycerol. *Microb Cell Fact* **15**: 55.

Muhr, A., Rechberger, E.M., Salerno, A., Reiterer, A., Schiller, M., Kwiecień, M., *et al.* (2013a) Biodegradable latexes from animal-derived waste: Biosynthesis and characterization of mcl-PHA accumulated by *Ps. citronellolis*. *React Funct Polym* **73**: 1391–1398.

Muhr, A., Rechberger, E.M., Salerno, A., Reiterer, A., Malli, K., Strohmeier, K., *et al.* (2013b) Novel description of mcl-PHA biosynthesis by *Pseudomonas chlororaphis* from animal-derived waste. *J Biotechnol* **165**: 45–51.

Nair, L.S., and Laurencin, C.T. (2007) Biodegradable polymers as biomaterials. *Prog Polym Sci* **32**: 762–798.

Narodoslawsky, M., Shazad, K., Kollmann, R., and Schnitzer, H. (2015) LCA of PHA production-identifying the ecological potential of bio-plastic. *Chem Biochem Eng Q* **29**: 299–305.

Poblete-Castro, I., Escapa, I.F., Jäger, C., Puchalka, J., Lam, C.M.C., Schomburg, D., *et al.* (2012) The metabolic response of *P. putida* KT2442 producing high levels of polyhydroxyalkanoate under single-and multiple-nutrient-limited growth: highlights from a multi-level omics approach. *Microb Cell Fact* **11**: 34.

Poblete-Castro, I., Binger, D., Rodrigues, A., Becker, J., dos Santos, V.A.M., and Wittmann, C. (2013) In-silico-driven metabolic engineering of *Pseudomonas putida* for enhanced production of poly-hydroxyalkanoates. *Metab Eng* **15**: 113–123.

Preusting, H., Nijenhuis, A., and Witholt, B. (1990) Physical characteristics of poly (3-hydroxyalkanoates) and poly (3-hydroxyalkenoates) produced by *Pseudomonas oleovorans* grown on aliphatic hydrocarbons. *Macromolecules* **23**: 4220–4224.

Rai, R.. (2010) Biosynthesis of polyhydroxyalkanoates and its medical applications (Doctoral dissertation, University of Westminster).

Rai, R., Keshavarz, T., Roether, J.A., Boccaccini, A.R., and Roy, I. (2011a) Medium chain length polyhydroxyalkanoates, promising new biomedical materials for the future. *Mater Sci Eng R-Rep* **72**: 29–47.

Rai, R., Yunos, D.M., Boccaccini, A.R., Knowles, J.C., Barker, I.A., Howdle, S.M., *et al.* (2011b) Poly-3-

hydroxyoctanoate P (3HO), a medium chain length poly-hydroxyalkanoate homopolymer from *Pseudomonas mendocina*. *Biomacromol* **12:** 2126–2136.

Rechendorff, K., Hovgaard, M.B., Foss, M., Zhdanov, V., and Besenbacher, F. (2006) Enhancement of protein adsorption induced by surface roughness. *Langmuir* **22:** 10885–10888.

Renard, E., Vergnol, G., and Langlois, V. (2011) Adhesion and proliferation of human bladder RT112 cells on functionalized polyesters. *AEMB* **32:** 214–220.

Ribeiro, P.L.L., da Silva, A.C.M.S., Menezes Filho, J.A., and Druzian, J.I. (2015) Impact of different by-products from the biodiesel industry and bacterial strains on the production, composition, and properties of novel polyhydroxyalkanoates containing achiral building blocks. *Ind Crops Prod* **69:** 212–223.

Sato, Y., and Murahara, M. (2004) Protein adsorption on PTFE surface modified by ArF excimer laser treatment. *J Adhes Sci Technol* **18:** 1545–1555.

Shrivastav, A., Mishra, S.K., Shethia, B., Pancha, I., Jain, D., and Mishra, S. (2010) Isolation of promising bacterial strains from soil and marine environment for polyhydroxyalkanoates (PHAs) production utilizing Jatropha biodiesel byproduct. *Int J Biol Macromol* **47:** 283–287.

Song, J.H., Jeon, C.O., Choi, M.H., Yoon, S.C., and Park, W. (2008) Polyhydroxyalkanoate (PHA) production using waste vegetable oil by Pseudomonas sp. strain DR2. *JMicrobiol Biotechnol* **18:** 1408–1415.

Teeka, J., Imai, T., Reungsang, A., Cheng, X., Yuliani, E., and Thiantanankul, J. (2012) Characterization of polyhydroxyalkanoates (PHAs) biosynthesis by isolated *Novosphingobium* sp. THA_AIK7 using crude glycerol. *J Ind Microbiol Biotechnol* **39:** 749–758.

Ulery, B.D., Nair, L.S., and Laurencin, C.T. (2011) Biomedical applications of biodegradable polymers. *J Polym Sci Part B Polym Phys* **49:** 832–864.

Valappil, S.P., Misra, S.K., Boccaccini, A.R., and Roy, I. (2006) Biomedical applications of polyhydroxyalkanoates, an overview of animal testing and in vivo responses. *Expert Rev Med Dev* **3:** 853–868.

Weatherburn, M.W. (1967) Phenol-hypochlorite reaction for determination of ammonia. *Anal Chem* **39:** 971–974.

Yang, S., Leong, K.F., Du, Z., and Chua, C.K. (2001) The design of scaffolds for use in tissue engineering. Part I. Traditional factors. *Tissue Eng* **7:** 679–689.

Permissions

All chapters in this book were first published in MB, by John Wiley & Sons Ltd.; hereby published with permission under the Creative Commons Attribution License or equivalent. Every chapter published in this book has been scrutinized by our experts. Their significance has been extensively debated. The topics covered herein carry significant findings which will fuel the growth of the discipline. They may even be implemented as practical applications or may be referred to as a beginning point for another development.

The contributors of this book come from diverse backgrounds, making this book a truly international effort. This book will bring forth new frontiers with its revolutionizing research information and detailed analysis of the nascent developments around the world.

We would like to thank all the contributing authors for lending their expertise to make the book truly unique. They have played a crucial role in the development of this book. Without their invaluable contributions this book wouldn't have been possible. They have made vital efforts to compile up to date information on the varied aspects of this subject to make this book a valuable addition to the collection of many professionals and students.

This book was conceptualized with the vision of imparting up-to-date information and advanced data in this field. To ensure the same, a matchless editorial board was set up. Every individual on the board went through rigorous rounds of assessment to prove their worth. After which they invested a large part of their time researching and compiling the most relevant data for our readers.

The editorial board has been involved in producing this book since its inception. They have spent rigorous hours researching and exploring the diverse topics which have resulted in the successful publishing of this book. They have passed on their knowledge of decades through this book. To expedite this challenging task, the publisher supported the team at every step. A small team of assistant editors was also appointed to further simplify the editing procedure and attain best results for the readers.

Apart from the editorial board, the designing team has also invested a significant amount of their time in understanding the subject and creating the most relevant covers. They scrutinized every image to scout for the most suitable representation of the subject and create an appropriate cover for the book.

The publishing team has been an ardent support to the editorial, designing and production team. Their endless efforts to recruit the best for this project, has resulted in the accomplishment of this book. They are a veteran in the field of academics and their pool of knowledge is as vast as their experience in printing. Their expertise and guidance has proved useful at every step. Their uncompromising quality standards have made this book an exceptional effort. Their encouragement from time to time has been an inspiration for everyone.

The publisher and the editorial board hope that this book will prove to be a valuable piece of knowledge for researchers, students, practitioners and scholars across the globe.

List of Contributors

Ludmilla Aristilde
Department of Biological and Environmental Engineering, College of Agriculture and Life Sciences, Cornell University, Ithaca, NY 14853, USA

Cédric Grangeteau, Sandrine Rousseaux, Hervé Alexandre and Michéle Guilloux-Benatier
Univ. Bourgogne Franche-Comte, AgroSup Dijon, PAM UMR A 02.102, F-21000 Dijon, France.
IUVV Equipe VAlMiS, rue Claude Ladrey, BP 27877, 21078 Dijon Cedex, France.

Chloé Roullier-Gall and Philippe Schmitt-Kopplin
Chair of Analytical Food Chemistry, Technische Universit€at M€unchen, Alte Akademie 10, 85354 Freising-Weihenstephan, Germany
Research Unit Analytical BioGeoChemistry, Department of Environmental Sciences, Helmholtz Zentrum M€unchen, Ingolstaedter Landstrasse 1, 85764 Neuherberg, Germany

Régis D. Gougeon
Univ. Bourgogne Franche-Comte, AgroSup Dijon, PAM UMR A 02.102, F-21000 Dijon, France
IUVV Equipe PAPC, rue Claude Ladrey, BP 27877, 21078 Dijon Cedex, France

Fernando Guzmán-Chávez, Oleksandr Salo, Yvonne Nygård and Arnold J. M. Driessen
Molecular Microbiology, Groningen Biomolecular Sciences and Biotechnology Institute, University of Groningen, Nijenborgh 7, 9747 AG Groningen, The Netherlands

Peter P. Lankhorst
DSM Biotechnology Center, Alexander Fleminglaan 1, 2613 AX Delft, The Netherlands

Roel A. L. Bovenberg
Synthetic Biology and Cell Engineering, Groningen Biomolecular Sciences and Biotechnology Institute, University of Groningen, Nijenborgh 7, 9747 AG Groningen, The Netherlands
DSM Biotechnology Center, Alexander Fleminglaan 1, 2613 AX Delft, The Netherlands

Patricia Lozano-Martínez, Rubén M. Buey, Rodrigo Ledesma-Amaro, Alberto Jiménez and José Luis Revuelta
Metabolic Engineering Group, Departamento de Microbiología y Genética, Universidad de Salamanca, Edificio Departamental, Campus Miguel de Unamuno, 37007 Salamanca, Spain

Lynne E. Macaskie, Iryna P. Mikheenko and Jacob B. Omajai
Schools of Biosciences

Alan J. Stephen and Joseph Wood
School of Chemical Engineering, University of Birmingham, Edgbaston, Birmingham, B15 2TT, UK

Jitka Makovcova, Vladimir Babak, Michal Slany and Lenka Cincarova
Departments of Food and Feed Safety

Pavel Kulich
Chemistry and Toxicology

Josef Masek
Pharmacology and Immunotherapy, Veterinary Research Institute, Brno, Czech Republic

Frank Persson and Britt-Marie Wilén
Division of Water Environment Technology, Department of Civil and Environmental Engineering, Chalmers University of Technology, SE-41296 Gothenburg, Sweden.

Carolina Suarez and Malte Hermansson
Department of Chemistry and Molecular Biology, University of Gothenburg, SE-40530 Gothenburg, Sweden

Elzbieta Plaza and Razia Sultana
Department of Sustainable Development, Environmental Science and Engineering (SEED), Royal Institute of Technology (KTH), Teknikringen 76, SE-100 44 Stockholm, Sweden.

Elena Piacenza, Alessandro Presentato, Joseph A. Lemire, Marc Demeter and Raymond J. Turner
Biofilm Research Group, Department of Biological Sciences, University of Calgary, 2500 University Dr NW, Calgary, AB T2N 1N4, Canada

Emanuele Zonaro, Giovanni Vallini and Silvia Lampis
Environmental Microbiology Laboratory, Department of Biotechnology, University of Verona, Strada Le Grazie 15, 37134 Verona, Italy

Timothy Sibanda, Ramganesh Selvarajan and Memory Tekere
Department of Environmental Sciences, College of Agriculture and Environmental Science, UNISA Florida Campus, Florida 1709, South Africa

Alan J. Stephen and Sophie A. Archer
Schools of Chemical Engineering

Rafael L. Orozco and Lynne E. Macaskie
School of Biosciences, University of Birmingham, Edgbaston, Birmingham, B15 2TT, UK

Alessandro Tanca, Cristina Fraumene, Daniela Pagnozzi and Maria Filippa Addis
Porto Conte Ricerche, Science and Technology Park of Sardinia, Tramariglio, Alghero, Italy.

Valeria Manghina, Antonio Palomba and Sergio Uzzau
Porto Conte Ricerche, Science and Technology Park of Sardinia, Tramariglio, Alghero, Italy.
Department of Biomedical Sciences, University of Sassari, Sassari, Italy

Marcello Abbondio and Massimo Deligios
Department of Biomedical Sciences, University of Sassari, Sassari, Italy

Lorena Fernández-Cabezón, Beatriz Galán and José L. García
Department of Environmental Biology, Centro de Investigaciones Biologicas, Consejo Superior de Investigaciones Científicas, Ramiro de Maeztu 9, 28040 Madrid, Spain

Mayra Cuéllar-Cruz and Daniela Lucio-Hernández
Departamento de Biología, División de Ciencias Naturales y Exactas, Campus Guanajuato, Universidad de Guanajuato, Noria Alta S/N, Col. Noria Alta, C.P. 36050, Guanajuato, Mexico.

Isabel Martínez-Ángeles and Abel Moreno
Departamento de Química de Biomacromoléculas, Instituto de Química, Universidad Nacional Autónoma de México, Av. Universidad 3000, Ciudad Universitaria, Ciudad de Mexico, 04510, México.

Nicola Demitri and Maurizio Polentarutti
Elettra – Sincrotone Trieste, S.S. 14 km 163.5 in Area Science Park, 34149, Basovizza – Trieste, Italy.

María J. Rosales-Hoz
Departamento de Química, Centro de Investigación y de Estudios Avanzados del I.P.N., Apdo. Postal 14-740, 07000, México, D.F., México

Helen Y. Buse and Vicente Gomez-Alvarez
Pegasus Technical Services, Inc c/o US EPA, 26 W Martin Luther King Drive NG-16, Cincinnati, OH 45268, USA

Pan Ji, Amy Pruden and Marc A. Edwards
Department of Civil and Environmental Engineering, Virginia Tech, Blacksburg, VA, USA.

Nicholas J. Ashbolt
School of Public Health, University of Alberta, Edmonton, AB T6G 2G7, Canada.

Pooja Basnett, Barbara Lukasiewicz, Elena Marcello and Ipsita Roy
Faculty of Science and Technology, University of Westminster, London, UK.

Harpreet K. Gura
Aarhus University, Denmark

Jonathan C. Knowles
Eastman Dental Institute, University College London, London, UK.
Department of Nanobiomedical Science & BK21 Plus NBM Global Research Center for Regenerative Medicine, Dankook University, Cheonan 330-714, Republic of Korea.

Index

A

Aerobic Ammonium Oxidation (AOB), 76

Anthropogenic Factors, 14

Antimicrobial Activity, 31, 85, 88, 90-91, 94, 96, 98, 113, 115

Ashbya Gossypii, 45, 47-48

Aureobasidium Pullulans, 14

B

Bacillus Mycoides, 85, 87-90, 96, 98

Bacterial Colonization, 60

Bacteroidales, 132, 136

Benzanthracene, 100, 103, 105, 107

Bio Se-nemo-s, 88-92, 94-97

Bio-precious Metal Materials, 50

Biodiesel Waste, 52, 184-189, 191, 194-197

Biofilm Formation, 60-61, 63, 67-69, 73, 80, 85-86, 90, 94-96, 99, 165, 182-183

Biofilms, 60-75, 79-81, 84-86, 88, 90-95, 97-98, 167, 169-175, 177-179, 181-183

Biofuel, 1-4, 6, 8-12, 47, 49, 53, 58

Biominerals, 149

Bioplastics, 184-185

Biosynthesis, 1, 3-4, 7-8, 10-11, 30-35, 37, 39-40, 43, 45, 48-49, 57-58, 98-99, 139, 149-150, 165, 167, 188, 197-199

Biotechnological Process, 138-139

Biotransformation, 98, 114, 139-143, 146-148

Bottlenecks, 116-117

C

Candida Species, 149-156, 159, 161-163, 167-168

Cdc Biofilm Reactor, 176, 179

Cell Homeostasis, 159

Clostridium Acetobutylicum, 1-2, 12-13

Confocal Laser Scanning
Microscopy (CLSM), 92-93, 97

D

Decomposition, 1, 99

Dna Extraction, 25, 29, 79, 81, 83, 102, 124-125, 127, 133-134, 136-137

Drinking Water Biofilm, 183

E

Enzyme Production, 103-104

F

Fermentation, 1, 10, 12, 14-15, 17, 19-25, 27-29, 45, 48, 51, 53, 103-104, 112, 116-122, 124, 130, 140, 184, 186, 195

Filamentous Fungus, 30, 45, 148

Fossil Oils, 54

Fuel Cells, 51-53, 55-58, 118, 121-122

G

Gene Deletion, 37-38, 46

Gram-negative Bacteria, 58, 60-69, 71-72, 148, 168

Green Catalysts, 50

H

Hansenula Anomala, 15

Hybridization, 74-75, 82

Hydrocarbon Degradation, 105, 107

Hydrocarbon Utilization, 100, 103-104

Hydrophobicity, 72, 191-192, 196

I

Industrial Production, 45, 140

L

Legionella Pneumophila, 169-171, 178, 181-183

M

Mannose-derived Carbons, 3

Metabolite Labelling, 1, 6, 10-11

Metallic Bionanocatalysts, 50

Microbial Hydrogen Production, 116

Microbial Life, 100

Microbial Metabolic Pathways, 128

Microbiome Diversity, 169, 173

Microorganisms, 14, 29, 45-46, 60-62, 67, 72-73, 75-76, 100-101, 105, 107, 109, 112-113, 117, 122, 124, 127, 138, 140, 149-150, 153, 157, 169, 183-184, 186

Mycobacterium Smegmatis, 138, 142-145

N

Natural Gemstones, 149

O

Overexpression, 30, 33-35, 37-38, 43-44, 46, 48

P

Partial Nitritation-anammox (PNA), 74

Penicillium Chrysogenum, 30, 34, 37, 39-40

Photosynthetic Bacteria, 119, 122

Physicochemical Properties, 115, 149

Phytosanitary Protection, 14, 16-17, 19-20, 25-26

Polyhydroxyalkanotes (PHAS), 184

Polymicrobial Growth, 61

Pseudomonas Aeruginosa, 85-86, 91, 93, 182-183

Pyrosequencing, 16-17, 21-23, 26, 135

R

Rhodobacter Sphaeroides, 51, 118, 120, 122

S

Saccharomyces Cerevisiae, 14, 27-29, 47-49, 165, 168

Salmonella Enterica, 60-61, 64-67, 69, 71-72, 136

Scanning Electron Microscopy, 61, 65-66, 69, 71, 163, 191-192, 194, 196-197

Senps, 85-97

Sheep Faecal Microbiota
Composition, 125

Sheep Farming, 124

Sorbicillin Biosynthesis, 30, 33, 40

Sorbicillinoids, 30-31, 33-35, 37, 39-40

Staphylococcus Aureus, 60-62, 64-66, 69, 71-73, 85, 91-93, 99

Substrate Availability, 74-75, 77

Surface Roughness, 191-192, 196, 198-199

Surface Topography, 191, 196

Synthetic Extreme Environments, 100-101, 112

T

Testosterone Production, 138, 147

V

Vermamoeba Vermiformis, 169, 171

W

Wine Microbiology, 14, 28-29